# MONOTONE ITERATIVE TECHNIQUES FOR DISCONTINUOUS NONLINEAR DIFFERENTIAL EQUATIONS

# PURE AND APPLIED MATHEMATICS

## A Program of Monographs, Textbooks, and Lecture Notes

## MONOGRAPHS AND TEXTBOOKS IN PURE AND APPLIED MATHEMATICS

1. *K. Yano,* Integral Formulas in Riemannian Geometry (1970)
2. *S. Kobayashi,* Hyperbolic Manifolds and Holomorphic Mappings (1970)
3. *V. S. Vladimirov,* Equations of Mathematical Physics (A. Jeffrey, ed.; A. Littlewood, trans.) (1970)
4. *B. N. Pshenichnyi,* Necessary Conditions for an Extremum (L. Neustadt, translation ed.; K. Makowski, trans.) (1971)
5. *L. Narici et al.,* Functional Analysis and Valuation Theory (1971)
6. *S. S. Passman,* Infinite Group Rings (1971)
7. *L. Dornhoff,* Group Representation Theory. Part A: Ordinary Representation Theory. Part B: Modular Representation Theory (1971, 1972)
8. *W. Boothby and G. L. Weiss, eds.,* Symmetric Spaces (1972)
9. *Y. Matsushima,* Differentiable Manifolds (E. T. Kobayashi, trans.) (1972)
10. *L. E. Ward, Jr.,* Topology (1972)
11. *A. Babakhanian,* Cohomological Methods in Group Theory (1972)
12. *R. Gilmer,* Multiplicative Ideal Theory (1972)
13. *J. Yeh,* Stochastic Processes and the Wiener Integral (1973)
14. *J. Barros-Neto,* Introduction to the Theory of Distributions (1973)
15. *R. Larsen,* Functional Analysis (1973)
16. *K. Yano and S. Ishihara,* Tangent and Cotangent Bundles (1973)
17. *C. Procesi,* Rings with Polynomial Identities (1973)
18. *R. Hermann,* Geometry, Physics, and Systems (1973)
19. *N. R. Wallach,* Harmonic Analysis on Homogeneous Spaces (1973)
20. *J. Dieudonné,* Introduction to the Theory of Formal Groups (1973)
21. *I. Vaisman,* Cohomology and Differential Forms (1973)
22. *B.-Y. Chen,* Geometry of Submanifolds (1973)
23. *M. Marcus,* Finite Dimensional Multilinear Algebra (in two parts) (1973, 1975)
24. *R. Larsen,* Banach Algebras (1973)
25. *R. O. Kujala and A. L. Vitter, eds.,* Value Distribution Theory: Part A; Part B: Deficit and Bezout Estimates by Wilhelm Stoll (1973)
26. *K. B. Stolarsky,* Algebraic Numbers and Diophantine Approximation (1974)
27. *A. R. Magid,* The Separable Galois Theory of Commutative Rings (1974)
28. *B. R. McDonald,* Finite Rings with Identity (1974)
29. *J. Satake,* Linear Algebra (S. Koh et al., trans.) (1975)
30. *J. S. Golan,* Localization of Noncommutative Rings (1975)
31. *G. Klambauer,* Mathematical Analysis (1975)
32. *M. K. Agoston,* Algebraic Topology (1976)
33. *K. R. Goodearl,* Ring Theory (1976)
34. *L. E. Mansfield,* Linear Algebra with Geometric Applications (1976)
35. *N. J. Pullman,* Matrix Theory and Its Applications (1976)
36. *B. R. McDonald,* Geometric Algebra Over Local Rings (1976)
37. *C. W. Groetsch,* Generalized Inverses of Linear Operators (1977)
38. *J. E. Kuczkowski and J. L. Gersting,* Abstract Algebra (1977)
39. *C. O. Christenson and W. L. Voxman,* Aspects of Topology (1977)
40. *M. Nagata,* Field Theory (1977)
41. *R. L. Long,* Algebraic Number Theory (1977)
42. *W. F. Pfeffer,* Integrals and Measures (1977)
43. *R. L. Wheeden and A. Zygmund,* Measure and Integral (1977)
44. *J. H. Curtiss,* Introduction to Functions of a Complex Variable (1978)
45. *K. Hrbacek and T. Jech,* Introduction to Set Theory (1978)
46. *W. S. Massey,* Homology and Cohomology Theory (1978)
47. *M. Marcus,* Introduction to Modern Algebra (1978)
48. *E. C. Young,* Vector and Tensor Analysis (1978)
49. *S. B. Nadler, Jr.,* Hyperspaces of Sets (1978)
50. *S. K. Segal,* Topics in Group Kings (1978)
51. *A. C. M. van Rooij,* Non-Archimedean Functional Analysis (1978)
52. *L. Corwin and R. Szczarba,* Calculus in Vector Spaces (1979)

53. *C. Sadosky,* Interpolation of Operators and Singular Integrals (1979)
54. *J. Cronin,* Differential Equations (1980)
55. *C. W. Groetsch,* Elements of Applicable Functional Analysis (1980)
56. *I. Vaisman,* Foundations of Three-Dimensional Euclidean Geometry (1980)
57. *H. I. Freedan,* Deterministic Mathematical Models in Population Ecology (1980)
58. *S. B. Chae,* Lebesgue Integration (1980)
59. *C. S. Rees et al.,* Theory and Applications of Fourier Analysis (1981)
60. *L. Nachbin,* Introduction to Functional Analysis (R. M. Aron, trans.) (1981)
61. *G. Orzech and M. Orzech,* Plane Algebraic Curves (1981)
62. *R. Johnsonbaugh and W. E. Pfaffenberger,* Foundations of Mathematical Analysis (1981)
63. *W. L. Voxman and R. H. Goetschel,* Advanced Calculus (1981)
64. *L. J. Corwin and R. H. Szcarba,* Multivariable Calculus (1982)
65. *V. I. Istrăţescu,* Introduction to Linear Operator Theory (1981)
66. *R. D. Järvinen,* Finite and Infinite Dimensional Linear Spaces (1981)
67. *J. K. Beem and P. E. Ehrlich,* Global Lorentzian Geometry (1981)
68. *D. L. Armacost,* The Structure of Locally Compact Abelian Groups (1981)
69. *J. W. Brewer and M. K. Smith, eds.,* Emily Noether: A Tribute (1981)
70. *K. H. Kim,* Boolean Matrix Theory and Applications (1982)
71. *T. W. Wieting,* The Mathematical Theory of Chromatic Plane Ornaments (1982)
72. *D. B.Gauld,* Differential Topology (1982)
73. *R. L. Faber,* Foundations of Euclidean and Non-Euclidean Geometry (1983)
74. *M. Carmeli,* Statistical Theory and Random Matrices (1983)
75. *J. H. Carruth et al.,* The Theory of Topological Semigroups (1983)
76. *R. L. Faber,* Differential Geometry and Relativity Theory (1983)
77. *S. Barnett,* Polynomials and Linear Control Systems (1983)
78. *G. Karpilovsky,* Commutative Group Algebras (1983)
79. *F. Van Oystaeyen and A. Verschoren,* Relative Invariants of Rings (1983)
80. *I. Vaisman,* A First Course in Differential Geometry (1984)
81. *G. W. Swan,* Applications of Optimal Control Theory in Biomedicine (1984)
82. *T. Petrie and J. D. Randall,* Transformation Groups on Manifolds (1984)
83. *K. Goebel and S. Reich,* Uniform Convexity, Hyperbolic Geometry, and Nonexpansive Mappings (1984)
84. *T. Albu and C. Năstăsescu,* Relative Finiteness in Module Theory (1984)
85. *K. Hrbacek and T. Jech,* Introduction to Set Theory: Second Edition (1984)
86. *F. Van Oystaeyen and A. Verschoren,* Relative Invariants of Rings (1984)
87. *B. R. McDonald,* Linear Algebra Over Commutative Rings (1984)
88. *M. Namba,* Geometry of Projective Algebraic Curves (1984)
89. *G. F. Webb,* Theory of Nonlinear Age-Dependent Population Dynamics (1985)
90. *M. R. Bremner et al.,* Tables of Dominant Weight Multiplicities for Representations of Simple Lie Algebras (1985)
91. *A. E. Fekete,* Real Linear Algebra (1985)
92. *S. B. Chae,* Holomorphy and Calculus in Normed Spaces (1985)
93. *A. J. Jerri,* Introduction to Integral Equations with Applications (1985)
94. *G. Karpilovsky,* Projective Representations of Finite Groups (1985)
95. *L. Narici and E. Beckenstein,* Topological Vector Spaces (1985)
96. *J. Weeks,* The Shape of Space (1985)
97. *P. R. Gribik and K. O. Kortanek,* Extremal Methods of Operations Research (1985)
98. *J.-A. Chao and W. A. Woyczynski, eds.,* Probability Theory and Harmonic Analysis (1986)
99. *G. D. Crown et al.,* Abstract Algebra (1986)
100. *J. H. Carruth et al.,* The Theory of Topological Semigroups, Volume 2 (1986)
101. *R. S. Doran and V. A. Belfi,* Characterizations of C*-Algebras (1986)
102. *M. W. Jeter,* Mathematical Programming (1986)
103. *M. Altman,* A Unified Theory of Nonlinear Operator and Evolution Equations with Applications (1986)
104. *A. Verschoren,* Relative Invariants of Sheaves (1987)
105. *R. A. Usmani,* Applied Linear Algebra (1987)
106. *P. Blass and J. Lang,* Zariski Surfaces and Differential Equations in Characteristic $p > 0$ (1987)
107. *J. A. Reneke et al.,* Structured Hereditary Systems (1987)

108. *H. Busemann and B. B. Phadke,* Spaces with Distinguished Geodesics (1987)
109. *R. Harte,* Invertibility and Singularity for Bounded Linear Operators (1988)
110. *G. S. Ladde et al.,* Oscillation Theory of Differential Equations with Deviating Arguments (1987)
111. *L. Dudkin et al.,* Iterative Aggregation Theory (1987)
112. T. Okubo, *Differential Geometry* (1987)
113. *D. L. Stancl and M. L. Stancl,* Real Analysis with Point-Set Topology (1987)
114. *T. C. Gard,* Introduction to Stochastic Differential Equations (1988)
115. *S. S. Abhyankar,* Enumerative Combinatorics of Young Tableaux (1988)
116. *H. Strade and R. Farnsteiner,* Modular Lie Algebras and Their Representations (1988)
117. *J. A. Huckaba,* Commutative Rings with Zero Divisors (1988)
118. *W. D. Wallis,* Combinatorial Designs (1988)
119. *W. Więsław,* Topological Fields (1988)
120. *G. Karpilovsky,* Field Theory (1988)
121. *S. Caenepeel and F. Van Oystaeyen,* Brauer Groups and the Cohomology of Graded Rings (1989)
122. *W. Kozlowski,* Modular Function Spaces (1988)
123. *E. Lowen-Colebunders,* Function Classes of Cauchy Continuous Maps (1989)
124. *M. Pavel,* Fundamentals of Pattern Recognition (1989)
125. *V. Lakshmikantham et al.,* Stability Analysis of Nonlinear Systems (1989)
126. *R. Sivaramakrishnan,* The Classical Theory of Arithmetic Functions (1989)
127. *N. A. Watson,* Parabolic Equations on an Infinite Strip (1989)
128. *K. J. Hastings,* Introduction to the Mathematics of Operations Research (1989)
129. *B. Fine,* Algebraic Theory of the Bianchi Groups (1989)
130. *D. N. Dikranjan et al.,* Topological Groups (1989)
131. *J. C. Morgan II,* Point Set Theory (1990)
132. *P. Biler and A. Witkowski,* Problems in Mathematical Analysis (1990)
133. *H. J. Sussmann,* Nonlinear Controllability and Optimal Control (1990)
134. *J.-P. Florens et al.,* Elements of Bayesian Statistics (1990)
135. *N. Shell,* Topological Fields and Near Valuations (1990)
136. *B. F. Doolin and C. F. Martin,* Introduction to Differential Geometry for Engineers (1990)
137. *S. S. Holland, Jr.,* Applied Analysis by the Hilbert Space Method (1990)
138. *J. Okniński,* Semigroup Algebras (1990)
139. *K. Zhu,* Operator Theory in Function Spaces (1990)
140. *G. B. Price,* An Introduction to Multicomplex Spaces and Functions (1991)
141. *R. B. Darst,* Introduction to Linear Programming (1991)
142. *P. L. Sachdev,* Nonlinear Ordinary Differential Equations and Their Applications (1991)
143. *T. Husain,* Orthogonal Schauder Bases (1991)
144. *J. Foran,* Fundamentals of Real Analysis (1991)
145. *W. C. Brown,* Matrices and Vector Spaces (1991)
146. *M. M. Rao and Z. D. Ren,* Theory of Orlicz Spaces (1991)
147. *J. S. Golan and T. Head,* Modules and the Structures of Rings (1991)
148. *C. Small,* Arithmetic of Finite Fields (1991)
149. *K. Yang,* Complex Algebraic Geometry (1991)
150. *D. G. Hoffman et al.,* Coding Theory (1991)
151. *M. O. González,* Classical Complex Analysis (1992)
152. *M. O. González,* Complex Analysis (1992)
153. *L. W. Baggett,* Functional Analysis (1992)
154. *M. Sniedovich,* Dynamic Programming (1992)
155. *R. P. Agarwal,* Difference Equations and Inequalities (1992)
156. *C. Brezinski,* Biorthogonality and Its Applications to Numerical Analysis (1992)
157. *C. Swartz,* An Introduction to Functional Analysis (1992)
158. *S. B. Nadler, Jr.,* Continuum Theory (1992)
159. *M. A. Al-Gwaiz,* Theory of Distributions (1992)
160. *E. Perry,* Geometry: Axiomatic Developments with Problem Solving (1992)
161. *E. Castillo and M. R. Ruiz-Cobo,* Functional Equations and Modelling in Science and Engineering (1992)
162. *A. J. Jerri,* Integral and Discrete Transforms with Applications and Error Analysis (1992)
163. *A. Charlier et al.,* Tensors and the Clifford Algebra (1992)

164. *P. Biler and T. Nadzieja*, Problems and Examples in Differential Equations (1992)
165. *E. Hansen*, Global Optimization Using Interval Analysis (1992)
166. *S. Guerre-Delabrière*, Classical Sequences in Banach Spaces (1992)
167. *Y. C. Wong*, Introductory Theory of Topological Vector Spaces (1992)
168. *S. H. Kulkarni and B. V. Limaye*, Real Function Algebras (1992)
169. *W. C. Brown*, Matrices Over Commutative Rings (1993)
170. *J. Loustau and M. Dillon*, Linear Geometry with Computer Graphics (1993)
171. *W. V. Petryshyn*, Approximation-Solvability of Nonlinear Functional and Differential Equations (1993)
172. *E. C. Young*, Vector and Tensor Analysis: Second Edition (1993)
173. *T. A. Bick*, Elementary Boundary Value Problems (1993)
174. *M. Pavel*, Fundamentals of Pattern Recognition: Second Edition (1993)
175. *S. A. Albeverio et al.*, Noncommutative Distributions (1993)
176. *W. Fulks*, Complex Variables (1993)
177. *M. M. Rao*, Conditional Measures and Applications (1993)
178. *A. Janicki and A. Weron*, Simulation and Chaotic Behavior of $\alpha$-Stable Stochastic Processes (1994)
179. *P. Neittaanmäki and D. Tiba*, Optimal Control of Nonlinear Parabolic Systems (1994)
180. *J. Cronin*, Differential Equations: Introduction and Qualitative Theory, Second Edition (1994)
181. *S. Heikkilä and V. Lakshmikantham*, Montotone Iterative Techniques for Discontinuous Nonlinear Differential Equations (1994)
182. *X. Mao*, Exponential Stability of Stochastic Differential Equations (1994)

*Additional Volumes in Preparation*

# MONOTONE ITERATIVE TECHNIQUES FOR DISCONTINUOUS NONLINEAR DIFFERENTIAL EQUATIONS

Seppo Heikkilä

*University of Oulu*
*Oulu, Finland*

V. Lakshmikantham

*Florida Institute of Technology*
*Melbourne, Florida*

CRC Press
Taylor & Francis Group
Boca Raton London New York

CRC Press is an imprint of the
Taylor & Francis Group, an **informa** business
A TAYLOR & FRANCIS BOOK

CRC Press
Taylor & Francis Group
6000 Broken Sound Parkway NW, Suite 300
Boca Raton, FL 33487-2742



ISBN-13: 978-0-824-79224-4 (hbk)



Publisher's Note

The publisher has gone to great lengths to ensure the quality of this reprint but points out that some imperfections in the original copies may be apparent.

**Library of Congress Cataloging-in-Publication Data**

Heikkilä, Seppo.
Monotone iterative techniques for discontinuous nonlinear differential equations / Seppo Heikkilä, V. Lakshmikantham.
p. cm. -- (Monographs and textbooks in pure and applied mathematics; 181)
Includes bibliographical references and index.
ISBN 0-8247-9224-6
1. Differential equations, Nonlinear--Numerical solutions. 2. Iterative methods (Mathematics). I. Lakshmikantham, V. II. Title. III. Series.
QA372.H47 1994
515'.355--dc20 94-2465
CIP

**Visit the Taylor & Francis Web site at**
**http://www.taylorandfrancis.com**

**and the CRC Press Web site at**
**http://www.crcpress.com**

# Preface

It is now well known that the method of upper and lower solutions coupled with the monotone iterative technique offers an effective and flexible mechanism for proving theoretical as well as constructive existence and comparison results for a variety of continuous nonlinear differential systems. The development of the corresponding theory for discontinuous nonlinear differential equations has taken different directions. To investigate only theoretical existence results, one can convert the given function which is discontinuous in the dependent variable into a set-valued map and then consider set-valued differential equations. Alternatively, one could employ Zorn's Lemma directly to prove existence results. However, the usual monotone iterative methods are no longer applicable if we wish to prove constructive existence results for discontinuous differential equations or for set-valued differential equations. In fact, owing to the intrinsic difficulties of relating the partial ordering with inclusion so as to reveal the monotone character of set-valued maps, no real progress is being made in this direction. A natural question is therefore whether one can develop some generalized monotone iterative method that is suitable to deal with discontinuous nonlinear problems to yield constructive results. The answer is positive and this is precisely what we attempt to describe in this monograph.

We shall develop in detail a generalized monotone iterative method in the context of partially ordered sets based on elementary set theory and using well-ordered chains of iterations. We then apply it to derive suitable fixed point theorems in ordered abstract spaces which are important tools in the investigation of existence and comparison results for discontinuous nonlinear problems.

Our aim in this book is to develop the basic theory, exhibit its common features and provide insight into certain discontinuous problems of ordinary and partial differential equations by means of systematic application of the developed techniques, including some problems at resonance.

Some of the characteristic features of the monograph are as follows: This is the first book that

(i) presents a systematic investigation of a generalized monotone iterative method in terms of upper and lower solutions that is suitable for the study of a variety of nonlinear discontinuous differential equations;
(ii) provides new existence and comparison results when the functions involved in the differential equations admit a threefold decomposition, namely, continuous, Lipschitz and discontinuous functions in the dependent variable;
(iii) includes the extensions of the method of upper and lower solutions and monotone iterative technique to Carathéodory systems in finite as well as infinite dimensional spaces;
(iv) contains classical monotone iterative methods as a special case and exhibits new results even in that framework;
(v) offers a comprehensive treatment of existence and comparison of extremal strong, weak or mild solutions to discontinuous differential equations in ordered Banach

spaces without requiring any kind of compactness hypotheses.

This book consists of five chapters treating first order, second order and partial differential equations of elliptic and parabolic type, as well as differential equations in Banach spaces. The necessary basic analysis of ordered spaces, generalized iteration ideas and required fixed point theorems are incorporated in the first chapter. Examples and special cases are given to demonstrate the theory throughout the book.

We wish to express our gratitude to Dr. S. Carl for all his contributions and corrections to Chapter 4, and to Dr. M. Kumpulainen for his careful reading of Chapter 1. We also want to thank Professors S. Leela, V. Mustonen and S. Seikkala for reading parts of the manuscript and for their comments. The preparation of the manuscript was facilitated by a grant of the Academy of Finland, and by loving care and support of Seija-Liisa Heikkilä. We are grateful to Donn Harnish for formatting the manuscript to the final form.

Seppo Heikkilä
V. Lakshmikantham

# Contents

# MONOTONE ITERATIVE TECHNIQUES FOR DISCONTINUOUS NONLINEAR DIFFERENTIAL EQUATIONS

# 1

# Methods and Prerequisites

## 1.0. INTRODUCTION

This chapter introduces the main ideas that are required in the development of a generalized monotone iterative technique, and therefore forms a basis for the remaining chapters. Moreover, we incorporate in this chapter other necessary concepts and tools needed for the later use.

In section 1.1 we present basic concepts relative to partially ordered sets, discuss a recursion principle and a generalized monotone iteration method, and study properties of chains, especially iteratively generated chains, well-ordered chains and monotone sequences in partially ordered sets as well as in ordered topological and metric spaces. Section 1.2 is devoted to derive fixed point theorems for nondecreasing and mixed monotone maps in ordered spaces utilizing the generalized monotone iterative technique. These fixed point results are mostly applied to equations in ordered Banach spaces and function spaces, the properties of which are introduced in section 1.3. Section 1.4 describes essential abstract calculus relative to measure, integration and differentiation of Banach-valued functions. This section also includes results on integral inequalities, presents some substitution theorems and fixed point results in function spaces. Finally, in section 1.5 we indicate the method of upper and lower solutions in the framework of Carathéodory systems of differential equations which are essential when we investigate existence results for discontinuous nonlinear differential equations. Here we consider initial value problems as well as periodic boundary value problems, relative to existence, uniqueness and extremality of their solutions, and derive comparison results which are required for our later use.

## 1.1. ANALYSIS IN ORDERED SPACES

We shall begin with introducing basic concepts related to partially ordered sets. After presenting a recursion principle and some of its applications, we shall give a detailed description of a generalized monotone iteration method in the context of partially ordered sets. For the applicability of this method to nonlinear problems, we shall also derive sufficient conditions for the existence of supremums and infimums of chains, as well as for countability of well-ordered chains and for the convergence of monotone sequences in ordered topological and metric spaces.

### 1.1.1. Introduction to theory of posets

Given a nonempty set $P$, we say that a relation $\leq$ is a *partial ordering* of $P$ if it is reflexive: $x \leq x$ for all $x \in P$, antisymmetric: $x \leq y$ and $y \leq x$ imply $x = y$, and transitive: $x \leq y$ and $y \leq z$ imply $x \leq z$. By a *poset* $P = (P, \leq)$ we mean a nonempty set $P$ equipped with a partial ordering relation $\leq$. If $x$, $y \in P$, denote $x < y$ when $x \leq y$ and $x \neq y$. The so obtained relation $<$, called *strict ordering* of $P$, is irreflexive: $x \not< x$ for all $x \in P$, and transitive. We shall use also notations $y \geq x$ for $x \leq y$ and $y > x$ for $x < y$. Given $a$, $b \in P$, denote $[a) = \{x \in P \mid a \leq x\}$,

$(b] = \{x \in P \mid x \leq b\}$ and $[a, b] = \{x \in P \mid a \leq x \leq b\}$. We say that $b \in P$ is an *upper bound* of a subset $A$ of a poset $P$ if $A \subseteq (b]$. If $A$ has an upper bound which belongs to $A$, we call it the *greatest element* or the *maximum* of $A$, and denote it by $\max A$. If $x < b$ for each $x \in A$, we call $b$ a *strict upper bound.* The notions of a (strict) lower bound and the least element of $A$, called also the minimum of $A$ and denoted by $\min A$, are defined similarly. If the set $B$ of the upper bounds of $A$ has the least element $b$, i.e. $b \in B$ and $B \subseteq [b)$, we say that $b$ is the *least upper bound*, or the *supremum* of $A$, and denote $b = \sup A$. A notion of the greatest lower bound, or the infimum of $A$, denoted by $\inf A$, is defined similarly. An element $x$ of $A$ is called *maximal* if there does not exist any element $y \in A$ such that $x < y$. If $A$ has the greatest element, it is also maximal. $A$ is said to be *order bounded* if $A$ has both a lower bound and an upper bound, or equivalently if $A$ is contained in an order interval $[a, b]$.

We say that a poset $P$ is a *lattice* if $\sup\{x, y\}$ and $\inf\{x, y\}$ exist for all $x$, $y \in P$. A subset $C$ of $P$ is said to be *a chain*, or *totally ordered* if $x \leq y$ or $y \leq x$ for all $x$, $y \in C$. For instance, the set $\mathbb{R}$ of the real numbers is totally ordered with respect to its natural ordering. $C$ is called *well-ordered* if each nonempty subset of $C$ has the least element. If each nonempty subset of $C$ has the greatest element, we say that $C$ is *inversely well-ordered.* A set $D$ together with a reflexive and transitive relation $\leq$ is called *directed* if for each pair $x$, $y \in D$ there exists $z \in D$ such that $x \leq z$ and $y \leq z$. Each well-ordered set is a chain, each chain is a lattice and each lattice is directed. Note that the empty set $\emptyset$ is a well-ordered subset of any poset. A subset $A$ of a set $C \subseteq P$ is said to be *cofinal* in $C$ if for each $x \in C$ there is $y \in A$ such that $x \leq y$.

A set $C$ is called *countable* if $C$ is either finite or there is a bijection from the set $\mathbb{N}$ of natural numbers onto $C$. We say that a sequence $(x_n)_{n=o}^{\infty}$ in a poset $P$ is *nondecreasing* (resp. *increasing*) if $x_n \leq x_m$ (resp. $x_n < x_m$) whenever $n$, $m \in \mathbb{N}$ and $n < m$. A nonincreasing (resp. decreasing) sequence is defined similarly by reversing the above inequalities between $x_n$ and $x_m$. A sequence is called *monotone* if it is nondecreasing or nonincreasing. Denote by $\sup_n x_n$ and $\inf_n x_n$ the possible supremum and infimum of $\{x_n\}_{n\in I\!N}$ in $P$.

### 1.1.2. A recursion principle and applications

The following lemma contains a recursion principle which has many interesting applications. For instance, it forms a basis to the generalized monotone iteration method introduced later on. In the proof we shall need only elementary tools of set theory.

**Lemma 1.1.1:** *If $\mathcal{D}$ is a set of subsets of a poset $P$ with $\emptyset \in \mathcal{D}$ and $f : \mathcal{D} \to P$, there is a unique well-ordered chain $C$ so that*

$$x \in C \text{ if and only if } x = f\{y \in C \mid y < x\}.$$

*If $f(C)$ exists, it is not a strict upper bound of $C$.*

*Proof.* To shorten notations, denote

$$C^{<x} = \{y \in C \mid y < x\}, \text{ when } x \in P \text{ and } C \subseteq P.$$

A chain $C$ of $P$ is called *conforming* if it is well-ordered and $x = f(C^{<x})$ for all $x \in C$. For instance, $\{f(\emptyset)\}$ is conforming. We shall first show that conforming chains satisfy

(a) *If $A$ and $B$ are conforming and $A \not\subseteq B$, then $B = A^{<x}$ with $x = \min(A \setminus B)$.*

The choice of $x$ implies that $A^{<x} \subseteq B$. To prove that $B = A^{<x}$, make a counter-hypothesis: $B \setminus A^{<x}$ is nonempty. Then $B \setminus A^{<x}$ has the least element $y$ and $B^{<y} \subseteq A^{<x}$. Strict inclusion holds, for otherwise $y = f(B^{<y}) = f(A^{<x}) = x$, which is impossible, since $y \in B$ and $x \notin B$. Thus $z = \min(A^{<x} \setminus B^{<y})$ exists and $A^{<z} \subseteq B^{<y}$. Since $z \in A^{<x} \subseteq B$, $z \notin B^{<y}$ and $y \in B$, then $y \leq z < x$. This and $B^{<y} \subseteq A^{<x}$ imply that $B^{<y} \subseteq A^{<y} \subseteq A^{<z}$. Therefore $B^{<y} = A^{<z}$, whence $y = f(B^{<y}) = f(A^{<z}) = z$. But then $y \in A$ and $y < x$, so that $y \in A^{<x}$, which contradicts the choice of $y$. Thus $B \setminus A^{<x}$ is empty and $B = A^{<x}$.

Property (a) of the conforming chains is now applied to prove

(b) *The union $C$ of all the conforming chains is well-ordered and satisfies $x \in C$ if and only if $x = f(C^{<x})$.*

To show that $C$ is well-ordered, let $Y$ be a nonempty subset of $C$. Choose a conforming chain $B$ so that $Y \cap B$ is nonempty, and denote $y = \min(Y \cap B)$. Given any $x \in Y$ there is a conforming chain $A$ such that $x \in A$. If $x \in B$, then $x \in Y \cap B$, whence $y \leq x$. If $x \notin B$, then $B \subseteq A^{<x}$ by (a), so that $y < x$. Thus $y \leq x$ for each $x \in Y$, whence $y = \min Y$ exists, and $C$ is well-ordered.

To prove that $x = f(C^{<x})$ for given $x \in C$, choose a conforming chain $B$ so that $x \in B$. Given $y \in C^{<x}$, there is a conforming chain $A$ such that $y \in A^{<x}$. If $B \subset A$, then $B = A^{<z}$ for some $z \in A$, by (a). Since $x \in B$, then $x < z$, whence $B^{<x} = (A^{<z})^{<x} = A^{<x}$. Thus $y$ is in $B^{<x}$. This holds also when $A \subseteq B$, so that $C^{<x} \subseteq B^{<x}$. Obviously, $B^{<x} \subseteq C^{<x}$, whence $C^{<x} = B^{<x}$. Because $B$ is conforming, then $x = f(B^{<x}) = f(C^{<x})$.

Conversely, if $x \in P$ and $x = f(C^{<x})$, then $C^{<x} \cup \{x\}$ is conforming, whence $x \in C$. Thus $x \in C$ iff $x = f(C^{<x})$.

To prove the uniqueness, let $B$ be another well-ordered chain in $P$ for which $x \in B$ iff $x = f(B^{<x})$. Since $B$ is conforming, then $B \subseteq C$. If $B \neq C$, it follows from (a) that $B = C^{<x}$ where $x = \min(C \setminus B)$. But then $f(B^{<x}) = f(C^{<x}) = x$ exists and $x \notin B$, which contradicts the defining property of $B$. Thus $B = C$.

Finally, if $f(C)$ exists, it cannot be a strict upper bound of $C$, for otherwise $C \cup \{f(C)\}$ would be a conforming chain and is not contained in $C$, which is impossible since $C$ is the union of all the conforming chains. □

Before applications let us consider which kind of elements the well-ordered chain $C$ defined in lemma 1.1.1 contains. $x_0 = f(\emptyset)$ is its least element, and the next possible elements of $C$ are of the form $x_n = f(\{x_0, \ldots, x_{n-1}\})$, as long as $x_n$ is defined and $x_{n-1} < x_n$. If $x_\omega = f(\{x_n\}_{n=0}^\infty)$ exists and is a strict upper bound of $\{x_n\}_{n=0}^\infty$, then $x_\omega$ is the next element of $C$, and so on.

Denote by $2^P$ the set of all subsets of $P$. We say that $x \in P$ is a *fixed point* of a multifunction $F\colon P \to 2^P$ if $x \in Fx$. As an application of lemma 1.1.1 and the axiom of choice we obtain

**Proposition 1.1.1:** *Given a poset $P$ and a multifunction $F\colon P \to 2^P$, assume that each well-ordered chain $C$ in $\bigcup F[P]$ has such an upper bound $x$ in $P$ that $x \leq y$ for some $y$ in $Fx$. Then $F$ has a maximal fixed point $x$, which is also a maximal element of $Fx$.*

*Proof.* Denote by $\mathcal{D}$ the set of all those well-ordered chains $C$ in $\bigcup F[P]$ which have a strict upper bound in $\bigcup F[P]$. Let $f\colon \mathcal{D} \to \bigcup F[P]$ be a choice function of such upper bounds, and let $C$ be the well-ordered chain for which $x \in C$ iff $x = f(C^{<x})$. By hypothesis $C$ has such an upper bound $x \in P$ that $x \leq y$ for some $y \in Fx$. Equality $x = y$ must hold, i.e. $x$ is a fixed point of $F$, for otherwise $f(C)$ would exist and would be a strict upper bound of $C$, which contradicts the last conclusion of lemma 1.1.1. By the same reason, $x$ is a maximal fixed point of $F$ and a maximal element of $Fx$. □

If $Fx \equiv P$, proposition 1.1.1 is reduced to the following result.

**Zorn's Lemma:** *If each well-ordered chain in $P$ has an upper bound in $P$, then $P$ has a maximal element.*

This form of Zorn's Lemma is due to Bourbaki (1949-1950). It implies the classical form, where all chains of $P$, not only well-ordered chains, are assumed to possess an upper bound in $P$. Converse holds as well, because of the following consequence of lemma 1.1.1 and the axiom of choice.

**Lemma 1.1.2:** *Each chain of any poset contains a well-ordered cofinal chain.*

*Proof.* Make a counter-hypothesis: there is a chain $P$ such that each of its well-ordered chain has a strict upper bound in $P$. $P$

is nonempty, whence we can choose an element $a$ from it. Let $\mathcal{D}$ consist of the empty set and all the well-ordered chains in $P$ whose least element is $a$. Choose a function $f\colon \mathcal{D} \to P$ which assigns to each nonempty $A \in \mathcal{D}$ its strict upper bound in $P$, and for which $f(\emptyset) = a$. Let $C$ be the well-ordered chain in $P$ such that $x \in C$ iff $x = f(C^{<x})$. Since $\emptyset \neq C \in \mathcal{D}$, the definition of $f$ implies that $f(C)$ exists and is a strict upper bound of $C$. But this contradicts the last conclusion of lemma 1.1.1, whence there is a well-ordered chain $W$ in $P$ which has no strict upper bound. In particular, $W$ is cofinal in $P$. □

We say that $x \in P$ is a *fixed point* of a mapping $F\colon P \to P$ if $x = Fx$. The single-valued version of proposition 1.1.1 reads as follows.

**Proposition 1.1.2:** *A mapping $F\colon P \to P$ has a maximal fixed point if each well-ordered chain in $F[P]$ has such an upper bound $x$ in $P$ that $x \leq Fx$.*

### 1.1.3. A generalized monotone iteration method

As an application of lemma 1.1.1 we shall now introduce a generalized monotone iteration method and apply it to prove a fixed point theorem for a mapping $G\colon P \to P$, where $P$ is a poset.

An ordinary iteration method is inadequate, for instance, when its application leads to a nonending string of increasing iteration sequences $(G^n a_i)_{n=o}^{\infty}$, where $a_o \in P$ is given and $a_i$ is the least upper bound of $\{G^n a_{i-1}\}_{n\in\mathbb{N}}$ for $i = 1, 2, \ldots$. Zorn's lemma provides "a precise way of expressing the fact that we can establish a never-ending string of such denumerable sets", as stated in Lang (1969), pp. 13-14. The use of Zorn's lemma, while it yields fixed point theorems having many interesting applications to differential equations (cf. Amann (1977), Stuart (1978)), however leans heavily on the axiom of choice, and thus gives no idea of how to construct solutions of the differential equations in question. An alternative method is to use monotone transfinite sequences of iterations (cf. Heikkilä (1990)). However, to define proper indexes, ordinals, of such transfinite sequences one needs also the axiom schema of replacement (cf. Krivine (1971), Levy (1979), Mendelson (1987)). Moreover, the collection of all the ordinals is not a set, whence the elementary set theory does not serve a sufficient background for the use of such generalized sequences.

The generalized monotone iterative method presented in the next theorem is based on the use of elementary set theory, and has no dependence on the axiom of choice or the axiom schema of replacement.

**Theorem 1.1.1:** *Given $G: P \to P$ and $a \in P$, there is a unique well-ordered chain $C$, called a well-ordered (w.o.) chain of $G$-iterations of $a$, in $P$ satisfying*
(I) $a = \min C$, *and* $a < x \in C$ *if and only if* $x = \sup G[C^{<x}]$.
*If $x_* = \sup G[C]$ exists and $a \leq x_* \leq Gx_*$, then $x_* = \max C$ and $x_*$ is a fixed point of $G$.*

*Proof.* The first assertion follows from lemma 1.1.1 when we define $f(\emptyset) = a$ and $f(U) = \sup G[U]$ if $\emptyset \neq U \subseteq P$ and $\sup G[U]$ exists. Assume now that $x_* = \sup G[C]$ exists and $a \leq x_* \leq Gx_*$, where $C$ is defined by (I). If $x \in C$ and $x \neq a$, then $x = \sup G[C^{<x}] \leq \sup G[C] = x_*$. Thus $x_*$ is an upper bound of $C$, whence lemma 1.1.1 implies that $x_* = \max C$. In particular, $Gx_* \leq \sup G[C] = x_*$. This and $x_* \leq Gx_*$ imply that $x_* = Gx_*$. □

The following result characterizes the w.o. chains of iterations.

**Lemma 1.1.3:** *Given $G: P \to P$ and $a \in P$, let $C$ be the w.o. chain of $G$-iterations of $a$. Then*

$$Gx = \min\{y \in C \mid x < y\} \quad \textit{for all } x \in C \;\; x < Gx. \tag{1.1.1}$$

*Proof.* Let $x \in C$, $x < Gx$, be given. Denoting $z = \min\{y \in C \mid x < y\}$, then $z = \sup G[C^{<z}] = \sup G[C^{<x} \cup \{x\}]$. If $x = a$, then $C^{<x} = \emptyset$, whence $z = Gx$. If $a < x$, then $x = \sup G[C^{<x}] < z = \sup G[C^{<x} \cup \{x\}]$, whence $z = Gx$ also in this case. Thus (1.1.1) holds. □

Denote $[a, b) = \{x \in P \mid a \leq x < b\}$ and $(a, b) = \{x \in P \mid a < x < b\}$ when $a, b \in P$. As an immediate consequence of theorem 1.1.1 and lemma 1.1.3 we obtain

**Corollary 1.1.1:** *Let $[a, b]$ be an order interval of a totally ordered space whose well-ordered chains have supremums. If $G\colon [a, b] \to [a, b]$ such that $y < Gy$ for all $y \in [a, b)$, and $Gb = b$, and if $C$ is the w.o. chain $C$ of $G$-iterations of $a$, then $[a, b]$ is the disjoint union of $C$ and the intervals $(x, Gx)$, $x \in C$.*

*Proof.* $G$ satisfies the hypotheses of theorem 1.1.1 with $P = [a, b]$. Thus $\max C$ is a fixed point of $G$, whence $b = \max C$. Let $t \in [a, b] \setminus C$ be given, and denote $z = \min\{y \in C \mid t < y\}$. Since $t < z$ and $z = \sup G[C^{<z}]$, there is $x \in C$, $x < z$, such that $t < Gx$. The choice of $z$ implies that $x < t$. Thus $t \in (x, Gx)$. Conversely, from (1.1.1) it follows that $(x, Gx) \subseteq [a, b] \setminus C$ for each $x \in C$, whence $[a, b] \setminus C$ is the union of the intervals $(x, Gx)$, $x \in C$. Since $P$ is totally ordered, it follows from (1.1.1) that these intervals are disjoint. □

From lemma 1.1.3 it follows that the least elements of the w.o. chain $C$ of $G$-iterations of $a$ are the elements of the iteration sequences $(G^n a_i)_{n=o}^{\infty}$ with $a_o = a$ and $a_i = \sup_n G^n a_{i-1}$, $i = 1, 2, \ldots$, as long as these sequences exist and are increasing. In particular, we have

**Corollary 1.1.2:** *Let $G$ in theorem 1.1.1 be nondecreasing in $[a)$. Define a sequence $(a_m)_{m=o}^{\infty}$ by $a_o = a$ and*

$$a_{m+1} = \sup_{n_m} G^{n_m}\big(\sup_{n_{m-1}} G^{n_{m-1}}(\cdots(\sup_{n_o} G^{n_o} a)\cdots)\big), \; m \in \mathbb{N}.$$

*Then* $a_m \leq x_*$ *for each* $m \in I\!N$, *and* $x_* = a_m$ *if and only if* $a_m = Ga_m$.

The purpose of the following example is to ensure that ordinary iterations are inadequate to describe w.o. chains of iterations, even in the case when $G\colon [0,1] \to [0,1]$ is a nondecreasing stepfunction.

**Example 1.1.1:** Given $a$, $b \in I\!R$, $a < b$, denote

$$P_o = P_o[a,b) = \{b - 2^{-n}(b-a) \mid n \in I\!N\}.$$

The points of $P_o[a,b)$ subdivide $[a,b)$ into infinite number of disjoint subintervals

$$[a_n, b_n) = [b - 2^{-n}(b-a), b - 2^{-n-1}(b-a)), \quad n \in I\!N.$$

If each of these subintervals are partitioned by $P_o$ we obtain a more dense partition of $[a,b)$, denote it by $P_o^2$. Thus

$$P_o^2 = P_o^2[a,b) = \bigcup_{n=o}^{\infty} P_o[a_n, b_n).$$

Continuing this process we obtain a sequence of partitions $P_o^m = P_o^m[a,b)$, $m = 1, 2, \ldots$, of $[a,b)$.

Consider now the interval $[0,1)$. Make first the partition $P_o[0,1)$. Each of the so obtained subintervals $[1-2^{-m}, 1-2^{-m-1})$, $m \in I\!N$, are then partitioned by $P_o^{m+1}$. The resulting partition of $[0,1)$ is denoted by $P_1$. Thus

$$P_1 = P_1[0,1) = \bigcup_{m=o}^{\infty} P_o^{m+1}[1-2^{-m}, 1-2^{-m-1}).$$

It is elementary matter to show that

$$P_o^{m+1}[1-2^{-m}, 1-2^{-m-1}) = \{c(n_o, \ldots, c_m) \mid n_o, \ldots, n_m \in \mathbb{N}\},$$

where

$$\begin{aligned} c(n_o, \ldots, n_m) =& 1 - 2^{-m-1} - \sum_{k=o}^{m} 2^{-k-m-2} \prod_{j=o}^{k} 2^{-n_j} \\ & - 2^{-2m-2} \prod_{j=o}^{m} 2^{-n_j}. \end{aligned} \qquad (1.1.2)$$

To analyze the set $P_1[0,1)$, notice first that each of its elements is a rational number. Denote $D = \{(n_o, \ldots, n_m) \mid m, n_o, \ldots, n_m \in \mathbb{N}\}$, and define equality and strict ordering in $D$ by $(n_o, \ldots, n_m) = (p_o, \ldots, p_q)$ iff $m = q$ and $n_j = p_j$ for all $j = 1, \ldots m$, and $(n_o, \ldots, n_m) < (p_o, \ldots, p_q)$ if $m = q$ and $n_i < p_i$ where $i = \min\{j \mid n_j \neq p_j\}$, or if $m < q$.

By using the properties of natural numbers it is easy to see that condition $s \leq t$ iff $s = t$ or $s < t$ defines a well-ordering relation on $D$.

Denoting by $0_m$ the null sequence of length $m$, we have for each $m \in \mathbb{N}$,

$$\begin{aligned} c(0_{m+1}) &= 1 - 2^{-m-1} - \sum_{k=o}^{m} 2^{-k-m-2} - 2^{-2m-2} \\ &= 1 - 2^{-m-1} - 2^{-m-1} + 2^{-2m-2} - 2^{-2m-2} = 1 - 2^{-m}. \end{aligned}$$

From (1.1.2) it follows that

$$c(n_o, \ldots, n_m) \to 1 - 2^{-m-1} = c(0_{m+2}) \quad \text{as } n_o, \ldots, n_m \to \infty.$$

Since $c(n_o, \ldots, n_m)$ is increasing with respect to each $n_j$, we then have for each $(n_o, \ldots, n_m) \in D$,

$$c(0_{m+1}) = 1 - 2^{-m} \leq c(n_o, \ldots, n_m) < c(0_{m+2}) \qquad \text{(a)}$$

To prove that $c(n_o, \ldots, n_m)$ increases when $(n_o, \ldots, n_m)$ increases with respect to the above defined well-ordering of $D$, assume that $(n_o, \ldots, n_m) < (p_o, \ldots, p_q)$. If $m < q$, it follows from (a) that

$$c(n_o, \ldots, n_m) < c(0_{m+2}) \leq c(0_{q+1}) \leq c(p_o, \ldots, p_q).$$

If $m = q$, there is an index $i$ such that $n_j = p_j$ for $j < i$ and $n_i < p_i$. If $i = m$, then

$$c(n_o, \ldots, n_m) < a(n_o, \ldots, n_{m-1}, p_m) \leq c(p_o, \ldots, p_q).$$

Assume now that $i < m$. If $n_{i+1}, \ldots, n_m \to \infty$, it follows from (1.1.2) that $c(n_o, \ldots, n_m)$ increases to the number

$$c = 1 - 2^{-m-1} - \sum_{k=o}^{i+1} 2^{-k-m-1} \prod_{j=o}^{k} 2^{-n_j}.$$

Since

$$c(n_o, \ldots, n_i + 1, 0_{m-i}) = 1 - 2^{-m-1} - \sum_{k=0}^{i-1} 2^{-k-m-2} \prod_{j=o}^{k} 2^{-n_j}$$

$$- \sum_{k=i}^{m} 2^{-k-m-3} \prod_{j=o}^{i} 2^{-n_j} - 2^{-2m-3} \prod_{j=o}^{i} 2^{-n_j},$$

we have

$$\begin{aligned} c - c(n_o, \dots, n_i + 1, 0_{m-i}) \\ = \sum_{k=i}^{m} 2^{-k-m-3} \prod_{j=o}^{i} 2^{-n_j} + 2^{-2m-3} \prod_{j=o}^{i} 2^{-n_j} - 2^{-i-m-2} \prod_{j=o}^{i} 2^{-n_j} \\ = 2^{-m-2} \prod_{j=o}^{i} 2^{-n_j} \left( \sum_{k=i}^{m} 2^{-k-1} + 2^{-m-1} - 2^{-i} \right) = 0. \end{aligned}$$

Thus $c(n_o, \dots, n_i + 1, 0_{m-i}) = c$. In particular,

$$c(n_o, \dots, n_i + 1, 0_{m-i}) = \sup\{c(n_o, \dots, n_m) \mid n_{i+1}, \dots, n_m \in I\!N\}.$$

Noticing that in the considered case

$$c(n_o, \dots, n_{i-1}, n_i + 1, 0_{m-i}) \le c(p_1, \dots, p_q),$$

it follows that $c(n_o, \dots, n_m) < c(p_o, \dots, p_q)$ also in this case.

The above reasoning shows that

$$c(n_o, \dots, n_m) < c(p_o, \dots, p_q) \text{ if } (n_o, \dots, n_m) < (p_o, \dots, p_q).$$

The set $C = \{c(t) \mid t \in D\}$ is then a well-ordered chain in the set $Q$ of rational numbers with $\min C = 0$ and $\sup C = 1$.

Let $G\colon Q \to Q$ be any mapping for which

$$Gc(n_o, \dots, n_m) = c(n_o, \dots, n_m + 1) \tag{1.1.3}$$

for each $(n_o, \dots, n_m) \in D$. In particular, $Gx$ is for $x \in C$ the least element of $C$ greater than $x$. If $G1 < 1$, it can be shown that $C$ is the w.o. chain of $G$-iterations of 0. If $G1 = 1$, then

$C \cup \{1\}$ is the w.o. chain of $G$-iterations of 0 with 1 its maximum, and also a fixed point of $G$. It is easy to see that

$$\sup_m(\sup_{n_m} G^{n_m}(\sup_{n_{m-1}} G^{n_{m-1}}(\cdots(\sup_{n_o} G^{n_o}0)\cdots)) = c(0,0,0),$$

whence ordinary iterations don't suffice to describe $C$. Moreover, we can continue the partition of $[0,1)$ with replacing $P_o$ above by $P_1$, and obtain a more dense partition process $P_2$. If $P_1$ is replaced by $P_2$, and so on, we obtain a sequence of partitions $P_m$ of $[0,1)$, each of which leads to a well-ordered chain of rational numbers.

### 1.1.4. Countable chains and monotone sequences

As we shall see, the result of corollary 1.1.1 has important applications in analysis in the case when we can prove that $C$ is countable. A sufficient condition for this is given in the following lemma.

**Lemma 1.1.4:** *A well-ordered chain $C$ in a poset $P$ is countable if its subchains possess countable cofinal chains.*

*Proof.* Make a counter-hypothesis: There is an uncountable well-ordered chain $C$ in $P$ whose each subchain has a countable cofinal chain. We may assume that $C^{<x}$ is countable for each $x \in C$, for otherwise we replace $C$ by $C^{<z}$, where $z$ is the least element of $C$ for which $C^{<z}$ is uncountable.

Assume first that $C$ does not have the maximum, and let $\{x_n\}_{n=o}^{\infty}$ be a cofinal chain in $C$. If $x \in C$, there is $y \in C$ such that $x < y$, since $\max C$ does not exist, and $n \in \mathbb{N}$ so that $y \leq x_n$, because $\{x_n\}_{n=o}^{\infty}$ is cofinal in $C$. This proves that $C = \bigcup_{n=o}^{\infty} C^{<x_n}$, which is countable, since each $C^{<x_n}$ is countable.

Assume next that $z = \max C$ exists. Since $z \in C$, then $C^{<z}$ is countable, whence also $C = C^{<z} \cup \{z\}$ is countable.

The above proof shows that in all cases $C$ is countable, which contradicts the counter-hypothesis, and hence proves the assertion. □

The following results are also needed in the sequel.

**Lemma 1.1.5:** *Let $(x_n)_{n=o}^{\infty}$ be a sequence in a poset $P$*

a) *If $(x_n)_{n=o}^{\infty}$ is totally ordered, it has a monotone subsequence.*

b) *If $(x_n)_{n=o}^{\infty}$ is nondecreasing (resp. nonincreasing), then it has the supremum (resp. the infimum) $x$ if and only if $x$ is the supremum (resp. the infimum) of some of its subsequences.*

*Proof.* To prove a), assume first that $(x_n)_{n=o}^{\infty}$ does not possess the greatest element. Since $(x_n)_{n=o}^{\infty}$ is totally ordered, then to each $x_n$ there is $m \in \mathbb{N}$ such that $x_n < x_m$. This implies that $(x_n)_{n=o}^{\infty}$ has an increasing subsequence.

Assume then that $(x_n)_{n=o}^{\infty}$ does not contain any increasing subsequence. Thus for each $m \in \mathbb{N}$ the sequence $(x_n)_{n=m}^{\infty}$ has no increasing subsequence, and hence has the greatest element by the first part of the proof. Let $x_{n_o}$ be the greatest element of $(x_n)_{n=o}^{\infty}$, and when $x_{n_k}$ is chosen, denote by $x_{n_{k+1}}$ the greatest element of $(x_n)_{n=n_k+1}^{\infty}$. This construction yields a nonincreasing subsequence $(x_{n_k})_{k=o}^{\infty}$.

The assertion b) is an easy consequence of the fact that each subsequence of a monotone sequence is cofinal. □

### 1.1.5. Ordered topological spaces

In order to apply the generalized monotone iteration method introduced in theorem 1.1.1 one has to know sufficient conditions for the existence of supremums of well-ordered chains. Such conditions can be derived, for instance, by using topological methods. We assume that the reader is familiar with basic concepts related to topology (cf. Nagata (1974), Narici and Beckenstein (1985)).

By an *ordered topological space* we mean a nonempty set $X$ equipped with a partial ordering $\leq$ and a topology such that the sets $[a)$ and $(a]$ are closed for each $a \in X$. This implies also that each order interval $[a, b]$ of $X$ is closed. We say that $x \in X$ is a *cluster point* of a sequence $(x_n)_{n=o}^{\infty}$ of $X$ if for each neighborhood $U$ of $x$ and for each $n \in I\!N$ there is $m \geq n$ such that $x_m \in U$. In particular, each subsequential limit is a cluster point. Moreover, if $(x_n)_{n=o}^{\infty}$ belongs to a subset $Y$ of $X$, then each cluster point of $(x_n)_{n=o}^{\infty}$ belongs to the closure $\overline{Y}$ of $Y$.

**Proposition 1.1.3:** *If a monotone sequence $(x_n)_{n=o}^{\infty}$ in an ordered topological space $X$ has a cluster point $x$, then $x = \sup_n x_n$ if $(x_n)_{n=o}^{\infty}$ is nondecreasing, and $x = \inf_n x_n$ if $(x_n)_{n=o}^{\infty}$ is nonincreasing.*

*Proof.* Assume that $x$ is a cluster point of a nondecreasing sequence $(x_n)_{n=o}^{\infty}$. Given $n \in I\!N$ and any open neighborhood $U$ of $x$, there is $m \geq n$ such that $x_m \in U$. Since $x_n \leq x_m$, then $[x_n) \cap U$ is nonempty. Thus $x \in \overline{[x_n)}$, and since $[x_n)$ is closed, then $x \in [x_n)$, i.e. $x_n \leq x$. This holds for each $n \in I\!N$, whence $x$ is an upper bound of $\{x_n\}_{n \in I\!N}$. If $z$ is an upper bound of $\{x_n\}_{n \in I\!N}$, then $\{x_n\}_{n \in I\!N}$ is contained in $(z]$. Since $(z]$ is closed and $x$ belongs to $\overline{\{x_n\}_{n \in I\!N}}$, then $x$ belongs to $(z]$, so that $x \leq z$. Thus $x = \sup_n x_n$. The proof that $x = \inf_n x_n$ if $x$ is a cluster point of a nonincreasing sequence $(x_n)_{n=o}^{\infty}$ is similar. □

**Corollary 1.1.3:** *A monotone sequence in an ordered topological space $X$ converges if its each subsequence has a cluster point.*

*Proof.* Assume that each subsequence of a nondecreasing sequence $(x_n)_{n=o}^{\infty}$ of $X$ has a cluster point. Let $x$ be a cluster point of $(x_n)_{n=o}^{\infty}$. Make a counter-hypothesis: $(x_n)_{n=o}^{\infty}$ does not converge to $x$. Thus there is a neighborhood $V$ of $x$ and a subsequence $(x_{n_k})_{k=o}^{\infty}$ of $(x_n)_{n=o}^{\infty}$ such that $x_{n_k} \notin V$ for each $k \in I\!N$. By a hypothesis, $(x_{n_k})_{k=o}^{\infty}$ has a cluster point $y$. It is also a cluster point of $(x_n)_{n=o}^{\infty}$, whence $y = x$ by proposition 1.1.3. But this contradicts the fact that $V$ does not contain any element of $(x_{n_k})_{k=o}^{\infty}$. Thus $(x_n)_{n=o}^{\infty}$ converges to $x$. The proof in the case when $(x_n)_{n=o}^{\infty}$ is nonincreasing is similar. □

**Lemma 1.1.6:** *Let $C$ be a well-ordered chain in an ordered topological space. If each subchain $D$ of $C$ contains a nondecreasing sequence which converges to* $\sup D$, *then $C$ is countable.*

*Proof.* By lemma 1.1.4 it suffices to show that each subchain of $C$ has a countable cofinal chain. Let $D$ be a subchain of $C$. By hypothesis $D$ contains a nondecreasing sequence $(x_n)_{n=o}^{\infty}$ which converges to $x = \sup D$. If $x \in D$, then $\{x\}$ is a countable cofinal subchain of $D$. Assume next that $x \notin D$. If $y \in D$, there is $n \in I\!N$ such that $y \leq x_n$, for otherwise $x_n \in (y]$ for each $n \in I\!N$, whence $x = \lim_n x_n \leq y$. But this is impossible, since $x = \sup D \notin D$. Thus $\{x_n\}_{n=o}^{\infty}$ is a countable cofinal chain in $D$. □

**Proposition 1.1.4:** *A chain $C$ in an ordered topological space $X$ has the supremum if $C$ has a relatively compact cofinal subset. If $X$ is first countable, then $C$ contains a nondecreasing sequence which converges to* $\sup C$.

*Proof.* Let $C$ be a chain in $X$, and assume that $B$ is a relatively compact cofinal subset of $C$. Since $B$ is a chain, then $\{[y) \cap \overline{B} \mid y \in B\}$ is a family of closed subsets of $\overline{B}$ possessing a finite intersection property. Since $\overline{B}$ is compact, then $\bigcap\{[y) \cap \overline{B} \mid y \in B\}$ contains a point $x$. In particular, $x \in [y)$ for each $y \in B$, whence $x$ is an upper bound of $B$. If $z$ is an upper bound of $B$, it follows from $x \in \overline{B} \subseteq \overline{(z]} = (z]$ that $x \leq z$, whence $x = \sup B$.

Given any $z \in C$ there is $y \in B$ such that $z \leq y \leq x$, whence $x$ is an upper bound of $C$. If $w$ is any upper bound of $C$, then it is also an upper bound of $B$, whence $x \leq w$. Thus $x = \sup C$.

If $x \in C$, then $x = \max C$, so that choosing $y_n = x$, $n \in I\!N$, we obtain a nondecreasing sequence $(y_n)_{n=o}^{\infty}$ in $C$, which converges to $x$. Assume next that $x \notin C$, and that $X$ is first countable. Let $\{U_n\}_{n \in I\!N}$ be an open base at $x$. We may assume that

$U_{n+1} \subseteq U_n$ for each $n \in I\!N$. Since $x \in \overline{C}$ by the above proof, we can pick for each $n \in I\!N$ an element $x_n$ from $U_n \cap C$. The so obtained sequence converges to $x$. Because $x \notin C$, then $x_n < x$ for each $n \in I\!N$. This implies that the sequence $(x_n)_{n=o}^{\infty}$ is totally ordered and cannot have the greatest element, whence it has an increasing subsequence by the proof of lemma 1.1.5, and it converges to $x$. □

**Lemma 1.1.7:** *If a chain $C$ in an ordered topological space $X$ has a separable cofinal subset $A$, and if each nondecreasing sequence of $A$ has a cluster point in $X$, then $C$ contains a nondecreasing sequence which converges to* $\sup C$.

*Proof.* Let $B = \{y_k\}_{k=o}^{\infty}$ be a dense subset of $A$. Denoting

(a) $x_n = \max\{y_k \mid 0 \le k \le n\}, \quad n \in I\!N,$

we obtain a nondecreasing sequence $(x_n)_{n=o}^{\infty}$ in $A$. By hypothesis each subsequence of $(x_n)_{n=o}^{\infty}$ has a cluster point, whence $(x_n)_{n=o}^{\infty}$ converges by proposition 1.1.3 and corollary 1.1.3 to its supremum $x$. This implies by (a) that $y_n \le x_n \le x$ for each $n \in I\!N$, so that $B$ is contained in $(x]$. Since $(x]$ is closed and $A \subseteq \overline{B}$, then $A \subseteq (x]$, whence $x$ is an upper bound of $A$, and hence an upper bound of $C$, since $A$ is cofinal in $C$. In fact, $x = \sup C$, for if $y$ is any upper bound of $C$, then $x_n \le y$ for each $n \in I\!N$, which ensures that $x \le y$. □

### 1.1.6. Ordered metric spaces

By an *ordered metric space* we mean a metric space $X = (X, d)$ equipped with a partial ordering $\le$ such that the intervals $(x]$ and $[x)$ are closed for each $x \in X$. We say that a subset $Y$ of a metric space $X$ is *bounded* if there is $M > 0$ such that $d(x, y) \le M$ for all $x$, $y \in Y$. It is well-known that a subset of a metric space is compact if and only if it is sequentially or countably compact, and that each compact set is closed and bounded.

**Proposition 1.1.5:** *Let $C$ be a chain in an ordered metric space $X$. If each nondecreasing sequence of $C$ has a cluster point, then $C$ contains a nondecreasing sequence which converges to* $\sup C$.

*Proof.* Let $x_o$ be an element of $C$. The assumptions imply that for each $n = 1, 2, \ldots$ there is an element $x_n$ in $C$ such that $x_{n-1} \leq x_n$ and $\operatorname{diam}([x_n) \cap C) < \frac{1}{n}$. For otherwise we can construct a nondecreasing sequence $(y_k)_{k=o}^{\infty}$ of $C$ such that $d(y_k, y_{k+1}) \geq \frac{1}{n}$ for some $n$ and for all $k \in \mathbb{N}$, whence $(y_k)_{k=o}^{\infty}$ diverges, which is impossible, since all the nondecreasing sequences of $C$ converge by corollary 1.1.3. Because the sequence $(x_n)_{n=o}^{\infty}$ is nondecreasing, then $x = \lim_n x_n$ exists.

To show that $x = \sup C$, let $y \in C$ be given. If $y \leq x_n$ for some $n \in \mathbb{N}$, then $y \leq \lim_n x_n = x$. On the other hand, if $x_n \leq y$ for each $n \in \mathbb{N}$, it follows from the above construction that $d(x_n, y) \leq \frac{1}{n}$ for each $n$, whence $y = \lim_n x_n = x$. Thus in both cases $y \leq x$, so that $x$ is an upper bound of $C$. If $z$ is any upper bound of $C$, then $x_n \leq z$ for each $n \in \mathbb{N}$, whence $x = \lim_n x_n \leq z$. Thus $x$ is the supremum of $C$. □

As a consequence of propositions 1.1.4 and 1.1.5 and lemmas 1.1.6 and 1.1.7 we obtain

**Proposition 1.1.6:** *A well-ordered chain $C$ of an ordered topological space $X$ is countable in the following cases.*

a) *$X$ is first countable and each subchain of $C$ is relatively compact.*
b) *Each subset of $C$ is separable and each nondecreasing sequence of $C$ has a cluster point.*
c) *$X$ is an ordered metric space and each nondecreasing sequence of $C$ has a cluster point.*

The next result gives a necessary and sufficient condition for the convergence of each monotone sequence of a chain in an ordered metric space.

**Proposition 1.1.7:** *If $C$ is a chain in an ordered metric space, then each monotone sequence of $C$ converges if and only if $C$ is relatively compact.*

*Proof.* If $C$ is relatively compact, then it is also relatively countably compact, whence each monotone sequence of $C$ converges by corollary 1.1.3.

Conversely, assume that all monotone sequences of $C$ converge, and let $(y_n)_{n=o}^{\infty}$ be a sequence in $\overline{C}$. Choose to each $n \in \mathbb{N}$ such an element $x_n$ from $C$ that $d(y_n, x_n) \leq \frac{1}{n}$. By lemma 1.1.5 the so obtained sequence $(x_n)_{n=o}^{\infty}$ of $C$ has a monotone subsequence $(x_{n_k})_{k=o}^{\infty}$. By hypothesis, $y = \lim_k x_{n_k}$ exists. Since $d(y_{n_k}, y) \leq \frac{1}{n_k} + d(x_{n_k}, y)$, it follows that $(y_{n_k})_{k=o}^{\infty}$ converges to $y$. Thus each sequence of $\overline{C}$ has a convergent subsequence, whence $\overline{C}$ is sequentially compact. Since $P$ is a metric space, then $\overline{C}$ is compact, i.e. $C$ is relatively compact. □

**Remark 1.1.1:** If one restricts the considerations in subsections 1.1.5 and 1.1.6 to well-ordered chains, then the axiom of choice is not needed in the proofs.

## 1.2. FIXED POINT THEOREMS

In this section we shall derive fixed point results for nondecreasing and mixed monotone mappings in ordered spaces, by using the generalized monotone iteration method introduced in theorem 1.1.1 and its dual method.

### 1.2.1. Fixed point results for nondecreasing mappings

We shall first introduce fixed point theorems in a poset $P$. Given a subset $W$ of $P$, we say that a mapping $G: P \to P$ is *nondecreasing* (resp. *nonincreasing*) in $W$ if $Gx \leq Gy$ (resp. $Gx \geq Gy$) whenever $x, y \in W$ and $x \leq y$. $G$ is called *monotone* if $G$ is nondecreasing or nonincreasing. We say that $x \in W$ is the *least fixed point* of $G$ in $W$ if $x = Gx$, and if $x \leq y$ whenever $y \in W$ and $y = Gy$. The greatest fixed point of $G$ in $W$ is defined similarly, by reversing the inequality. If both least and greatest fixed point of $G$ in $W$ exist, we call them *extremal fixed points* of $G$ in $W$.

As a consequence of theorem 1.1.1 we obtain

**Theorem 1.2.1:** *Given $G: P \to P$ and $a \in P$, assume that $x_* = \sup G[C]$ exists, where $C$ is the w.o. chain of $G$-iterations of $a$. If $a \leq Ga$ and if $G$ is nondecreasing in $[a)$, then $x_*$ is the least fixed point of $G$ in $[a)$ and*

$$x_* = \max C = \min\{y \in [a) \mid Gy \leq y\}.$$

*Proof.* The proof of theorem 1.1.1 implies that $a \leq x \leq x_*$ for each $x \in C$. Since $G$ is nondecreasing in $[a)$, it follows that $Gx \leq Gx_*$ whenever $x \in C$. Thus $x_* = \sup G[C] \leq Gx_*$, whence $x_*$ is by theorem 1.1.1 a fixed point of $G$ in $[a)$, and $x_* = \max C$.

To show that $x_* = \min\{y \in [a) \mid Gy \leq y\}$, let $y \in [a)$ satisfy $Gy \leq y$. Since $G$ is nondecreasing in $[a, y]$, then $G[a, y] \subseteq [a, y]$. This implies that $C \subseteq [a, y]$, for otherwise, if $x$ is the least element of $C$ such that $x \notin [a, y]$, then $C^{<x} \subseteq [a, y]$, whence $G[C^{<x}] \subseteq [a, y]$, so that $x = \sup G[C^{<x}] \in [a, y]$, a contradiction. Thus $C \subseteq [a, y]$, which implies that $a \leq x_* = \max C \leq y$. This holds for each $y \in [a)$ for which $Gy \leq y$, whence $x_* = \min\{y \in [a) \mid Gy \leq y\}$. In particular, $x_*$ is the least fixed point of $G$ in $[a)$. □

From the definition of the w.o. chain $C$ of $G$-iterations of $a$ (cf. theorem 1.1.1) it follows that also $G[C]$ is a well-ordered chain if $x \leq Gx$ for each $x \in C$. As a consequence of theorems 1.1.1 and 1.2.1 we then obtain

**Corollary 1.2.1:** *Given $G: P \to P$, assume that* $\sup G[C]$ *exists whenever $C$ and $G[C]$ are well-ordered chains in $P$.*

a) *If $x \leq Gx$ for each $x \in P$, then $G$ has a fixed point.*
b) *If $a \leq Ga$ and $G$ is nondecreasing in $[a)$, then $G$ has the least fixed point $x_*$ in $[a)$.*

For the sake of completeness we shall state dual results of theorems 1.1.1 and 1.2.1.

**Proposition 1.2.1:** *Given $G: P \to P$ and $b \in P$, assume that $x^* = \inf G[C]$ exists, where $C$ is the inversely well-ordered (i.w.o.) chain of $G$- iterations of $b$, i.e.*

$b = \max C$, *and* $b > x \in C$ *iff* $x = \inf G[\{y \in C \mid y > x\}]$.

a) *If $Gx^* \leq x^* \leq b$, then $x^*$ is a fixed point of $G$ in $(b]$.*

b) *If $Gb \leq b$ and $G$ is nondecreasing in $(b]$, then $x^* = \max\{y \in (b] \mid y \leq Gy\}$, and $x^*$ is the greatest fixed point of $G$ in $(b]$.*

As a consequence of theorem 1.2.1, lemma 1.1.7 and propositions 1.1.4 and 1.1.5 we obtain

**Proposition 1.2.2:** *Let $P$ be a subset of an ordered topological space $X$. A mapping $G: P \to P$ has the least fixed point if*

(G1) *$G[P]$ has a lower bound $a$ in $P$ and $G$ is nondecreasing in $[a) \cap P$,*

*and if one of the following assumptions hold.*

a) *$\overline{G[P]}$ is a compact subset of $P$,*

b) *$(Gx_n)_{n=o}^{\infty}$ converges in $P$ whenever $(x_n)_{n=o}^{\infty}$ is a nondecreasing sequence in $P$, and $\overline{G[P]}$ is metrizable or its chains are separable.*

*Proof.* Let $C$ be the w.o. chain of $G$-iterations of $a$. If the hypothesis a) holds, it follows from proposition 1.1.4 that $G[C]$ has the supremum in $P$. Assume next that b) holds, and let $(y_n)_{n=o}^{\infty}$ be a nondecreasing sequence in $G[C]$. The property (I) of $C$ given in theorem 1.1.1 implies that $(y_n)_{n=o}^{\infty} = (Gx_n)_{n=o}^{\infty}$, where $(x_n)_{n=o}^{\infty}$ is a nondecreasing sequence of $C$. This ensures by condition b) that $(y_n)_{n=o}^{\infty}$ converges in $P$. Thus $G[C]$ has by proposition 1.1.5 and lemma 1.1.7 the supremum in $P$.

The above proof implies that $G$ satisfies the hypotheses of theorem 1.2.1, which implies the assertion. □

The following two theorems and their consequences derived

in section 1.4 play a central role when we develop the theory of discontinuous differential equations.

**Theorem 1.2.2:** *Let $Y$ be a subset of an ordered metric space $X$, $[a,b]$ a nonempty order interval in $Y$, and $G\colon [a,b] \to [a,b]$ a nondecreasing mapping. If $(Gx_n)_{n=o}^{\infty}$ converges in $Y$ whenever $(x_n)_{n=o}^{\infty}$ is a monotone sequence in $[a,b]$, then the w.o. chain of $G$-iterations of $a$ has the maximum $x_*$, and the i.w.o. chain of $G$-iterations of $b$ has the minimum $x^*$, and*

$$x_* = \min\{y \mid Gy \leq y\}, \quad x^* = \max\{y \mid y \leq Gy\}. \tag{1.2.1}$$

*In particular, $x_*$ and $x^*$ are the extremal fixed points of $G$.*

*Proof.* The assertions follow from theorem 1.2.1 and proposition 1.2.2 b) and their duals. □

**Theorem 1.2.3:** *Let $Y$ be a subset of an ordered metric space $X$, $[a,b]$ a nonempty order interval in $Y$, and $G\colon [a,b] \to [a,b]$ a nondecreasing mapping. If $\sup G[C]$ and $\inf G[\bar{C}]$ exist in $Y$ when $C$ is the w.o. chain of $G$-iterations of $a$ and $\bar{C}$ is the i.w.o. chain of $G$-iterations of $b$, then $x_* = \max C$ is the least fixed point of $G$, and $x^* = \min \bar{C}$ is the greatest fixed point of $G$, and (1.2.1) holds.*

*Proof.* The assertions follow from theorem 1.2.1 and proposition 1.2.1. □

The following result is a direct consequence of corollary 1.1.2, its dual and proposition 1.1.3.

**Corollary 1.2.2:** *Let the hypotheses of theorem 1.2.2 hold. Define sequences $(a_m)_{m=o}^{\infty}$ and $(b_m)_{m=o}^{\infty}$ by $a_o = a$, $b_o = b$ and*

$$a_{m+1} = \lim_{n_m} G^{n_m}(\lim_{n_{m-1}} G^{n_{m-1}}(\cdots(\lim_{n_o} G^{n_o}a)\cdots)),$$

*and*

$$b_{m+1} = \lim_{n_m} G^{n_m}(\lim_{n_{m-1}} G^{n_{m-1}}(\cdots(\lim_{n_o} G^{n_o}b)\cdots)).$$

*Then the extremal fixed points $x_*$ and $x^*$ of $G$ exists. Moreover,*

a) $a_m \leq x_* \leq x^* \leq b_m$ *for each* $m \in \mathbb{N}$.

b) $x_* = a_m$ *if and only if* $a_m = Ga_m$. *This holds for* $m > 0$ *if* $G$ *is left continuous at* $a_m$.

c) $x^* = b_m$ *if and only if* $b_m = Gb_m$. *This holds for* $m > 0$ *if* $G$ *is right continuous at* $b_m$.

d) *If* $G$ *is left continuous, then* $x_* = a_1 = \lim_n G^n a$.

e) *If* $G$ *is right continuous, then* $x^* = b_1 = \lim_n G^n b$.

**Remarks 1.2.1:** If condition (G1) is replaced in proposition 1.2.2 by $x \leq Gx$ for each $x \in P$, then the existence of at least one fixed point of $G$ is ensured by theorem 1.1.1.

The use of the w.o. (resp. i.w.o.) chains of $G$-iterations in the above proofs allows us to restrict the assumptions to the range of $G$. This restriction is useful for instance, when we derive in subsection 1.4.8 fixed point results in ordered spaces of continuous functions. Moreover, in view of proposition 1.1.6 and corollary 1.2.2 these iteration chains are practically in all applications countable, and are reduced to ordinary sequences of monotone iterations when $G$ is right or left continuous. The axiom of choice or any of its variant, e.g. Zorn's lemma, is not needed in the proofs of this and the next subsection.

### 1.2.2. Coupled fixed points of mixed monotone operators

Given a poset $P$, we say that a mapping $A: P \times P \to P$ is *mixed monotone* if $A(\cdot, z)$ is nondecreasing and $A(z, \cdot)$ is nonincreasing for each $z \in P$. We say that $A$ is *order bounded* if there exist $\underline{v}$, $\bar{w} \in P$ such that $\underline{v} \leq A(v, w) \leq \bar{w}$ for all $(v, w) \in P \times P$. A point $(v, w)$ of $P \times P$ is called a *coupled fixed point of* $A$ if it satisfies

$$v = A(v, w), \qquad w = A(w, v). \tag{1.2.2}$$

A coupled fixed point $(v_*, w^*)$ of $A$ is called *extremal* if $v_* \le w^*$ and $v_* \le v$, $w \le w^*$ for any coupled fixed point $(v, w)$ of $A$.

**Proposition 1.2.3:** *Let $P$ be a poset. A mixed monotone and order bounded mapping $A\colon P \times P \to P$ has the extremal coupled fixed point if $A[C]$ has the supremum and the infimum in $P$ whenever $C$ is a well-ordered or an inversely well-ordered chain in $Y = P \times P$ with respect to the partial ordering defined by*

$$(v_1, w_1) \le (v_2, w_2) \ \textit{ if and only if } v_1 \le v_2 \textit{ and } w_2 \le w_1. \tag{1.2.3}$$

*Proof.* The mixed monotonicity of $A$ implies that the equation

$$G(v, w) = (A(v, w), A(w, v)) \tag{1.2.4}$$

defines a nondecreasing mapping $G\colon Y \to Y$. Due to the order boundedness of $A$, the range $G[Y]$ of $G$ has a lower bound $a = (\underline{v}, w)$ in $Y$. Let $C$ be the w.o. chain of $G$-iterations of $a$. The set $C' = \{(w, v) \mid (v, w) \in C\}$ is an inversely well-ordered chain in $Y$. From the hypotheses it follows that $v_* = \sup A[C]$ and $w^* = \inf A[C']$ exist in $P$. Moreover, it is easy to see that $(v_*, w^*) = \sup G[C]$ in $Y$. Thus $(v_*, w^*)$ is by theorem 1.2.1 the least fixed point of $G$ in $Y$. This implies by (1.2.4) that $(v_*, w^*)$ is a coupled fixed point of $A$.

If $(v, w)$ is a coupled fixed point of $A$, it follows from (1.2.2) and (1.2.4) that $(v, w)$ is a fixed point of $G$ in $Y$. Because $(v_*, w^*)$ is the least one, then $(v_*, w^*) \le (v, w)$, i.e. $v_* \le v$ and $w \le w^*$. In particular, we can choose $v = w^*$ and $w = v_*$, whence $v_* \le w^*$. Since $(w, v)$ is also a coupled fixed point of $A$, it follows that $v_* \le v$, $w \le w^*$. Thus $(v_*, w^*)$ is extremal. □

The previous result is now applied to derive existence results for extremal fixed points of mixed monotone operators in ordered topological spaces.

**Proposition 1.2.4:** *Let $P$ be a nonempty subset of an ordered topological space. An order bounded and mixed monotone mapping $A\colon P \times P \to P$ has the extremal coupled fixed point if $\overline{A[P \times P]}$ is a compact subset of $P$.*

*Proof.* Denote $Y = P \times P$, and let $G\colon Y \to Y$ be defined by (1.2.4). The given hypotheses imply that $G[Y]$ has a lower bound $a$ in $Y$, that $G$ is nondecreasing with respect to the partial ordering defined by (1.2.3), and that $\overline{G[Y]}$ is a compact subset of $Y$. Thus $G$ has by proposition 1.2.2 a) the least fixed point $x_* = (v_*, w^*)$. As shown in the proof of proposition 1.2.3, $(v_*, w^*)$ is the extremal coupled fixed point of $A$. □

**Theorem 1.2.4:** *Given a nonempty subset $P$ of an ordered metric space $X = (X, d)$ and an order bounded and mixed monotone mapping $A\colon P \times P \to P$, assume that*

(A1) *$(A(v_j, w_j))_{j=1}^\infty$ converges whenever $(v_j)_{j=1}^\infty$ and $(w_j)_{j=1}^\infty$ are sequences in $P$, one being nondecreasing and the other nonincreasing.*

*Then $A$ has the extremal coupled fixed point.*

*Proof.* Note first that $Y = P \times P$ is an ordered topological space with respect to the partial ordering given by (1.2.3) and the metric $d((v_1, w_1), (v_2, w_2)) = d(v_1, v_2) + d(w_1, w_2)$. Let $G\colon Y \to Y$ be defined by (1.2.4). If $(x_j)_{j=1}^\infty = ((v_j, w_j))_{j=1}^\infty$ is a nondecreasing sequence in $Y$, then $(v_j)_{j=1}^\infty$ is nondecreasing and $(w_j)_{j=1}^\infty$ nonincreasing in $P$. The hypothesis (A1) implies that the sequences $(A(v_j, w_j))_{j=1}^\infty$ and $(A(w_j, v_j))_{j=1}^\infty$ converge in $P$. Thus the sequence $(Gx_j)_{j=1}^\infty$ converges in $Y$. This implies by proposition 1.2.2 b) that $G$ has the least fixed point $(v_*, w^*)$. From the proof of proposition 1.2.3 it follows that $(v_*, w^*)$ is the extremal coupled fixed point of $A$. □

In the case when $P$ is a subset of an ordered topological space where all the monotone sequences converge, and $A\colon P \times P \to P$ is continuous, then also $G$, given by (1.2.4), is continuous in the product topology, whence the well-ordered chain $C$ of $G$-iterations of $a = (\underline{v}, \bar{w})$ is either the iteration sequence $(G^n a)_{n=0}^{\infty}$ or its finite initial sequence. Thus, if $A$ is continuous, its extremal coupled fixed point $(v_*, w^*)$ in the results derived above can be obtained by $v_* = \lim_{j\to\infty} v_j$ and $w^* = \lim_{j\to\infty} w_j$, where $(v_j)_{j=0}^{\infty}$ and $(w_j)_{j=0}^{\infty}$ are defined by

$$v_{j+1} = A(v_j, w_j), \quad w_{j+1} = A(w_j, v_j), \ j \in \mathbb{N}, \ v_0 = \underline{v}, \ w_0 = \bar{w}.$$

## 1.3. ORDERED NORMED AND FUNCTION SPACES

The previous fixed point results are applied mostly to operators in ordered Banach spaces and function spaces. For this purpose we shall introduce in this section basic concepts of these spaces. As for a more detailed study of such spaces see section 5.8.

### 1.3.1. Ordered vector spaces and normed spaces

Unless otherwise stated, only vector spaces and normed spaces over the field $\mathbb{R}$ of real numbers are considered here. A subset $K$ of a vector space $E$ is called a *positive cone* in $E$ if

$K + K \subseteq K$, $K \cap (-K) = \{0\}$ and $cK \subseteq K$ for each $c \geq 0$.

For instance, the set $\mathbb{R}_+$ of nonnegative real numbers form a positive cone in $\mathbb{R}$.

It is easy to see that the relation

$$x \leq y \text{ if and only if } y - x \in K \tag{1.3.1}$$

defines a partial ordering '$\leq$' in $E$. We say then that $E$ is *ordered by* $K$, and that $K$ is the *order cone* of $E$. From (1.3.1) it follows that $K = \{y \in E \mid 0 \leq y\}$.

By an *ordered normed space* we mean a real vector space $E$, equipped with a norm $\|\cdot\| : E \to \mathbb{R}$, and ordered by a positive cone $K$ which is closed in the strong topology of $E$.

It is elementary to verify that the following properties hold.

**Lemma 1.3.1:** *Let $E$ be an ordered normed space.*

a) *$x \leq y$ if and only if $-y \leq -x$.*

b) *If $x \leq y$, then $x + z \leq y + z$ for each $z \in E$.*

c) *$x \leq y$ and $c \geq 0$ imply that $cx \leq cy$ and $-cy \leq -cx$.*

d) *If $\sup A$ exists in $E$ and $c \geq 0$, then $\sup(cA) = c \sup A$ and $\inf(-cA) = -c \sup A$.*

e) *If $A \subseteq E$ and $\sup A$ exists, then $\sup(z + A) = z + \sup A$ for each $z \in E$.*

f) *If $x_n \to x$ and $y_n \to y$ in $E$ and $x_n \leq y_n$ for $n$ large enough, then $x \leq y$.*

The order cone $K$ of an ordered normed space $E = (E, \|\cdot\|)$ is called *normal* if the norm $\|\cdot\|$ of $E$ is *semimonotone* in $K$, i.e. there is $\gamma > 0$ such that

$$y, z \in K \text{ and } y \leq z \quad \text{imply } \|y\| \leq \gamma \|z\|. \tag{1.3.2}$$

$K$ is said to be *regular* if each nondecreasing and order bounded sequence of $K$ is strongly convergent. If each nondecreasing and bounded sequence of $K$ is strongly convergent, we say that $K$ is *fully regular.*

**Lemma 1.3.2:** *If $E$ is an ordered normed space, then normality, regularity and/or full regularity of the order cone $K$ of $E$ remain valid if the norm is replaced by any equivalent norm. If $K$ is normal, there is a norm of $E$, equivalent to the original one, which is monotone in $K$, i.e. (1.3.2) holds when $\gamma = 1$.*

*Proof.* The first assertion is an immediate consequence of definitions. The second assertion is proved in Amann (1976) (see also Guo and Lakshmikantham (1988), Schaefer (1966)). □

The following result holds for all finite-dimensional ordered normed spaces.

**Proposition 1.3.1:** *The order cone of any finite-dimensional ordered normed space is fully regular.*

*Proof.* Let $E$ be a finite-dimensional vector space, and let $\{e_1, \ldots, e_m\}$ be a base of $E$. Since all the norms of $E$ are equivalent, we can choose the norm

$$\|y\|_\infty = \max\{|y_1|, \ldots, |y_m|\}, \ y = \sum_{i=1}^{m} y_i e_i \in E.$$

Let $E$ be ordered by any closed cone $K$, and let $(x_n)_{n=o}^{\infty}$ be a bounded and nondecreasing sequence in $K$. Then all the coordinate sequences of $(x_n)_{n=o}^{\infty}$ are bounded in $\mathbb{R}$. This and Bolzano-Weierstrass Theorem imply that each subsequence of $(x_n)_{n=o}^{\infty}$ has a strongly convergent subsequence. Thus $(x_n)_{n=o}^{\infty}$ is strongly convergent by corollary 1.1.3. □

If $(x_n)_{n=o}^{\infty}$ is a monotone and (order) bounded sequence in $E$, then $(x_n - x_o)_{n=o}^{\infty}$ or $(x_o - x_n)_{n=o}^{\infty}$ is nondecreasing and (order) bounded sequence in $K$. Thus we have

**Lemma 1.3.3:** *Let $K$ be the order cone of an ordered normed space $E$. If $K$ is regular (resp. fully regular), then each order bounded (resp. bounded) and monotone sequence of $E$ is strongly convergent.*

As a consequence of proposition 1.1.5 and lemma 1.3.3 we obtain the following results.

**Proposition 1.3.2:** *If $E$ is an ordered normed space with regular order cone, then each order bounded chain $C$ of $E$ contains a nondecreasing (resp. a nonincreasing) sequence which converges strongly to* $\sup C$ *(resp.* $\inf C$*).*

**Proposition 1.3.3:** *If $E$ is an ordered normed space $E$ with fully regular order cone, then each bounded or order bounded chain $C$ of $E$ contains a nondecreasing sequence which converges strongly to* $\sup C$ *and a nonincreasing sequence which converges strongly to* $\inf C$.

### 1.3.2. Ordered Banach spaces

We say that a normed space $E$ is a *Banach space* if it is complete, i.e. if all its Cauchy sequences are strongly convergent. It is well-known that the continuous linear functionals $f: E \to \mathbb{R}$ form a Banach space $E'$ with respect to the norm $\|f\| = \sup\{|f(x)| \mid x \in E,\ \|x\| \leq 1\}$, also in the case when $E$ is not complete.

We call a normed space $E$ *reflexive* if to each $T \in (E')'$ there corresponds $x \in E$ such that $T(f) = f(x)$ for all $f \in E'$. Each reflexive normed space is a Banach space. Finite-dimensional normed spaces and Hilbert spaces are reflexive, as well as uniformly convex Banach spaces (cf. Narici and Beckenstein (1985), Yoshida (1974)).

By an *ordered Banach space* we mean a Banach space ordered by a closed positive cone $K$. As for relations between the concepts of normality, regularity and full regularity of $K$ we have

**Proposition 1.3.4:** *Let $E$ be an ordered Banach space with order cone $K$.*

a) *$K$ is normal if it is regular, and $K$ is regular if it is fully regular.*

b) *The converse of a) holds if $E$ is reflexive.*

*Proof.*

a) Cf. Deimling (1985), Krasnosel'skii (1964).

b) Cf. Sun (1984), Guo and Lakshmikantham (1988). □

We say that a sequence $(y_n)_{n=o}^{\infty}$ of a normed space $E$ converges *weakly* to $y \in E$ if $f(y_n) \to f(y)$ for each $f \in E'$. This is equivalent to the convergence in the weak topology of $E$ (cf. Narici and Beckenstein (1985), Yoshida (1974)).

**Proposition 1.3.5:** *Let $B$ be a bounded subset of a reflexive Banach space $E$.*

a) *$B$ is weakly relatively compact and weakly relatively sequentially compact.*

b) *The weak closure of $B$ is formed by weak limits of its sequences.*

*Proof.*

*a) Cf. Narici and Beckenstein (1985).*

*b) Cf. Browder (1976).* □

*As a consequence of propositions 1.1.4 and 1.3.5 and lemma 1.1.6 we obtain*

**Proposition 1.3.6:** *Let $C$ be a bounded chain in a reflexive ordered Banach space.*

a) *$C$ contains a nondecreasing sequence which converges weakly to* $\sup C$.

b) *If $C$ is well-ordered, it is countable.*

*Proof.* From proposition 1.3.5 a) it follows that $C$ is weakly relatively compact, whence the proof of proposition 1.1.4 implies that $x = \sup C$ exists and belongs to the weak closure of $C$. Proposition 1.3.5 b) implies the existence of a sequence of $C$ which converges weakly to $\sup C$. Since all subchains of $C$ are also weakly relatively compact, then the assertion b) follows from lemma 1.1.6. □

We say that a normed space $(E_1, \|\cdot\|_1)$ is *continuously embedded* in a normed space $(E_2, \|\cdot\|_2)$ if $E_1 \subseteq E_2$, and if there is $c > 0$ such that

$$\|x\|_2 \le c\,\|x\|_1 \quad \text{for all} \quad x \in E_1. \tag{1.3.3}$$

Let $E_2$ be an ordered normed space with order cone $K$, and $E_1$ a normed space which is continuously embedded in $E_2$. Since $E_1 \cap K$ is a positive cone in $E_1$, then the partial ordering of $E_2$ induces a partial ordering in $E_1$. This partial ordering does not necessarily make $E_1$ an ordered normed space, since $E_1 \cap K$ may not be closed in $E_1$. However, the following result holds.

**Proposition 1.3.7:** *Let $E_1$ be a reflexive Banach space which is continuously embedded in an ordered normed space $E_2$ with fully regular order cone. If a bounded subset $C$ of $E_1$ is a chain with respect to the partial ordering of $E_2$, then $C$ contains a nondecreasing sequence which converges weakly in $E_1$ and strongly in $E_2$ to* $\sup C$, *and a nonincreasing sequence which converges weakly in $E_1$ and strongly in $E_2$ to* $\inf C$.

*Proof.* Let $C$ be a bounded subset of $E_1$, and assume that $C$ is a chain with respect to the partial ordering of $E_2$. Since $E_1$ is continuously embedded in $E_2$, then $C$ is a bounded chain in $E_2$. Because the order cone of $E_2$ is fully regular, then $C$ contains by proposition 1.3.3 a nondecreasing sequence $(x_n)_{n=o}^{\infty}$ which converges strongly in $E_2$ to $x = \sup C$. Since $(x_n)_{n=o}^{\infty}$ is a bounded sequence in a reflexive Banach space $E_1$, then it has by proposition 1.3.5 a subsequence $(x_{n_k})_{k=o}^{\infty}$ which has a weak limit $y$ in $E_1$. On the other hand, $(x_{n_k})_{k=o}^{\infty}$ converges strongly, and hence also weakly, to $x$ in $E_2$. If $f \in E_2'$, then denoting $g = f|E_1$, we have $g \in E_1'$ and

$$f(y) = g(y) = \lim_k g(x_{n_k}) = \lim_k f(x_{n_k}) = f(x).$$

This implies that $y = x$. In particular, $x = \sup C \in E_1$.

The above result implies that the chain $-C$ contains a nondecreasing sequence $(y_k)_{k=o}^{\infty}$ which converges weakly in $E_1$ and strongly in $E_2$ to $\sup(-C)$. Thus $(-y_k)_{k=o}^{\infty}$ is a nonincreasing sequence in $C$ which converges weakly in $E_1$ and strongly in $E_2$ to $\inf C$. □

### 1.3.3. Ordered spaces of continuous functions

In this subsection we shall study existence of supremums and infimums of chains in the space $C(X, E)$ of continuous functions $x: X \to E$, where $X$ is a topological space and $E$ an ordered metric or normed space, the partial ordering of $C(X, E)$ being defined by

$$x \leq y \text{ if and only if } x(t) \leq y(t) \text{ for each } t \in X. \tag{1.3.4}$$

Assume first that $E = (E, d)$ is an ordered metric space. We say that a subset $C$ of $C(X, E)$ is *equicontinuous* if for each $t \in X$ and for each $\epsilon > 0$ there exists a neighborhood $U$ of $t$ such that

$$d(x(s), x(t)) \leq \epsilon \text{ for all } x \in C \text{ and } s \in U. \tag{1.3.5}$$

**Proposition 1.3.8:** *If $C$ is an equicontinuous chain in $C(X, E)$, and if each nondecreasing sequence of $C$ converges pointwise, then $C$ has the supremum in $C(X, E)$.*

*Proof.* From proposition 1.1.5 it follows that $x^*(t) = \sup_{x \in C} x(t)$ exists in $E$ for each $t \in X$. To prove that the so obtained function $x^*: X \to E$ is continuous, let $t \in X$ be given. Since $C$ is equicontinuous, then to each $\epsilon > 0$ there corresponds such a neighborhood $U$ of $t$ that (1.3.5) holds. Let $s \in U$ be fixed. By proposition 1.1.5 there exists a nondecreasing sequence $(x_n)_{n=o}^{\infty}$ in $C$ such that $(x_n(s))_{n=o}^{\infty}$ converges in $E$ to $x^*(s)$, and a nondecreasing sequence $(y_n)_{n=o}^{\infty}$ in $C$ such that $(y_n(t))_{n=o}^{\infty}$ converges in $E$ to $x^*(t)$. Denoting $z_n = \max\{x_n, y_n\}$, $n \in \mathbb{N}$, we obtain a nondecreasing sequence $(z_n)_{n=o}^{\infty}$ of the elements of $C$. By hypothesis it converges pointwise. Since $x_n(s) \leq z_n(s) \leq x^*(s)$ and $y_n(t) \leq z_n(t) \leq x^*(t)$, it then follows that $z_n(s) \to x^*(s)$ and $z_n(t) \to x^*(t)$ as $n \to \infty$. This and (1.3.5) imply when $n \to \infty$ in $d(x^*(s), x^*(t)) \leq d(x^*(s), z_n(s)) + d(z_n(s), z_n(t)) + d(z_n(t), x^*(t))$

that

$$d(x^*(s), x^*(t)) \le \epsilon.$$

This holds for each $s \in U$, whence $x^*$ is continuous at $t$. Thus $x^* \in C(X, E)$. From the definition of $x^*$ it follows that $x^*$ is the supremum of $C$ in $C(X, E)$. □

Proposition 1.3.8 and its dual imply.

**Proposition 1.3.9:** *If $E$ is an ordered normed space with regular (resp. fully regular) order cone, then each pointwise order bounded (resp. bounded) and equicontinuous chain $C$ of $C(X, E)$ has the supremum $x^*$ and the infimum $x_*$ in $C(X, E)$. Moreover, if there is a function $\psi\colon X \times X \to \mathbb{R}_+$ such that*

$$\|x(t) - x(s)\| \le \psi(t, s) \quad \text{for all } x \in C \text{ and } \ t, s \in X, \tag{1.3.6}$$

*then for all $t, s \in X$,*

$$\|x^*(t) - x^*(s)\| \le \psi(t, s) \text{ and } \|x_*(t) - x_*(s)\| \le \psi(t, s). \tag{1.3.7}$$

*Proof.* The hypotheses imply that each monotone sequence of $C$ converges pointwise, so that the first assertion follows from proposition 1.3.8 and its dual. If (1.3.6) holds, it follows from the proof of proposition 1.3.8, with $\epsilon$ replaced by $\psi(t, s)$, that the first inequality of (1.3.7) hold. The validity of the second inequality can be proved similarly. □

**Proposition 1.3.10:** *Let $E$ be a reflexive Banach space which is ordered by a closed cone or via continuous embedding in an ordered Banach space with fully regular order cone. If $C$ is a pointwise bounded and equicontinuous chain in $C(X, E)$, then $x^* = \sup C$ and $x_* = \inf C$ exist in $C(X, E)$. Moreover, if (1.3.6) holds, then also (1.3.7) holds.*

*Proof.* Let $C$ be a pointwise bounded and equicontinuous chain in $C(X,E)$. The given hypotheses imply by propositions 1.3.6 and 1.3.7 that $x^*(t) = \sup_{x \in C} x(t)$ exists in $E$ for each $t \in X$. To prove that the so obtained function $x^* : X \to E$ is the supremum of $C$, it suffices to show its continuity. Let $t \in X$ and $\epsilon > 0$ be given. By the equicontinuity hypothesis there is such a neighborhood $U$ of $t$ that

$$\|x(s) - x(t)\| \leq \epsilon \quad \text{whenever } s \in U \text{ and } x \in C. \tag{a}$$

Let $s \in U$ be fixed. By propositions 1.3.6 and 1.3.7 there exists a nondecreasing sequence $(x_n)_{n=o}^{\infty}$ in $C$ such that $(x_n(s))_{n=o}^{\infty}$ converges weakly in $E$ to $x^*(s)$, and a nondecreasing sequence $(y_n)_{n=o}^{\infty}$ in $C$ such that $(y_n(t))_{n=o}^{\infty}$ converges weakly in $E$ to $x^*(t)$. Denoting $z_n = \max\{x_n, y_n\}$, $n \in \mathbb{N}$, we obtain a nondecreasing sequence $(z_n)_{n=o}^{\infty}$ in $C$. The sequences $(z_n(s))_{n=o}^{\infty}$ and $(z_n(t))_{n=o}^{\infty}$ are nondecreasing and bounded sequences in $E$. Since $E$ is reflexive, we can pick such a subsequence $(z_{n_k})_{k=o}^{\infty}$ of $(z_n)_{n=o}^{\infty}$ that $(z_{n_k}(s))_{k=o}^{\infty}$ converges weakly in $E$ to $x^*(s)$, and that $(z_{n_k}(t))_{k=o}^{\infty}$ converges weakly in $E$ to $x^*(t)$. Thus $(z_{n_k}(s) - z_{n_k}(t))_{k=o}^{\infty}$ converges in $E$ weakly to $x^*(s) - x^*(t)$, so that by (a) and (5.8.7)

$$\|x^*(s) - x^*(t)\| \leq \liminf_{k \to \infty} \|z_{n_k}(s) - z_{n_k}(t)\| \leq \epsilon.$$

This holds for each $s \in U$, which shows that $x^*$ is continuous at $t$. Thus $x^* \in C(X,E)$, whence $x^*$ is the supremum of $C$ in $C(X,E)$. The existence of $\inf C$ follows from the above proof and from the fact that $\inf C = -\sup(-C)$.

If (1.3.6) holds, then replacing $\epsilon$ in the above proof by $\psi(t,s)$, we obtain the first inequality of (1.3.7). The second one can be proved similarly. □

As for further results concerning ordered Banach spaces and ordered function spaces see section 5.8.

## 1.4. RESULTS OF NONLINEAR ANALYSIS

In this section we shall first introduce basic concepts of measure and integration theory of Banach-valued functions, present some classes of measurable functions and prove substitution theorems which are needed in the sequel. We shall also prove the fundamental theorem of calculus for Banach-valued functions. Finally we shall present fixed point theorems for operators in function spaces. These fixed point theorems are basic tools in our study of discontinuous differential equations.

### 1.4.1. Measurability and integrability

We assume that the reader is familiar with the basics of the measure and integration theory of real-valued functions (see, e.g., McShane (1974), Munroe (1959), Royden (1968)). We shall now consider the case when the values of functions in question are in a Banach space $E$.

Let $(\Omega, \mathcal{A}, \mu)$ be a *measure space*, i.e. $\Omega$ is a nonempty set, $\mathcal{A}$ is a $\sigma$-algebra of (measurable) subsets of $\Omega$, and $\mu\colon \mathcal{A} \to I\!R_+ \cup \{\infty\}$ is a positive measure. A measurable subset $Z$ of $\Omega$ is called *($\mu$-) null set* if $\mu(Z) = 0$. We say that a property **P** holds *almost everywhere* (a.e.) in $\Omega$, or for *almost all* (a.a.) $t \in \Omega$, if there is a null set $Z$ in $\Omega$ such that **P** holds for all $t \in \Omega \setminus Z$.

By a *step function* we mean a function $y\colon \Omega \to E$ which has only finite number of values, and $y^{-1}(\{z\})$ is of finite measure for each $z \neq 0$. Each step function can be represented in the form $y = \sum_{i=1}^{n} \chi_{A_i} y_i$, where each $y_i$ is a nonzero element of $E$, the sets $A_i \in \mathcal{A}$, $i = 1, \ldots, n$, are disjoint and of finite measure, and $\chi_A$ denotes the characteristic function of $A$. Define

$$\int_\Omega y\, d\mu = \sum_{i=1}^{n} \mu(A_i) y_i$$

when $y$ is a step function.

We say that a function $y\colon \Omega \to E$ is *$\mu$-measurable* if there is a sequence $(y_n)_{n=1}^{\infty}$ of step functions such that $\lim_{n\to\infty} y_n(t) = y(t)$ for a.a. $t \in \Omega$. In the following we shall identify two $\mu$-measurable functions if they coincide in the complement of a null set.

A function $y\colon \Omega \to E$ is called *$\mu$-integrable* if there is a sequence $(y_n)_{n=1}^{\infty}$ of step functions such that $\lim_{n\to\infty} y_n(t) = y(t)$ for a.a. $t \in \Omega$, and $\int_\Omega |y_n - y_m| d\mu \to 0$ as $n, m \to \infty$, where $|y|$ denotes the function $t \mapsto \|y(t)\|$.

The sequence $(y_n)_{n=1}^{\infty}$ which has the properties above is called an *approximating sequence* of $y$.

If $y\colon \Omega \to E$ is $\mu$-integrable, and $(y_n)_{n=1}^{\infty}$ is an approximating sequence of $y$, define

$$\int_\Omega y = \int_\Omega y\, d\mu = \lim_{n\to\infty} \int_\Omega y_n d\mu. \tag{1.4.1}$$

It can be shown that the value of the integral in (1.4.1) is independent of the choice of the approximating sequence $(y_n)_{n=1}^{\infty}$. Moreover, each $\mu$-integrable function $y\colon \Omega \to E$ is also $\mu$-measurable.

The set $L^1(\Omega, E)$ of all the $\mu$-integrable functions $y\colon \Omega \to E$ is a vector space with respect to addition and scalar multiplication of functions, and complete with respect to the seminorm $\|\cdot\|_1$ defined by

$$\|y\|_1 = \int_\Omega |y|\, d\mu. \tag{1.4.2}$$

Moreover, $\|y\|_1 = 0$ if and only if $y(t) = 0$ for a.a. $t \in \Omega$. This and the identification of a.e. equal functions imply that $\|\cdot\|_1$ is a norm, and that $(L^1(\Omega, E), \|\cdot\|)$ is a Banach space.

We say that $x\colon \Omega \to E$ is *essentially bounded* if there is $M > 0$ such that $\|x(t)\| \le M$ for a.a. $t \in \Omega$. If $\|x(t)\| \le M$ for all $t \in \Omega$, we say that $x$ is *bounded*.

Assume now that $\Omega$ is a Lebesgue measurable subset of $\mathbb{R}^m$, $\mathcal{A}$ the $\sigma$-algebra of Lebesgue measurable subsets of $\Omega$, and $E$ a Banach space. We say that $y$ is *strongly measurable* if $y$ is $\mu$-measurable, where $\mu$ denotes the Lebesgue measure on $\mathbb{R}^m$.

Each strongly measurable function $y\colon \Omega \to E$ is (a.e. equal to) a measurable function, i.e. $y^{-1}[A]$ is Lebesgue measurable whenever $A \subseteq E$ is open. Converse holds if $E$ is separable (cf. Lang (1969)). $y\colon \Omega \to E$ is called *Bochner integrable* if $y$ is $\mu$-integrable, i.e. if $y \in L^1(\Omega, E)$. In the case when $E = \mathbb{R}^n$ and $\mu(\Omega) < \infty$ it can be shown (cf. Mikusinski (1978)) that $y\colon \Omega \to \mathbb{R}^n$ is Bochner integrable if and only if it is Lebesgue integrable.

If $\Omega$ is a compact real interval $[a, b]$, denote

$$\int_\Omega y = \int_a^b y(t)dt = -\int_b^a y(t)dt.$$

## 1.4.2. Measurability of right or left regulated functions

To provide further examples of measurable and integrable functions we shall now define and study right regulated, left regulated and regulated functions. They are important also in future applications.

If $J$ is a real interval and $X$ a metric space, we say that a function $y\colon J \to X$ is *right regulated* if $y$ has right-hand limit $y(t+) = \lim_{\delta \to 0+} y(t + \delta)$ at each point $t$ of $J$ different from the possible right end point. A left regulated function is defined similarly. $y\colon J \to X$ is called *regulated* if is right and left regulated. We say that $g\colon J \to X$ is *countably stepped* if it has a countable number of values, assuming each value on a countable union of intervals. If $g$ has a finite number of values and obtains each value on a finite union of intervals, we call $g$ a *stepped* function. It is well-known (cf. Brown and Page (1970), Dieudonné (1960)) that if $J = [a, b] \subseteq \mathbb{R}$, and $y\colon J \to X$ is regulated, then $y$ has a countable number of discontinuity points, and $y$ can be uniformly approximated by stepped functions. By using the generalized iteration method we shall prove something similar to right regulated functions.

**Lemma 1.4.1:** *Let $X$ be a metric space and $J = [a, b) \subset \mathbb{R}$. If $y\colon J \to X$ is right regulated, then the set of the discontinuity* *points of $y$ is countable. Moreover, $y$ can be uniformly approximated by countably stepped functions.*

*Proof.* For each positive integer $n$ define $G_n: [a,b] \to [a,b]$ by

$$G_n(x) = \sup\{u \in (x,b] \mid d(y(s),y(t)) \le \frac{1}{n} \text{ for all } s,\ t \in (x,u)\}$$

when $x \in [a,b)$, and $G_n(b) = b$. Since $x < G_n(x)$ for all $x \in [a,b)$, then each $G_n$ satisfies the hypotheses of corollary 1.1.1. This and proposition 1.1.6 imply that the w.o. chain $C_n$ of $G_n$-iterations of $a$ is countable and $[a,b] \setminus C_n$ is a disjoint union of open intervals $(x, G_n(x))$, $x \in C_n$.

Given $\epsilon > 0$, choose $n \in I\!N$ such that $\frac{1}{n} \le \epsilon$. If $t \in [a,b] \setminus \bigcup_{n=1}^{\infty} C_n$, there is $x \in C_n$ such that $t \in (x, G_n(x))$. From the definition of $G_n$ it follows that $d(y(s),y(t)) \le \frac{1}{n} \le \epsilon$ for each $s \in (x, G_n(x))$. Thus all the discontinuity points of $y$ belong to the countable set $\bigcup_{n=1}^{\infty} C_n$.

Choose from each interval $(x, G_n(x))$, $x \in C_n$, its midpoint $t(x)$, and define

$$y_n(t) = \begin{cases} y(t(x)), & t \in (x, G_n(x)),\ x \in C_n \setminus \{b\}, \\ y(t), & t \in C_n \setminus \{b\}. \end{cases}$$

$y_n: J \to X$ is for each $n$ countably stepped and $d(y_n(t), y(t)) \le \frac{1}{n}$ for all $t \in J$. Thus the sequence $(y_n)_{n=1}^{\infty}$ converges uniformly to $y$ on $J$, implying the last assertion. □

If $(a,b)$ is a nonempty open real interval, there exists a decreasing sequence $(a_n)_{n=o}^{\infty}$ in $(a,b)$ which converges to $a$. Then $(a,b) = \bigcup_{n=o}^{\infty}[a_n, b) = \bigcup_{n=1}^{\infty}[a_n, a_{n-1}) \cup [a_o, b)$. The dual reasoning shows that $(a,b)$ can be represented as a countable union of disjoint half-open intervals of the form $(u,v]$. From lemma 1.4.1 and its dual it then follows.

**Corollary 1.4.1:** *Let $J$ be a countable union of bounded real intervals $J_i$ and $X$ a metric space. If $y\colon J \to X$ is right or left regulated in the interiors of $J_i$, then $y$ has a countable number of discontinuities, and can be uniformly approximated by countably stepped functions.*

For instance, any open subset of $\mathbb{R}$ is a countable union of open intervals.

**Lemma 1.4.2:** *If $J$ is a real interval and $X$ a complete metric space, then the uniform limit of right regulated functions is right regulated.*

*Proof.* Assume that sequence $(y_n)_{n=o}^{\infty}$ of right regulated functions converge uniformly on $J$ to $y\colon J \to X$. Given $x \in J$, $x \neq \max J$ and $\epsilon > 0$, choose $n \in \mathbb{N}$ so that

$$d(y_n(t), y(t)) \leq \frac{\epsilon}{3} \text{ for all } t \in J, \tag{a}$$

and $u > x$ such that

$$d(y_n(s), y_n(t)) \leq \frac{\epsilon}{3} \text{ for all } s,\ t \in (x, u). \tag{b}$$

From (a) and (b) it follows that

$$d(y(s), y(t)) \leq \epsilon \text{ for all } s,\ t \in (x, u).$$

This shows that $y$ is right regulated. □

It is well-known (cf. Munroe (1959)) that a bounded real function is Riemann integrable on a compact real interval if and only if the set of its discontinuity points is a null set. Since a countable set of real numbers is a null set, then corollary 1.4.1, lemma 1.4.2 and its dual imply.

**Corollary 1.4.2:** *A bounded real function $f$ is Riemann integrable on a compact interval $J$ if $f$ is the uniform limit of a sequence of functions which are right or left regulated in the interior of $J$.*

In the case when $X$ is an ordered Banach space we obtain

**Proposition 1.4.1:** *If $E$ is an ordered Banach space whose order cone is regular, then each monotone function $y\colon J \to E$ is regulated.*

*Proof.* Assume first that $y$ is nondecreasing. Let $t$ be an inner point of $J$, and let $t_o \in J$ be so chosen that $t_o < t$. Denoting $t_n = t - \frac{t-t_o}{n}$, $n = 1, 2, \dots$, then $(y(t_n))_{n=o}^{\infty}$ is a nondecreasing sequence in $X$, and $y(t_o) \leq y(t_n) \leq y(t)$ for each $n \in I\!N$. Thus $(y(t_n))_{n=o}^{\infty}$ is also order-bounded. Since the order cone of $E$ is regular, then

$$z = \lim_{n\to\infty} y(t_n) = \sup_n y(t_n)$$

exists. Given $\epsilon > 0$ there is $n \in I\!N$ such that $\|y(t_n) - z\| < \epsilon$. Because $y$ is nondecreasing, then

$$y(t_n) \leq y(s) \leq z \text{ for each } \ s \in (t_n, t).$$

Since the order cone of $E$ is also normal, there is $c > 0$ such that

$$\|z - y(s)\| \leq c\|y(t_n) - z\| < c\epsilon \text{ for each } \ s \in (t_n, t).$$

Thus $\lim_{s\to t-} y(s) = z$, which shows that $y(t-) = z$.

The proof of the existence of $y(t+)$ is similar, as well as the proof in the case when $y$ is nonincreasing, and when $t$ is a possible end point of $J$. □

**Example 1.4.1:** Define functions $g_m : I\!R \to I\!R$, $m = 1, 2, \dots$, by

$$g_m(t) = \sum_{k=1}^{\infty} \frac{[2 + [\sqrt[m]{kt}] - \sqrt[m]{kt}]}{(km)^2} \sin\left(\frac{1}{1 + [\sqrt[m]{kt}] - \sqrt[m]{kt}}\right),$$

where $[x]$ denotes the greatest integer $\leq x$. It is easy to see that each $g_m$ is right regulated. Denoting $f_n = \sum_{m=1}^{n} g_m$ we obtain a sequence $(f_n)_{n=1}^{\infty}$ which converges uniformly to a real function $f$. Corollaries 1.4.1 and 1.4.2 imply then that $f$ has a countable number of discontinuities, that $f$ can be uniformly approximated by countably stepped functions, and that $f$ is Riemann integrable on each compact interval. A closer analysis shows that $f$ is neither right continuous nor has the left-hand limit at $\frac{n}{\sqrt[m]{k}}$ for any integer $n$ and for any positive integers $k$, $m$. Thus $f$ is not regulated, monotone or of bounded variation on any interval.

### 1.4.3 Almost right and almost left regulated functions

We shall now consider a weaker concept of regulated functions. If $J$ is a real interval, and $E$ a normed space, we say that a function $y: J \to E$ is *almost right regulated* if for each $x \in J$ distinct from the possible right end point of $J$, and for each $\epsilon > 0$ there exists $u \in J$, $x < u$, and a null set $Z$ in $(x, u)$ such that $\|y(s) - y(t)\| \leq \epsilon$ for all $s$, $t \in (x, u) \backslash Z$. An almost left regulated function is defined similarly.

If $y$ is a.e. equal to a right (resp. left) regulated function, then $y$ is almost right (resp. left) regulated. Since the Dirichlet function $y(t) = \begin{cases} 1 & \text{if } t \text{ is irrational,} \\ 0 & \text{if } t \text{ is rational,} \end{cases}$ is a.e. equal to a constant function, it is almost right regulated, as well as almost left regulated. Note also that the Dirichlet function has neither the right-hand nor the left-hand limit at any point.

**Proposition 1.4.2:** *Let $E$ be a normed space and $a, b \in \mathbb{R}$, $a < b$. If $y\colon [a,b) \to E$ is almost right regulated, then $y$ is strongly measurable.*

*Proof.* Given a positive integer $n$, let $G_n : [a,b) \to [a,b]$ be a choice function which chooses to each $x \in [a,b)$ a point $G_n(x) \in (x,b]$ such that

$$\|y(s) - y(t)\| \le \frac{1}{n} \quad \text{for all } s,\, t \in (x, G_n(x)) \setminus Z_{x,n}, \tag{a}$$

where $Z_{x,n}$ is a null set in $(x, G_n(x))$. By defining $G_n(b) = b$ we obtain a function $G_n : [a,b] \to [a,b]$ which satisfies the hypotheses of corollary 1.1.1. This and proposition 1.1.6 imply that the w.o. chain $C_n$ of $G_n$-iterations of $a$ is countable, and that $[a,b]$ is the disjoint union of $C_n$ and the open intervals $(x, G_n(x))$, $x \in C_n$. Since $C_n$ is countable, then the set $Z_n = \bigcup_{x \in C_n} Z_{x,n}$ is a null set in $[a,b]$. Choose from each interval $(x, G_n(x))$, $x \in C_n$, a point $s_{x,n}$, and define

$$y_n(t) = \sum_{x \in C_n} (\chi_{(x,G_n(x))}(t) y(s_{x,n}) + \chi_{\{x\}}(t) y(t)), \quad t \in [a,b). \tag{b}$$

$y_n : [a,b) \to X$ is a countably stepped function. If $t \in [a,b) \setminus Z_n$, there is exactly one $x \in C_n$ such that either $t \in (x, G_n(x))$ or $t = x$. From (a) and (b) it follows that we have in both cases $\|y_n(t) - y(t)\| \le \frac{1}{n}$. Thus

$$\|y_n(t) - y(t)\| \le \frac{1}{n} \quad \text{for all } \ t \in [a,b) \setminus Z_n. \tag{c}$$

We shall show next that $y_n$ is strongly measurable. If $C_n$ is finite, then $y_n$ is a step function, and hence strongly measurable. If $C_n$

is infinite, then it has a representation as a sequence $(a_j^n)_{j=o}^{\infty}$. Define for each $m \in I\!N$ a function $g_m\colon [a,b) \to E$ by

$$g_m(t) = \begin{cases} y_n(t), & t \in \bigcup_{j=o}^{m}(a_j^n, G_n(a_j^n)), \\ 0, & t \in [a,b) \setminus \bigcup_{j=o}^{m}(a_j^n, G_n(a_j^n)). \end{cases}$$

Each $g_m$ is a step function and the sequence $(g_m)_{m=o}^{\infty}$ converges pointwise in $[a,b) \setminus Z_n$ to $y_n$. Thus $y_n$ is strongly measurable.

Assume now that the above construction is carried out for each positive integer $n$. The set $Z = \bigcup_{n=1}^{\infty} Z_n$ is a null set in $[a,b]$, and (c) implies that the sequence $(y_n)_{n=1}^{\infty}$ of the countably stepped functions, defined by (b), converges uniformly on $[a,b] \setminus Z$ to $y$. Since each $y_n$ is strongly measurable, it follows that $y$ is strongly measurable. □

The above proof differs from that of lemma 1.4.1 in the sense that we used the axiom of choice. As a consequence of proposition 1.4.2 and its dual we obtain

**Corollary 1.4.3:** *Let $J$ be a countable union of order bounded real intervals $J_i$. A function $y$ from $J$ to a normed space $E$ is strongly measurable if it is a.e. pointwise limit of a sequence of functions which are almost right or almost left regulated in the interiors of $J_i$.*

### 1.4.4. Substitution theorems

Let $\Omega$ be a measurable space and $V$ a subset of a Banach space $E$. We shall first consider measurability of the superposition operator $f(\cdot, y(\cdot))$, where $f\colon \Omega \times V \to E$ and $y\colon \Omega \to V$.

**Theorem 1.4.1:** *Let $J$ be a compact real interval. Given $f\colon J \times V \to E$, assume that $E$ is ordered by a regular order cone, and that there is a null set $Z$ in $J$ such that*

(a) $f(\cdot, x)$ *is right continuous or left continuous in* $J \setminus Z$ *for each* $x \in V$;

(b) $f(t, \cdot)$ *is nondecreasing in* $V$ *for each* $t \in J \setminus Z$.

*If* $y \colon J \to V$ *is nondecreasing, then* $f(\cdot, y(\cdot))$ *is strongly measurable.*

*Proof.* Assume right continuity in (a). If $J = [t_0, t_1]$ and $y \colon J \to V$ is nondecreasing, it follows from (b) that for each $s \in J \setminus Z$ the function $t \mapsto f(s, y(t))$ is nondecreasing on $J$. Then it is also regulated by proposition 1.4.1, and thus strongly measurable.

For each $n = 1, 2, \ldots$ choose a partition $P_n = \{t_j^n\}_{j=0}^n$ of $J$ such that

$$\max_{1 \le j \le n} |t_j^n - t_{j-1}^n| \le \frac{2(t_1 - t_0)}{n} \quad \text{and} \quad t_j^n \in J \setminus Z$$

for all $j = 1, \ldots, n-1$. The functions $f_n \colon J \to E$, defined by

$$f_n(t) = \sum_{j=1}^{n-1} f(t_j^n, y(t)) \chi_{(t_{j-1}^n, t_j^n]}(t), \quad t \in J, \; n = 1, 2, \ldots,$$

are strongly measurable. Let $t \in (t_0, t_1) \setminus Z$ be given. For $n$ large enough there exists $t_{j_n}^n \in P_n$ such that

$$f_n(t) = f(t_{j_n}^n, y(t)), \quad \text{and } 0 \le t_{j_n}^n - t < \frac{2(t_1 - t_0)}{n}.$$

Since $s \mapsto f(s, y(t))$ is right continuous in $J \setminus Z$, then

$$\lim_{n \to \infty} f_n(t) = f(t, y(t)).$$

This holds for each $t \in (t_0, t_1) \setminus Z$, whence $f(\cdot, y(\cdot))$ is strongly measurable on $J$. □

In the next result we don't assume the regularity of $K$ or the monotonicity of $f(t,\cdot)$.

**Theorem 1.4.2:** *Assume that $E$ is an ordered Banach space with order cone $K$, that $V \subseteq E$ is open, that $J = [a,b]$, and that $f\colon J \times V \to E$ satisfies one of the the following conditions.*

(1) *For each $(t,x) \in [a,b) \times V$ there is $z \in E$ so that $\|f(t+h, x+k) - z\| \to 0$ as $h \to 0+$ and $k \to 0$ in $K$.*

(2) *For each $(t,x) \in (a,b] \times V$ there is $z \in E$ so that $\|f(t-h, x-k) - z\| \to 0$ as $h \to 0+$ and $k \to 0$ in $K$.*

*If $y\colon J \to V$ is nondecreasing and continuous, then $f(\cdot, y(\cdot))$ is strongly measurable, and has at most countable number of discontinuities on $J$.*

*Proof.* The hypothesis (1) implies that if $y\colon J \to V$ is nondecreasing and continuous, then the function $t \mapsto f(t, y(t))$ is right regulated on $[a,b)$. The assertions follow then from lemma 1.4.1 and proposition 1.4.2. The proof when (2) holds is similar. □

In the next result $E$ need not to be ordered.

**Theorem 1.4.3:** *Let $(\Omega, \mathcal{A}, \mu)$ be a measure space, $U$ a measurable subset of $\Omega$, and $V$ a subset of a Banach space $E$. Assume that $f\colon U \times V \to E$ is a Carathéodory function, i.e.*

(c1) *$f(\cdot, z)$ is $\mu$-measurable for each $z \in V$;*

(c2) *$f(t,\cdot)$ is continuous in $V$ for a.a. $t \in U$.*

*If $y\colon U \to V$ is $\mu$-measurable, then $f(\cdot, y(\cdot))$ is $\mu$-measurable.*

*Proof.* Let $y\colon U \to V$ be $\mu$-measurable, and let $(y_n)_{n=o}^{\infty}$ be a sequence of step functions which converges pointwise a.e. on $U$ to $y$. From (c1) it follows that $f(\cdot, y_n(\cdot))$ is $\mu$-measurable for each $n \in I\!N$. Condition (c2) implies that

$$\lim_{n\to\infty} f(t, y_n(t)) = f(t, y(t)) \quad \text{for a.a.} \quad t \in U.$$

This implies that $f(\cdot, y(\cdot))$ is $\mu$-measurable. □

Denote by $L(E)$ the Banach space of all the continuous linear mappings $T\colon E \to E$ with the norm $\|T\| = \sup\{\|Th\| \mid \|h\| \leq 1\}$.

**Corollary 1.4.4:** *Let $U$ be a Lebesgue measurable subset of $\mathbb{R}^n$, and let $A\colon U \to L(E)$ be given. If $A(\cdot)z$ is strongly measurable for each $z \in E$, and if $y\colon U \to E$ is strongly measurable, then $A(\cdot)y(\cdot)$ is strongly measurable.*

*Proof.* The given hypotheses imply that $(t, z) \mapsto A(t)z$ is a Carathéodory function, whence the assertion follows from theorem 1.4.3. □

Let $(\Omega, \mathcal{A}, \mu)$ be a measure space, $E$ a Banach space and $\mathcal{B}$ the $\sigma$-algebra of the Borel sets of $E$. Denote by $\mathcal{A} \times \mathcal{B}$ the smallest $\sigma$-algebra containing all products $A \times B$ with $A \in \mathcal{A}$ and $B \in \mathcal{B}$. We say that $f\colon \Omega \times E \to E$ is a *standard function* if there is a null set $Z \in \mathcal{A}$ such that for each $V \in \mathcal{B}$

$$\{(t, x) \in (\Omega \setminus Z) \times E \mid f(t, x) \in V\} \in \mathcal{A} \times \mathcal{B}.$$

The following result is due to I.V. Shragin (1979).

**Theorem 1.4.4:** *If $E$ is separable and $\mu$ is complete and $\sigma$-finite, then $f(\cdot, y(\cdot))$ is measurable whenever $f\colon \Omega \times E \to E$ is a standard function and $y\colon \Omega \to E$ is measurable.*

Measurability can be replaced in theorem 1.4.4 by $\mu$-measurability. In particular, we have

**Corollary 1.4.5:** *Let $\Omega$ be a Lebesgue measurable subset of $\mathbb{R}^n$. If $f\colon \Omega \times \mathbb{R}^m \to \mathbb{R}^m$ is a Borel measurable function and $y\colon \Omega \to \mathbb{R}^m$ is measurable, then $f(\cdot, y(\cdot))$ is measurable.*

**Example 1.4.2:** Let $D$ be a subset of $[0, 1]$ which is not Lebesgue measurable. Define

$$f(t, x) = \begin{cases} 1 & \text{if either } x > t, \text{ or } x = t \text{ and } x \in D, \\ 0 & \text{if either } x < t, \text{ or } x = t \text{ and } x \notin D. \end{cases}$$

For the so obtained function $f\colon [0,1] \times I\!R \to I\!R$ the functions $f(\cdot, x)$ and $f(t, \cdot)$ have at most one discontinuity point (at $x = t$), and $f(t, \cdot)$ is nondecreasing for all $t \in [0,1]$. The function $y(t) = t$ is continuous and nondecreasing, but $f(\cdot, y(\cdot)) = \chi_D$ is not measurable.

### 1.4.5. $L^p$-spaces of real valued functions

Let $(\Omega, \mathcal{A}, \mu)$ be a measure space and $1 \leq p < \infty$. Denote by $L^p(\Omega) = L^p(\Omega, I\!R)$ the set of those $\mu$-measurable functions $u\colon \Omega \to I\!R$ for which $\int_\Omega |u|^p d\mu < \infty$. Identifying a.e. equal functions it is easy to see that $L^p(\Omega, I\!R)$ is a Banach space with respect to the norm $\|u\|_p = (\int_\Omega |u|^p d\mu)^{\frac{1}{p}}$. The set $L^p(\Omega, I\!R_+) = L^p_+(\Omega = \{u \in L^p(\Omega, I\!R) \mid u(t) \geq 0$ for a.a. $t \in \Omega\}$ is a closed cone in $L^p(\Omega, I\!R)$, inducing a partial ordering $' \leq'$, equivalent to a.e. pointwise ordering in $L^p(\Omega, I\!R)$. Let $u, v \in L^p(\Omega, I\!R_+)$, $u \leq v$ be given. Since

$$\begin{aligned}\|v\|_p^p &= \int_\Omega (u + (v - u))^p d\mu \geq \int_\Omega u^p d\mu + \int_\Omega (v - u)^p d\mu \\ &= \|u\|_p^p + \|v - u\|_p^p,\end{aligned}$$

it follows that $L^p(\Omega, I\!R_+)$ is a normal order cone in $L^p(\Omega)$. If $1 < p < \infty$, it can be shown that $L^p(\Omega, I\!R)' = L^{\frac{p}{p-1}}(\Omega, I\!R)$. In particular, $L^p(\Omega, I\!R)$ is reflexive. From proposition 1.3.4 it then follows that $L^p(\Omega, I\!R_+)$ is also regular and fully regular order cone of $L^p(\Omega, I\!R)$ when $1 < p < \infty$. This follows from the dominated convergence theorem when $p = 1$.

Consider next the case $p = \infty$. Denote by $L^\infty(\Omega, I\!R)$ the space of (the equivalence classes of) those $\mu$-measurable functions $u\colon \Omega \to I\!R$, which are essentially bounded in $\Omega$, i.e. whose *essential supremums*

$$\begin{aligned}\|u\|_\infty &= \text{ess sup}\{|u(t)| \mid t \in \Omega\} \\ &= \inf\{c \geq 0 \mid |u(t)| \leq c \text{ for a.a. } t \in \Omega\}\end{aligned}$$

are finite. $(L^\infty(\Omega, \mathbb{R}), \|\cdot\|_\infty)$ is a Banach space and the set $L^\infty(\Omega, \mathbb{R}_+)$ of its nonnegative-valued elements is its closed and normal cone.

**Example 1.4.3:** If the set $\mathbb{N}$ is endowed with the counting measure $\mu$, then $L^p(\mathbb{N})$ can be identified for $1 \le p < \infty$ with the space $l^p$ of those sequences $x = (x_n)_{n=o}^\infty$ of real numbers for which $\|x\|_p = (\sum_{n=o}^\infty |x_n|^p)^{\frac{1}{p}}$ is finite.

$L^\infty(\mathbb{N})$ is equal to the space $l^\infty$ of all bounded sequences $x = (x_n)_{n=o}^\infty$ of $\mathbb{R}$, and $\|x\|_\infty = \sup_{n\in\mathbb{N}} |x_n|$. The partial ordering of $L^p(\mathbb{N})$, defined by $L_+^p(\mathbb{N})$ equals to that, defined in $l^p$ by

$$(x_n)_{n=o}^\infty \le (y_n)_{n=o}^\infty \text{ if and only if } x_n \le y_n \text{ for each } n \in \mathbb{N}.$$

### 1.4.6. Integral inequalities

Next we shall derive some inequalities for integrals of functions with values in ordered Banach spaces.

**Proposition 1.4.3:** *Let $E$ be an ordered Banach space and $K$ its order cone. If $y\colon \Omega \to E$ is $\mu$-integrable and $y(t) \in K$ a.e. on $\Omega$, then*

$$0 \le \int_A y\,d\mu \le \int_B y\,d\mu \quad \textit{whenever } A,\ B \in \mathcal{A},\ A \subseteq B. \tag{1.4.3}$$

*Proof.* The first inequality in (1.4.3) is equivalent to $\int_A y d\mu \in K$. To prove this, note first that $\int_A y d\mu = 0$ if $\mu(A) = 0$.

Assume now that $0 < \mu(A) < \infty$, and make a counter-hypothesis:

$$v = \frac{1}{\mu(A)} \int_A y d\mu \notin K.$$

Since $K$ is closed and convex, there exist (see Narici and Beckenstein (1985), p. 164) $T \in E'$ and $c \in \mathbb{R}$ such that

$$Tu \geq c \text{ for all } u \in K \text{ and } Tv < c, \tag{a}$$

whence

$$T(\int_A y\,d\mu) = T(\mu(A)v) = \mu(A)Tv < c\mu(A). \tag{b}$$

On the other hand, since $y(t) \in K$ for a.a. $t \in A$, it follows from (a) that $Ty(t) \geq c$ for a.a. $t \in A$, whence

$$T(\int_A y\,d\mu) = \int_A (T \circ y)\,d\mu \geq \int_A c\,d\mu = c\mu(A),$$

contradicting (b). Thus $v \in K$, which implies that $0 \leq \int_A y\,d\mu$.

If $\mu(A) = \infty$, there exist measurable subsets $A_n$ of $\Omega$ with $\mu(A_n) < \infty$ such that

$$\|\int_A y\,d\mu - \int_{A_n} y\,d\mu\| < \frac{1}{n}, \quad n = 1, 2, \ldots.$$

Since $\int_{A_n} y\,d\mu \in K$ for each $n \in \mathbb{N}$ and $K$ is closed, then $\int_A y\,d\mu \in K$ also in this case.

The second inequality in (1.4.3) follows from

$$\int_B y\,d\mu - \int_A y\,d\mu = \int_{B \setminus A} y\,d\mu \in K.$$

□

As an immediate consequence of proposition 1.4.3 we obtain

**Corollary 1.4.6:** *Let $E$ be an ordered Banach space. If $x, y\colon \Omega \to E$ are $\mu$-integrable and $x(t) \le y(t)$ for a.a. $t \in \Omega$, then*

$$\int_A x\,d\mu \le \int_A y\,d\mu \quad \textit{whenever } A \in \mathcal{A}. \tag{1.4.4}$$

### 1.4.7. Bochner integrability and a.e. differentiability

Let $E = (E, \|\cdot\|)$ be a Banach space and $J = [a, b]$, $a < b$. We say that a mapping $x\colon J \to E$ is *absolutely continuous* if for each $\epsilon > 0$ there corresponds such a $\delta > 0$ that for any sequence $[a_j, b_j]$, $j = 1, \ldots, n$, of disjoint subintervals of $J$ with $\sum_{j=1}^n (b_j - a_j) < \delta$ we have $\sum_{j=1}^n \|x(b_j) - x(a_j)\| < \epsilon$. $x$ is called *Lipschitz continuous* if there is $M > 0$ such that $\|x(s) - x(t)\| \le M|s - t|$ for all $s, t \in J$. $x$ is said to be of *strong bounded variation* if

$$V_J(x) = \sup\{\sum \|x(t_{j+1} - x(t_j)\| \mid \{t_j\} \text{ is a partition of } J\}$$

is finite. Each Lipschitz continuous function is absolutely continuous, and each absolutely continuous function is of strong bounded variation.

We say that $x\colon J \to E$ is *differentiable* at a point $t \in (a, b)$ if there is $z \in E$ such that

$$\left\| \frac{x(s) - x(t)}{s - t} - z \right\| \to 0 \tag{1.4.5}$$

as $s \to t$. Denote $z = x'(t)$, and call it the *derivative of $x$* at $t$. If $t \in [a, b)$ and (1.4.5) holds as $s \to t+$, we say that $z$ is the *right-hand derivative* of $x$ at $t$, and denote $z = x'_+(t)$. The *left-hand derivative* $x'_-(t)$ of $x$ at $t \in (a, b]$ is defined similarly. We say that $x$ is *differentiable on* $J$ if $x$ is differentiable at each point of $(a, b)$, and if $x'_+(a)$ and $x'_-(b)$ exist. If $x'(t)$ exists for a.a. $t \in J$, we say that $x$ is *a.e. differentiable on* $J$. If $x\colon J \to \mathbb{R}$ is a.e. differentiable, we define $x'(t) = 0$ at those points $t \in J$ where $x$ is not differentiable.

The following result is proved, for instance, in Royden (1968).

**Lemma 1.4.3:** *If $u\colon J \to \mathbb{R}$ is absolutely continuous, then it is a.e. differentiable, $u'$ is Lebesgue integrable and*

$$u(t) = u(t_o) + \int_{t_o}^{t} u'(s)ds \quad \textit{for all} \quad t_o, t \in J.$$

*If $v\colon J \to \mathbb{R}$ is Lebesgue integrable, $t_o \in J$ and $u(t) = \int_{t_o}^{t} v(s)ds$, $t \in J$, then $u$ is absolutely continuous and $u'(t) = v(t)$ for a.a. $t \in J$.*

We shall also need the following result, which is proved in McShane (1974).

**Theorem 1.4.5:** *If $f \in L^{\infty}([a,b], \mathbb{R})$ and $u\colon [c,d] \to [a,b]$ is absolutely continuous, then*

$$\int_{c}^{d} f(u(t))u'(t)\,dt = \int_{u(c)}^{u(d)} f(v)\,dv.$$

*This holds also when $f \in L^{1}([a,b], \mathbb{R})$ and $u\colon [c,d] \to [a,b]$ is absolutely continuous and monotone.*

The next theorem extends the fundamental theorem of calculus, presented in lemma 1.4.3, to the case when the values of the functions in question are in a Banach space $E$.

**Theorem 1.4.6:** *If $x, y\colon J \to E$ and $t_o \in J$, then the following conditions are equivalent.*

a) *$x$ is absolutely continuous and a.e. differentiable, and*

$$x'(t) = y(t) \quad \textit{for a.a.} \quad t \in J. \tag{1.4.6}$$

b) $y$ *is Bochner integrable and*

$$x(t) = x(t_o) + \int_{t_o}^{t} y(s)ds \quad \textit{for all} \quad t \in J. \tag{1.4.7}$$

*Proof.* Assume first that $x$ is absolutely continuous, and that $x'(t) = y(t)$ for all $t \in J \setminus Z$, where $Z$ is a null set in $J$. To show that $y$ is strongly measurable, denote

$$h_n = \frac{b-a}{2^n} \quad \text{and} \quad t_n^k = a + k\,h_n,$$

for $n \in \mathbb{N}$ and $k = 0, 1, \dots, 2^n$, and define step functions $y_n : J \to E$ by $y_n(b) = 0$, $n \in \mathbb{N}$, and

$$y_n(t) = \frac{x(t_n^k) - x(t_n^{k-1})}{h_n}, \; t_n^{k-1} \le t < t_n^k, \; k = 1, \dots, 2^n. \tag{a}$$

Let $t \in J \setminus Z$ be given. For each $n \in \mathbb{N}$ there corresponds such a $k_n \in \{1, \dots, 2^n\}$ that $t_n^{k_n - 1} \le t < t_n^{k_n}$. Because $x'(t)$ exists, then in the formulae

$$x(t_n^{k_n}) = x(t) + x'(t)(t_n^{k_n} - t) + |t_n^{k_n} - t|\, u_n,$$

and

$$x(t_n^{k_n-1}) = x(t) + x'(t)(t_n^{k_n-1} - t) + |t_n^{k_n-1} - t|\, v_n$$

$\|u_n\|$ and $\|v_n\| \to 0$ when $n \to \infty$. From these formulae it follows that

$$x(t_n^{k_n}) - x(t_n^{k_n-1}) = x'(t)h_n + |t_n^{k_n} - t|u_n - |t_n^{k_n-1} - t|v_n.$$

Thus for each $n \in I\!N$,

$$\|y_n(t) - y(t)\| = \|\frac{x(t_n^{k_n}) - x(t_n^{k_n - 1})}{h_n} - x'(t)\| \le \|u_n\| + \|v_n\|,$$

whence

$$\lim_{n \to \infty} y_n(t) = y(t) \quad \text{for each } \ t \in J \setminus Z. \tag{b}$$

This proves that $y$ is strongly measurable.

Since $x$ is absolutely continuous, it is also of strong bounded variation. From (a) it follows that for all $n \in I\!N$ we have

$$\int_J \|y_n(t)\| dt = \sum_{k=1}^{2^n} \|x(t_n^k) - x(t_n^{k-1})\| \le V_J(x).$$

This and (b) imply that $y$ is Bochner integrable.

Next we shall show that

$$\frac{d}{dt} \int_{t_o}^t y(s) ds = y(t) \text{ for a.a. } \ t \in J. \tag{1.4.8}$$

The mapping $t \mapsto \|y(t)\|$ is Lebesgue integrable, whence the function $u(t) = \int_{t_o}^t \|y(s)\| ds$, $t \in J$, is absolutely continuous by lemma 1.4.3, and

$$\|\int_t^{t+h} y(s) ds\| \le |\int_t^{t+h} \|y(s)\| ds| = |u(t+h) - u(t)|, \ t, \ t+h \in J.$$

This implies that the mapping $t \mapsto \int_{t_o}^t y(s) ds$ is absolutely continuous. Since

$$\|\frac{1}{h} \int_t^{t+h} y(s) ds - y(t)\| \le \frac{1}{h} \int_t^{t+h} \|y(s) - y(t)\| ds$$

whenever $t$, $t+h \in J$, $h \neq 0$, then the assertion (1.4.8) is proved when we show that

$$\lim_{h\to 0} \frac{1}{h} \int_t^{t+h} \|y(s) - y(t)\| ds = 0 \text{ for a.a. } t \in J. \tag{c}$$

Since $y$ is strongly measurable, there exists such a sequence $(y_n)_{n=o}^{\infty}$ of step functions $y_n \colon J \to E$ and a null set $Z_o$ in $J$ that

$$\lim_{n\to\infty} y_n(t) = y(t) \quad \text{for } t \in J \setminus Z_o. \tag{d}$$

Since each $y_n$ has only a finite number of values, then the set $V = \bigcup y_n[J]$ is countable, i.e. $V = \{x_k\}_{k=1}^{\infty}$. For each $k$ the function $v(t) = \|y(t) - x_k\|$ is Lebesgue integrable, whence by lemma 1.4.3 there exists such a null set $Z_k$ in $J$ that

$$\lim_{h\to 0} \frac{1}{h} \int_t^{t+h} \|y(s) - x_k\| ds = \|y(t) - x_k\| \text{ for all } t \in J \setminus Z_k. \tag{e}$$

Denote $Z = \bigcup_{k=0}^{\infty} Z_k \cup \{a, b\}$. Given $t \in J \setminus Z$ and $\epsilon > 0$, there exists by (d) and the definition of $V$ such $x_k \in V$ that

$$\|y(t) - x_k\| < \frac{\epsilon}{3}. \tag{f}$$

By (e) there exists $\delta > 0$ such that

$$\left|\frac{1}{h} \int_t^{t+h} \|y(s) - x_k\| ds - \|y(t) - x_k\|\right| < \frac{\epsilon}{3} \text{ for } 0 < |h| < \delta. \tag{g}$$

Hence, for $0 < |h| < \delta$ it follows from (f) and (g) that

$$\frac{1}{h} \int_t^{t+h} \|y(s) - y(t)\| ds < \epsilon.$$

Thus (c) holds, which implies (1.4.8).

Next we shall show that $x$ satisfies the integral equation (1.4.7). Define

$$z(t) = x(t) - x(t_o) - \int_{t_o}^{t} y(s)ds, \quad t \in J.$$

$z$ is absolutely continuous, and $z'(t) = x'(t) - y(t) = 0$ for a.a. $t \in J$. Given $t$, $t + h \in J$, $h \neq 0$, we have

$$|\|z(t+h)\| - \|z(t)\|| \leq \|z(t+h) - z(t)\|$$

and

$$\left|\frac{\|z(t+h)\| - \|z(t)\|}{h}\right| \leq \left\|\frac{z(t+h) - z(t)}{h}\right\|.$$

Thus also the function $u(t) = \|z(t)\|$, $t \in J$ is absolutely continuous and $u'(t) = 0$ for a.a. $t \in J$. Lemma 1.4.3 implies then that

$$u(t) = u(t_o) + \int_{t_o}^{t} u'(s)ds = u(t_o) = 0 \quad \text{for all} \quad t \in J,$$

whence by the definitions of $u$ and $z$ we see that (1.4.7) holds.

Conversely, assume that $y$ is Bochner integrable, and that (1.4.7) holds. The above proof shows that (1.4.8) is valid. From (1.4.7) and (1.4.8) it follows that $x$ is absolutely continuous and $x'(t) = y(t)$ for a.a. $t \in J$, whence (1.4.6) holds. □

The following results are also needed in the sequel.

**Lemma 1.4.4:** *Let $E$ be an ordered Banach space and $K$ its order cone. If $x\colon [a,b] \to E$ is absolutely continuous and a.e. differentiable, and if $x'(t) \in K$ for a.a. $t \in [a,b]$, then $x$ is nondecreasing.*

*Proof.* From theorem 1.4.6 and corollary 1.4.6 it follows that

$$x(t) - x(\bar{t}) = \int_{\bar{t}}^{t} x'(s)\,ds \in K, \ a \le \bar{t} \le t \le b,$$

which implies the assertion. □

**Corollary 1.4.7:** *Let $E$ be an ordered Banach space and $K$ its order cone. If $y$, $z\colon [a,b] \to E$ are Bochner integrable and $y(t) \le z(t)$ for a.a. $t \in [a,b]$, then for all $t \in [a,b]$,*

$$0 \le \int_a^t z(s)\,ds - \int_a^t y(s)\,ds \le \int_a^b z(s)\,ds - \int_a^b y(s)\,ds.$$

*Proof.* Define $x(t) = \int_a^t (z(s) - y(s))\,ds$, $a \le t \le b$. Since $x$ is absolutely continuous and $x'(t) = z(t) - y(t) \in K$ for a.a. $t \in [a,b]$, then $x$ is by lemma 1.4.4 nondecreasing. Thus we have $0 = x(a) \le x(t) \le x(b)$ for all $t \in [a,b]$, which concludes the proof. □

An absolutely continuous function from a real interval to a Banach space $E$ is not necessarily a.e. differentiable, as we see from

**Example 1.4.4:** Choose $E = (c_o) = \{x = (x_n)_{n=1}^{\infty} \mid \lim_n x_n = 0\}$, with the norm $\|x\| = \max_n\{|x_n|\}$, and $J = [0, 2\pi]$. Define $x\colon J \to (c_o)$ by

$$x(t) = (\frac{1}{n}\sin(nt))_{n=1}^{\infty}, \ t \in J.$$

$x$ is absolutely continuous, since

$$\|x(t) - x(s)\| = \max_n |\frac{1}{n}\sin(nt) - \frac{1}{n}\sin(ns)| \le |t - s|, \ s, t \in J.$$

If $x$ is differentiable at $t \in J$, it is easy to see that $x'(t) = (\cos(nt))_{n=1}^{\infty}$. But $\lim_n \cos(nt) \neq 0$ for all $t \in J$, whence $x'(t)$ does not belong to $(c_o)$ for any $t \in J$.

However, it can be shown (cf. Kufner John and Fučik (1977)).

**Lemma 1.4.5:** *If a Banach space $E$ is reflexive or is separable and a dual of a Banach space, then each mapping $x\colon [a,b] \to E$ which is of strong bounded variation is a.e. differentiable.*

### 1.4.8. Fixed point results in function spaces

In this subsection we shall first present fixed point results in ordered spaces of absolutely continuous functions. They will be among our basic tools in the study of discontinuous differential equations. For the study of Carathéodory type of differential equations we shall also present fixed point results for mappings in the spaces of continuous functions.

Given a compact real interval $J$ and an ordered Banach space $E$, denote by $C(J,E)$ the space of continuous functions $x\colon J \to E$. Define a norm and a partial ordering in $C(J,E)$ by

$$\begin{aligned} &\|x\|_o = \max\{\|x(t)\| \mid t \in J\}, \quad \text{and} \\ &x \le y \quad \text{if and only if } x(t) \le y(t) \text{ for each } \ t \in J. \end{aligned} \tag{1.4.9}$$

By $AC(J,E)$ we denote the space of all the absolutely continuous functions $x\colon J \to E$. Assume that $AC(J,E)$ is equipped with the partial ordering induced by that of $C(J,E)$ given above.

As a consequence of theorem 1.2.2 we obtain

**Theorem 1.4.7:** *Given a nonempty order interval $[\alpha,\beta]$ in $AC(J,E)$ and a nondecreasing mapping $G\colon [\alpha,\beta] \to [\alpha,\beta]$, assume there is $v \in AC(J,\mathbb{R})$ such that*

$$\|Gx(s)-Gx(t)\| \le |v(s)-v(t)| \quad \textit{for } x \in [\alpha,\beta], \ s, t \in J, \tag{1.4.10}$$

*and that the order cone of $E$ is regular. Then the w.o. chain of $G$-iterations of $\alpha$ has the maximum $x_*$, the i.w.o. chain of $G$-iterations of $\beta$ has the minimum $x^*$, and*

$$x_* = \min\{x \mid Gx \leq x\}, \quad x^* = \max\{x \mid x \leq Gx\}. \tag{1.4.11}$$

*In particular, $x_*$ and $x^*$ are the extremal fixed points of $G$.*

*Proof.* Let $(x_n)_{n=o}^{\infty}$ be a monotone sequence in $[\alpha, \beta]$. Since $G$ is nondecreasing, then $(Gx_n(t))_{n=o}^{\infty}$ is for each $t \in J$ a monotone sequence in $[\alpha(t), \beta(t)]$. Because the order cone of $E$ is regular, then

$$x(t) = \lim_{n\to\infty} Gx_n(t) \text{ exists in } [\alpha(t), \beta(t)] \text{ for each } t \in J. \tag{a}$$

From (1.4.10) it follows that for all $s, t \in J$ and $n \in \mathbb{N}$,

$$\|Gx_n(s) - Gx_n(t)\| \leq |v(s) - v(t)|. \tag{b}$$

Thus the sequence $(Gx_n)_{n=o}^{\infty}$ is equicontinuous, whence the convergence in (a) is uniform. Moreover, (b) implies as $n \to \infty$ that

$$\|x(s) - x(t)\| \leq |v(s) - v(t)| \text{ for all } s, t \in J \text{ and } n \in \mathbb{N}. \tag{c}$$

Because $v \in AC(J, \mathbb{R})$, it follows from (c) that $x \in AC(J, E)$.

The above proof shows that $G$ satisfies the hypotheses of theorem 1.2.2 when $X = C(J, E)$ with the metric induced by the norm $\|\cdot\|_o$ and $Y = AC(J, E)$, whence the assertions follow from theorem 1.2.2. □

As an immediate consequence to theorem 1.4.7 we have

**Proposition 1.4.4:** *Given a nonempty order interval* $[\alpha, \beta]$ *in* $AC(J, E)$ *and a nondecreasing mapping* $G\colon [\alpha, \beta] \to [\alpha, \beta]$, *assume there is* $\gamma \in L^1(J, \mathbb{R}_+)$ *such that*

$$\|(Gx)'(t)\| \leq \gamma(t) \quad \text{for all } x \in [\alpha, \beta] \text{ and for a.a } t \in J. \tag{1.4.12}$$

*If the order cone of E is regular, then the conclusions of theorem 1.4.7 hold.*

*Proof.* From (1.4.12) it follows that (1.4.10) holds when

$$v(t) = \int_{t_o}^{t} \gamma(s)\, ds, \quad t \in J, \tag{1.4.13}$$

where $t_o$ is a point of $J$. Thus $G$ satisfies the hypotheses of theorem 1.4.7. □

The results of theorem 1.4.7 and proposition 1.4.4 can be applied in the case when $E = \mathbb{R}^m$, endowed with the coordinate-wise partial ordering, and also when $E = l^p$, $1 \leq p < \infty$, equipped with the coordinate-wise partial ordering, since it equals to that induced by the regular order cone $l^p_+$. These special cases will be used in the study of finite and infinite systems of discontinuous differential equations. Because $l^p_+$ is not regular when $p = \infty$, then another proof to the conclusions of theorem 1.4.7 is needed in this case.

**Proposition 1.4.5:** *Given a nonempty order interval* $[\alpha, \beta]$ *in* $AC(J, l^p)$, $1 \leq p \leq \infty$, *and a nondecreasing mapping* $G\colon [\alpha, \beta] \to [\alpha, \beta]$, *assume there exists* $v \in AC(J, \mathbb{R})$ *such that (1.4.10) holds when* $\|\cdot\| = \|\cdot\|_p$. *Then the conclusions of theorem 1.4.7 hold.*

*Proof.* In view of the above remark it suffices to consider the case when $p = \infty$. Let $C$ be a chain in $G[\alpha, \beta]$. Define for each $t \in J$

$$\begin{aligned} &y(t) = (y_1(t), y_2(t), \dots), \quad \text{where} \\ &y_i(t) = \sup\{x_i(t) \mid (x_1, x_2, \dots) \in C\}. \end{aligned} \tag{a}$$

Since $C \subset [\alpha, \beta]$, then $y_i(t) \in [\alpha_i(t), \beta_i(t)]$ for each $t \in J$ and $i = 1, 2, \ldots$, whence $y(t) \in l^\infty$ for each $t \in J$. Let $s, t \in J$ and $i \in \{1, 2, \ldots\}$ be given. Choose nondecreasing sequences $(y_n)$ and $(z_n)$ from $C$ so that $\lim_n (y_n)_i(s) = y_i(s)$ and $\lim_n (z_n)_i(t) = y_i(t)$. Denoting $x_n = \max\{y_n, z_n\}$, then

$$\lim_n (x_n)_i(s) = y_i(s) \quad \text{and} \quad \lim_n (x_n)_i(t) = y_i(t). \tag{b}$$

Because $x_n \in G[\alpha, \beta]$ for each $n \in \mathbb{N}$, it follows from (1.4.10) that

$$|(x_n)_i(s) - (x_n)_i(t)| \leq |v(s) - v(t)| \quad \text{for all} \quad n \in \mathbb{N}. \tag{c}$$

(b) and (c) imply that

$$|y_i(s) - y_i(t)| \leq |v(s) - v(t)|. \tag{d}$$

Since (d) holds for each $i = 1, 2, \ldots$ and for all $s, t \in J$, it follows that

$$\|y(s) - y(t)\|_\infty \leq |v(s) - v(t)| \quad \text{for all} \quad s, t \in J.$$

In particular, $y \in AC(J, l^\infty)$. Since $y_i(t) \in [\alpha_i(t), \beta_i(t)]$, for each $t \in J$ and $i = 1, 2, \ldots$, then $y \in [\alpha, \beta]$. From (a) is follows that $y = \sup C$.

The proof that $z = \inf C$ exists in $[\alpha, \beta]$ and satisfies (1.4.10) is similar. Thus $G$ satisfies the hypotheses of theorem 1.2.3 when $X = C(J, l^\infty)$ and $Y = AC(J, l^\infty)$, whence the conclusions follow from theorem 1.2.3. □

As for the existence of extremal coupled fixed points of mixed monotone mappings in ordered spaces of absolutely continuous mappings, we have the following consequence of theorem 1.2.4.

**Theorem 1.4.8:** *Let $[\alpha, \beta]$ be a nonempty order interval in $AC(J, E)$ and $A\colon [\alpha, \beta] \times [\alpha, \beta] \to [\alpha, \beta]$ a mixed monotone mapping. Assume there is $v \in AC(J, \mathbb{R})$ such that for all $x, y \in [\alpha, \beta]$*

*and* $s,\ t \in J$,

$$\|A(x,y)(s) - A(x,y)(t)\| \le |v(s) - v(t)|, \tag{1.4.14}$$

*and that the order cone of E is regular. Then A has the extremal coupled fixed points.*

*Proof.* Let $(v_n)_{n=o}^{\infty}$ and $(w_n)_{n=o}^{\infty}$ be sequences in $[\alpha,\beta]$, one being nondecreasing and the other one nonincreasing. Because $A$ is mixed monotone, then $(A(v_n,w_n))_{n=o}^{\infty}$ is a monotone sequence in $[\alpha,\beta]$. From (1.4.14) it follows that for all $s,\ t \in J$ and $n \in \mathbb{N}$,

$$\|A(v_n,w_n)(s) - A(v_n,w_n)(t)\| \le |v(s) - v(t)|.$$

Thus the reasoning used in the proof of theorem 1.4.7 shows that $(A(v_n,w_n))_{n=o}^{\infty}$ converges uniformly to a function $x \in AC(J,E)$. Consequently, $A$ satisfies the hypotheses of theorem 1.2.4 when $X = C(J,E)$ with the metric induced by the norm $\|\cdot\|_o$ and $P = AC(J,E)$, so that the assertions follow from theorem 1.2.4. □

As a special case of theorem 1.4.8 we obtain

**Proposition 1.4.6:** *Let $[\alpha,\beta]$ be a nonempty order interval in $AC(J,E)$ and $A\colon [\alpha,\beta]\times[\alpha,\beta] \to [\alpha,\beta]$ a mixed monotone mapping. If there is $\gamma \in L^1(J,\mathbb{R}_+)$ such that*

$$\|A(x,y)'(t)\| \le \gamma(t), \tag{1.4.15}$$

*for all $x,\ y \in [\alpha,\beta]$ and for a.a. $t \in J$, and if the order cone of E is regular, then A has the extremal coupled fixed points.*

*Proof.* The hypotheses of theorem 1.4.8 hold when

$$v(t) = \int_{t_o}^{t} \gamma(s)\,ds, \quad t \in J,$$

which implies the assertion. □

In the case when $E = l^p$ we have

**Proposition 1.4.7:** *Let $[\alpha, \beta]$ be a nonempty order interval in $AC(J, l^p)$, $1 \le p \le \infty$, and $A\colon [\alpha, \beta] \times [\alpha, \beta] \to [\alpha, \beta]$ a mixed monotone mapping. Assume there is $v \in AC(J, \mathbb{R})$ such that*

$$\|A(x, y)(s) - A(x, y)(t)\|_p \le |v(s) - v(t)| \tag{1.4.16}$$

*for all $x$, $y \in [\alpha, \beta]$ and $s$, $t \in J$. Then $A$ has the extremal coupled fixed points.*

*Proof.* The conclusion follows from theorem 1.4.8 when $1 \le p < \infty$. If $p = \infty$, the reasoning similar to that used in the proof of proposition 1.4.5 shows that the hypotheses of proposition 1.2.3 hold when $P = [\alpha, \beta]$. □

Next we shall derive a fixed point result for a mapping from $C(J, E)$ into itself, where $J = [0, T]$ and $E$ is a Banach space. If $x \in C(J, E)$, denote $|x| = t \mapsto \|x(t)\|$. We shall show that $F\colon C(J, E) \to C(J, E)$ has a unique fixed point if the following condition holds.

(F) There is $b \in C(J, \mathbb{R}_+)$ and such a nondecreasing mapping $Q\colon [0, b] \to [0, b]$ that $Qb(t) < b(t)$ and $Q^n b(t) \to 0$ for each $t \in J$ and

$$|Fy - F\bar{y}| \le Q|y - \bar{y}| \text{ whenever } |y - \bar{y}| \le b. \tag{1.4.17}$$

We shall first prove an auxiliary result.

**Lemma 1.4.6:** *Given $F\colon C(J, E) \to C(J, E)$, assume that condition (F) holds. Then $F$ has at most one fixed point which, if* *it exists, is the uniform limit of the sequence* $(F^n y_o)_{n=o}^{\infty}$ *for each* $y_o \in C(J,E)$. *Moreover, if there is* $y_o \in C(J,E)$ *and* $u \in [0,b]$ *such that*

$$|y_o - Fy_o| + Qu \le u, \tag{1.4.18}$$

*then* $F$ *has a fixed point* $x$ *and*

$$|x - y_o| \le u. \tag{1.4.19}$$

*Proof.* Assume that $x \in C(J,E)$ and $x = Fx$. Given $y_o \in C(J,E)$, denote

$$x_i = x + \frac{i}{m}(y_o - x), \quad i = 0, \ldots, m \ge \frac{\|y_o - x\|_o}{\min_{t\in J}(b(t) - Qb(t))}.$$

Because

$$|x_{i-1} - x_i| = \frac{1}{m}|y_o - x| \le \frac{1}{m}\|y_o - x\|_o \le b - Qb \le b$$

for each $i = 1, \ldots, m$, it follows from (1.4.17) by induction that

$$|F^n x_{i-1} - F^n x_i| \le Q^n b, \quad i = 1, \ldots, m, \ n \in \mathbb{N}.$$

Thus

$$|x - F^n y_o| = |F^n x - F^n y_o| \le \sum_{i=1}^{m} |F^n x_{i-1} - F^n x_i| \le m\, Q^n b \ \text{(a)}$$

for each $n \in \mathbb{N}$. Since $(Q^n b)_{n=o}^{\infty}$ is a nonincreasing sequence in $C(J, \mathbb{R}_+)$ and converges pointwise on $J$ to 0, it follows from Dini's theorem that this convergence is uniform on $J$. This and (a) imply that $F^n y_o \to x$ uniformly on $J$.

If $\bar{x}$ is also a fixed point of $F$, then the choice $y_o = \bar{x}$ above yields $F^n\bar{x} \to x$. But $F^n\bar{x} = \bar{x}$ for all $n \in I\!N$, so that $x = \bar{x}$, which proves the uniqueness assertion.

Assume next that $y_o \in C(J, E)$ and $u \in [0, b]$ satisfy (1.4.18). Denote

$$W = \{y \in C(J, E) \mid |y - y_o| \le u\}.$$

Since $Q$ is nondecreasing it follows from (1.4.17) and (1.4.18) that if $y \in W$, then

$$\begin{aligned}|Fy - y_o| &\le |Fy_o - y_o| + |Fy - Fy_o| \\ &\le |Fy_o - y_o| + Q|y - y_o| \le |Fy_o - y_o| + Qu \le u,\end{aligned}$$

whence $Fy \in W$. Thus $F[W] \subseteq W$, so that the assertion

$$|F^{m+n}y_o - F^m y_o| \le Q^m u \quad \text{for all} \quad n \in I\!N \tag{b}$$

holds when $m = 0$. Because $Q$ is nondecreasing, it follows from (1.4.17) and (b) that for all $n \in I\!N$,

$$|F^{m+n+1}y_o - F^{m+1}y_o| \le Q|F^{m+n}y_o - F^m y_o| \le Q^{m+1}u.$$

Thus (b) holds for all $m \in I\!N$. Since $0 \le Q^n u \le Q^n b$ for each $n \in I\!N$, then $Q^n u \to 0$ uniformly on $J$. From (b) it then follows that $(F^n y_o)_{n=o}^{\infty}$ is a Cauchy sequence in $C(J, E)$ with respect to the norm $\|\cdot\|_o$, and thus converges in $C(J, E)$. Denote $x = \lim_{n\to\infty} F^n y_o$. From (b) we obtain when $n \to \infty$, that

$$|x - F^m y_o| \le Q^m u \quad \text{for all} \quad m \in I\!N.$$

This implies that $x$ belongs to $W$, and by (1.4.17) that

$$|Fx - F^{m+1}y_o| \le Q^{m+1}u \quad \text{for all} \quad m \in I\!N.$$

Thus $Fx = \lim_{n\to\infty} F^n y_o$, so that $x = Fx$. Because $W$ is a closed subset of $C(J,E)$, then $x \in W$, whence (1.4.19) holds. □

**Theorem 1.4.9:** *If $F\colon C(J,E) \to C(J,E)$ satisfies condition (F), then for each $y_o \in C(J,E)$ the sequence $(F^n y_o)_{n=o}^{\infty}$ converges uniformly on $J$ to a unique fixed point of $F$.*

*Proof.* Given $y_o \in C(J,E)$, denote $x_o = y_o - Fy_o$, and

$$x_i = \frac{(m-i)x_o}{m}, \quad i = 0,\ldots,m \geq \frac{\|x_o\|_o}{\min_{t\in J}(b(t) - Qb(t))}.$$

This implies that

$$|x_{i-1} - x_i| = \frac{1}{m}|x_o| \leq \frac{1}{m}\|x_o\|_o \leq b - Qb$$

for each $i = 1,\ldots,m$. Denoting $F_1 = y \mapsto x_1 + Fy$, we have

$$|F_1 y - F_1 \bar{y}| = |Fy - F\bar{y}|$$

for all $y$, $\bar{y} \in C(J,E)$, whence $F_1$ satisfies condition (F). Moreover,

$$|y_o - F_1 y_o| = |x_o - x_1| \leq b - Qb,$$

so that $u = b$ satisfies (1.4.18) with $F$ replaced by $F_1$. Thus $F_1$ satisfies the hypotheses of lemma 1.4.6, whence $F_1$ has a fixed point $y_1$.

Replacing $x_o$, $x_1$ and $y_o$ above by $x_1$, $x_2$ and $y_1$, respectively, we see that the operator $F_2 = y \mapsto x_2 + Fy$ has a fixed point. Repeating this argument $m$ times it follows that $F_m = y \mapsto x_m + Fy$ has a fixed point. But $x_m = 0$, whence $F$ has a fixed point $x$. In view of lemma 1.4.6, $x$ is the only fixed point of $F$ and $F^n y_o \to x$ uniformly on $J$ for each choice of $y_o \in C(J,E)$. □

In particular, we have

**Proposition 1.4.8:** *Given $F\colon C(J,E) \to C(J,E)$, assume that*

$$|Fy - F\bar{y}| \le Q|y - \bar{y}| \quad \textit{for all} \quad y,\, \bar{y} \in C(J,E),$$

*where $Q\colon C(J,\mathbb{R}_+) \to C(J,\mathbb{R}_+)$ is nondecreasing, and for each $u_o \in \mathbb{R}_+$ there is $b \in C(J,\mathbb{R}_+)$ such that $u_o + Qb \le b$ and $Q^n b \to 0$ pointwise in $J$. Then for each $y_o \in C(J,E)$ the sequence $(F^n y_o)_{n=o}^{\infty}$ converges uniformly in $J$ to a unique fixed point $x$ of $F$. Moreover, (1.4.19) holds for each $u \in C(J,\mathbb{R}_+)$ which satisfies (1.4.18).*

## 1.5. METHOD OF UPPER AND LOWER SOLUTIONS

We shall devote most of this section to the study of the method of upper and lower solutions, which plays a prominent role in our study of differential equations. Since our results concern discontinuous differential equations, we are obliged to work in the framework of Carathéodory systems and in the space $AC(J,E)$ of absolutely continuous functions from a real interval $J = [0,T]$ into an ordered Banach space $E$.

Assume that the norm $\|\cdot\|_o$ and the partial ordering $\le$ are defined in $C(J,E)$ by (1.4.9). It is easy to see that the order interval $[\alpha,\beta] = \{x \in AC(J,E) \mid \alpha \le x \le \beta\}$ is for all $\alpha, \beta \in AC(J,E)$ a convex subset of $C(J,E)$.

In the cases of scalar differential equations and their finite systems we have $E = \mathbb{R}^m$. We shall assume that $\mathbb{R}^m$ is equipped with componentwise ordering and with the norm

$$\|x\| = \max\{|x_i| \mid 1 \le i \le m\}, \; x = (x_1,\ldots,x_m) \in \mathbb{R}^m.$$

### 1.5.1. Initial value problems of Carathéodory type

We shall first concentrate our attention to the scalar initial value problems (IVP for short) where we shall develop important ideas involved. Let us begin by considering the IVP

$$x' = f(t, x), \qquad x(0) = x_o, \tag{1.5.1}$$

where $f\colon J \times \mathbb{R} \to \mathbb{R}$ is a Carathéodory function and $J = [0, T]$. A function $y \in AC(J, \mathbb{R})$ is said to be an *upper solution* of (1.5.1) if

$$y'(t) \geq f(t, y(t)) \text{ for a.a. } t \in J, \text{ and } y(0) \geq x_o,$$

and a *lower solution* if the reversed inequalities are satisfied. If equalities hold, we say that $y$ is a *solution* of (1.5.1) on $J$. A solution $y \in AC(J, \mathbb{R})$ of (1.5.1) is called *maximal* if $x(t) \leq y(t)$ on $J$ for each solution $x$ of (1.5.1), and *minimal* if the reverse inequality holds. If both minimal and maximal solutions exist we call them *extremal solutions* of (1.5.1).

If we know the existence of lower and upper solutions $y$, $z$ of (1.5.1) such that $y \leq z$, we can prove the existence of the extremal solutions of (1.5.1) in the order interval $[y, z]$, provided that $f$ satisfies the classical Carathéodory conditions. This is the content of the next result.

**Theorem 1.5.1:** *Let $y$, $z \in AC(J, \mathbb{R})$ be lower and upper solutions of (1.5.1) such that $y \leq z$. Let $f$ be a Carathéodory function in $\Omega = \{(t, x) \mid y(t) \leq x \leq z(t),\ t \in J\}$. If there is $m \in L^1(J, \mathbb{R}_+)$ such that*

$$|f(t, x)| \leq m(t) \quad \textit{for all } x \in [y(t), z(t)] \textit{ and for a.a. } t \in J,$$

*then the IVP (1.5.1) has the extremal solutions in the order interval $[y, z]$.*

*Proof.* Let $p(t, x) = \max\{y(t), \min\{x, z(t)\}\}$, $(t, x) \in \Omega$, and define

$$\tilde{f}(t, x) = f(t, p(t, x)) + r(t, x), \text{ where } r(t, x) = \frac{p(t, x) - x}{1 + x^2}.$$

Then $\tilde{f}$ defines a Carathéodory extension of $f$ to $J \times I\!R$, and there exists $M \in L^1(J, I\!R_+)$ such that

$$|\tilde{f}(t,x)| \leq M(t) \quad \text{for all } x \in I\!R \text{ and for a.a. } \ t \in J.$$

Therefore, by the classical Carathéodory existence theorem (cf. Carathéodory (1948)), the IVP

$$x' = \tilde{f}(t,x), \qquad x(0) = x_o \tag{a}$$

has a solution on $J$. Moreover, denoting by $S$ the set of all the solutions of (a), it can be shown (cf. Walter (1970)) that the equations

$$x^*(t) = \sup\{x(t) \mid x \in S\}, \ x_*(t) = \inf\{x(t) \mid x \in S\}, \ t \in J$$

define the maximal solution $x^*$ and the minimal solution $x_*$ of (a) on $J$ (cf. also Coddington and Levinson (1955)).

To complete the proof we shall show that each solution $x$ of (a) satisfies $y(t) \leq x(t) \leq z(t)$ on $J$. Given a solution $x$ of (a) suppose that there exist $t_1$, $t_2 \in J$, $t_1 < t_2$, with

$$y(t_1) = x(t_1) \ \text{ and } \ y(t) > x(t), \ t_1 < t < t_2.$$

Then, letting $v(t) = y(t) - x(t)$, we have for a.a. $t \in (t_1, t_2)$

$$v'(t) \leq f(t, y(t) - f(t, y(t)) - \frac{y(t) - x(t)}{1 + x^2(t)} \leq 0,$$

so that $v(t) \leq v(t_1) = 0$ for $t \in (t_1, t_2)$, which is a contradiction. Thus $y(t) \leq x(t)$ on $J$. One can similarly prove that $x(t) \leq z(t)$ on $J$. Consequently, each solution $x$ of (a) satisfies $y(t) \leq x(t) \leq z(t)$ on $J$, and obviously $x \in AC(J, I\!R)$. By the definition of

$p(t, x)$ and $r(t, x)$ we then see that each solution of (a) is also a solution of (1.5.1) in $[y, z]$ and vice versa, whence $x_*$ and $x^*$ are the extremal solutions of (1.5.1) in $[y, z]$. □

A basic result relative to upper and lower solutions is the following standard comparison result (cf. Lakshmikantham and Leela (1969), Walter (1970)).

**Lemma 1.5.1:** *Given a Carathéodory function* $f: J \times \mathbb{R} \to \mathbb{R}$, *assume there is* $m \in L^1(J, \mathbb{R}_+)$ *such that*

$$|f(t, x)| \leq m(t) \quad \textit{for all } x \in \mathbb{R} \textit{ and for a.a. } t \in J.$$

a) *If $y$ is a lower solution and $z$ the maximal solution of (1.5.1), then $y \leq z$.*
b) *If $y$ is the minimal solution and $z$ an upper solution of (1.5.1), then $y \leq z$.*

The next result offers a construction of lower and upper solutions $y$, $z$ of (1.5.1) such that $y(t) \leq z(t)$ on $J$.

**Lemma 1.5.2:** *Let $f_1$, $f_2$, $f: J \times \mathbb{R} \to \mathbb{R}$ satisfy the hypotheses of lemma 1.5.1, and assume that*

$$f_1(t, x) \leq f(t, x) \leq f_2(t, x) \tag{1.5.2}$$

*for a.a. $t \in J$ and for all $x \in \mathbb{R}$. Let $y$ be any solution of*

$$y' = f_1(t, y), \qquad y(0) = y_o \leq x_o,$$

*and $z$ be the maximal solution of*

$$z' = f_2(t, z), \qquad z(0) = z_o \geq x_o,$$

*on $J$. Then $y$, $z$ are lower and upper solutions of (1.5.1) satisfying $y \leq z$.*

*Proof.* Because of (1.5.2) we get

$$y' \leq f_2(t,y), \qquad y(0) \leq x_o \leq z(0).$$

Hence, by lemma 1.5.1 we have $y \leq z$. Clearly, $y$, $z$ are lower and upper solutions of (1.5.1) on $J$. □

The following result will also be used to formulate growth conditions for the function $f$, which ensure existence of upper and lower solutions of (1.5.1).

**Lemma 1.5.3:** *If $p \in L^1(J, \mathbb{R}_+)$, $\psi$, $\frac{1}{\psi} \in L^\infty_{loc}(\mathbb{R}_+, \mathbb{R}_+)$, and $\int_o^\infty \frac{dv}{\psi(v)} = \infty$, then for each $w_o \in \mathbb{R}_+$ the IVP*

$$w' = p(t)\,\psi(w), \quad w(0) = w_o \tag{1.5.3}$$

*has a unique solution $w$ on $J$. Moreover, if $\psi$ is nondecreasing and if $u \in AC(J, \mathbb{R})$ satisfies the inequality*

$$u(t) \leq w_o + \int_o^t p(s)\,\psi(u(s))\,ds \tag{1.5.4}$$

*on $J$, then $u \leq w$. This holds in particular for any lower solution $u$ of (1.5.3).*

*Proof.* Let $w_o \in \mathbb{R}_+$ be given. The hypotheses imply that the equation

$$\int_{w_o}^{w(t)} \frac{dv}{\psi(v)} = \int_o^t p(s)ds \tag{a}$$

defines a nondecreasing function $w\colon J \to I\!R_+$ for which $w(0) = w_o$. If $M$ =ess sup $\{\psi(v) \mid w(0) \le v \le w(T)\}$, and $0 \le \bar{t} \le t \le T$, we have

$$\frac{w(t) - w(\bar{t})}{M} \le \int_{w(\bar{t})}^{w(t)} \frac{dv}{\psi(v)} = \int_{\bar{t}}^{t} p(s)ds.$$

This implies that $w$ is also absolutely continuous. By theorem 1.4.5 we can then rewrite (a) as

$$\int_o^t \frac{w'(s)ds}{\psi(w(s))} = \int_o^t p(s)ds, \quad t \in J,$$

which implies that $w$ is a solution of (1.5.3) on $J$.

To prove uniqueness, let $w$ be a solution of (1.5.3) on $J$. Since both $p$ and $\psi$ are nonnegative-valued, then $w$ is absolutely continuous and nondecreasing. This implies by theorem 1.4.5 that

$$\int_{w_o}^{w(t)} \frac{dv}{\psi(v)} = \int_o^t \frac{w'(s)\,ds}{\psi(w(s))} = \int_o^t p(s)\,ds \text{ for all } t \in J.$$

Thus the only solution $w$ of (1.5.3) is given by (a).

Finally, let $u \in AC(J, I\!R)$ satisfy (1.5.4) on $J$, and let $z$ be the solution of (1.5.3) with $z(0) = \max\{\|u\|_o, \|w\|_o\}$. Obviously, $u \le z$ and

$$z(t) \ge w_o + \int_o^t p(s)\,\psi(z(s))\,ds. \tag{b}$$

Define a mapping $G\colon [u, z] \to AC(J, I\!R)$ by

$$Gx(t) = w_o + \int_o^t p(s)\,\psi(x(s))\,ds. \tag{c}$$

Since $\psi$ is nondecreasing, it follows from (1.5.4), (b) and (c) that $G$ is nondecreasing, $u \le Gu$ and $Gz \le z$. Thus $G\colon [u,z] \to [u,z]$, and for each $x \in [u,z]$

$$|(Gx)'(t)| \le \psi(z(T))\,p(t), \qquad \text{for a.a. } t \in J. \tag{d}$$

The above proof implies that $G$ satisfies the hypotheses of proposition 1.4.4, whence $G$ has a fixed point $x$ in $[u,z]$. From the definition (c) of $G$ it follows that $x$ is a solution of the IVP (1.5.3). But $w$ is the only solution of (1.5.3), whence $x = w \in [u,z]$. This implies that $u \le w$, and concludes the proof. □

We shall need also the following comparison and uniqueness result.

**Lemma 1.5.4:** *Assume that $f\colon J \times \mathbb{R} \to \mathbb{R}$ satisfies for a.a. $t \in J$ and for all $x_1, x_2 \in \mathbb{R}$, $x_1 \ge x_2$,*

$$f(t,x_1) - f(t,x_2) \le g(t, x_1 - x_2), \tag{1.5.5}$$

*where $g\colon J \times \mathbb{R}_+ \to \mathbb{R}_+$ and the IVP*

$$x' = g(t,x), \qquad x(0) = 0 \tag{1.5.6}$$

*has the zero-function as the only lower solution on $J$. If $y$ is a lower solution and $z$ an upper solution of (1.5.1) on $J$, then $y \le z$. In particular, (1.5.1) has at most one solution on $J$.*

*Proof.* Let $y$ be a lower solution and $z$ an upper solution of (1.5.1). Denoting $m(t) = \max\{y(t) - z(t), 0\}$, $t \in J$, it follows from (1.5.5) and (1.5.6) that $m$ is a lower solution of (1.5.6), and hence $m(t) \equiv 0$, so that $y \le z$. In particular, if $y$ and $z$ are solutions of the IVP (1.5.1), then the above result implies that $y \le z$ and $z \le y$, whence $y = z$. This proves the uniqueness assertion. □

As a special case of lemma 1.5.4 we have

**Corollary 1.5.1:** *The conclusions of lemma 1.5.4 hold if there exists* $p \in L^1(J, I\!R_+)$ *such that*

$$f(t, x_1) - f(t, x_2) \le p(t)(x_1 - x_2) \tag{1.5.7}$$

*for a.a.* $t \in J$ *and for all* $x_1, x_2 \in I\!R$, $x_1 \ge x_2$.

*Proof.* If $m \in AC(J, I\!R_+)$ satisfies $m'(t) \le p(t)m(t)$ for a.a. $t \in J$, then $m(t) \le m(0)\exp(\int_o^t p(s)\,ds)$, $t \in J$. Thus $m(t) \equiv 0$ is the only lower solution of (1.5.6) when $g(t, x) = p(t)\,x$, $(t, x) \in J \times I\!R_+$. □

The first step in the study of the IVP (1.5.1) is often to replace it by such an operator equation that it can be solved, for instance by using known fixed point results, and that the so obtained solutions are also solutions of the IVP (1.5.1). The next lemma gives a method for such a replacement. Because of our needs we prove the result for Banach-valued functions, by applying theorem 1.4.6.

**Lemma 1.5.5:** *If* $E$ *is a Banach space,* $f\colon J \times E \to E$ *and* $x_o \in E$, *then* $x\colon J \to E$ *is a solution of the IVP (1.5.1) on* $J$ *if and only if* $x$ *satisfies for any* $p \in L^1(J, I\!R)$ *the integral equation*

$$\begin{aligned} x(t) &= e^{-P(t)}[x_o + \int_o^t e^{P(s)}(f(s, x(s)) + p(s)x(s))ds], \\ P(t) &= \int_o^t p(s)\,ds, \quad t \in J. \end{aligned} \tag{1.5.8}$$

*Proof.* Let $p \in L^1(J, I\!R)$ be given. Assume first that $x\colon J \to E$ is a solution of (1.5.1). By definition, $x$ is absolutely continuous and a.e. differentiable. Thus $x'$ is by theorem 1.4.6 Bochner integrable on $J$. Applying (1.5.1) we obtain

$$\begin{aligned}
&e^{-P(t)}[x_o + \int_o^t e^{P(s)}(f(s,x(s)) + p(s)x(s))\,ds] \\
&= e^{-P(t)}[x_o + \int_{t_o}^t e^{P(s)}(x'(s) + p(s)x(s))ds] \\
&= e^{P(t)}[x_o + e^{P(t)}x(t) - x_o] = x(t).
\end{aligned}$$

This holds for all $t \in J$, whence $x$ satisfies (1.5.8).

Conversely, assume that $x\colon J \to E$ is a solution of (1.5.8). From theorem 1.4.6 it follows that $x$ is absolutely continuous and a.e. differentiable on $J$. By differentiation we obtain

$$x'(t) = -p(t)x(t) + f(t,x(t)) + p(t)x(t) = f(t,x(t))$$

for a.a. $t \in J$. Moreover, the insertion $t = 0$ in (1.5.8) gives $x(0) = x_o$, whence $x$ is a solution of the IVP (1.5.1). □

**Corollary 1.5.2:** *If $E$ is a Banach space, $p \in L^1(J,\mathbb{R})$ and $q \in L^1(J,E)$, then the linear IVP*

$$x' + p(t)x = q(t), \qquad x(0) = x_o$$

*has a unique solution*

$$x(t) = e^{-P(t)}[x_o + \int_o^t e^{P(s)}q(s)ds], \; P(t) = \int_o^t p(s)\,ds, \quad t \in J.$$

*Proof.* The assertion follows from lemma 1.5.5 when $f(t,x) = q(t) - p(t)x$. □

### 1.5.2. Periodic boundary value problems

A fundamental comparison result relative to periodic boundary value problems (PBVP for short) is.

**Lemma 1.5.6:** *Given $f: J \times \mathbb{R} \to \mathbb{R}$, assume there is $p \in L^1(J, \mathbb{R})$ with $\int_o^T p(s)\, ds > 0$ such that $f(t,x) + p(t)\, x$ is nonincreasing in $x$. If $y$, $z \in AC(J, \mathbb{R})$ satisfy the inequalities*

$$\begin{aligned} &y' \le f(t,y) \quad \text{a.e. on } J,\ y(0) \le y(T) \quad \text{and} \\ &z' \ge f(t,z) \quad \text{a.e. on } J,\ z(0) \ge z(T), \end{aligned} \tag{1.5.9}$$

*then*

$$y(t) \le z(t) \quad \text{on } J. \tag{1.5.10}$$

*Proof.* If (1.5.10) is false, then either (i) $w(t) = y(t) - z(t) > 0$ on $J$, or there exist $t_1$, $t_2 \in J$, $t_1 < t_2$ such that (ii) $w(t_1) = 0$ and $w(t) > 0$ on $(t_1, t_2)$, or (iii) $w(t_2) = 0$ and $w(t) > 0$ on $(t_1, t_2)$.

Suppose that (i) holds. Then, denoting $P(t) = \int_o^t p(s)\, ds$, the given hypotheses imply that

$$w(T) \le w(0)\, e^{-P(T)}.$$

Since $y(0) \le y(T)$ and $z(0) \ge z(T)$, it follows that

$$w(T) \le w(T)\, e^{-P(T)},$$

or equivalently, $w(T) \le 0$, which is a contradiction. If (ii) holds, we get

$$w(t) \le w(t_1)\, e^{P(t_1) - P(t)} = 0, \quad t \in (t_1, t_2),$$

a contradiction. Finally, if (iii) holds, then $w(t) > 0$ for all $t \in [0, t_2)$, for otherwise there would be $t_o \in [0, t_2)$ such that $w(t_o) = 0$ and $w(t) > 0$ on $(t_o, t_2)$. But then (ii) would hold with $t_1$

replaced by $t_o$, which is impossible. In particular, $w(0) > 0$, which by (1.5.9) implies that $w(T) > 0$. Since $w(t_2) = 0$ and $w$ is continuous, there is $t_3 \in [t_2, T)$ such that $w(t_3) = 0$ and $w(t) > 0$ on $(t_3, T)$. But then (ii) would hold with $t_1$, $t_2$ replaced by $t_3$, $T$, which is impossible. Thus (iii) cannot hold either, whence (1.5.10) is true. □

As an immediate consequence of lemma 1.5.6 we obtain

**Corollary 1.5.3:** *If $f\colon J \times \mathbb{R} \to \mathbb{R}$, $p \in L^1(J, \mathbb{R})$ with $\int_o^T p(s)\,ds > 0$, and if $f(t,x) + p(t)\,x$ is nonincreasing in $x$, then the PBVP*

$$x' = f(t,x), \qquad x(0) = x(T) \tag{1.5.11}$$

*can have at most one solution.*

The next result shows that Carathéodory conditions ensure the existence of a solution to (1.5.11) between $y$, $z$ satisfying (1.5.9).

**Theorem 1.5.2:** *Assume that $f\colon J \times \mathbb{R} \to \mathbb{R}$ and $y$, $z \in AC(J, \mathbb{R})$ satisfy (1.5.9), and that $y \le z$. If $f$ is a Carathéodory function on $\Omega = \{(t,x) \mid t \in J,\ y(t) \le x \le z(t)\}$, and if there is $m \in L^2(J, \mathbb{R})$ such that*

$$|f(t,x)| \le m(t) \quad \text{for a.a. } t \in J \text{ and for all } \ x \in [y(t), z(t)],$$

*then the PBVP (1.5.11) has a solution in the order interval $[y, z]$.*

*Proof.* For the proof of theorem 1.5.2 in the case when $f$ is continuous, see Lakshmikantham and Leela (1983). It is not difficult to modify the proof when $f$ is a Carathéodory function, following the proof in Lakshmikantham and Leela (1983) and the proof of theorem 1.5.1. We omit the details. □

The PBVP (1.5.11) can be converted to an integral equation by the similar manner as the IVP (1.5.1).

**Lemma 1.5.7:** *Let $E$ be a Banach space and $f\colon J\times E\to E$. A function $x\colon J\to E$ is a solution of the PBVP (1.5.11) on $J$ if and only if $x$ satisfies for any $p\in L^1(J,\mathbb{R})$, with $P(t)=\int_o^t p(s)\,ds$ nonzero at $t=T$, the integral equation*

$$\begin{aligned} x(t) &= e^{-P(t)}\int_o^t e^{P(s)}(f(s,x(s))+p(s)x(s))\,ds \\ &+ \frac{e^{-P(t)}}{e^{P(T)}-1}\int_o^T e^{P(s)}(f(s,x(s))+p(s)x(s))\,ds. \end{aligned} \tag{1.5.12}$$

*Proof.* Let $p\in L^1(J,\mathbb{R}_+)$ satisfy $P(T)=\int_o^T p(s)ds\neq 0$. Assume first that $x\colon J\to E$ is a solution of (1.5.11). By definition, $x$ is absolutely continuous and a.e. differentiable, whence $x'$ is by theorem 1.4.6 Bochner integrable. Applying (1.5.11) we obtain

$$\begin{aligned} &\int_o^t e^{P(s)}(f(s,x(s))+p(s)x(s))\,ds \\ &+\frac{1}{e^{P(T)}-1}\int_o^T e^{P(s)}(f(s,x(s))+p(s)x(s))\,ds \\ &=\int_o^t e^{P(s)}(x'(s)+p(s)x(s))\,ds \\ &+\frac{1}{e^{P(T)}-1}\int_o^T e^{P(s)}(x'(s)+p(s)x(s))\,ds \\ &= e^{P(t)}x(t)-x(0)+\frac{1}{e^{P(T)}-1}(e^{P(T)}(x(T)-x(0)) \\ &= e^{P(t)}x(t), \qquad t\in J. \end{aligned}$$

Thus $x$ is a solution of the integral equation (1.5.12). Conversely, assume that $x$ satisfies (1.5.12), or equivalently, the integral equation

$$e^{P(t)}x(t)=\int_o^t e^{P(s)}(f(s,x(s))+p(s)x(s))\,ds$$

$$+\frac{1}{e^{P(T)}-1}\int_o^T e^{P(s)}(f(s,x(s))+p(s)x(s))\,ds.$$

By differentiation we obtain

$$e^{P(t)}(x'(t)+p(t)x(t)) = e^{P(t)}(f(t,x(t))+p(t)x(t)) \quad \text{for a.a. } t \in J.$$

Thus $x$ satisfies the differential equation of (1.5.11) a.e. on $J$. Moreover, the insertion $t = 0$ and $t = T$ in (1.5.12) gives $x(0) = x(T)$, whence $x$ is a solution of the PBVP (1.5.11). □

**Corollary 1.5.4:** *If $E$ is a Banach space, $p \in L^1(J, I\!R)$, $P(t) = \int_o^t p(s)\,ds$, $P(T) \neq 0$ and $q \in L^1(J, E)$, then the linear PBVP*

$$x' + p(t)x = q(t), \qquad x(0) = x(T) \tag{1.5.13}$$

*has a unique solution*

$$x(t) = e^{-P(t)}\left(\int_o^t e^{P(s)}q(s)\,ds + \int_o^T \frac{e^{P(s)}q(s)}{e^{P(T)}-1}ds\right). \tag{1.5.14}$$

*Proof.* The assertion follows from lemma 1.5.7 when we choose $f(t,x) = q(t) - p(t)x$. □

## 1.6. NOTES AND COMMENTS

The definitions and concepts concerning partially ordered sets are standard (see, e.g., Birkhoff (1973), Hrbacek and Jech (1978), Krivine (1971), Levy (1979), Mendelson (1987), and Schaefer (1966)). The contents of subsections 1.1.2 and 1.1.3 are adapted from Heikkilä (1994a) (see also Carl, Heikkilä and Kumpulainen (1993), Heikkilä (1990b), and Lewin (1991)). Subsections 1.1.4 and 1.1.5 are based on Bakhtin (1972), Heikkilä (1990b), Heikkilä and Hu (1994), and Nagata (1974).

The fixed point results of section 1.2 are adapted from Carl and Heikkilä (1993), Carl, Heikkilä and Kumpulainen (1993), and Heikkilä, Kumpulainen and Lakshmikantham (1992). As for the special cases and related results see Abian and Brown (1961), Amann (1977), Birkhoff (1973), Chen (1991), Dunford and Schwartz (1958), Guo and Lakshmikantham (1988), Heikkilä (1988), Heikkilä (1989a), Heikkilä (1992), Heikkilä, Lakshmikantham and Sun (1992), Höft and Höft (1976), Kolibiar (1982), Shmitson (1971), Sun (1991), Sun and Sun (1986, 1989), and Tarski (1955).

The definitions and basic properties of ordered normed spaces given in section 1.3 are found from Deimling (1985), Guo and Lakshmikantham (1988), Krasnosel'skii (1964), Krein and Rutman (1950), and Zeidler (1985). Proposition 1.3.7 is taken from Heikkilä (1992). The results of subsection 1.3.3 are new. Definitions and basic properties on measure and integration theory introduced in section 1.4 are adapted from Hille (1972), Hille and Phillips (1957), Lang (1969), and Yoshida (1974). The results of subsections 1.4.2 and 1.4.3 are based on Heikkilä (1994a). Theorem 1.4.1 is from Heikkilä (1990d), theorem 1.4.2 is new, and theorem 1.4.3 is from Hille and Phillips (1957). Proposition 1.4.3 and corollaries 1.4.6 and 1.4.7 and lemma 1.4.4 are from Heikkilä (1989d). Example 1.4.2 is taken from Appel (1988). The proof of theorem 1.4.6 is based on Hille and Phillips (1957), and Mikusinski (1978). The results of subsection 1.4.8 are formulated from more general fixed point results for the needs of future applications. As for theorem 1.4.9, see Heikkilä (1989b), and Seikkala (1978).

The results of section 1.5 extend the well-known results concerning the applications of the method of upper and lower solutions to IVP's and PBVP's of first order differential equations with continuous nonlinearities (see for ex. Lakshmikantham and Leela (1969)) to these problems of Carathéodory type. Theorem

1.5.1 and and lemma 1.5.6 are taken from Heikkilä and Lakshmikantham (1994a). Lemma 1.5.3 is a generalization of Gronwall's lemma to discontinuous case. Lemma 1.5.4 is adapted from Walter (1970), while lemma 1.5.7 and corollary 1.5.4 are from Heikkilä (1991).

# 2

# First Order Differential Equations

## 2.0. INTRODUCTION

This chapter is devoted to applications of the generalized monotone iterative technique developed in chapter 1 to first order discontinuous nonlinear differential equations in the framework of upper and lower solutions. We shall derive existence and extremality results for first order scalar differential equations and systems, and study the dependence of the solutions on the data.

We shall begin section 2.1 by proving existence of the extremal solutions for discontinuous scalar initial value problem in a general form, where the dependence on the unknown function is decomposed into continuous and discontinuous parts. The method of upper and lower solutions coupled with the generalized monotone iteration method is employed to obtain existence results in the order interval generated by upper and lower solutions. Since, in general, such an order interval need not contain all the solutions, we also offer sufficient conditions to construct an order interval within which we obtain the extremals of all the solutions. Results on the dependence of the extremal solutions on data are also included, as well as some special cases where no continuity hypotheses are needed. As a byproduct we obtain new results which imply existence of solutions when the functions involved in the IVP permits a decomposition of several types.

In section 2.2 we generalize most of the results of section 2.1 to mixed monotone IVP's, where one is required to consider coupled quasisolutions, and to find their extremals between upper and lower coupled quasisolutions. Of course, when the given problem satisfies a uniqueness assumption, these coupled quasisolutions reduce to the unique solution. In general, extremal coupled quasisolutions sandwich the solution set and provide effective bounds for the solutions.

In section 2.3 we concentrate our attention to the study of periodic boundary value problems in the same spirit. The main results of sections 2.1 and 2.2 are extended to PBVP's. We note however that the notion of extremal solutions for PBVP's is to be understood differently than of IVP's, and also one needs to utilize Lyapunov-Schmidt method for considering periodic problems and problems at resonance in general.

Section 2.4 extends several theorems discussed in section 2.1 to finite systems of IVP's. As is usual, such extensions demand of imposing quasimonotone properties on the systems involved. As a byproduct, we obtain also existence and comparison results for higher order IVP's involving discontinuous nonlinearities. Results derived in section 2.5 extend the results of section 2.2 to corresponding finite mixed monotone systems. Section 2.6 discusses PBVP's for discontinuous systems, containing also mixed monotone systems.

The results derived for finite systems of discontinuous nonlinear differential equations are extended in sections 2.7-2.9 to infinite systems in $l^p$-spaces. Section 2.7 deals with extremal solutions of infinite systems of IVP's. Mixed monotone systems are considered in section 2.8, while section 2.9 describes the existence of extremal solutions of infinite systems of PBVP's. Examples are provided throughout to demonstrate the theory discussed.

## 2.1. SCALAR INITIAL VALUE PROBLEMS

We shall consider in this section the IVP

$$x' = f(t, x, x), \qquad x(0) = x_o, \tag{2.1.1}$$

where $f\colon J \times \mathbb{R}^2 \to \mathbb{R}$, $J = [0, T]$.

In the following considerations $Z$ denotes a null set in $J$ with respect to the Lebesgue measure $\mu$. Unless otherwise is stated, measurability of a function is considered as $\mu$-measurability. Let us first introduce a set of conditions for $f\colon J \times \mathbb{R}^2 \to \mathbb{R}$.

(A0) $\alpha, \beta \in AC(J, \mathbb{R})$, $\alpha \le \beta$, $\alpha' \le f(t, \alpha, \alpha)$ and $\beta' \ge f(t, \beta, \beta)$ on $J \setminus Z$.

(A1) There is $N \in L^1(J, \mathbb{R}_+)$ such that $|f(t, x, y)| \le N(t)$ for all $t \in J \setminus Z$ and $x$, $y \in [\alpha(t), \beta(t)]$.

(A2) $f(\cdot, x, y(\cdot))$ is measurable for all $x \in \mathbb{R}$ and $y \in AC(J, \mathbb{R})$.

(A3) $f(t, \cdot, y)$ is continuous for all $t \in J \setminus Z$ and $y \in \mathbb{R}$.

(A4) There is $M \in L^1(J, \mathbb{R}_+)$ and a nondecreasing function $\varphi \in C(\mathbb{R}, \mathbb{R})$ such that $f(t, x, y) + M(t)\,\varphi(y)$ is nondecreasing in $y$ for all $t \in J \setminus Z$ and $x \in \mathbb{R}$.

(A5) $|f(t, x, y)| \le p(t)\, h(|x|, |y|)$ for all $t \in J \setminus Z$ and $x$, $y \in \mathbb{R}$, where $p \in L^1(J, \mathbb{R}_+)$, $h\colon \mathbb{R}_+^2 \to (0, \infty)$ is nondecreasing in both of its arguments and $\int_o^\infty \frac{dv}{h(v,v)} = \infty$.

It turns out that conditions (A0)–(A4) imply the existence of the extremal solutions of (2.1.1) in the order interval $[\alpha, \beta]$ of $AC(J, \mathbb{R})$ if $x_o \in [\alpha(0), \beta(0)]$. If (A2)–(A5) hold, we shall show that (2.1.1) has for each $x_o \in \mathbb{R}$ the extremals among all its solutions, and that they are nondecreasing with respect to $x_o$ and $f$. We shall also consider the special cases

$$f(t, x, y) = f_1(t, x) + f_2(t, y), \quad \text{and} \quad f(t, x, y) = q(x)\, g(t, y).$$

In the latter case we shall allow $f$ to be discontinuous in all of its arguments.

### 2.1.1. Existence of extremal solutions

We shall begin by proving our main result concerning the existence of the extremal solutions of (2.1.1) between assumed upper and lower solutions.

**Theorem 2.1.1:** *If the hypotheses (A0)–(A4) hold, then the IVP (2.1.1) has the extremal solutions in the order interval* $[\alpha, \beta]$ *for each* $x_o \in [\alpha(0), \beta(0)]$.

*Proof.* Let $x_o \in [\alpha(0), \beta(0)]$ and $y \in [\alpha, \beta]$ be given. Consider the IVP

$$x' = F(t, x; y(t)), \qquad x(0) = x_o, \tag{2.1.2}$$

where

$$F(t, x; y(t)) = f(t, x, y(t)) + M(t)[\varphi(y(t)) - \varphi(x)]. \tag{2.1.3}$$

From (A2) and (A3) it follows that $(t, x) \mapsto F(t, x; y(t))$ is a Carathéodory function in $\Omega = \{(t, x) \mid t \in J,\, x \in [\alpha(t), \beta(t)]\}$ for every $y \in [\alpha, \beta]$. By (A4) and (2.1.3) we get

$$\begin{aligned} F(t, \alpha(t); y(t)) - f(t, \alpha(t), \alpha(t)) &= f(t, \alpha(t), y(t)) \\ + M(t)\left(\varphi(y(t)) - \varphi(\alpha(t))\right) &- f(t, \alpha(t), \alpha(t)) \geq 0 \end{aligned}$$

on $J \setminus Z$. This and (A0) imply that

$$\alpha' \leq f(t, \alpha, \alpha) \leq F(t, \alpha; y) \quad \text{on} \quad J \setminus Z.$$

Similarly, $\beta' \geq F(t, \beta; y)$ on $J \setminus Z$. Condition (A1) implies that for all $x \in [\alpha(t), \beta(t)]$ and $t \in J \setminus Z$,

$$|F(t, x; y(t))| \leq K\, M(t) + N(t), \tag{a}$$

where $K = \varphi(\max \beta) - \varphi(\min \alpha)$. Hence, by theorem 1.5.1 the IVP (2.1.2) has for each $y \in [\alpha, \beta]$ the extremal solutions in $[\alpha, \beta]$.

We now define a map $G\colon [\alpha, \beta] \to [\alpha, \beta]$ by

$$Gy = x, \quad y \in [\alpha, \beta], \tag{2.1.4}$$

where $x$ is the maximal solution of (2.1.2) in $[\alpha, \beta]$. Because

$$(Gy)'(t) = F(t, Gy(t); y(t)) \text{ for all } y \in [\alpha, \beta] \text{ and for a.a. } t \in J,$$

it follows by (a) that

$$|(Gy)'(t)| \leq N(t) + K\,M(t)$$

for all $y \in [\alpha, \beta]$ and for a.a. $t \in J$. To prove that $G$ is nondecreasing, let $y_1$, $y_2 \in [\alpha, \beta]$, $y_1 \leq y_2$ be given, and suppose that $x_1$ and $x_2$ are the corresponding maximal solutions of

$$x_i' = F(t, x_i; y_i(t)), \qquad x_i(0) = x_o, \quad i = 1,\, 2$$

in $[\alpha, \beta]$. Since $y_1 \leq y_2$, we have by (A4) and (2.1.3)

$$x_1' \leq F(t, x_1; y_2(t)) \text{ a.e. on } J,$$

and hence by lemma 1.5.1 it follows that $x_1 \leq x_2$, thus proving that $Gy_1 \leq Gy_2$.

The above proof shows that $G$ satisfies the hypotheses of proposition 1.4.4. Thus the i.w.o. chain of $G$-iterations of $\beta$, defined in proposition 1.2.1, has the minimum $x^*$, which is the greatest fixed point of $G$. From the definition of $G$ it follows that $x^*$ is also a solution of the IVP (2.1.1) in $[\alpha, \beta]$.

If $x$ is any solution of (2.1.1) in $[\alpha, \beta]$, then it satisfies also the IVP (2.1.2) with $y = x$. But $Gx$ is the maximal solution of

(2.1.2) with $y = x$, whence $x \le Gx$. This implies by (1.4.11) that $x \le x^*$. Thus $x^*$ is the maximal solution of the IVP (2.1.1) in $[\alpha, \beta]$.

The similar reasoning shows that the equation (2.1.4), which assigns to each $y \in [\alpha, \beta]$ the minimal solution $x$ of the IVP (2.1.2) in $[\alpha, \beta]$, defines an operator $G\colon [\alpha, \beta] \to [\alpha, \beta]$ which satisfies the hypotheses of proposition 1.4.4. Thus the w.o. chain of $G$-iterations of $\alpha$, defined in theorem 1.1.1, has the maximum $x_*$, which is the least fixed point of $G$, and also a solution of the IVP (2.1.1) in $[\alpha, \beta]$. Moreover, $Gx \le x$ for any solution $x$ of (2.1.1) in $[\alpha, \beta]$, whence (1.4.11) implies that $x_*$ is the least of all the solutions of (2.1.1) in $[\alpha, \beta]$. □

The following corollaries are important in themselves.

**Corollary 2.1.1:** *Let $f\colon J \times \mathbb{R}^2 \to \mathbb{R}$ be defined by*

$$f(t, x, y) = f_1(t, x) + f_2(t, y), \quad t \in J, \ x, \ y \in \mathbb{R}, \tag{2.1.5}$$

*where $f_1(t, x)$ is a Carathéodory function and $f_2(t, y)$ is a standard function and nondecreasing in $y$. If (A0) holds, and if there exists $N \in L^1(J, \mathbb{R}_+)$ such that*

$$\sup\{|f_i(t, y)| \mid y \in [\alpha(t), \beta(t)]\} \le N(t) \quad \textit{for a.a.} \quad t \in J, \ i = 1, 2,$$

*then the conclusion of theorem 2.1.1 is valid.*

**Corollary 2.1.2:** *If (2.1.5) and (A0) hold, if $f_1$ and $f_2$ are continuous in $\Omega = \{(t, x) \mid t \in J, \ x \in [\alpha(t), \beta(t)]\}$, and if $f_2(t, \cdot)$ is nondecreasing in $[\alpha(t), \beta(t)]$ for all $t \in J$, then the conclusion of theorem 2.1.1 is valid.*

The results of theorem 2.1.1 and corollaries 2.1.1 and 2.1.2 hold also when conditions (A2)–(A4) are restricted to hold in the set $D = \{(t, x, y) \mid t \in J, \ x, \ y \in [\alpha(t), \beta(t)]\}$.

The next result provides a condition which implies the validity of (A0) and (A1).

**Proposition 2.1.1:** *Assume that*

$$|f(t,x,y)| \le H(t,|x|,|y|) \quad \textit{for} \quad x,\, y \in \mathbb{R}, \quad t \in J \setminus Z, \tag{2.1.6}$$

*where* $H\colon J \times \mathbb{R}_+^2 \to \mathbb{R}_+$, $H(t,u,v)$ *is nondecreasing in* $u$ *and in* $v$ *and the IVP*

$$w' = H(t,w,w), \qquad w(0) = |x_o| \tag{2.1.7}$$

*has an upper solution* $w$ *on* $J$. *Then (A0) and (A1) hold for* $\alpha = -w$, $\beta = w$ *and* $N = w'$.

*Proof.* If $w$ is an upper solution of (2.1.7), we obtain by (2.1.6) that

$$\sup\{|f(t,x,y)| \mid |x|,\, |y| \le w(t)\} \le H(t,w(t),w(t)) \le w'(t) \tag{a}$$

for a.a. $t \in J$. From (a) it follows that

$$w' \ge f(t,w,w), \quad \text{and } -w' \le f(t,-w,-w) \text{ a.e. on } J. \tag{b}$$

The assertions are then direct consequences of (a) and (b). □

Conditions which ensure validity of (A0) and (A1) on a subinterval of $J$ are given in the following proposition.

**Proposition 2.1.2:** *Given* $f\colon J \times \mathbb{R}^2 \to \mathbb{R}$ *and a null set* $Z$ *in* $J$, *assume there is* $M > 0$ *and* $N \in L^1(J,\mathbb{R}_+)$ *such that*

$$\sup\{|f(t,x,y)| \mid |x|,\, |y| \le M\} \le N(t) \quad \textit{for } t \in J \setminus Z.$$

*Given* $x_o \in (-M, M)$ *and* $J_o = [0, c]$, *where* $c \in (0, T]$ *is so chosen that* $\int_o^c N(s)\,ds \le M - |x_o|$, *then (A0) and (A1) hold when* $J$ *is restricted to* $J_o$, *for*

$$\beta(t) = |x_o| + \int_o^t N(s)\,ds, \quad t \in J_o \text{ and } \alpha = -\beta.$$

*Proof.* The above choices ensure that $\beta(t) \le M$ on $J_o$, whence

$$\sup\{|f(t,x,y)| \mid |x| \le \beta(t),\ |y| \le \beta(t)\} \le N(t) = \beta'(t) \tag{a}$$

for a.a. $t \in J_o$. The assertion follows easily from (a). □

### 2.1.2. Existence of extremals among all solutions

The result of theorem 2.1.1 ensures the existence of the extremal solutions of the IVP (2.1.1) within the order interval $[\alpha, \beta]$, provided that $\alpha(0) \le x_o \le \beta(0)$. However, it does not guarantee that all the solutions of (2.1.1) are in that order interval or that any solution of (2.1.1) even exists if $x_o \notin [\alpha(0), \beta(0)]$. It turns out that (A2)–(A5) are sufficient conditions for the existence of the least and the greatest of all the solutions of (2.1.1), for any fixed initial value $x_o$. Moreover, we obtain an order interval which contains all the solutions of (2.1.1) when $x_o$ is fixed.

**Theorem 2.1.2:** *Given* $f: J \times \mathbb{R}^2 \to \mathbb{R}$ *and a null set* $Z$ *of* $J$, *assume that conditions (A2)–(A5) hold. Then for each fixed* $x_o \in \mathbb{R}$ *the IVP (2.1.1) has the extremal solutions, and all the solutions of (2.1.1) belong to the order interval* $[\alpha, \beta]$, *given by*

$$\alpha(t) = x_o + |x_o| - w(t), \quad \beta(t) = x_o - |x_o| + w(t), \quad t \in J, \tag{2.1.8}$$

*where* $w \in AC(J, \mathbb{R})$ *is the solution of the IVP*

$$w' = p(t)\,h(w, w), \qquad w(0) = |x_o|. \tag{2.1.9}$$

*Proof.* Let $x_o \in I\!R$ be given. Condition (A5) implies by lemma 1.5.3 an existence and uniqueness of $w$. Since $[\alpha, \beta] \subseteq [-w, w]$, it follows from (A5) that for all $x$, $y \in [\alpha, \beta]$

$$|f(t, x(t), y(t))| \leq p(t)\, h(w(t), w(t)) = w'(t) \quad \text{for a.a.} \quad t \in J. \qquad \text{(a)}$$

This implies that $\alpha$ and $\beta$ given by (2.1.8) are lower and upper solutions of (2.1.1) and that (A1) holds for $N = w'$. Since $f$ satisfies also the hypotheses (A2)–(A4) in $J \times I\!R^2$, then the IVP (2.1.1) has by theorem 2.1.1 the extremal solutions $x_*$ and $x^*$ in $[\alpha, \beta]$.

If $x$ is a solution of (2.1.1), then it satisfies also the integral equation

$$x(t) = x_o + \int_o^t f(s, x(s), x(s))ds, \quad t \in J. \qquad \text{(b)}$$

This implies by (A5) that

$$|x(t)| \leq |x_o| + \int_o^t p(s)\, h(|x(s)|, |x(s)|)\, ds, \quad t \in J, \qquad \text{(c)}$$

whence $|x(t)| \leq w(t)$ on $J$ by lemma 1.5.3. Applying (a), (b) and (A5) we obtain

$$|x(t) - x_o| \leq \int_o^t p(s)\, h(w(s), w(s))\, ds = w(t) - |x_o|,$$

so that $x \in [\alpha, \beta]$. Since $x_*$ and $x^*$ are the extremal solutions of (2.1.1) in $[\alpha, \beta]$, then $x \in [x_*, x^*]$. But $x$ was an arbitrary solution of (2.1.1), whence $x_*$ and $x^*$ are the least and the greatest of all the solutions of the IVP (2.1.1), and all the solutions of (2.1.1) are contained in $[\alpha, \beta]$. □

**Remark 2.1.1:** If we don't assume in condition (A5) the divergence of the integral $\int_o^\infty \frac{dv}{h(v,v)}$, then for given $x_o \in \mathbb{R}$ we obtain an existence of the extremal solutions of (2.1.1) on the interval $J_o = [0, c]$, where $c \in (0, T]$ is so chosen that

$$\int_o^c p(s)\, ds < \int_{|x_o|}^\infty \frac{dv}{h(v,v)}.$$

If $f$ is defined in $J \times V \times V$, where $V \subset \mathbb{R}$, and if conditions (A2)–(A5) hold, then the extremal solutions of (2.1.1) exist in such a subinterval $J_o = [0, c]$ of $J$ that $[\alpha(t), \beta(t)] \subseteq V$ for each $t \in J_o$, where $\alpha$, $\beta$ are given by (2.1.8). This holds, for instance if $w(c) - |x_o| < d = \inf\{|y - x_o| \mid y \notin V\}$, or equivalently, if

$$\int_o^c p(s)\, ds < \int_{|x_o|}^{|x_o|+d} \frac{dv}{h(v,v)}.$$

By theorem 1.4.4 and corollary 1.4.5 the assumption (A2) holds if $f$ is a standard function or a Borel function. In the case when $f$ is nonnegative-valued and $f(t, x, y)$ is nondecreasing in $y$ one can restrict in the proofs of theorems 2.1.1 and 2.1.2 to seek the extremal solutions among nondecreasing functions of the order interval $[x_o, \beta]$ respectively. This restriction can be done also for functions $x$, $y$ in condition (A1). As a consequence of theorem 2.1.2 we then obtain

**Proposition 2.1.3:** *Given $f\colon J \times \mathbb{R}^2 \to \mathbb{R}$ and a null set $Z$ in $J$, assume that $f$ satisfies condition (A5) and*

(A6) *$f(t, \cdot, y)$ is continuous and $f(t, y, \cdot)$ is nondecreasing for all $y \in \mathbb{R}$ and $t \in J \setminus Z$,*

*and that one of the following conditions holds for all $x$, $y \in \mathbb{R}$.*

(A7) *$f(\cdot, x, y)$ is right continuous or left continuous in $J \setminus Z$.*
(A8) *$\lim_{h,k,l \to 0+} f(t + h, x + k, y + l)$ exists for all $t \in [0, T)$.*
(A9) *$\lim_{h,k,l \to 0-} f(t + h, x + k, y + l)$ exists for all $t \in (0, T]$.*

*Then the IVP (2.1.1) has for each $x_o \in \mathbb{R}$ the extremal solutions. Moreover, if (A8) or (A9) holds, then for each solution $x$ of (2.1.1), $x'(t) = f(t, x(t), x(t))$ in the complement of a countable subset of $J$.*

*Proof.* Let $x \in \mathbb{R}$ be given. Since $f(t,x,y)$ is nondecreasing in $y$, then (A7) implies by theorem 1.4.1 that $f(\cdot, x, y(\cdot))$ is measurable whenever $y \in C(J, \mathbb{R})$ is nondecreasing. This holds by theorem 1.4.2 also when (A8) or (A9) is valid. Thus the existence of the extremal solutions of (2.1.1) follows from the above remarks. If (A8) or (A9) holds, then for each solution $x$ of (2.1.1) the function $f(\cdot, x(\cdot), x(\cdot))$ is right or left regulated, and hence has by lemma 1.4.1 and its dual only countable number of discontinuities. This implies the last assertion. □

## 2.1.3. Case when $f$ is discontinuous in all its arguments

The so far we have allowed the function $f(t,x,y)$ in (2.1.1) to be discontinuous in $t$ and in $y$, but always assumed continuity in $x$. This continuity can also be dropped in certain cases as we see from the following result.

**Theorem 2.1.3:** *Let $f\colon J \times \mathbb{R}^2 \to \mathbb{R}$ be defined by*

$$f(t,x,y) = q(x)\, g(t,y), \quad t \in J,\ x,\ y \in \mathbb{R}, \tag{2.1.10}$$

*where $q$, $\frac{1}{q} \in L^{\infty}_{loc}(\mathbb{R}, \mathbb{R}_+)$, $g\colon J \times \mathbb{R} \to \mathbb{R}$ is a standard function and $g(t, \cdot)$ is nondecreasing for a.a. $t \in J$.*

a) *If $f$ satisfies condition (A0), then (2.1.1) has the extremal solutions in $[\alpha, \beta]$.*

b) *If condition (A5) holds for $f$, then (2.1.1) has for each $x_o \in \mathbb{R}$ the extremal solutions.*

*Proof.* a) Let $\alpha$, $\beta$ be as in condition (A0). If $y \in [\alpha, \beta]$, then

$$\frac{\alpha'(t)}{q(\alpha(t))} \le g(t, \alpha(t)) \le g(t, y(t)) \le g(t, \beta(t)) \le \frac{\beta'(t)}{q(\beta(t))}$$

for a.a. $t \in J$. This, the given hypotheses and theorem 1.4.5 imply that $g(\cdot, y(\cdot)) \in L^1(J, \mathbb{R})$, and that the equation

$$\int_{x_0}^{Gy(t)} \frac{dv}{q(v)} = \int_0^t g(s, y(s))ds, \quad t \in J \tag{a}$$

defines a nondecreasing mapping $G\colon [\alpha, \beta] \to [\alpha, \beta]$. Denote $M$ =ess sup $\{q(v) \mid \min G\alpha \leq v \leq \max G\beta\}$. If $y \in [\alpha, \beta]$ and $0 \leq \bar{t} \leq t \leq T$, then

$$\begin{aligned}\frac{|Gy(t) - Gy(\bar{t})|}{M} &\leq \left| \int_{Gy(\bar{t})}^{Gy(t)} \frac{dv}{q(v)} \right| = \left| \int_{\bar{t}}^{t} g(s, y(s))\, ds \right| \\ &\leq \int_{\bar{t}}^{t} (|g(s, \alpha(s))| + |g(s, \beta(s))|)\, ds.\end{aligned}$$

Hence, if $0 \leq \bar{t} \leq t \leq T$, then

$$|Gy(t) - Gy(\bar{t})| \leq M \int_{\bar{t}}^{t} (|g(s, \alpha(s))| + |g(s, \beta(s))|)\, ds. \tag{b}$$

The above proof shows that $G$ satisfies the hypotheses of theorem 1.4.7, whence the w.o. chain of $G$-iterations of $\alpha$ has the maximum $x_*$, which is the least fixed point of $G$, and the i.w.o. chain of $G$-iterations of $\beta$ has the minimum $x^*$, which is the greatest fixed point of $G$.

From (2.1.1), (2.1.10) and theorem 1.4.5 it follows that $x \in [\alpha, \beta]$ is a solution of (2.1.1) if and only if it satisfies the integral equation

$$\int_{x_0}^{x(t)} \frac{dv}{g(v)} = \int_0^t g(s, x(s))ds, \quad t \in J,$$

or equivalently, by (a), $x$ is a fixed point of $G$. Since $x_*$ and $x^*$ are the extremal fixed points of $G$, then they are the extremal solutions of (2.1.1) in $[\alpha, \beta]$.

b) If condition (A5) holds, then the proof of theorem 2.1.2 shows that $\alpha$, $\beta$ given by (2.1.8) are lower and upper solutions of (2.1.1), and that all the solutions of (2.1.1) belong to $[\alpha, \beta]$. By the above proof the IVP (2.1.1) has extremal solutions $x_*$ and $x^*$ in $[\alpha, \beta]$, whence they are the least and the greatest of all the solutions of (2.1.1) on $J$. □

### 2.1.4. Dependence on the data

Consider now the dependence of the solutions of the IVP (2.1.1) on the initial value $x_o$ and on the function $f$.

As an auxiliary result we shall first prove.

**Lemma 2.1.1:** *Let $f: J \times \mathbb{R}^2 \to \mathbb{R}$ satisfy the hypotheses (A2)–(A5), and let $x_*$ and $x^*$ be the extremal solutions of the IVP (2.1.1).*

a) *If $x$ is a lower solution of (2.1.1), then $x \leq x^*$.*

b) *If $x$ is an upper solution of (2.1.1), then $x_* \leq x$.*

*Proof.* a) Let $x$ be a lower solution of the IVP (2.1.1). Denote $w_o = \max\{|x_o|, \|x\|_o\}$, and let $w$ be the solution of the IVP

$$w' = p(t)\, h(w, w), \qquad w(0) = w_o. \tag{a}$$

By choosing $\alpha = x$, $\beta = w$ and $N = h(w(T), w(T))\, p$, it follows that $f$ satisfies the hypotheses of theorem 2.1.1. Thus the IVP (2.1.1) has a solution $z$ in $[x, w]$. But $x^*$ is the greatest of all the solutions of (2.1.1), whence $z \leq x^*$, so that $x \leq x^*$.

The proof of the assertion b) is similar. □

**Proposition 2.1.4:** *Let $f, \hat{f}: J \times \mathbb{R}^2 \to \mathbb{R}$ satisfy the hypotheses (A2)–(A5). Given $x_o, \hat{x}_o \in \mathbb{R}$, let $x_*$ be the minimal solution of the IVP (2.1.1), and let $\hat{x}^*$ be the maximal solution of the IVP*

$$x' = \hat{f}(t, x, x), \qquad x(t_o) = \hat{x}_o. \tag{2.1.11}$$

*If $x_o \leq \hat{x}_o$ and $f(t,x,y) \leq \hat{f}(t,x,y)$ for a.a. $t \in J$ and for all $x$, $y \in I\!R$, then $x_* \leq x \leq \hat{x}^*$ on $J$ for any solution $x$ of (2.1.1) or (2.1.11).*

*Proof.* Let $x$ be a solution of (2.1.1) or (2.1.11). If $x$ is a solution of (2.1.1), it is a lower solution of (2.1.11), whence $x \leq \hat{x}^*$ on $J$ by lemma 2.1.1. If $x$ is a solution of (2.1.11), then $x \leq \hat{x}^*$, because $\hat{x}^*$ is the maximal solution of (2.1.11) on $J$. The proof that $x_* \leq x$ on $J$ is similar. □

As an immediate consequence of proposition 2.1.4 we obtain

**Corollary 2.1.3:** *If conditions (A2)–(A5) hold, then the extremal solutions of the IVP (2.1.1) are nondecreasing with respect to $x_o$ and to $f$.*

**Proposition 2.1.5:** *Let $f$, $\hat{f}$: $J \times I\!R^2 \to I\!R$ be such that $y \mapsto f(t,y,y)+q(t)y$ or $y \mapsto \hat{f}(t,y,y)+q(t)y$ is nondecreasing for some $q \in L^1(J, I\!R)$ and for a.a. $t \in J$. Given $x_o$, $\hat{x}_o \in I\!R$, $x_o < \hat{x}_o$, assume that $x$ is a solution of the IVP (2.1.1) on $J$, and that $\hat{x}$ is a solution of the IVP (2.1.11) on $J$. If $f(t,x,y) \leq \hat{f}(t,x,y)$ for a.a. $t \in J$ and for all $x$, $y \in I\!R$, then $x(t) < \hat{x}(t)$ for each $t \in J$.*

*Proof.* Denoting $Q(t) = \int_o^t q(s)\,ds$, $t \in J$, it follows from lemma 1.5.5 that

$$e^{Q(t)}\,x(t) = x_o + \int_o^t e^{Q(s)}[f(s,x(s),x(s)) + q(s)x(s)]\,ds, \quad \text{(a)}$$

and

$$e^{Q(t)}\hat{x}(t) = x_o + \int_o^t e^{Q(s)}[\hat{f}(s,\hat{x}(s),\hat{x}(s)) + q(s)\hat{x}(s)]\,ds. \quad \text{(b)}$$

If the conclusion does not hold, there is $c \in (0,T]$ such that $x(c) = \hat{x}(c)$ and $x(t) \leq \hat{x}(t)$ for each $t \in [0,c]$. This and the given hypotheses imply that for $0 \le t \le c$,

$$f(s, x(s), x(s)) + q(s)x(s) \le \hat{f}(s, \hat{x}(s), \hat{x}(s)) + q(s)\hat{x}(s). \qquad \text{(c)}$$

From (a), (b) and (c) it follows that

$$\begin{aligned} x_o - \hat{x}_o &= e^{Q(c)}\hat{x}(c) - \hat{x}(0) - (e^{Q(c)}x(c) - x(0)) \\ &= \int_o^c e^{Q(s)}[\hat{f}(s, \hat{x}(s), \hat{x}(s)) + q(s)\hat{x}(s) \\ &\quad - f(s, x(s), x(s)) - q(s)x(s)]\, ds \ge 0. \end{aligned}$$

But then $x_o \ge \hat{x}_o$, which contradicts the choice of these initial values. Thus $x(t) < \hat{x}(t)$ for each $t \in J$. □

**Proposition 2.1.6:** *Assume that the functions $q\colon \mathbb{R} \to \mathbb{R}_+$ and $g\colon J \times \mathbb{R} \to \mathbb{R}$ are as in theorem 2.1.3, and let $\hat{q}\colon \mathbb{R} \to \mathbb{R}_+$ and $\hat{g}\colon J \times \mathbb{R} \to \mathbb{R}$ satisfy the hypotheses given for $q$ and $g$ in theorem 2.1.3, respectively. Given $x_o$, $\hat{x}_o \in \mathbb{R}$, $x_o < \hat{x}_o$, let $x$ be any solution of the IVP*

$$x' = q(x)\, g(t, x), \qquad x(0) = x_o \tag{2.1.12}$$

*on $J$, and let $\hat{x}$ be any solution of the IVP*

$$x' = \hat{q}(x)\, \hat{g}(t, x), \qquad x(t_o) = \hat{x}_o \tag{2.1.13}$$

*on $J$. Then $x(t) < \hat{x}(t)$ for each $t \in J$ in the following cases.*

a) $0 \le g(t, x) \le \hat{g}(t, x)$ *for a.a.* $t \in J$ *and for all* $x \in \mathbb{R}$, *and* $q(v) \le \hat{q}(v)$ *for a.a.* $v \in \mathbb{R}$.
b) $g(t, x) \le \hat{g}(t, x)$ *for a.a.* $t \in J$ *and for all* $x \in \mathbb{R}$, *and* $q(v) = \hat{q}(v)$ *for a.a.* $v \in \mathbb{R}$.

*Proof.* Make a counter-hypothesis: there is $c \in (0, T]$ such that $x(c) = \hat{x}(c)$, and $x(t) \leq \hat{x}(t)$ in $J_o = [0, c]$. From the proof of theorem 2.1.3 it follows that

$$\int_{x_o}^{x(c)} \frac{dv}{q(v)} = \int_o^c g(s, x(s))ds, \quad \int_{\hat{x}_o}^{x(c)} \frac{dv}{\hat{q}(v)} = \int_o^c \hat{g}(s, \hat{x}(s))ds.$$

Since $x \leq \hat{x}$ in $J_o$, and since $g(t, \cdot)$ and $\hat{g}(t, \cdot)$ are both nondecreasing for a.a. $t \in J$, then in both the cases a) and b) we have

$$\int_{x_o}^{x(c)} \frac{dv}{q(v)} = \int_o^c g(s, x(s))ds \leq \int_o^c \hat{g}(s, \hat{x}(s))ds = \int_{\hat{x}_o}^{x(c)} \frac{dv}{\hat{q}(v)}.$$

But then $\hat{x}_o \leq x_o$, which contradicts the assumption that $x_o < \hat{x}_o$. This implies that the assertion is true. □

### 2.1.5. Special cases

As a consequence of theorem 2.1.2 we obtain

**Proposition 2.1.7:** *Given for each $j = 1, \ldots, n$, a bounded and continuous function $q_j \colon \mathbb{R} \to \mathbb{R}_+$, a standard function $g_j \colon J \times \mathbb{R} \to \mathbb{R}$ and $\mu_j$, $\nu_j \in L^1(J, \mathbb{R}_+)$, assume there exists a nondecreasing function $\varphi \in C(\mathbb{R}, \mathbb{R})$ such that $y \mapsto g_j(t, y) + \nu_j(t)\varphi(y)$ is nondecreasing for a.a. $t \in J$, and a nondecreasing function $\psi : \mathbb{R}_+ \to (0, \infty)$ with $\int_o^\infty \frac{dv}{\psi(v)} = \infty$, such that $|g_j(t, y)| \leq \mu_j(t)\,\psi(|y|)$ for a.a. $t \in J$ and for all $y \in \mathbb{R}$. Then the IVP*

$$x' = \sum_{j=1}^n q_j(x) g_j(t, x), \qquad x(0) = x_o \tag{2.1.14}$$

*has for each $x_o \in \mathbb{R}$ the extremal solutions, which are nondecreasing with respect to $x_o$ and $g_j$.*

*Proof.* Define a function $f\colon J \times \mathbb{R}^2 \to \mathbb{R}$ by

$$f(t,x,y) = \sum_{j=1}^{n} q_j(x) g_j(t,y), \quad t \in J,\ x,\ y \in \mathbb{R}.$$

If $y \in C(J, \mathbb{R})$, then the hypotheses given for $g_j$ imply by theorem 1.4.4 that each $g_j(\cdot, y(\cdot))$ is measurable. Thus $f(\cdot, x, y(\cdot))$ is measurable for each $x \in \mathbb{R}$, whence (A2) is valid. Since

$$|f(t,x+k,y) - f(t,x,y)| \le \psi(|y|) \sum_{j=1}^{n} |q_j(x+k) - q_j(x)| \mu_j(t)$$

for a.a. $t \in J$ and for all $x$, $k$, $y \in \mathbb{R}$, it follows that $f(t,\cdot,y)$ is continuous for all $y \in \mathbb{R}$ and for a.a. $t \in J$. Thus condition (A3) holds.

If $x$, $y$, $\bar{y} \in \mathbb{R}$ and $y \le \bar{y}$, we have for a.a. $t \in J$

$$\begin{aligned} f(t,x,\bar{y}) - f(t,x,y) &= \sum_{j=1}^{n} q_j(x)(g_j(t,\bar{y}) - g_j(t,y)) \\ &\ge -\sum_{j=1}^{n} q_j(x)\nu_j(t)(\varphi(\bar{y}) - \varphi(y)) \\ &\ge -\sum_{j=1}^{n} K\nu_j(t)(\varphi(\bar{y}) - \varphi(y)), \end{aligned}$$

where $K = \sup\{q_j(x) \mid x \in \mathbb{R},\ 1 \le j \le n\}$. This implies that condition (A4) holds with $M = K\sum_{j=1}^{n} \nu_j$. Condition (A5) holds with $p = \sum_{j=1}^{n} \mu_j$ and $h(u,v) = K\psi(v)$. Thus all the conditions (A2)–(A5) hold, whence (2.1.14) has by theorem 2.1.2 extremal solutions. The last assertion follows from proposition 2.1.4. □

From theorem 2.1.3 it follows

**Proposition 2.1.8:** *Assume that $q$ and $g$ satisfy the hypotheses of theorem 2.1.3 and that $g: J \times \mathbb{R} \to \mathbb{R}$ is a Carathéodory function. Then the sequence $(y_n)_{n=o}^{\infty}$ of functions $y_n: J \to \mathbb{R}$, defined by*

$$\int_{x_o}^{y_{n+1}(t)} \frac{dv}{q(v)} = \int_o^t g(s, y_n(s))ds, \quad t \in J, \; n \in \mathbb{N}, \qquad (2.1.15)$$

*converges uniformly on $J$ to the minimal (resp. the maximal) solution of (2.1.12) in the order interval $[\alpha, \beta]$ if $y_o = \alpha$ (resp. $y_o = \beta$).*

*Proof.* The given hypotheses imply that the equation

$$\int_{x_o}^{Gy(t)} \frac{dv}{q(v)} = \int_o^t g(s, y(s))ds, \quad t \in J \qquad \text{(a)}$$

defines a nondecreasing operator $G: [\alpha, \beta] \to [\alpha, \beta]$. By choosing $y_o = \alpha$, it follows from (2.1.15) and (a) that $(y_n)_{n=o}^{\infty} = (G^n\alpha)_{n=o}^{\infty}$. Thus the sequence $(y_n)_{n=o}^{\infty}$ is nondecreasing and belongs to $[\alpha, \beta]$. The proof of theorem 2.1.3 implies also that $(y_n)_{n=o}^{\infty}$ converges uniformly on $J$ to a function $x_* \in [\alpha, \beta]$. Since $g$ is a Carathéodory function, it follows from (2.1.15) as $n \to \infty$ that $x = x_*$ satisfies the integral equation

$$\int_{x_o}^{x(t)} \frac{dv}{q(v)} = \int_o^t g(s, x(s))ds, \quad t \in J. \qquad \text{(b)}$$

Thus $x_*$ is a fixed point of $G$, and hence, by the proof of theorem 2.1.3, a solution of (2.1.12).

If $x \in [\alpha, \beta]$ is a solution of (2.1.12), it is also a fixed point of $G$. Since $y_o = \alpha \leq x$ and $G$ is nondecreasing, it follows by induction that $y_n = G^n\alpha \leq x$ for each $n \in \mathbb{N}$. Thus $x_* = \lim_n y_n \leq x$, whence $x_*$ is the minimal solution of (2.1.12) in $[\alpha, \beta]$.

The proof that $(y_n)_{n=o}^{\infty}$ converges to the maximal solution of (2.1.12) in $[\alpha,\beta]$ when $y_o=\beta$ is similar to the above one. □

If $q(v)=1$ a.e. on $\mathbb{R}$, then (2.1.15) can be rewritten as

$$y_{n+1}(t)=x_o+\int_{t_o}^{t} g(s,y_n(s))ds, \quad t\in J,\ n\in\mathbb{N}. \tag{2.1.16}$$

Thus we obtain

**Corollary 2.1.4:** *Assume that $q\colon\mathbb{R}\to(0,\infty)$, that $q(v)=1$ a.e. on $\mathbb{R}$, that $g\colon J\times\mathbb{R}\to\mathbb{R}$ is a Carathéodory function, that $g(t,\cdot)$ is nondecreasing on $\mathbb{R}$ for a.a. $t\in J$, and that the IVP (2.1.12) has a lower solution $\alpha$ and an upper solution $\beta$ on $J$ such that $\alpha\le\beta$. Then the sequence $(y_n)_{n=o}^{\infty}$ of the successive approximations $y_n\colon J\to\mathbb{R}$, defined by (2.1.16), converges uniformly on $J$ to the minimal (resp. the maximal) solution of the IVP (2.1.12) in the order interval $[\alpha,\beta]$ if $y_o=\alpha$ (resp. $y_o=\beta$).*

In the study of differential equations in Banach spaces we shall need the following result.

**Proposition 2.1.9:** *Assume that $g\colon J\times\mathbb{R}_+\to\mathbb{R}_+$ is a Carathéodory function, that $g(t,\cdot)$ is nondecreasing on $\mathbb{R}$ for a.a. $t\in J$, that the IVP*

$$x'=g(t,x), \qquad x(0)=x_o \tag{2.1.17}$$

*has for some $x_o=r_o>0$ an upper solution, and that the zero-function is the only solution of (2.1.17) when $x_o=0$. Then the IVP (2.1.17) has for all $x_o\in[0,r_o]$ the minimal solution $x(\cdot,x_o)$, it is nondecreasing with respect to $x_o$, and $x(t,\frac{1}{n})\to0$ uniformly over $t\in J$ as $n\to\infty$.*

*Proof.* Let $x_o\in[0,r_o]$ be given, and let $\beta$ be an upper solution of the IVP (2.1.17). Because the zero function is a lower solution of (2.1.17), it follows from corollary 2.1.4 that the sequence $(y_n)_{n=o}^{\infty}$ of the successive approximations (2.1.16) with $y_o(t) \equiv 0$ converges uniformly on $J$ to the minimal solution $x(\cdot, x_o)$ of (2.1.17).

Assume now that $0 \leq \hat{x}_o \leq x_o$, and let $(\hat{y}_n)_{n=0}^{\infty}$ be the sequence of the successive approximations defined by $\hat{y}_o(t) \equiv 0$ and

$$\hat{y}_{n+1}(t) = \hat{x}_o + \int_{t_o}^{t} g(s, \hat{y}_n(s))ds, \quad t \in J, \ n \in I\!N. \tag{a}$$

Since $g(t, \cdot)$ is nondecreasing for a.a. $t \in J$, it follows from (2.1.16) and (a) by induction that $\hat{y}_n \leq y_n$ for each $n \in I\!N$. This implies as $n \to \infty$ that $x(\cdot, \hat{x}_o) \leq x(\cdot, x_o)$.

To prove the last assertion, choose $k \geq \frac{1}{r_o}$ and denote $x_n = x(\cdot, \frac{1}{n})$, $n \geq k$. The so obtained sequence $(x_n)_{n=k}^{\infty}$ is nonincreasing by the above proof, and bounded below by the zero function, whence

$$x(t) = \lim_{n \to \infty} x_n(t) \tag{b}$$

exists for each $t \in J$. Since

$$x_n(t) = \frac{1}{n} + \int_{o}^{t} g(s, x_n(s))\, ds, \quad t \in J, \ n \geq k, \tag{c}$$

then

$$0 \leq x_n(t) - x_m(t) \leq \frac{1}{n} - \frac{1}{m} + \int_{o}^{T} (g(s, x_n(s)) - g(s, x_m(s)))\, ds$$

$$= x_n(T) - x_m(T)$$

whenever $0 \leq t \leq T$ and $k \leq n \leq m$, so that the convergence in (b) is uniform. In particular, $x \in C(J, I\!R_+)$. Because

$$\lim_{n \to \infty} g(t, x_n(t)) = g(t, x(t)) \quad \text{for a.a.} \quad t \in J,$$

it follows from (c) by the dominated convergence theorem, as $n \to \infty$, that

$$x(t) = \int_o^t g(s, x(s))\, ds, \quad t \in J.$$

But this means that $x$ is a solution of the IVP (2.1.17) with $x_o = 0$. This implies by a hypothesis that $x(t) \equiv 0$, so that $x_n(t) = x(t, \frac{1}{n}) \to 0$ uniformly over $t \in J$ as $n \to \infty$. □

As an immediate consequence of proposition 2.1.7 we have

**Corollary 2.1.5:** *Assume that for each $j = 1, \ldots, n$, $p_j \colon J \to \mathbb{R}_+$ is Lebesgue integrable, $q_j \colon \mathbb{R} \to \mathbb{R}_+$ is bounded and continuous, $g_j \colon \mathbb{R} \to \mathbb{R}$, and that there is a nondecreasing function $\varphi \in C(\mathbb{R}, \mathbb{R})$ such that $g_j + \varphi$ is nondecreasing, and a nondecreasing function $\psi \colon \mathbb{R}_+ \to (0, \infty)$ with $\int_o^\infty \frac{dv}{\psi(v)} = \infty$, such that $|g_j(y)| \le \psi(|y|)$ for all $y \in \mathbb{R}$. Then the IVP*

$$x' = \sum_{j=1}^{n} p_j(t) q_j(x) g_j(x), \qquad x(t_o) = x_o$$

*has for each $x_o \in \mathbb{R}$ extremal solutions, which are nondecreasing with respect to $x_o$ and $g_j$.*

As a consequence of theorem 2.1.3 we obtain

**Corollary 2.1.6:** *If $q, \frac{1}{q} \in L^\infty_{loc}(\mathbb{R}, \mathbb{R}_+)$, $p \in L^1(J, \mathbb{R}_+)$, if $g \colon \mathbb{R} \to \mathbb{R}$ is nondecreasing and $\int_o^\infty \frac{dv}{h(v,v)} = \infty$, where $h(u, v) = 1 +$ ess sup$\{|q(x)| \mid |x| \le u\} \max\{|g(v)|, |g(-v)|\}$, then the IVP*

$$x' = p(t) q(x) g(x), \qquad x(0) = x_o$$

*has for each $x_o \in \mathbb{R}$ the extremal solutions, and they are nondecreasing with respect to $x_o$ and $g$.*

**Remark 2.1.2:** Given $n \in \mathbb{N}$, let $\varphi_n\colon \mathbb{R} \to \mathbb{R}$ be the odd extension of the function

$$\varphi_n(v) = (v + \exp_n(0)) \prod_{j=1}^{n} \log_j(v + \exp_n(0)), \quad v \in \mathbb{R}_+, \tag{2.1.18}$$

where $\log_n$ and $\exp_n$ denote $n$-fold iterated logarithm and exponential functions, respectively, and $\exp_o(0) = 0$. Given positive numbers $\gamma$, $\mu$ and $\nu$ and positive integers $m$, $n$ and $j$, define functions $\varphi\colon \mathbb{R} \to \mathbb{R}$, $h\colon \mathbb{R}_+^2 \to \mathbb{R}_+$ and $\psi\colon \mathbb{R}_+ \to \mathbb{R}_+$ by

$$\varphi = \varphi_m, \; h(x, y) = \varphi_n(\mu x + \nu y) + \gamma, \quad \text{and} \quad \psi(v) = \varphi_j(v).$$

These functions satisfy the hypotheses given for $\varphi$ and $h$ in conditions (A4) and (A5), and for $\psi$ in corollary 2.1.5.

**Example 2.1.1:** Denote $D = \{(n_o, \ldots, n_m) \mid m, n_o, \ldots, n_m \in \mathbb{N} \text{ and } n_o > 0\}$, and define for each $(n_o, \ldots, n_m) \in D$ a rational number $a(n_o, \ldots, n_m)$ by

$$a(n_o, \ldots, n_m) = M + 2^{-m-1} + \sum_{k=o}^{m} 2^{-k-m-2} \prod_{j=o}^{k} 2^{-n_j}$$
$$+ \sum_{j=0}^{m} 2^{-2m-2} \prod_{j=o}^{m} 2^{-n_j}, \tag{2.1.19}$$

where $M \geq 0$. As for the properties of these numbers see example 1.1.1. Given $p \in L^1(J, \mathbb{R}_+)$, denote $P(t) = \int_o^t p(s)\, ds, \; t \in J$, and define a function $f\colon J \times \mathbb{R} \to \mathbb{R}$ by

$$f(t, y) = \begin{cases} M\,p(t), & 0 \leq y \leq M\,P(t), \\ (M+1)p(t), & y > (M+1)P(t), \\ a(n_o, \ldots, n_m + 1)p(t), & a(n_o, \ldots, n_m + 1)P(t) \\ & \quad < y \leq a(n_o, \ldots, n_m)P(t), \\ -f(t, -y), & y < 0. \end{cases}$$

It is easy to show that the hypotheses (A2)–(A5) hold. For instance, if $y \in AC(J, \mathbb{R})$, there is a partition of $J$ into a countable number of subintervals such that $f(t, y(t)) = \text{const}\cdot p(t)$ in each subinterval, whence $f(\cdot, y(\cdot))$ is measurable. Thus the IVP

$$x' = f(t, x), \qquad x(0) = x_o \tag{2.1.20}$$

has for each $x_o \in \mathbb{R}$ extremal solutions.

In the case when $x_o = 0$, also condition (A0) holds with

$$\beta(t) = (M + 1)P(t), \ t \in J, \qquad \alpha = -\beta.$$

Define a mapping $G$ in $[\alpha, \beta]$ by

$$Gy(t) = \int_o^t f(s, y(s))\, ds, \ t \in J.$$

Because $f(t, \cdot)$ is nondecreasing then also $G$ is. Condition (A0) implies that $\alpha \le G\alpha$ and $G\beta \le \beta$, whence $G\colon [\alpha, \beta] \to [\alpha, \beta]$. It is easy to see that

$$Ga(n_o, \ldots, n_m)P(t) = a(n_o, \ldots, n_m + 1)P(t)$$

for all $t \in J$ and $(n_o, \ldots, n_m) \in D$. In view of this one can show that the w.o. chain of $G$-iterations of $\alpha$ is $\{a(n_o, \ldots, n_m)\alpha \mid (n_o, \ldots, n_m) \in D\}$, and that the i.w.o. chain of $G$-iterations of $\beta$ is $\{a(n_o, \ldots, n_m)\beta \mid (n_o, \ldots, n_m) \in D\}$. Thus the least and the greatest fixed points of $G$ are

$$x_*(t) = \sup\{a(n_o, \ldots, n_m)\alpha(t) \mid (n_o, \ldots, n_m) \in D\} = -M\,P(t),$$

and

$$x^*(t) = \inf\{a(n_o, \ldots, n_m)\beta(t) \mid (n_o, \ldots, n_m) \in D\} = M\,P(t).$$

These functions are also the extremal solutions of the IVP (2.1.20) when $x_o = 0$.

If $g$ in theorem 2.1.3 or in corollary 2.1.6 is nonnegative-valued, then condition $\frac{1}{q} \in L^\infty_{loc}(I\!R, I\!R_+)$ can be weakened by theorem 1.4.5 to $\frac{1}{q} \in L^1_{loc}(I\!R, I\!R_+)$. The following counter-example shows that this hypothesis, as well as the condition (A4) cannot be dropped in the discontinuous case.

**Example 2.1.2:** Let $m$ be a positive number. Define a function $q\colon I\!R \to I\!R$ by $q(y) = \begin{cases} m, & \text{if } y = 0, \\ 0, & \text{if } y \neq 0. \end{cases}$
Let $J = [0, T]$, $T > 0$ be given. Consider the IVP

$$x' = q(x), \qquad x(0) = x_o. \tag{a}$$

Obviously, $q \in L^\infty(I\!R, I\!R_+)$. If $x\colon J \to I\!R$ is continuous, then the set $A = \{t \in J \mid x(t) = 0\}$ is closed, and hence Lebesgue measurable. Thus, the set $\{t \in J \mid q(x(t)) < r\}$ is $J$ if $r > m$, $A$ if $0 < r \leq m$ and $\emptyset$ if $r \leq 0$, whence $q(x(\cdot))$ is measurable. But the IVP (a) has no solution on $J$ when $x_o = 0$. To see this, note first that $x\colon J \to I\!R$ is a solution of (a) on $J$ if and only if $x$ is a solution of the integral equation

$$x(t) = \int_o^t q(x(s))ds, \qquad t \in J. \tag{b}$$

Each possible solution of (b) on $J$ is absolutely continuous and nondecreasing on $J$ and vanishes at 0. If $x$ is such a function, denote $r = \sup\{s \in J \mid x(t) = 0 \text{ on } [0, s]\}$. If $r = 0$, then $x(t) > 0$ for each $t \in (0, T]$. But then $\int_o^t q(x(s))ds = 0$ for each $t \in J$, whence $x$ does not satisfy (b) on $J$. If $r > 0$, then $x(t) = 0$ for each $t \in [0, r)$, whence $\int_o^t q(x(s))ds = mt$ for each $t \in [0, r)$, so that $x$ is not a solution of (b) on $J$. This implies that (b), and hence also (a), has no solution on any interval $J = [0, T]$, $T > 0$.

## 2.2. MIXED MONOTONE IVP'S

In this section we shall consider the following the scalar IVP

$$x' = f(t, x, x, x), \qquad x(0) = x_o, \tag{2.2.1}$$

where $f\colon J \times \mathbb{R}^3 \to \mathbb{R}$, $J = [0, T]$, by assuming that $f(t, x, y, z)$ is nondecreasing in $y$ and nonincreasing in $z$. Since we don't impose any continuity hypotheses with respect to these variables, we cannot expect the solvability of (2.2.1) in the sense discussed above (see example 2.2.1). Therefore we shall make the following definitions.

The functions $y$, $z \in AC(J, \mathbb{R})$ are said to be *coupled quasisolutions* of (2.2.1) if

$$\begin{aligned} y'(t) &= f(t, y(t), y(t), z(t)) \text{ for a.a. } t \in J,\ y(0) = x_o, \\ z'(t) &= f(t, z(t), z(t), y(t)) \text{ for a.a. } t \in J,\ z(0) = x_o. \end{aligned} \tag{2.2.2}$$

We shall first consider the existence of coupled quasisolutions of the IVP (2.2.1) in the case when there is a null set $Z$ in $J$ such that $f\colon J \times \mathbb{R}^3 \to \mathbb{R}$ satisfies some of the following hypotheses.

(B0) $\alpha$, $\beta \in AC(J, \mathbb{R})$, $\alpha \le \beta$, $\alpha' \le f(t, \alpha, \alpha, \beta)$ and $\beta' \ge f(t, \beta, \beta, \alpha)$ on $J \setminus Z$.

(B1) There is $N \in L^1(J, \mathbb{R}_+)$ such that $|f(t, x, y, z)| \le N(t)$ for all $t \in J \setminus Z$ and $x$, $y$, $z \in [\alpha(t), \beta(t)]$.

(B2) $f(\cdot, x, y(\cdot), z(\cdot))$ is measurable for all $y$, $z \in AC(J, \mathbb{R})$ and $x \in \mathbb{R}$.

(B3) $f(t, \cdot, y, z)$ is continuous for all $y$, $z \in \mathbb{R}$ and $t \in J \setminus Z$.

(B4) $f(t, y, \cdot, z)$ is nondecreasing and $f(t, y, z, \cdot)$ is nonincreasing for all $y$, $z \in \mathbb{R}$ and $t \in J \setminus Z$.

(B5) $f(t, x_1, y, z) - f(t, x_2, y, z) \le g(t, x_1 - x_2)$ for all $x_1$, $x_2$, $y$, $z \in \mathbb{R}$, $x_1 \ge x_2$, and $t \in J \setminus Z$, where $g\colon J \times \mathbb{R}_+ \to \mathbb{R}_+$ and $u(t) \equiv 0$ is the only lower solution of the IVP $u' = g(t, u)$, $u(0) = 0$.

(B6) $|f(t,x,y,z)| \le p(t)\,h(|x|,|y|,|z|)$ for all $t \in J \setminus Z$ and $x$, $y$, $z \in \mathbb{R}$, where $p \in L^1(J,\mathbb{R}_+)$, $h\colon \mathbb{R}_+^3 \to (0,\infty)$ is nondecreasing in all its arguments and $\int_o^\infty \frac{dv}{h(v,v,v)} = \infty$.

We shall show that conditions (B0)–(B4) imply the existence of coupled quasisolutions of (2.2.1) in the order interval $[\alpha,\beta]$ of $AC(J,\mathbb{R})$ if $x_o \in [\alpha(0),\beta(0)]$. If also (B5) holds, there exist *extremal* coupled quasisolutions $y$, $z$ in $[\alpha,\beta]$, i.e. $y \le u, v \le z$ for all coupled quasisolutions $u$, $v$ of (2.2.1) in $[\alpha,\beta]$. If (B2)–(B6) hold, it turns out that (2.2.1) has for each $x_o \in \mathbb{R}$ the extremals among all its coupled quasisolutions.

We shall also consider the special cases

$$f(t,x,y,z) = f_1(t,x) + f_2(t,y) + f_3(t,z),$$

and

$$f(t,x,y,z) = q(x)\,g(t,y,z).$$

In the latter case $f$ can be discontinuous in all of its arguments.

### 2.2.1. Existence of coupled quasisolutions

We shall first consider an existence of coupled quasisolutions of the IVP (2.2.1) between $\alpha$ and $\beta$ given by (B0).

**Theorem 2.2.1:** *If there is a null set $Z$ in $J$ so that the hypotheses (B0)–(B4) hold, then the IVP (2.2.1) has coupled quasisolutions in the order interval $[\alpha,\beta]$ for each $x_o \in [\alpha(0),\beta(0)]$.*

*Proof.* Given $x_o \in [\alpha(0),\beta(0)]$ and $y$, $z \in [\alpha,\beta]$, consider the IVP

$$x' = f(t,x,y(t),z(t)), \qquad x(0) = x_o. \tag{2.2.3}$$

From (B2) and (B3) it follows that $(t,x) \mapsto f(t,x,y(t),z(t))$ is a Carathéodory function in $\Omega = \{(t,x) \mid x \in [\alpha(t),\beta(t),\ t \in J\}$. By (B4) we get

$$f(t,\alpha(t),\alpha(t),\beta(t)) \le f(t,\alpha(t),y(t),z(t))$$

and

$$f(t,\beta(t),\beta(t),\alpha(t)) \geq f(t,\beta(t),y(t),z(t))$$

on $J \setminus Z$. This and (B0) imply that

$$\alpha' \leq f(t,\alpha,y(t),z(t)) \text{ and } \beta' \geq f(t,\beta,y(t),z(t)) \text{ on } J \setminus Z.$$

Condition (B1) implies that

$$|f(t,x,y(t),z(t))| \leq N(t) \text{ for } x \in [\alpha(t),\beta(t)],\ t \in J \setminus Z. \quad \text{(a)}$$

Hence, by theorem 1.5.1 the IVP (2.2.3) has for all $y$, $z \in [\alpha,\beta]$ the extremal solutions in $[\alpha,\beta]$.

We now define a map $A\colon [\alpha,\beta] \times [\alpha,\beta] \to [\alpha,\beta]$ by

$$A(y,z) = x, \quad y,\ z \in [\alpha,\beta], \tag{2.2.4}$$

where $x$ is the maximal solution of (2.2.3) in $[\alpha,\beta]$. We shall show that $A$ is mixed monotone. Let $y_1$, $y_2$, $z \in [\alpha,\beta]$, $y_1 \leq y_2$, be given, and suppose that $x_1$ and $x_2$ are the corresponding maximal solutions of

$$x_i' = f(t,x_i,y_i(t),z(t)), \qquad x_i(0) = x_o,\ i = 1,\ 2,$$

in $[\alpha,\beta]$. Since $y_1 \leq y_2$, we have by (B4)

$$x_1' \leq f(t,x_1,y_2(t),z(t)) \text{ a.e. on } J,$$

which implies by lemma 1.5.1 that $x_1 \leq x_2$. Thus $A(y_1,z) \leq A(y_2,z)$, so that $A(y,z)$ is nondecreasing in $y$. Similarly, it can be shown that $A(y,z)$ is nonincreasing in $z$, whence $A$ is mixed monotone. From the definition of $A$ and from (a) it follows that

$$|(A(y,z)'(t)| \leq N(t) \text{ for all } y,\ z \in [\alpha,\beta] \text{ and for a.a. } t \in J.$$

The above proof implies that $A$ satisfies the hypotheses of proposition 1.4.6, whence $A$ has the extremal coupled fixed points $y$, $z$. In particular,

$$y = A(y, z), \qquad z = A(z, y). \tag{2.2.5}$$

From (2.2.2)–(2.2.5) it then follows that the functions $y$, $z$ are coupled quasisolutions of the IVP (2.2.1) in $[\alpha, \beta]$. □

The coupled quasisolutions $y$, $z$ were obtained in the proof of theorem 2.2.1 as extremal coupled fixed points of a mixed monotone operator defined by (2.2.4). But this does not guarantee that $y$ and $z$ are *extremal coupled quasisolutions of* (2.2.1) in $[\alpha, \beta]$ in the sense that $y \le v$, $w \le z$ whenever $v$, $w$ are coupled quasisolutions of (2.2.1) in $[\alpha, \beta]$. Next we shall show that condition (B5) ensures this extremality.

**Proposition 2.2.1:** *If $f\colon J \times \mathbb{R}^3 \to \mathbb{R}$ satisfies conditions (B0)–(B5), then the IVP (2.2.1) has the extremal coupled quasisolutions in $[\alpha, \beta]$.*

*Proof.* Conditions (B0)–(B4) imply that the IVP (2.2.3) has for each choice of $y$, $z \in [\alpha, \beta]$ a solution in $[\alpha, \beta]$, and condition (B5) implies by lemma 1.5.4 that this solution is uniquely determined. This and the definition (2.2.4) of $A$ imply that $v$, $w$ are coupled quasisolutions of (2.2.1) in $[\alpha, \beta]$ if and only if they are coupled fixed points of $A$. But $A$ has by the proof of theorem 2.2.1 the extremal coupled fixed points $y$, $z$, whence $y \le v$, $w \le z$ for all coupled quasisolutions $v$, $w$ in $[\alpha, \beta]$. □

As an immediate consequence of theorem 2.2.1, proposition 2.2.1 and corollary 1.5.1 we obtain

**Corollary 2.2.1:** *Let $f$ in theorem 2.2.1 be defined by*

$$f(t, x, y, z) = f_1(t, x) + f_2(t, y) + f_3(t, z), \tag{2.2.6}$$

*where $f_1$ is a Carathéodory function, $f_2$, $f_3$ are standard functions, $f_2(t, y)$ is nondecreasing in $y$ and $f_3(t, z)$ is nonincreasing* *in* $z$. *If (B0) holds, and if there exists* $N \in L^1(J, \mathbb{R}_+)$ *such that*

$$\sup\{|f_j(t,y)| \mid y \in [\alpha(t), \beta(t)]\} \leq N(t)$$

*for a.a.* $t \in J$ *and for* $j = 1, 2, 3$, *then the IVP (2.2.1) has for each* $x_o \in [\alpha(0), \beta(0)]$ *coupled quasisolutions in* $[\alpha, \beta]$. *Moreover, if there is* $p \in L^1(J, \mathbb{R}_+)$ *such that*

$$f_1(t, x_1) - f_1(t, x_2) \leq p(t)(x_1 - x_2)$$

*for all* $x_1$, $x_2 \in \mathbb{R}$, $x_1 \geq x_2$, *and for a.a.* $t \in J$, *then (2.2.1) has the extremal coupled quasisolutions in* $[\alpha, \beta]$.

The results of theorem 2.2.1, proposition 2.2.1 and corollary 2.2.1 hold also when conditions (B2)–(B4) are restricted to hold in the set $D = \{(t, x, y, z) \mid t \in J,\ x,\ y,\ z \in [\alpha(t), \beta(t)]\}$.

The next result provides a mean to find $\alpha$, $\beta \in AC(J, \mathbb{R})$ and $N \in L^1(J, \mathbb{R}_+)$ such that (B0) and (B1) hold.

**Proposition 2.2.2:** *Given* $f\colon J \times \mathbb{R}^3 \to \mathbb{R}$ *and a null set* $Z$ *of* $J$, *assume that*

$$|f(t,x,y,z)| \leq H(t, |x|, |y|, |z|) \quad \text{for } x,\ y,\ z \in \mathbb{R} \text{ and } t \in J \setminus Z,$$

*where* $H\colon J \times \mathbb{R}^3_+ \to \mathbb{R}_+$, $H(t,u,v,w)$ *is nondecreasing in* $u$, $v$ *and* $w$, *and that the IVP*

$$w' = H(t, w, w, w), \qquad w(0) = |x_o|$$

*has an upper solution* $w$ *on* $J$. *Then (B0) and (B1) hold with* $\alpha = -w$, $\beta = w$ *and* $N = w'$.

*Proof.* The given hypotheses imply that

$$\begin{aligned} &\sup\{|f(t,x,y,z)| \mid |x| \leq w(t),\ |y| \leq w(t)\ |z| \leq w(t)\} \\ &\leq H(t, w(t), w(t), w(t)) \leq w'(t) \end{aligned} \tag{a}$$

for a.a. $t \in J$. From (a) it follows that

$$w' \geq f(t, w, w, -w) \quad \text{and} \quad -w' \leq f(t, -w, -w, w) \text{ a.e. on } J. \quad \text{(b)}$$

The assertions are then direct consequences of (a) and (b). □

### 2.2.2. Existence of extremal coupled quasisolutions

The next result gives sufficient conditions for the existence of the extremal coupled quasisolutions among all the coupled quasisolutions of the IVP (2.2.1) for any fixed initial value $x_o$, and also an order interval which contains all the coupled quasisolutions of (2.2.1).

**Theorem 2.2.2:** *Given $f\colon J \times \mathbb{R}^3 \to \mathbb{R}$ and a null set $Z$ of $J$, assume that conditions (B2)–(B6) hold. Then the IVP (2.2.1) has for each $x_o \in \mathbb{R}$ the extremal coupled quasisolutions, and all the coupled quasisolutions of (2.2.1) belong to the order interval $[\alpha, \beta]$, given by*

$$\alpha(t) = x_o + |x_o| - w(t), \quad \beta(t) = x_o - |x_o| + w(t), \quad t \in J, \quad (2.2.7)$$

*where $w \in AC(J, \mathbb{R})$ is the solution of the IVP*

$$w' = p(t)\, h(w, w, w), \qquad w(0) = |x_o|. \qquad (2.2.8)$$

*Proof.* Let $x_o \in \mathbb{R}$ be given. Since $[\alpha, \beta] \subseteq [-w, w]$, it follows from (B6) that for all $x, y, z \in [\alpha, \beta]$ and for a.a. $t \in J$,

$$|f(t, x(t), y(t), z(t))| \leq p(t)\, h(w(t), w(t), w(t)) = w'(t). \qquad \text{(a)}$$

This ensures that the hypotheses (B0) and (B1) hold for $\alpha$ and $\beta$ given by (2.2.7) and for $N = w'$. Since $f$ satisfies also the hypotheses (B2)–(B5), then the IVP (2.2.1) has by proposition 2.2.1 the extremal coupled quasisolutions $y$, $z$ in $[\alpha, \beta]$.

If $u$, $v$ are coupled quasisolutions of (2.2.1), then they satisfy also the integral equations

$$u(t) = x_o + \int_o^t f(s, u(s), u(s), v(s))ds, \quad t \in J, \tag{b}$$

and

$$v(t) = x_o + \int_o^t f(s, v(s), v(s), u(s))ds, \quad t \in J. \tag{c}$$

Denoting $x(t) = \max\{|u(t)|, |v(t)|\}$, $t \in J$, it follows from (b), (c) and (B6) that

$$x(t) \le |x_o| + \int_o^t p(s)h(x(s), x(s), x(s))\, ds, \quad t \in J, \tag{d}$$

Since $w$ is the solution of (2.2.8), it follows from lemma 1.5.3 that $x(t) \le w(t)$ on $J$. This implies by (a), (b) and (B6) that

$$|u(t) - x_o| \le \int_o^t p(s)\, h(w(s), w(s), w(s))\, ds = w(t) - |x_o|,$$

so that $u \in [\alpha, \beta]$. From (a), (c) and (B6) it follows by the similar reasoning that $v \in [\alpha, \beta]$. Since $y$ and $z$ are the extremal coupled quasisolutions of solutions of (2.2.1) in $[\alpha, \beta]$, then $u$, $v \in [y, z]$. But $u$, $v$ were arbitrary coupled quasisolutions of (2.2.1), whence $y$ and $z$ are extremal coupled quasisolutions of the IVP (2.2.1), and all the coupled quasisolutions of (2.2.1) are contained in $[\alpha, \beta]$. □

**Remark 2.2.1:** By theorem 1.4.4 and corollary 1.4.5 the assumption (B2) holds if $f$ is a standard function or a Borel function.

If $f$ is defined in $J \times V \times V \times V$, where $V \subset I\!R$, and if conditions (B2)–(B6) hold, then the extremal solutions of (2.2.1) exist in such a subinterval $J_o = [0, c]$ of $J$ that $[\alpha(t), \beta(t)] \subseteq V$ for each $t \in J_o$, where $\alpha$, $\beta$ are given by (2.2.7). This holds, for instance, if $w(c) - |x_o| < d = \inf\{|y - x_o| \mid y \notin V\}$, or equivalently, if

$$\int_o^c p(s)\, ds < \int_{|x_o|}^{|x_o|+d} \frac{dv}{h(v, v, v)}.$$

**Example 2.2.1:** Define functions $p$, $h$: $I\!R \to I\!R_+$ by

$$p(t) = \begin{cases} 1, & t \text{ is irrational,} \\ 2, & t \text{ is rational,} \end{cases} \quad \text{and} \quad h(x) = \begin{cases} 1, & x \leq 0, \\ 0, & x > 0. \end{cases}$$

Consider the IVP (2.2.1), where $f\colon J \times I\!R^3 \to I\!R$ is defined by

$$f(t, x, y, z) = p(t)(1 + x)h(-y)h(z), \quad t \in J,\ x,\ y,\ z \in I\!R. \quad \text{(a)}$$

It is easy to see that conditions (B2)–(B6) hold, whence the IVP (2.2.1) has for each $x_o \in I\!R$ extremal coupled quasisolutions $y$, $z$. In the case when $x_o = 0$ the least possible candidate for $y$ is $y(t) \equiv 0$. Routine calculations show that for this $y$ (2.2.2) holds if and only if $z(t) = e^t - 1$, $t \in J$. It is also easy to see that these $y$, $z$ are the only coupled quasisolutions of (2.2.1) in the considered case.

To show that the IVP (2.2.1) does not possess an ordinary solution on any interval $J = [0, T]$, $T > 0$, when $x_o = 0$ and $f$ is given by (a), make a counter-hypothesis: there is $T > 0$ such that (2.2.1) has a solution $x$ on $J$ when $x_o = 0$. The definition of $f$ and lemma 1.5.5 imply that $x$ satisfies the integral equation

$$x(t) = \int_o^t p(s)(1 + x(s))h(-x(s))h(x(s))\, ds,\ t \in J. \quad \text{(b)}$$

In particular, $x$ is nondecreasing and $x(t) \geq 0$ for each $t \in J$. Assume first that $x(t) \equiv 0$ in $(0, c)$ for some $c \in (0, T]$. In view of (b) we then have

$$x(t) = \int_o^t p(s)\, ds = t, \quad t \in (0, c),$$

which is impossible, since we assumed that $x(t) \equiv 0$ in $(0, c)$. Thus $x(t) > 0$ for each $t \in (0, T]$. But then $h(x(t)) \equiv 0$ in $(0, T]$, whence (b) implies that $x(t) \equiv 0$, a contradiction.

### 2.2.3. Discontinuous case of $f$

The following result is a counterpart to theorem 2.1.3.

**Theorem 2.2.3:** *Let $f\colon J \times I\!R^3 \to I\!R$ be defined by*

$$f(t, x, y, z) = q(x)\, g(t, y, z), \quad t \in J, \ x, \ y, \ z \in I\!R, \tag{2.2.9}$$

*where $q$, $\frac{1}{q} \in L^\infty_{loc}(I\!R, I\!R_+)$, $g\colon J \times I\!R^2 \to I\!R$ is a standard function, and $g(t, \cdot, y)$ is nondecreasing and $g(t, y, \cdot)$ is nonincreasing for a.a. $t \in J$ and for all $y \in I\!R$.*

a) *If (B0) holds, then for each $x_o \in [\alpha(0), \beta(0)]$ the IVP (2.2.1) has the extremal coupled quasisolutions in $[\alpha, \beta]$.*

b) *If (B6) holds, then the IVP has the extremal coupled quasisolutions for each $x_o \in I\!R$.*

*Proof.* a) Let $\alpha$, $\beta \in AC(J, I\!R)$ be as in condition (B0) and let $x_o \in [\alpha(0), \beta(0)]$ be given. If $y$, $z \in [\alpha, \beta]$, then

$$\frac{\alpha'(t)}{q(\alpha(t))} \leq g(t, \alpha(t), \beta(t)) \leq g(t, y(t), z(t))$$
$$\leq g(t, \beta(t), \alpha(t)) \leq \frac{\beta'(t)}{q(\beta(t))}$$

for a.a. $t \in J$. In view of this, theorem 1.4.5 and the given hypotheses, the equation

$$\int_{x_o}^{A(y,z)(t)} \frac{dv}{q(v)} = \int_o^t g(s, y(s), z(s))ds, \quad t \in J \tag{a}$$

defines a mapping $A\colon [\alpha,\beta]\times[\alpha,\beta] \to [\alpha,\beta]$. Moreover, $A(y,z)$ is nondecreasing in $y$ and nonincreasing in $z$ because $g$ is. Denote $M$ =ess $\sup\{q(x) \mid \min\alpha \le x \le \max\beta\}$. If $y$, $z \in [\alpha,\beta]$ and $0 \le \bar{t} \le t \le T$, then

$$\begin{aligned}&\frac{|A(y,z)(t) - A(y,z)(\bar{t})|}{M}\\ &\le \left|\int_{A(y,z)(\bar{t})}^{A(y,z)(t)} \frac{dv}{q(v)}\right| = \left|\int_{\bar{t}}^{t} g(s,y(s),z(s))\,ds\right|\\ &\le \int_{\bar{t}}^{t} \max\{|g(s,\alpha(s),\beta(s))|,\ |g(s,\beta(s),\alpha(s))|\}\,ds.\end{aligned}$$

Hence for all $y$, $z \in [\alpha,\beta]$

$$\begin{aligned}&|A(y,z)(t) - A(y,z)(\bar{t})|\\ &\le M\int_{\bar{t}}^{t} \max\{|g(s,\alpha(s),\beta(s))|, |g(s,\beta(s),\alpha(s))|\}\,ds\end{aligned} \tag{b}$$

whenever $0 \le \bar{t} \le t \le T$.

The above proof shows that $A$ satisfies the hypotheses of theorem 1.4.8, whence $A$ has the extremal coupled fixed points $y$, $z$. From theorem 1.4.5 it follows that $u$, $v \in [\alpha,\beta]$ are coupled quasisolutions of (2.2.1), with $f$ given by (2.2.9) if and only if

$$\int_{x_o}^{u(t)} \frac{dx}{q(x)} = \int_o^t g(s,u(s),v(s))ds, \ t \in J$$

and

$$\int_{x_o}^{v(t)} \frac{dx}{q(x)} = \int_o^t g(s, v(s), u(s))ds, \; t \in J,$$

which is by (a) equivalent that $u$, $v$ are coupled fixed points of $A$. Since $y$, $z$ are the extremal coupled fixed points of $A$, then they are the extremal coupled quasisolutions of (2.2.1) in $[\alpha, \beta]$.

b) Assume next that condition (B6) holds, and let $x_o \in \mathbb{R}$ be given. The proof of theorem 2.2.2 implies that (B0) holds with $\alpha$, $\beta$ defined by (2.2.7), and that all the coupled quasisolutions of (2.2.1) are contained in $[\alpha, \beta]$. Moreover, (2.2.1) has by the above proof extremal coupled quasisolutions on $[\alpha, \beta]$, which are then extremals of all coupled quasisolutions of (2.2.1). □

### 2.2.4. Special cases

As a consequence of theorem 2.2.2 we obtain

**Proposition 2.2.3:** *Given for each $j = 1, \ldots, n$ a bounded and Lipschitz continuous function $q_j \colon \mathbb{R} \to \mathbb{R}_+$ and a bounded and standard function $g_j \colon J \times \mathbb{R}^2 \to \mathbb{R}$, such that $g_j(t, \cdot, z)$ is nondecreasing and $g_j(t, y, \cdot)$ is nonincreasing for a.a. $t \in J$, then the IVP*

$$x' = \sum_{j=1}^{n} q_j(x) g_j(t, x, x), \qquad x(0) = x_o \tag{2.2.10}$$

*has for each $x_o \in \mathbb{R}$ the extremal coupled quasisolutions.*

*Proof.* Define a function $f \colon J \times \mathbb{R}^3 \to \mathbb{R}$ by

$$f(t, x, y, z) = \sum_{j=1}^{n} q_j(x) g_j(t, y, z), \quad t \in J, \; x, \, y, z \in \mathbb{R}. \tag{a}$$

Condition (B6) holds because all the functions in the right-hand side of (a) are bounded. If $y$, $z \in AC(J, \mathbb{R})$, then the hypotheses given for $g_j$ imply by theorem 1.4.4 that each $g_j(\cdot, y(\cdot), z(\cdot))$ is measurable. Thus $f(\cdot, x, y(\cdot), z(\cdot))$ is measurable for each $x \in \mathbb{R}$, whence (B2) is valid. Since

$$|f(t, x + k, y, z) - f(t, x, y, z)|$$
$$\leq \max\{\sup |g_j| \mid 1 \leq j \leq n\} \sum_{j=1}^{n} |q_j(x + k) - q_j(x)|$$

for a.a. $t \in J$ and for all $x$, $k$, $y, z \in \mathbb{R}$, it follows that $f(t, \cdot, y, z)$ is Lipschitz continuous for all $y$, $z \in \mathbb{R}$ and for a.a. $t \in J$. Thus conditions (B3) and (B5) hold. Since each $g_j(t, y, z)$ is nondecreasing in $y$ and nonincreasing in $z$, then condition (B4) is valid. Thus the conclusion follows from theorem 2.2.2. □

From theorem 2.2.3 it follows

**Proposition 2.2.4:** *If the functions $q$, $g$ satisfy the hypotheses of theorem 2.2.3, and if $g$ is a Carathéodory function, then for each $x_o \in [\alpha(0), \beta(0)]$ the sequences $(y_n)_{n=o}^{\infty}$ and $(z_n)_{n=o}^{\infty}$ of functions $y_n$, $z_n \colon J \longrightarrow \mathbb{R}$, defined by $y_o = \alpha$, $z_o = \beta$,*

$$\int_{x_o}^{y_{n+1}(t)} \frac{dv}{q(v)} = \int_o^t g(s, y_n(s), z_n(s))ds, \ n \in \mathbb{N}, \qquad (2.2.11)$$

*and*

$$\int_{x_o}^{z_{n+1}(t)} \frac{dv}{q(v)} = \int_o^t g(s, z_n(s), y_n(s))ds, \ n \in \mathbb{N}, \qquad (2.2.12)$$

*converge uniformly on $J$ to the extremal coupled quasisolutions of*

$$x' = q(x)\, g(t, x, x), \qquad x(0) = x_o \qquad (2.2.13)$$

*in the order interval* $[\alpha, \beta]$.

*Proof.* The given hypotheses imply that the equation

$$\int_{x_o}^{A(y,z)(t)} \frac{dv}{q(v)} = \int_o^t g(s, y(s), z(s))ds, \quad t \in J \tag{a}$$

defines a mixed monotone operator $A\colon [\alpha, \beta] \to [\alpha, \beta]$. From (2.2.11), (2.2.12) and (a) it follows that $(y_n)_{n=o}^{\infty}$ and $(z_n)_{n=o}^{\infty}$ satisfy $y_o = \alpha$, $z_o = \beta$, and

$$y_{n+1} = A(y_n, z_n), \; z_{n+1} = A(z_n, y_n), \; n \in I\!N. \tag{b}$$

Thus the sequence $(y_n)_{n=o}^{\infty}$ is nondecreasing, the sequence $(z_n)_{n=o}^{\infty}$ is nonincreasing and both are contained in $[\alpha, \beta]$. The proof of theorem 2.2.3 implies also that both the sequences $(y_n)_{n=o}^{\infty}$ and $(z_n)_{n=o}^{\infty}$ converge uniformly on $J$ to functions $y$, $z \in [\alpha, \beta]$, respectively. Since $g$ is a Carathéodory function, it follows from (2.2.11) and (2.2.12) as $n \to \infty$ that $y$, $z$ satisfy the integral equations

$$\int_{x_o}^{y(t)} \frac{dv}{q(v)} = \int_o^t g(s, y(s), z(s))ds, \quad t \in J,$$

and

$$\int_{x_o}^{z(t)} \frac{dv}{q(v)} = \int_o^t g(s, z(s), y(s))ds, \quad t \in J.$$

Thus $y$, $z$ is are coupled fixed points of $A$, and hence, by the proof of theorem 2.2.3, coupled quasisolutions of (2.2.13).

If $u$, $v \in [\alpha, \beta]$ are coupled quasisolutions of (2.2.13), they are also coupled fixed points of $A$ by the proof of theorem 2.2.3. Since $y_o = \alpha \le u$, $v \le \beta = z_o$ and $A$ is mixed monotone, it follows from (b) by induction that $y_n \le u$, $v \le z_n$ for each $n \in I\!N$. Thus $y = \lim_n y_n \le u$, $v \le \lim_n z_n = z$. This proves that $y$, $z$ are the extremal coupled quasisolutions of (2.2.13) in $[\alpha, \beta]$. □

If $q(v) = 1$ a.e. on $\mathbb{R}$, then (2.2.11) and (2.2.12) can be rewritten as

$$y_{n+1}(t) = x_o + \int_o^t g(s, y_n(s), z_n(s))ds, \quad t \in J, \ n \in \mathbb{N}, \quad (2.2.14)$$

and

$$z_{n+1}(t) = x_o + \int_o^t g(s, z_n(s), y_n(s)ds, \quad t \in J, \ n \in \mathbb{N}. \quad (2.2.15)$$

Thus we obtain

**Corollary 2.2.2:** *Assume that $q\colon \mathbb{R} \to (0, \infty)$, that $q(v) = 1$ a.e. on $\mathbb{R}$, that $g\colon J \times \mathbb{R}^2 \to \mathbb{R}$ is a Carathéodory function, and that $g(t, \cdot, y)$ is nondecreasing and $g(t, y, \cdot)$ is nonincreasing on $\mathbb{R}$ for a.a. $t \in J$. If there exist $\alpha$, $\beta \in AC(J, \mathbb{R})$ such that*

$$\alpha \leq \beta, \ \alpha' \leq g(t, \alpha, \beta), \quad \text{and } \beta' \geq g(t, \beta, \alpha) \text{ a.e. on } J,$$

*then the sequence $(z_n)_{n=o}^{\infty}$ and $(z_n)_{n=o}^{\infty}$ of the successive approximations defined by (2.2.14) and (2.2.15) converge uniformly on $J$ to the extremal coupled quasisolutions $y$, $z$ of the IVP (2.2.13) in the order interval $[\alpha, \beta]$ if $y_o = \alpha$, $z_o = \beta$ and $\alpha(0) \leq x_o \leq \beta(0)$.*

As a special case of proposition 2.2.3 we have

**Corollary 2.2.3:** *Assume that for each $j = 1, \ldots, n$, $p_j\colon J \to \mathbb{R}_+$ is Lebesgue integrable, $q_j\colon \mathbb{R} \to \mathbb{R}_+$ is bounded and Lipschitz continuous, $g_j\colon \mathbb{R} \to \mathbb{R}_+$ is bounded and nondecreasing, and $h_j\colon \mathbb{R} \to \mathbb{R}_+$ is bounded and nonincreasing. Then the IVP*

$$x' = \sum_{j=1}^{n} p_j(t) q_j(x) g_j(x) h_j(x), \qquad x(t_o) = x_o$$

*has for each $x_o \in \mathbb{R}$ extremal coupled quasisolutions.*

As a consequence of theorem 2.2.3 we obtain

**Corollary 2.2.4:** *If $q, \frac{1}{q} \in L^\infty(\mathbb{R}, \mathbb{R}_+)$, $p \in L^1(J, \mathbb{R}_+)$, if $g \colon \mathbb{R} \to \mathbb{R}_+$ is bounded and nondecreasing and if $h \colon J \to \mathbb{R}_+$ is bounded and nonincreasing, then the IVP*

$$x' = p(t)q(x)g(x)h(x), \qquad x(0) = x_o$$

*has for each $x_o \in \mathbb{R}$ the extremal coupled quasisolutions.*

## 2.3. FIRST ORDER PERIODIC BVP'S

In this section we shall study the existence of extremal solutions and extremal coupled quasisolutions of first order periodic boundary value problems.

### 2.3.1. Existence of extremal solutions

Consider the PBVP

$$x' = f(t, x, x), \qquad x(0) = x(T), \tag{2.3.1}$$

where $f \colon J \times \mathbb{R}^2 \to \mathbb{R}$, $J = [0, T]$. Let us first list a set of assumptions for the function $f$.

(C0) $\alpha, \beta \in AC(J, \mathbb{R})$, $\alpha \le \beta$, $\alpha(0) \le \alpha(T)$, $\beta(0) \ge \beta(T)$, $\alpha' \le f(t, \alpha, \alpha)$ and $\beta' \ge f(t, \beta, \beta)$ on $J \setminus Z$.

(C1) $|f(t, x, y)| \le N(t)$ for all $t \in J \setminus Z$ and $x, y \in [\alpha(t), \beta(t)]$.

(C2) $f(\cdot, x(\cdot), y(\cdot))$ is measurable for all $x, y \in [\alpha, \beta]$.

(C3) $f(t, x, y) + M(t)\, y$ is nondecreasing in $y \in [\alpha(t), \beta(t)]$ for all $t \in J \setminus Z$ and $x \in [\alpha(t), \beta(t)]$.

(C4) $f(t,x,y)-p(t)\,x$ is continuous and nonincreasing in $x \in [\alpha(t),\beta(t)]$ for all $t \in J \setminus Z$ and $y \in [\alpha(t),\beta(t)]$.

(C5) $f(t,x,y)+p(t)\,x$ is nondecreasing in $x \in [\alpha(t),\beta(t)]$ for all $t \in J \setminus Z$ and $y \in [\alpha(t),\beta(t)]$.

(C6) $N_1(t)+p_1(t)x+p_2(t)y \le f(t,x,y) \le N_2(t)+p_1(t)x+p_2(t)y$ for all $t \in J \setminus Z$ and $x$, $y \in \mathbb{R}$.

We are going to prove that the PBVP (2.3.1) has the extremal solutions in the order interval $[\alpha,\beta]$ if there is a null set $Z$ in $J$ such that either conditions (C0)–(C4) hold for some $N$, $M$, $p \in L^2(J,\mathbb{R})$ or conditions (C0)–(C3) and (C5) hold for some $N$, $M$, $p \in L^1(J,\mathbb{R})$. If (C0) and (C1) are replaced by (C6) with $N_i$, $p_i \in L^2(J,\mathbb{R})$ (resp. $L^1(J,\mathbb{R})$) with $\int_o^T (p_1(s)+p_2(s))\,ds < 0$, then these extremal solutions are the least and the greatest of all the solutions of (2.3.1). Moreover, in all the above cases the extremal solutions in question are nondecreasing with respect to $f$.

**Theorem 2.3.1:** *Assume there exist $N$, $M$, $p \in L^2(J,\mathbb{R})$ and a null set $Z$ in $J$ such that conditions (C0)–(C4) hold. Then there exist the extremal solutions to PBVP (2.3.1) in the order interval $[\alpha,\beta]$ of $AC(J,\mathbb{R})$.*

*Proof.* We may choose the functions $M$ and $p$ in conditions (C3) and (C4) so that $\int_o^T (M(s)-p(s))\,ds > 0$. Let $y \in [\alpha,\beta]$ be given. Consider the PBVP

$$x' = F(t,x;y(t)), \qquad x(0)=x(T), \tag{2.3.2}$$

where

$$F(t,x;y(t)) = f(t,x,y(t)) + M(t)(y(t)-x)). \tag{2.3.3}$$

From (C2) and (C4) it follows that $(t,x) \mapsto F(t,x;y(t))$ is a Carathéodory function in $\Omega = \{(t,x) \mid t \in J,\ x \in [\alpha(t),\beta(t)]\}$. Applying (C0), (C3) and (2.3.3) it is easy to see that

$$\alpha' \le F(t,\alpha;y(t)) \text{ and } \beta' \ge F(t,\beta;y(t)) \text{ on } J \setminus Z.$$

Condition (C1) implies that for all $x \in [\alpha(t), \beta(t)]$ and $t \in J \setminus Z$,

$$|F(t, x; y(t))| \leq K\, M(t) + N(t), \tag{a}$$

where $K = \max \beta - \min \alpha$. Hence, by theorem 1.5.2 the PBVP (2.3.2) has a solution $x$ in $[\alpha, \beta]$. In view of condition (C4) we see that the function $F(t, x; y(t)) + (M(t) - p(t))\, x$ is nonincreasing in $x \in [\alpha(t), \beta(t)]$ for all $t \in J \setminus Z$, whence $x$ is by corollary 1.5.3 the only solution of (2.3.2).

Define a map $G\colon [\alpha, \beta] \to [\alpha, \beta]$ by

$$Gy = x, \quad y \in [\alpha, \beta], \tag{2.3.4}$$

where $x$ is the solution of (2.3.2) in $[\alpha, \beta]$. To prove that $G$ is nondecreasing, let $y_1$, $y_2 \in [\alpha, \beta]$, $y_1 \leq y_2$, be given, and suppose that $x_1$ and $x_2$ are the corresponding solutions of

$$x_i' = F(t, x_i; y_i(t)), \qquad x_i(0) = x_i(T), \quad i = 1, 2 \tag{2.3.5}$$

in $[\alpha, \beta]$. Since $y_1 \leq y_2$, we have by (C3) and (2.3.3)

$$x_1' \leq F(t, x_1; y_2(t)) \text{ a.e. on } J, \qquad x_1(0) = x_1(T).$$

Thus the hypotheses of lemma 1.5.6 hold when $y = x_1$ and $z = x_2$, whence $x_1 \leq x_2$, thus proving that $Gy_1 \leq Gy_2$.

From (2.3.3)–(2.3.4) and (a) it follows that

$$|(Gy)'(t)| \leq N(t) + KM(t) \text{ for all } y \in [\alpha, \beta] \text{ and for a.a. } t \in J.$$

Thus $G$ satisfies the hypotheses of proposition 1.4.4, whence $G$ has the least fixed point $x_*$ and the greatest fixed point $x^*$. From the definition of $G$ it follows that $x_*$ and $x^*$ are also solutions of the PBVP (2.3.1) in $[\alpha, \beta]$.

If $x$ is any solution of (2.3.1) in $[\alpha, \beta]$, then it satisfies also the PBVP (2.3.2) with $y = x$. But this means that $x$ is a fixed point of $G$, whence $x_* \leq x \leq x^*$. Thus $x_*$ is the minimal solution and $x^*$ is the maximal solution of the PBVP (2.3.1) in $[\alpha, \beta]$. □

If condition (C4) is replaced in theorem 2.3.1 by condition (C5), we obtain

**Theorem 2.3.2:** *If there exist $N$, $M$, $p \in L^1(J, I\!R)$ and a null set $Z$ in $J$ such that conditions (C0)–(C3) and (C5) hold, then the PBVP (2.3.1) has the extremal solutions in $[\alpha, \beta]$.*

*Proof.* We can choose the functions $M$, $p \in L^1(J, I\!R)$ in conditions (C3) and (C5) so that $\int_0^T (M(s) + p(s))\, ds > 0$. Denote $q = M + p$, and define an operator $G$ in $[\alpha, \beta]$ by

$$\begin{aligned} Gx(t) =& e^{-Q(t)} \int_0^t e^{Q(s)} g(s, x(s))\, ds \\ &+ \frac{e^{-Q(t)}}{e^{Q(T)} - 1} \int_0^T e^{Q(s)} g(s, x(s))\, ds, \quad t \in J, \end{aligned} \tag{2.3.6}$$

where

$$g(t, x) = f(t, x, x) + q(t)\, x, \quad \text{and} \quad Q(t) = \int_0^t q(s)\, ds. \tag{2.3.7}$$

From (C0) and (2.3.7) it follows that

$$e^{Q(t)} g(t, \alpha(t)) \geq e^{Q(t)} (\alpha'(t) + q(t)\alpha(t)) \quad t \in J \setminus Z. \tag{a}$$

In view of (a), (C0) and (2.3.6) we obtain

$$\begin{aligned}
e^{Q(t)}G\alpha(t) &\geq \int_o^t e^{Q(s)}(\alpha'(s)) + q(s)\alpha(s))\,ds \\
&+ \frac{1}{e^{Q(T)}-1}\int_o^T e^{Q(s)}(\alpha'(s) + q(s)\alpha(s))\,ds \\
&= e^{Q(t)}\alpha(t)) - \alpha(0) + \frac{1}{e^{Q(T)}-1}(e^{Q(T)}(\alpha(T)) - \alpha(0)) \\
&\geq e^{Q(t)}\alpha(t), \quad t \in J.
\end{aligned}$$

Thus $\alpha \leq G\alpha$. Similarly, it can be shown that $G\beta \leq \beta$. From (C3), (C5) and (2.3.7) it follows that $g(t,\cdot)$ is nondecreasing in $[\alpha,\beta]$ for all $t \in J \setminus Z$. This and the definition (2.3.6) of $G$ imply that $G$ is nondecreasing. Obviously, $Gx \in AC(J,\mathbb{R})$ for each $x \in [\alpha,\beta]$, whence (2.3.6) defines a nondecreasing operator $G\colon [\alpha,\beta] \to [\alpha,\beta]$.

From (2.3.6) and (2.3.7) it follows that

$$(Gy)'(t) = f(t,y(t),y(t)) + q(t)(y(t) - Gy(t))$$

for all $y \in [\alpha,\beta]$ and for a.a. $t \in J$. This and (C1) imply that

$$|(Gy)'(t)| \leq N(t) + Kq(t) \quad \text{for all } y \in [\alpha,\beta] \text{ and for a.a. } t \in J,$$

where $K = 2\max\{\|\alpha\|_o, \|\beta\|_o\}$. Thus $G$ satisfies the hypotheses of proposition 1.4.4, whence $G$ has the least fixed point $x_*$ and the greatest fixed point $x^*$. From the definitions of $g$ and $G$ and lemma 1.5.7 it follows that $x_*$ and $x^*$ are also solutions of the PBVP (2.3.1) in $[\alpha,\beta]$.

If $x$ is any solution of (2.3.1) in $[\alpha,\beta]$, then it is by lemma 1.5.7 a fixed point of $G$, whence $x_* \leq x \leq x^*$. Thus $x_*$ is the minimal solution and $x^*$ is the maximal solution of the PBVP (2.3.1) in $[\alpha,\beta]$. □

Consider next the existence of the extremal solutions of (2.3.1) among all its solutions.

**Proposition 2.3.1:** *Assume there exist* $M$, $p$, $p_1$, $p_2$, $N_1$, $N_2 \in L^2(J, \mathbb{R})$, *with* $\int_o^T (p_1(s) + p_2(s))\, ds < 0$, *and a null set* $Z$ *in* $J$ *such that condition (C6) holds, and that (C2)–(C4) hold when* $\alpha$, $\beta$ *are given by*

$$\alpha(t) = e^{P(t)}[\int_o^t e^{-P(s)} N_1(s) ds + \int_o^T \frac{e^{-P(s)} N_1(s)}{e^{-P(T)} - 1} ds], \quad (2.3.8)$$

$$\beta(t) = e^{P(t)}[\int_o^t e^{-P(s)} N_2(s) ds + \int_o^T \frac{e^{-P(s)} N_2(s)}{e^{-P(T)} - 1} ds], \quad (2.3.9)$$

*where* $P(t) = \int_o^t (p_1(s) + p_2(s))\, ds$, $t \in J$. *Then there exist the extremal solutions to PBVP (2.3.1), and all the solutions of (2.3.1) lie within the order interval* $[\alpha, \beta]$.

*Proof.* From corollary 1.5.4 it follows that $\alpha$ and $\beta$, given by (2.3.8) and (2.3.9), are the unique solutions of the PBVP's

$$\alpha' = N_1(t) + (p_1(t) + p_2(t))\alpha(t), \quad \alpha(0) = \alpha(T), \quad \text{(a)}$$

and

$$\beta' = N_2(t) + (p_1(t) + p_2(t))\beta(t), \quad \beta(0) = \beta(T). \quad \text{(b)}$$

In view of (a), (b) and (C6) we see that condition (C0) holds, and that condition (C1) holds with

$$N(t) = |N_1(t)| + |N_2(t)| + (|p_1(t)| + |p_2(t)|)(\|\alpha\|_o + \|\beta\|_o).$$

Thus the PBVP (2.3.1) has the extremal solutions $x_*$ and $x^*$ in $[\alpha, \beta]$.

If $x$ is a solution of (2.3.1), it follows from (2.3.1) and (C6) that

$$x' \geq N_1(t) + (p_1(t) + p_2(t))x(t) \text{ a.e. on } J, \; x(0) = x(T), \quad \text{(c)}$$

and

$$x' \le N_2(t) + (p_1(t) + p_2(t))x(t) \text{ a.e. on } J, \; x(0) = x(T). \quad \text{(d)}$$

Hence, (a), (b), (c) and (d) imply by lemma 1.5.6 that $x \in [\alpha, \beta]$. In particular, $x \in [x_*, x^*]$, whence $x_*$ and $x^*$ are the extremals among all the solutions of (2.3.1). □

By the similar reasoning it follows from theorem 2.3.2 and lemmas 1.5.6 and 1.5.7.

**Proposition 2.3.2:** *Assume there exist $M$, $p$, $p_1$, $p_2$, $N_1$, $N_2 \in L^1(J, \mathbb{R})$ with $\int_o^T (p_1(s) + p_2(s))\,ds < 0$ and a null set $Z$ in $J$ such that (C6) holds, and that conditions (C2) (C3) and (C5) are valid with $\alpha$, $\beta$ defined by (2.3.8) and (2.3.9). Then the conclusions of proposition 2.3.1 hold.*

**Example 2.3.1:** Denote $D = \{(n_o, \ldots, n_m) \mid m, n_o, \ldots, n_m \in \mathbb{N} \text{ and } n_o > 0\}$, and define for each $(n_o, \ldots, n_m) \in D$ a rational number $a(n_o, \ldots, n_m)$ by

$$a(n_o, \ldots, n_m) = c + 2^{-m-1} + \sum_{k=0}^{m} 2^{-k-m-2} \prod_{j=o}^{k} 2^{-n_j}$$

$$+ \sum_{j=0}^{m} 2^{-2m-2} \prod_{j=o}^{m} 2^{-n_j}.$$

where $c \ge 0$. Given $J = [0, 1]$ and $p \in L^1(J, \mathbb{R}_+)$, denote $P(t) = 1 + \int_o^t p(s)\,ds$, $t \in J$. Define a function $g \colon J \times \mathbb{R} \to \mathbb{R}$ by

$$g(t,x) = \begin{cases} cp(t), & 0 \le x \le cP(t), \\ (c+1)p(t), & x > (c+1)P(t), \\ a(n_o, \ldots, n_m + 1)p(t), & a(n_o, \ldots, n_m + 1)P(t) \\ & < x \le a(n_o, \ldots, n_m)P(t), \\ -g(t, -x), & x < 0. \end{cases}$$

Choose $q \in L^1(J, \mathbb{R}_+)$ so that

$$\int_o^T q(s)\,ds > \log P(T),$$

and define a function $f\colon J \times \mathbb{R} \to \mathbb{R}$ by

$$f(t,x,y) = g(t, ye^{Q(t)})e^{-Q(t)} - q(t)y, \quad t \in J,\ x,\ y \in \mathbb{R},$$

where $Q(t) = \int_o^t q(s)\,ds,\ t \in J$. Defining

$$\beta(t) = (c+1)P(t)e^{-Q(t)}, \quad \alpha(t) = -\beta(t),\ t \in J,$$

it is easy to see that conditions (C0)–(C3) and (C5) hold, whence the PBVP (2.3.1) has by theorem 2.3.2 the extremal solutions in $[\alpha, \beta]$.

As for the dependence of the extremal solutions of (2.3.1) on $f$ we have

**Proposition 2.3.3:** *If the hypotheses of theorem 2.3.1 or theorem 2.3.2 hold, then the extremal solutions of the PBVP (2.3.1) in $[\alpha, \beta]$ are nondecreasing with respect to $f$.*

*Proof.* Let $f,\ \hat{f}\colon J \times \mathbb{R}^2 \to \mathbb{R}$ satisfy

$$f(t,x,y) \le \hat{f}(t,x,y) \text{ for a.a. } t \in J \text{ and for all } x,\ y \in \mathbb{R}. \quad \text{(a)}$$

Assume first that the hypotheses of theorem 2.3.1 hold for $f$ and $\hat{f}$. Let $x_*$ be the minimal solution of (2.3.1) in $[\alpha, \beta]$, and let $\hat{x}_*$ be the minimal solution of the PBVP

$$x' = \hat{f}(t,x,x), \qquad x(0) = x(T). \quad \text{(b)}$$

If $F(t, x; y(t))$ is defined by (2.3.3), it follows from (a), (b) and (2.3.3) that

$$\hat{x}_*'(t) \geq F(t, \hat{x}_*(t); \hat{x}_*(t)) \quad \text{a.e. on} \quad J, \quad \hat{x}_*(0) = \hat{x}_*(T). \tag{c}$$

From the definition (2.3.4) of the operator $G$ it follows that

$$(G\hat{x}_*)'(t) = F(t, G\hat{x}_*(t); \hat{x}_*(t)) \quad \text{a.e. on} \quad J, \quad G\hat{x}_*(0) = G\hat{x}_*(T). \tag{d}$$

Because $(t, x) \mapsto F(t, x; \hat{x}_*(t))$ satisfies the hypotheses of lemma 1.5.6, then (c) and (d) imply that $G\hat{x}_* \leq \hat{x}_*$. Since $G$ satisfies the hypotheses of theorem 1.4.7, then $x_* \leq \hat{x}_*$.

The proof that $x^* \leq \hat{x}^*$, where $x^*$ denotes the maximal solution of (2.3.1) in $[\alpha, \beta]$ and $\hat{x}^*$ is the maximal solution of (b) in $[\alpha, \beta]$, is similar.

Assume next that (a) holds for the functions $f$, $\hat{f}$ which satisfy the hypotheses of theorem 2.3.2, and let $G\colon [\alpha, \beta] \to [\alpha, \beta]$ be defined by (2.3.6), where $g$ is given by (2.3.7). If $x_*$ and $\hat{x}_*$ are as above, it follows from lemma 1.5.7 that

$$\begin{aligned} \hat{x}_*(t) &= e^{-Q(t)} \int_o^t e^{Q(s)} \hat{g}(s, \hat{x}_*(s))\, ds \\ &+ \frac{e^{-Q(t)}}{e^{Q(T)} - 1} \int_o^T e^{Q(s)} \hat{g}(s, \hat{x}_*(s))\, ds, \quad t \in J, \end{aligned} \tag{e}$$

where

$$\hat{g}(t, x) = \hat{f}(t, x, x) + q(t)\, x, \quad \text{and} \quad Q(t) = \int_o^t q(s)\, ds. \tag{f}$$

From (a), (f) and (2.3.7) it follows that

$$g(t, x) \leq \hat{g}(t, x), \quad \text{for a.a. } t \in J \text{ and for all } x \in [\alpha(t), \beta(t)],$$

whence (2.3.6) and (e) imply that $G\hat{x}_* \leq \hat{x}_*$. Since $G$ satisfies the hypotheses of theorem 1.4.7 and $x_*$ is the least fixed point of $G$, it follows that $x_* \leq \hat{x}_*$.

Similarly, it can be shown that $x^* \leq \hat{x}^*$, where $x^*$ denotes the maximal solution of (2.3.1) in $[\alpha, \beta]$ and $\hat{x}^*$ is the maximal solution of (b) in $[\alpha, \beta]$. □

The following corollaries are direct consequences of theorem 2.3.1 and proposition 2.3.3.

**Corollary 2.3.1:** *Let $f: J \times \mathbb{R}^2 \to \mathbb{R}$ satisfy conditions (C0) and (C1). If $f(t, x, y)$ is in $D = \{(t, x, y) \mid t \in J,\ x, y \in [\alpha(t), \beta(t)]\}$ Borel measurable, continuous and nonincreasing in $x$ and nondecreasing in $y$, then (2.3.1) has the extremal solutions in $[\alpha, \beta]$, and they are nondecreasing with respect to $f$.*

**Corollary 2.3.2:** *If $f: J \times \mathbb{R}^2 \to \mathbb{R}$ satisfies condition (C0), if $f(t, x, y)$ is in $D = \{(t, x, y) \mid t \in J,\ x, y \in [\alpha(t), \beta(t)]\}$ continuous, nonincreasing in $x$ and nondecreasing in $y$, then the PBVP (2.3.1) has the extremal solutions in $[\alpha, \beta]$, and they are nondecreasing with respect to $f$.*

The proof of proposition 2.3.3 can also be used to verify the following result.

**Proposition 2.3.4:** *If the hypotheses of proposition 2.3.1 or proposition 2.3.2 hold, then the extremal solutions of the PBVP (2.3.1) are nondecreasing with respect to $f$.*

### 2.3.2. Existence of extremal coupled quasisolutions

Consider next the existence of the extremal coupled quasisolutions of the PBVP

$$x' = f(t, x, x, x), \qquad x(0) = x(T), \tag{2.3.10}$$

in the case when $f\colon J \times \mathbb{R}^3 \to \mathbb{R}$ satisfies some of the following hypotheses.

(D0) $\alpha,\ \beta \in AC(J,\mathbb{R})$, $\alpha \le \beta$, $\alpha(0) \le \alpha(T)$, $\beta(0) \ge \beta(T)$, and $\alpha' \le f(t,\alpha,\alpha,\beta)$ and $\beta' \ge f(t,\beta,\beta,\alpha)$ on $J \setminus Z$.

(D1) $\sup\{|f(t,x,y,z)| \mid x,\ y,\ z \in [\alpha(t),\beta(t)]\} \le N(t)$ for all $t \in J \setminus Z$.

(D2) $f(\cdot,x(\cdot),y(\cdot),z(\cdot))$ is measurable for all $x,\ y,\ z \in [\alpha,\beta]$.

(D3) $f(t,x,y,z) - p(t)x$ is continuous and nonincreasing in $x \in [\alpha,\beta]$ for all $t \in J \setminus Z$ and $y,\ z \in [\alpha(t),\beta(t)]$.

(D4) $f(t,y,\cdot,z)$ is nondecreasing and $f(t,y,z,\cdot)$ is nonincreasing for all $y,\ z \in \mathbb{R}$ and $t \in J \setminus Z$.

Because of the mixed monotonicity of $f$ with respect to its last two arguments the PBVP (2.3.10) need not to possess extremal solutions in $[\alpha,\beta]$. Thus we need a generalized concept of solutions to (2.3.10), where $f\colon J \times \mathbb{R}^3 \to \mathbb{R}$. The functions $y,\ z \in AC(J,\mathbb{R})$ is said to be *coupled quasisolutions* of (2.3.10) if

$$\begin{aligned} &y'(t) = f(t,y(t),y(t),z(t)) \text{ for a.a. } t \in J,\\ &z'(t) = f(t,z(t),z(t),y(t)) \text{ for a.a. } t \in J,\\ &y(0) = y(T), \quad \text{and} \quad z(0) = z(T). \end{aligned} \tag{2.3.11}$$

**Theorem 2.3.3:** *If there is a null set $Z$ in $J$ and $N, p \in L^2(J,\mathbb{R})$ such that the hypotheses (D0)–(D4) hold, then the PBVP (2.3.10) has coupled quasisolutions in the order interval $[\alpha,\beta]$ of $AC(J,\mathbb{R})$.*

*Proof.* Denote $M(t) = |p(t)| + 1$, $t \in J$. Given $y,\ z \in [\alpha,\beta]$, consider the PBVP

$$x' = F(t,x;y(t),z(t)), \qquad x(0) = x(T), \tag{2.3.12}$$

where

$$F(t,x;y(t),z(t)) = f(t,x,y(t),z(t)) + M(t)(y(t) - x)). \tag{2.3.13}$$

From (D2) and (D3) it follows that $(t,x) \mapsto F(t,x;y(t),z(t))$ satisfies the Carathéodory conditions in $\Omega = \{(t,x) \mid t \in J,\ x \in [\alpha(t),\beta(t)]\}$. Applying (D0), (D3), (D4) and (2.3.13) it is easy to see that

$$\alpha' \leq F(t,\alpha;y(t),z(t)) \quad \text{and} \quad \beta' \geq F(t,\beta;y(t),z(t)) \text{ on } J \setminus Z.$$

Condition (D1) implies that for all $x \in [\alpha(t),\beta(t)]$ and $t \in J \setminus Z$,

$$|F(t,x;y(t),z(t))| \leq K\,M(t) + N(t), \tag{a}$$

where $K = \max\beta - \min\alpha$. Hence, by theorem 1.5.2 the PBVP (2.3.12) has a solution $x$ in $[\alpha,\beta]$. In view of condition (D3) we see that the function $F(t,x;y(t),z(t)) + (M(t) - p(t))\,x$ is nonincreasing in $x \in [\alpha(t),\beta(t)]$ for all $t \in J \setminus Z$, whence $x$ is by corollary 1.5.3 the only solution of (2.3.12).

Define a map $A\colon [\alpha,\beta] \times [\alpha,\beta] \to [\alpha,\beta]$ by

$$A(y,z) = x, \quad y,\, z \in [\alpha,\beta], \tag{2.3.14}$$

where $x$ is the solution of (2.3.12) in $[\alpha,\beta]$. To prove that $A$ is mixed monotone, let $y_1,\, y_2,\, z_1,\, z_2 \in [\alpha,\beta]$, $y_1 \leq y_2$, $z_2 \leq z_1$ be given, and suppose that $x_1$ and $x_2$ are the corresponding solutions of

$$x_i' = F(t,x_i;y_i(t),z_i(t)), \qquad x_i(0) = x_i(T), \quad i = 1,\, 2,$$

in $[\alpha,\beta]$. Since $y_1 \leq y_2$ and $z_2 \leq z_1$, we have by (D4) and (2.3.13)

$$x_1' \leq F(t,x_1;y_2(t),z_2(t)) \text{ a.e. on } J, \qquad x_1(0) = x_1(T).$$

Thus the hypotheses of lemma 1.5.6 hold when $y = x_1$ and $z = x_2$, whence $x_1 \leq x_2$, thus proving that $A(y_1,z_1) \leq A(y_2,z_2)$.

From (2.3.12)–(2.3.14) and (a) it follows that

$$|(A(y,z)'(t)| \leq N(t) + K\,M(t)$$

for all $y$, $z \in [\alpha, \beta]$ and for a.a. $t \in J$.

In view of the above proof $A$ satisfies the hypotheses of proposition 1.4.6, whence $A$ has the extremal coupled fixed points $y$, $z$. From (2.3.11)–(2.3.14) it follows that $u$, $v \in [\alpha, \beta]$ are coupled quasisolutions of (2.3.10) if and only if they are coupled fixed points of $A$. Thus $y \leq u$, $v \leq z$ for all coupled quasisolutions $u$, $v$ of (2.3.10) in $[\alpha, \beta]$. □

The reasoning similar to that used in the proof of proposition 2.3.1 implies.

**Proposition 2.3.5:** *Given $f\colon J \times \mathbb{R}^3 \to \mathbb{R}$ and a null set $Z$ of $J$, assume there exist $N_1$, $N_2$, $p_1$, $p_2$, $p_3 \in L^2(J, \mathbb{R})$ with $\int_o^T (p_1(s) + p_2(s) + p_3(s))\,ds < 0$ such that*

$$\begin{aligned} &N_1(t) + p_1(t)x + p_2(t)y + p_3(t)z \leq f(t,x,y,z) \\ &\leq N_2(t) + p_1(t)x + p_2(t)y + p_3(t)z \end{aligned}$$

*for all $t \in J \setminus Z$ and $x$, $y$, $z \in \mathbb{R}$. If $\alpha$, $\beta$ are defined by (2.3.8) and (2.3.9), where $P(t) = \int_o^t (p_1(s) + p_2(s) + p_3(s))\,ds$, $t \in J$, and if conditions (D2)–(D4) hold, then the PBVP (2.3.10) has the extremal coupled quasisolutions, and all the coupled quasisolutions of (2.3.10) belong to the order interval $[\alpha, \beta]$.*

## 2.4. FINITE DIFFERENTIAL SYSTEMS

In this section we shall consider the existence of extremal solutions of the differential system

$$x_i' = f_i(t, x, x), \qquad x_i(0) = x_{oi}, \quad i = 1, \ldots, m, \tag{2.4.1}$$

on $J = [0, T]$, where $f = (f_1, \ldots, f_m) \colon J \times \mathbb{R}^{2m} \to \mathbb{R}^m$. We shall assume that $\mathbb{R}^m$ is normed and partially ordered by

$$\|x\| = \max\{|x_i| \mid i = 1, \ldots, m\}$$

and

$$x \le y \text{ if and only if } x_i \le y_i \text{ for each } i = 1, \ldots, m,$$

when $x = (x_1, \ldots, x_m)$, $y = (y_1, \ldots, y_m) \in \mathbb{R}^m$. Many of the results derived in section 2.1 for the scalar IVP will now be extended to the system (2.4.1). It is well-known that certain monotone properties are required when we deal with systems of inequalities, and therefore we shall now define the needed property. Given a subset $V$ of $\mathbb{R}^m$, we say that a function $g = (g_1, \ldots, g_m) \colon V \to \mathbb{R}^m$ is *quasimonotone nondecreasing* if for each $i = 1, \ldots, m$, $g_i(x) \le g_i(y)$ whenever $x$, $y \in V$, $x \le y$ and $x_i = y_i$, and *quasimonotone nonincreasing* if $g_i(x) \ge g_i(y)$ whenever $x$, $y \in V$, $x \le y$ and $x_i = y_i$. $g$ is called *quasimonotone* if it is quasimonotone nondecreasing or quasimonotone nonincreasing.

To avoid repetition we shall first introduce a list of conditions for the function $f$.

(A0) $\alpha$, $\beta \in AC(J, \mathbb{R}^m)$, $\alpha(t) \le \beta(t)$ on $J$, and $\alpha' \le f(t, \alpha, \alpha)$, $\beta' \ge f(t, \beta, \beta)$ on $J \setminus Z$.

(A1) There is $N \in L^1(J, \mathbb{R}_+)$ such that $\|f(t, x, y)\| \le N(t)$ for all $x$, $y \in [\alpha(t), \beta(t)]\}$ and $t \in J \setminus Z$.

(A2) $f(\cdot, x(\cdot), y(\cdot))$ is measurable for all $x$, $y \in AC(J, \mathbb{R}^m)$.

(A3) $f(t, \cdot, y)$ and $f(t, y, \cdot)$ are quasimonotone nondecreasing for all $t \in J \setminus Z$ and $y \in \mathbb{R}^m$.

(A4) For each $i = 1, \ldots, m$ there is $M_i \in L^1(J, \mathbb{R}_+)$ and a nondecreasing function $\varphi_i \in C(\mathbb{R}, \mathbb{R})$ such that $f_i(t, x, y) + M_i(t)\,\varphi_i(y_i)$ is continuous in $x_i$ and nondecreasing in $y_i$ for all $x = (x_1, \ldots, x_m)$, $y = (y_1, \ldots, y_m) \in \mathbb{R}^m$ and $t \in J \setminus Z$.

(A5) $\|f(t, x, y)\| \le p(t)\,h(\|x\|, \|y\|)$ for all $t \in J \setminus Z$ and $x, y \in \mathbb{R}^m$, where $p \in L^1(J, \mathbb{R}_+)$, $h\colon \mathbb{R}_+^2 \to (0, \infty)$ is nondecreasing in both of its arguments and $\int_o^\infty \frac{dv}{h(v,v)} = \infty$.

By assuming that $Z$ is in these conditions a null set in $J$, we shall prove that conditions (A0)–(A4) imply the existence of the extremal solutions of (2.4.1) in the order interval $[\alpha, \beta]$ of $AC(J, \mathbb{R}^m)$ whenever $x_o = (x_{o1}, \ldots, x_{om}) \in [\alpha(0), \beta(0)]$. If (A2)–(A5) hold, we shall show that the system (2.4.1) has for each $x_o = (x_{o1}, \ldots, x_{om}) \in \mathbb{R}^m$ the extremals among all its solutions, and that they are nondecreasing with respect to $x_o$ and $f$. In the special case when the components of $f$ are of the form $f_i(t, x, y) = q_i(x_i)\,g_i(t, y)$ we don't assume $f$ to be continuous in any of its arguments.

### 2.4.1. Existence of extremal solutions

The main result of this section is the following theorem which extends theorem 2.1.1 to differential systems.

**Theorem 2.4.1:** *Assume there is a null set $Z$ in $J$ so that (A0)–(A4) hold. Then the system (2.4.1) has the extremal solutions in the order interval $[\alpha, \beta]$ for each $x_o = (x_{o1}, \ldots, x_{om}) \in [\alpha(0), \beta(0)]$.*

*Proof.* Let $x_o = (x_{o1}, \ldots, x_{om}) \in [\alpha(0), \beta(0)]$, $y = (y_1, \ldots, y_m) \in [\alpha, \beta]$, and $i \in \{1, \ldots, m\}$ be fixed. The given hypotheses imply that the equation

$$\begin{aligned} F_i(t, x; y(t)) &= M_i(t)(\varphi_i(y_i(t)) - \varphi_i(x)) \\ &+ f_i(t, (y_1(t), \ldots, y_{i-1}(t), x, y_{i+1}(t), \ldots, y_m(t)), y(t)) \end{aligned} \tag{a}$$

defines a Carathéodory function $(t,x) \mapsto F_i(t,x;y(t))$ in the set $\Omega_i = \{(t,x) \in J \times I\!R \mid t \in J,\ x \in [\alpha_i(t), \beta_i(t)]\}$. From (A0), (A3) and (A4) it follows that

$$\begin{aligned}\alpha_i'(t) &\le f_i(t, \alpha(t), \alpha(t)) + M_i(t)\,(\varphi_i(\alpha_i(t)) - \varphi_i(\alpha_i(t)))\\ &\le f_i(t, (y_1(t), \dots, y_{i-1}(t), \alpha_i(t), y_{i+1}(t), \dots, y_m(t)), y(t))\\ &+ M_i(t)\,(\varphi_i(y_i(t)) - \varphi_i(\alpha_i(t))) = F_i(t, \alpha_i(t); y(t))\end{aligned}$$

for all $t \in J \setminus Z$. Similarly, it can be shown that

$$\beta_i'(t) \ge F_i(t, \beta_i(t); y(t)) \quad \text{on} \quad J \setminus Z.$$

Condition (A1) implies that

$$|F_i(t,x;y(t))| \le K_i\, M_i(t) + N(t) \tag{b}$$

for all $x \in [\alpha_i(t), \beta_i(t)]$ and $t \in J \setminus Z$, where $K_i = \varphi_i(\max \beta_i) - \varphi_i(\min \alpha_i)$. Hence, by theorem 1.5.1, each IVP of the system

$$x_i' = F_i(t, x_i; y(t)), \qquad x_i(0) = x_{oi}, \quad i = 1, \dots, m, \tag{2.4.2}$$

has for each $y \in [\alpha, \beta]$ the maximal solution $x_i$ in $[\alpha_i, \beta_i]$. Since the system (2.4.2) is uncoupled, then $x = (x_1, \dots, x_m)$ is its maximal solution in $[\alpha, \beta]$.

We now define a map $G = (G_1, \dots, G_m)\colon [\alpha, \beta] \to [\alpha, \beta]$ by

$$G_i y = x_i, \quad y \in [\alpha, \beta], \quad i = 1, \dots, m, \tag{2.4.3}$$

where $x = (x_1, \dots, x_m)$ is the maximal solution of (2.4.2) in $[\alpha, \beta]$. To prove that $G$ is nondecreasing, let $y$, $\bar{y} \in [\alpha, \beta]$, $\bar{y} \le y$ be given, and suppose that $x = (x_1, \dots, x_m)$ and $\bar{x} = (\bar{x}_1, \dots, \bar{x}_m)$ are the corresponding maximal solutions of (2.4.2) and

$$\bar{x}_i' = F_i(t, \bar{x}_i; \bar{y}(t)), \qquad \bar{x}_i(0) = x_{oi},\ i = 1, \dots, m,$$

in $[\alpha, \beta]$, respectively. Since $\bar{y} \leq y$, we have by (A3) and (A4) for each $i = 1, \ldots, m$

$$\bar{x}_i' \leq F_i(t, \bar{x}_i; y(t)) \quad \text{a.e. on } J,$$

and hence by lemma 1.5.1 it follows that $\bar{x}_i(t) \leq x_i(t)$ on $J$, thus proving that $G_i\bar{y} \leq G_i y$. This holds for each $i = i, \ldots, m$, whence $G\bar{y} \leq Gy$.

Because

$$(G_i y)'(t) = F_i(t, G_i y(t); y(t))$$

for all $y \in [\alpha, \beta]$ and for a.a. $t \in J$, $i = 1, \ldots, m$, it follows by (b) that

$$\|(Gy)'(t)\| \leq N(t) + K\,M(t) \quad \text{for all } y \in [\alpha, \beta] \text{ and for a.a. } t \in J,$$

where $K = \max\{K_1, \ldots, K_m\}$ and $M = \max\{M_1, \ldots, M_m\}$.

By the above proof $G$ satisfies the hypotheses of proposition 1.4.4, whence $G$ has the greatest fixed point $x^* = (x_1^*, \ldots, x_m^*)$, which by definitions of $F_i$ and $G_i$ is also a solution of the system (2.4.1) in $[\alpha, \beta]$.

If $x = (x_1, \ldots, x_m)$ is any solution of (2.4.1) in $[\alpha, \beta]$, then it satisfies also the system (2.4.2) with $y = x$. But $Gx$ is the maximal solution of (2.4.2) with $y = x$, whence $x \leq Gx$. This implies by (1.4.11) that $x \leq x^*$. Thus $x^* = (x_1^*, \ldots, x_m^*)$ is the maximal solution of the system (2.4.1) in $[\alpha, \beta]$.

The similar reasoning shows that the rule which assigns to each $y \in [\alpha, \beta]$ the minimal solution $x$ of the system (2.4.2) in $[\alpha, \beta]$, defines an operator $G\colon [\alpha, \beta] \to [\alpha, \beta]$ which satisfies the hypotheses of proposition 1.4.4. Thus $G$ has the least fixed point $x_*$, which is also a minimal solution of the system (2.4.1) in $[\alpha, \beta]$. □

The following consequence of theorem 2.4.1 is a generalization to corollary 2.1.1.

**Corollary 2.4.1:** *Let $f$ in theorem 2.4.1 be defined by*

$$f(t,x,y) = f^1(t,x) + f^2(t,y), \ t \in J, \ x, \ y \in \mathbb{R}^m,$$

*where $f^1(t,x)$ is a Carathéodory function and quasimonotone in $x$, and $f^2(t,y)$ is a standard function and nondecreasing in $y$. If (A0) holds, and if there is $N \in L^1(J, \mathbb{R})$ such that*

$$\sup\{\|f^j(t,y)\| \mid x, \ y \in [\alpha(t), \beta(t)]\} \leq N(t)$$

*for a.a. $t \in J$, $j = 1, 2$, then the conclusion of theorem 2.4.1 is valid.*

The results of theorem 2.4.1 and corollary 2.4.1 hold also when the conditions (A2)–(A4) are restricted to hold in the set $D = \{(t,x,y) \mid t \in J, \ x, \ y \in [\alpha(t), \beta(t)]\}$.

The next result is an extension of proposition 2.1.1 to systems, and the proof is similar.

**Proposition 2.4.1:** *Given $f \colon J \times \mathbb{R}^{2m} \to \mathbb{R}^m$ and a null set $Z$ of $J$, assume that*

$$\|f(t,x,y)\| \leq H(t, \|x\|, \|y\|) \ \textit{for all } x, \ y \in \mathbb{R}^m \ \textit{and} \ \ t \in J \setminus Z,$$

*where $H \colon J \times \mathbb{R}^2_+ \to \mathbb{R}_+$, $H(t,u,v)$ is nondecreasing in $u$ and in $v$, and the IVP*

$$w' = H(t,w,w), \qquad w(0) = \|x_o\|$$

*has an upper solution $w$ on $J$. Then (A0) and (A1) hold with $\alpha = (-w, \ldots, -w)$, $\beta = (w, \ldots, w)$ and $N = w'$.*

Theorem 2.1.2 can also be extended to the differential systems.

**Theorem 2.4.2:** *Given $f = (f_1, \ldots, f_m)\colon J \times \mathbb{R}^{2m} \to \mathbb{R}^m$ and a null set $Z$ of $J$, assume that $f$ satisfies conditions (A2)–(A5). Then for each fixed $x_o = (x_{o1}, \ldots, x_{om}) \in \mathbb{R}^m$ the system (2.4.1) has the extremal solutions, and all the solutions of (2.4.1) belong to the order interval $[\alpha, \beta]$, given by*

$$\alpha_i(t) = x_{oi} + \|x_o\| - w(t), \;\; \beta_i(t) = x_{oi} - \|x_o\| + w(t), \quad (2.4.4)$$

*where $w \in AC(J, \mathbb{R})$ is the solution of the IVP*

$$w' = p(t)\, h(w, w), \qquad w(0) = \|x_o\|. \quad (2.4.5)$$

*Proof.* Let $x_o = (x_{o1}, \ldots, x_{om}) \in \mathbb{R}^m$ be given. Since $[\alpha_i, \beta_i] \subseteq [-w, w]$ for each $i = 1, \ldots, m$, it follows from (A5) that for all $x$, $y \in [\alpha, \beta]$, and for a.a. $t \in J$,

$$|f_i(t, x(t), y(t))| \leq p(t)\, h(w(t), w(t)) = w'(t). \quad \text{(a)}$$

This ensures that the hypotheses (A0) and (A1) of theorem 2.4.1 are valid for $\alpha$ and $\beta$ given by (2.4.4) and for $N_i = w'$. Since $f$ satisfies also the hypotheses (A2)–(A4), then system (2.4.1) has by theorem 2.4.1 the extremal solutions $x_*$ and $x^*$ in $[\alpha, \beta]$.

If $x = (x_1, \ldots, x_m)$ is a solution of (2.4.1), then it satisfies also the integral equations

$$x_i(t) = x_{oi} + \int_o^t f_i(s, x(s), x(s))ds, \quad t \in J,\ i = 1, \ldots, m. \quad \text{(b)}$$

The equation

$$u(t) = \|x_o\| + \int_o^t \|f(s, x(s), x(s))\|\, ds, \quad t \in J, \quad \text{(c)}$$

defines a function $u \in AC(J, \mathbb{R}_+)$. Because (b) and (c) imply that $\|x(t)\| \leq u(t)$ on $J$, it follows from (A5) that

$$u'(t) = \|f(t, x(t), x(t))\| \leq p(t)\, h(\|x(t)\|, \|x(t)\|)$$
$$\leq p(t)\, h(u(t), u(t)) \text{ for a.a. } t \in J.$$

Thus $u$ is a lower solution of the IVP (2.4.5). Since $w$ is the solution of (2.4.5), it follows from lemma 1.5.3 that $u(t) \leq w(t)$ on $J$. This implies by (a), (b) and (c) that

$$|x_i(t) - x_{oi}| \leq u(t) - \|x_o\| \leq w(t) - \|x_o\| \tag{d}$$

for each $i = 1, \ldots, m$. From (d) and (2.4.4) we see that $x \in [\alpha, \beta]$. Since $x_*$ and $x^*$ are the extremal solutions of (2.4.1) in $[\alpha, \beta]$, then $x \in [x_*, x^*]$. But $x$ was an arbitrary solution of (2.4.1), whence $x_*$ and $x^*$ are the least and the greatest of all the solutions of the IVP (2.4.1), and all the solutions of (2.4.1) in $[\alpha, \beta]$. □

By theorem 1.4.4 and corollary 1.4.5 the assumption (A2) holds if $f$ is a standard function or a Borel function. In the case when the values of $f$ are in $\mathbb{R}^m_+$ and $f(t, x, y)$ is nondecreasing in $y$, it is sufficient that (A1) is valid for nondecreasing functions $x$, $y$ of the order interval $[x_o, \beta]$. In such a case theorems 1.4.1 and 1.4.2 imply that the so restricted condition (A1) hold if one of the following hypotheses is valid for all $x = (x_1, \ldots, x_m)$ and $y = (y_1, \ldots, y_m) \in \mathbb{R}^m$.

(a) $f(\cdot, x, y)$ is right continuous or left continuous in $J \setminus Z$.

(b) $f(t + h, x_1 + k_1, \ldots, x_m + k_m, y_1 + l_1, \ldots, y_m + l_m)$ has a limit when $h, k_i, l_i \to 0+$, for all $t \in [0, T)$

(c) $f(t + h, x_1 + k_1, \ldots, x_m + k_m, y_1 + l_1, \ldots, y_m + l_m)$ has a limit when $h, k_i, l_i \to 0-$, for all $t \in (0, T]$

Moreover, if either of conditions (b) or (c) holds, then theorem 1.4.2 ensures that each solution of the system (2.4.1) satisfy the differential equations of that system in the complement of a countable set.

**Example 2.4.1:** Let $g = (g_1, \ldots, g_m)\colon J \times I\!R^m \to I\!R^m$ be a bounded Carathéodory function, and assume that $g(t, \cdot)$ is quasi-monotone nondecreasing for a.a. $t \in J$. Let $C_i$, $i = 1, \ldots, m$, be nonempty, bounded and well-ordered sets in $I\!R_+$, for instance, $C_i = \{i + c(n_o, \ldots, n_k) \mid (n_o, \ldots, n_k) \in D\}$, where $c(n_o, \ldots, n_k)$ is defined by (1.1.2). Denote for each $j = 1, \ldots, m$

$$z_j(v) = \begin{cases} v & \text{if } v \le \min C_j \text{ or } v \ge \sup C_j, \\ \min\{z \in C_j \mid v < z\} & \text{if } \min C_j < v < \sup C_j. \end{cases}$$

Define functions $f_i\colon J \times I\!R^{2m} \to I\!R$, $i = 1, \ldots, m$, by

$$f_i(t, x, (y_1, \ldots, y_m)) = g_i(t, x) \sum_{j=1}^{i} z_j(y_j).$$

It is easy to see that $f = (f_1, \ldots, f_m)$ satisfies conditions (A2)–(A5), whence the system (2.4.1) has for each $x_o = (x_{o1}, \ldots, x_{om})$ the extremal solutions.

## 2.4.2. Systems with no continuous variables

Consider now the system

$$x_i' = q_i(x_i)\, g_i(t, x), \qquad x_i(0) = x_{oi}, \quad i = 1, \ldots, m. \tag{2.4.6}$$

As a generalization to theorem 2.1.3 we shall prove.

**Theorem 2.4.3:** *Given $q_i$, $\frac{1}{q_i} \in L^\infty_{loc}(I\!R, I\!R_+)$, and a standard function $g = (g_1, \ldots, g_m)\colon J \times I\!R^m \to I\!R^m$, assume that $g(t, \cdot)$ is nondecreasing for a.a. $t \in J$. If the system (2.4.6) has a lower solution $\alpha$ and an upper solution $\beta$ such that $\alpha \le \beta$, then (2.4.6) has the extremal solutions in the order interval $[\alpha, \beta]$.*

*Proof.* The given hypotheses imply that the equations

$$\int_{x_{oi}}^{G_i x(t)} \frac{dv}{q_i(v)} = \int_o^t g_i(s, x(s))ds, \quad t \in J, \quad i = 1, \dots, m \tag{a}$$

define a nondecreasing mapping $G = (G_1, \dots, G_m) \colon [\alpha, \beta] \to [\alpha, \beta]$. Denote $M_i = \text{ess sup}\,\{q_i(v) \mid \min G_i\alpha \leq v \leq \max G_i\beta\}$. If $x \in [\alpha, \beta]$ and $0 \leq \bar{t} \leq t \leq T$, it can be shown as in the proof of theorem 2.1.3 that

$$|G_i x(t) - G_i x(\bar{t})| \leq M_i \int_{\bar{t}}^t (|g_i(s, \alpha(s))| + |g_i(s, \beta(s))|)\, ds, \tag{b}$$

whenever $0 \leq \bar{t} \leq t \leq T$ and $1 \leq i \leq m$. Thus $G$ satisfies the hypotheses of theorem 1.4.7, whence $G$ has the least fixed point $x_*$ and the greatest fixed point $x^*$.

From (2.4.6) it follows by theorem 1.4.5 that $x = (x_1, \dots, x_m) \in [\alpha, \beta]$ is a solution of (2.4.6) if and only if

$$\int_{x_{oi}}^{x_i(t)} \frac{dv}{q_i(v)} = \int_o^t g_i(s, x(s))ds, \quad t \in J, \ i = 1, \dots, m.$$

By (a) this is equivalent to the fact that $x = (x_1, \dots, x_m)$ is a fixed point of $G = (G_1, \dots, G_m)$. Since $x_*$ and $x^*$ are the extremal fixed points of $G$, then they are also the least and the greatest of all the solutions of (2.4.6) in $[\alpha, \beta]$. □

### 2.4.3. Dependence on the data and special cases

Consider first the dependence of the solutions of the system (2.4.1) on the initial value $x_o$ and on the function $f$.

As a straightforward generalization of lemma 2.1.1 we have

**Lemma 2.4.1:** *Let $f: J \times \mathbb{R}^{2m} \to \mathbb{R}^m$ satisfy the hypotheses (A2)–(A5), let $x_o = (x_{o1} \ldots, x_{om}) \in \mathbb{R}^m$ be given, and let $x_*$ and $x^*$ be the minimal and the maximal solutions of the system (2.4.1) on $J$.*

a) *If $x$ is a lower solution of (2.4.1), then $x \le x^*$.*

b) *If $x$ is an upper solution of (2.4.1), then $x_* \le x$.*

As a direct application of lemma 2.4.1 we obtain

**Proposition 2.4.2:** *Let $f, \hat{f}: J \times \mathbb{R}^{2m} \to \mathbb{R}^m$ satisfy the hypotheses (A2)–(A5). Given $x_o, \hat{x}_o \in \mathbb{R}^m$, let $x_*$ be the minimal solution of the system (2.4.1), and let $\hat{x}^*$ be the maximal solution of the system*

$$x_i' = \hat{f}_i(t, x, x), \qquad x_i(0) = \hat{x}_{oi}, \quad i = 1, \ldots, m. \tag{2.4.7}$$

*If $x_o \le \hat{x}_o$ and $f(t,x,y) \le \hat{f}(t,x,y)$ for a.a. $t \in J$ and for all $x, y \in \mathbb{R}^m$, then $x_* \le x \le \hat{x}^*$ for any solution $x$ of (2.4.1) or (2.4.7).*

In particular, we have

**Corollary 2.4.2:** *If conditions (A2)–(A5) hold, then the extremal solutions of the system (2.4.1) exist and are nondecreasing with respect to $x_{oi}$ and $f_i$, $i = 1, \ldots, m$.*

As a consequence of theorem 2.4.3 we obtain

**Proposition 2.4.3:** *Assume that $q_i, \frac{1}{q_i} \in L^\infty(\mathbb{R}, \mathbb{R}_+)$, $i = 1, \ldots, m$, that $g: J \times \mathbb{R}^m \to \mathbb{R}^m$ is a standard function and $g(t, \cdot)$ is nondecreasing for a.a. $t \in J$, and that there exist $p \in L^1(J, \mathbb{R}_+)$ and a nondecreasing function $\psi: \mathbb{R}_+ \to (0, \infty)$ with $\int_o^\infty \frac{dv}{\psi(v)} = \infty$ such that*

$$|g_i(t,x)| \le p(t)\psi(\|x\|)$$

*for all $x \in \mathbb{R}^m$ and for a.a. $t \in J$, $i = 1, \ldots, m$. Then the system (2.4.6) has for each $x_o = (x_{o1}, \ldots, x_{om})$ the extremal solutions, which are nondecreasing with respect to $x_{oi}$ and $g_i$, $i = 1, \ldots, m$.*

*Proof.* Let $x_o \in \mathbb{R}^m$ be given. Denoting $M = \max\{$ ess sup $q_i \mid 1 \le i \le m\}$ it is easy to see that the function

$$f(t, x, y) = (q_1(x_1)g_1(t, y), \ldots, q_m(x_m)g_m(t, y))$$

satisfies condition (A5) with $h(u, v) = M\,\psi(v)$. From the proof of theorem 2.4.2 it follows an existence of $\alpha, \beta \in AC(J, \mathbb{R}^m)$ such that (A0) holds, and that all the solutions of (2.4.6) belong to $[\alpha, \beta]$. Theorem 2.4.3 implies that the system (2.4.6) has the extremal solutions in $[\alpha, \beta]$, whence they are extremals among all the solutions of (2.4.6).

To prove that the extremal solutions of (2.4.6) are nondecreasing with respect to $x_{oi}$ and $g_i$, assume that the functions $\hat{g}_i \colon J \times \mathbb{R}^m \to \mathbb{R}$, $i = 1, \ldots, m$, satisfy the hypotheses given for $g_i$ above. Assume also that

$$x_{oi} \le \hat{x}_{oi}, \quad \text{and } g_i(t, y) \le \hat{g}_i(t, y)$$

for all $y \in \mathbb{R}^m$ and for a.a. $t \in J$, $i = 1, \ldots, m$. Denoting by $\hat{x}^*$ the maximal solutions of the system

$$x_i' = q_i(x_i)\,\hat{g}_i(t, x), \qquad x_i(0) = \hat{x}_{oi}, \quad i = 1, \ldots, m, \tag{b}$$

and noticing that $x^*$ is a lower solution of (b), it follows by the similar reasoning as used in the proof of lemma 2.1.1 that $x^* \le \hat{x}^*$. Similarly, since the minimal solution $\hat{x}_*$ is an upper solution of (2.4.6), then $x_* \le \hat{x}_*$, which concludes the proof. □

In the study of higher order differential equations we shall need the following consequence of theorem 2.4.2 and proposition 2.4.2.

**Proposition 2.4.4:** *Given $a_{ij} \in L^1(J, I\!R)$, $i, j = 1, \dots, m$, with $a_{ij}(t) \geq 0$ a.e. on $J$ when $i \neq j$, and a function $f\colon J \times I\!R^{2m} \to I\!R^m$ which satisfies conditions (A2)-(A5), then the system*

$$x_i' = \sum_{j=1}^{m} a_{ij}(t)x_j + f_i(t, x, x), \quad x_i(0) = x_{oi}, \ 1 \leq i \leq m \quad (2.4.8)$$

*has for each $x_o = (x_{o1}, \dots, x_{om}) \in I\!R^m$ extremal solutions, which are nondecreasing with respect to $x_{oi}$ and $f_i$, $i = 1, \dots, m$.*

*Proof.* Define the function $g\colon J \times I\!R^{2m} \to I\!R^m$ by

$$\bar{f}(t, x, y) = A(t)x + f(t, x, y), \quad t \in J, \ x, \ y \in I\!R^m,$$

where $A(t) = (a_{ij}(t))_{i,j=1}^{m}$, $t \in J$. It is easy to see that $\bar{f}$ satisfies the hypotheses (A2)-(A4) given for $f$, and that

$$\|\bar{f}(t, x, y)\| \leq \bar{p}(t)\,\bar{h}(\|x\|, \|y\|) \ \text{ for all } \ x, \ y \in I\!R^m, \ t \in J \setminus Z,$$

where

$$\bar{p}(t) = p(t) + \max\{|a_{ij}(t)| \mid 1 \leq i, j \leq m\}, \ \bar{h}(u, v) = h(u, v) + u.$$

Obviously, $\bar{p} \in L^1(J, I\!R_+)$, and $\bar{h}(u, v)$ is nondecreasing in $u$ and in $v$. Because the integral $\int_o^\infty \frac{dv}{h(v,v)}$ diverges, it is elementary to verify that $\int_o^\infty \frac{dv}{h(v,v)+v}$ diverges, too. Thus $\bar{f}$ satisfies also condition (A5), whence the system (2.4.8) has by theorem 2.4.2 the extremal solutions, which are by proposition 2.4.2 nondecreasing in $x_o$ and in $f$. □

By similar reasoning it follows from proposition 2.4.3.

**Proposition 2.4.5:** *Let the hypotheses of proposition 2.4.3 hold, let $q_i$ : s be continuous, and assume that $a_{ij} \in L^1(J, \mathbb{R})$, $i, j = 1, \ldots, m$, with $a_{ij}(t) \geq 0$ a.e. on $J$ when $i \neq j$. Then the system*

$$x_i' = \sum_{j=1}^{m} a_{ij}(t)x_j + q_i(x_i)g_i(t,x), \quad x_i(0) = x_{oi}, \quad 1 \leq i \leq m \quad (2.4.9)$$

*has for each $x_o = (x_{o1}, \ldots, x_{om})$ the extremal solutions, which are nondecreasing with respect to $x_{oi}$ and $g_i$, $i = 1, \ldots, m$.*

As a consequence of proposition 2.4.5 we obtain

**Corollary 2.4.3:** *Assume that $q_i$, $\frac{1}{q_i} : \mathbb{R} \to \mathbb{R}_+$ are bounded and continuous and $p_i \in L^1(J, \mathbb{R}_+)$ for each $i = 1, \ldots, m$, and that $g = (g_1, \ldots, g_m) \colon \mathbb{R}^m \to \mathbb{R}_+^m$ is nondecreasing and $\int_o^\infty \frac{dv}{\psi(v)} = \infty$, where $\psi(v) = 1 + \max\{\|g(x)\| \mid \|x\| \leq v\}$. If $a_{ij} \in L^1(J, \mathbb{R})$, $i, j = 1, \ldots, m$, with $a_{ij}(t) \geq 0$ a.e. on $J$ when $i \neq j$, then the system*

$$x_i' = \sum_{j=1}^{m} a_{ij}(t)x_j + p_i(t)q_i(x_i)g_i(x), \quad x_i(0) = x_{oi}, \quad i = 1, \ldots, m$$

*has for each $x_o = (x_{o1}, \ldots, x_{om})$ the extremal solutions, which are nondecreasing with respect to $x_{oi}$ and $g_i$, $i = 1, \ldots, m$.*

From theorem 2.4.3 it follows

**Proposition 2.4.6:** *Assume that $q_i$ and $g_i$ satisfy the hypotheses of theorem 2.4.3, and that $g \colon J \times \mathbb{R}^m \to \mathbb{R}^m$ is a Carathéodory function. Then the sequence $(y_n)_{n=o}^\infty$ of functions $y_n \colon J \to \mathbb{R}$, defined by*

$$\int_{x_{oi}}^{(y_{n+1})_i(t)} \frac{dv}{q_i(v)} = \int_o^t g_i(s, y_n(s))ds, \quad t \in J, \ n \in \mathbb{N}, \quad (2.4.10)$$

*converges uniformly on $J$ to the minimal (resp. the maximal) solution of the system (2.4.6) in the order interval $[\alpha, \beta]$ if $y_o = \alpha$ (resp. $y_o = \beta$).*

*Proof.* The given hypotheses imply that the equation

$$\int_{x_{oi}}^{G_i y(t)} \frac{dv}{q_i(v)} = \int_o^t g_i(s, y(s))ds, \quad t \in J, \ i = 1, \ldots, m, \qquad \text{(a)}$$

defines a nondecreasing operator $G = (G_1, \ldots, G_m)\colon [\alpha, \beta] \to [\alpha, \beta]$. By choosing $y_o = \alpha$, it follows from (2.4.10) and (a) that $(y_n)_{n=o}^{\infty} = (G^n \alpha)_{n=o}^{\infty}$. Thus the sequence $(y_n)_{n=o}^{\infty}$ is nondecreasing and belongs to $[\alpha, \beta]$. Since $G$ satisfies the hypotheses of theorem 1.4.7, then the proof of theorem implies also that $(y_n)_{n=o}^{\infty}$ converges uniformly on $J$ to a function $x_* \in [\alpha, \beta]$. Since $g$ is a Carathéodory function, it follows from (2.4.10) as $n \to \infty$ that for each $i = 1, \ldots, m$ the $i$:th component $x_{*i}$ of $x_*$ satisfies the integral equation

$$\int_{x_{oi}}^{x_{*i}(t)} \frac{dv}{q_i(v)} = \int_o^t g_i(s, x_*(s))ds, \quad t \in J. \qquad \text{(b)}$$

Thus $x_*$ is a fixed point of $G$, and hence, by the proof of theorem 2.4.3, a solution of (2.4.6).

If $x = (x_1, \ldots, x_m) \in [\alpha, \beta]$ is a solution of the system (2.4.6), it is also a fixed point of $G$. Since $y_o = \alpha \le x$ and $G$ is nondecreasing, it follows by induction that $y_n = G^n \alpha \le x$ for each $n \in \mathbb{N}$. Thus $x_* = \lim_n y_n \le x$, whence $x_* = (x_{*1}, \ldots x_{*m})$ is the least solution of the system (2.4.6) in $[\alpha, \beta]$.

The proof that $(y_n)_{n=o}^{\infty}$ converges to the greatest solution of (2.4.6) in $[\alpha, \beta]$ when $v_o = \beta$ is similar to the above one. □

In the case when $q_i(v) = 1$ a.e. on $\mathbb{R}$ in (2.4.10), it can be rewritten as

$$y_{n+1}(t) = y_o + \int_{t_o}^t g(s, y_n(s))ds, \quad t \in J, \ n \in \mathbb{N}. \qquad (2.4.11)$$

Thus we obtain

**Corollary 2.4.4:** *Assume that* $q_i \colon I\!R \to (0,\infty)$, *that* $q_i(v) = 1$ *a.e. on* $I\!R$ *for each* $i = 1,\ldots,m$, *that* $g\colon J \times I\!R^m \to I\!R^m$ *is a Carathéodory function, and that* $g(t,\cdot)$ *is nondecreasing on* $I\!R^m$ *for a.a.* $t \in J$. *If the system*

$$x_i' = g_i(t,x), \qquad x_i(0) = x_{oi}, \quad i = 1,\ldots,m \tag{2.4.12}$$

*has a lower solution* $\alpha$ *and an upper solution* $\beta$ *on* $J$ *such that* $\alpha \leq \beta$, *then the sequence* $(y_n)_{n=o}^{\infty}$ *of the successive approximations* $y_n \colon J \to I\!R$, *defined by (2.4.11) converges uniformly on* $J$ *to the minimal (resp. the maximal) solution of the system (2.4.6) in the order interval* $[\alpha,\beta]$ *if* $y_o = \alpha$ *(resp.* $y_o = \beta$*).*

**Remark 2.4.1:** The quasimonotonicity of $f(t,x,y)$ in $x$ in proposition 2.4.4, the monotonicity of $g$ in proposition 2.4.5 and in corollary 2.4.3, as well as the nonpositivity of $a_{ij}$:s for $i \neq j$ cannot be dropped from the hypotheses, as we see from

**Example 2.4.2:** Given $a > 0$, consider the differential system

$$\begin{aligned} x_1' = x_1 - a\,x_2, &\quad x_1(0) = 0, \\ x_2' = b, &\quad x_2(0) = c. \end{aligned} \tag{a}$$

The unique solution of (a) is $x(t) = (ab\,t + (ab+ac)(1-e^t), b\,t + c)$. Thus the first coordinate of the solution $x$ of (a) decreases from 0 when $b$ or $c$ increases from 0.

### 2.4.4. Applications to higher order IVP's

We shall now apply the results of previous subsections by proving existence and comparison results for extremal solutions of higher order IVP's.

Consider first the following $m$:th order IVP

$$\begin{aligned} &y^{(m)} + a_m(t)y^{(m-1)} + \cdots + a_1(t)y = g(t, \underline{y}, \underline{y}), \\ &y(0) = x_{o1},\ y'(0) = x_{o2},\ \ldots,\ y^{(m-1)} = x_{om}, \end{aligned} \tag{2.4.13}$$

where $\underline{y} = (y, y', \ldots, y^{(m-1)})$. By a *solution* of (2.4.13) we mean a function $y\colon J \to \mathbb{R}$ with $y^{(m-1)} \in AC(J, \mathbb{R})$, which satisfies the differential equation of (2.4.13) a.e. on $J$, and all the initial conditions of (2.4.13). If $y\colon J \to \mathbb{R}$ so that $y^{(m-1)} \in AC(J, \mathbb{R})$, then $\underline{y} \in AC(J, \mathbb{R}^m)$. We say that a solution $y_*$ of (2.4.13) is its *minimal solution* if $\underline{y}_* \leq \underline{y}$ for each solution $y$ of (2.4.13), and the *maximal solution* if the reversed inequality holds.

As an application of proposition 2.4.4 we obtain

**Theorem 2.4.4:** *Given $g\colon J \times \mathbb{R}^{2m} \to \mathbb{R}$ and a null set $Z$ of $J$, assume that*

(g1) *$g(\cdot, x, y(\cdot))$ is measurable for $x \in \mathbb{R}^m$, $y \in AC(J, \mathbb{R}^m)$.*

(g2) *$g(t, (x_1, \ldots, x_m), y)$ is continuous in $x_m$ and nondecreasing in $x_i$, $i < m$ for all $y \in \mathbb{R}^m$ and $t \in J \setminus Z$.*

(g3) *There is $M \in L^1(J, \mathbb{R}_+)$ and a nondecreasing function $\varphi \in C(\mathbb{R}, \mathbb{R})$ such that $y \mapsto g(t, x, y) + M(t)\,\varphi(y_m)$ is nondecreasing for all $x \in \mathbb{R}^m$ and $t \in J \setminus Z$.*

(g4) *$|g(t, x, y)| \leq p(t)\,h(\|x\|, \|y\|)$ for all $t \in J \setminus Z$ and $x$, $y \in \mathbb{R}^m$, where $p \in L^1(J, \mathbb{R}_+)$, $h(u, v)$ is nondecreasing in $u$, $v$, $h(0,0) > 0$ and $\int_o^\infty \frac{dv}{h(v,v)} = \infty$.*

*If $a_j \in L^1(J, \mathbb{R})$, $j = 1, \ldots, m$, and $a_j(t) \leq 0$ for a.a. $t \in J$ when $j < m$, then for each $x_o = (x_{o1}, \ldots, x_{om}) \in \mathbb{R}^m$ the IVP (2.4.13) has the extremal solutions, which are nondecreasing up to their $m-1$:th derivatives with respect to $g$ and $x_{oi}$, $i = 1, \ldots, m$.*

*Proof.* let $x_{oi} \in \mathbb{R}$, $i = 1, \ldots, m$, be given. Denoting

$$x = (x_1, \ldots, x_m) = \underline{y} = (y, y', \ldots, y^{(m-1)}),$$

it is easy to see that $y$ is a solution of the IVP (2.4.13) if and only if $x$ is a solution of the system

$$x_i' = \sum_{j=1}^{m} a_{ij}(t)x_j + f_i(t,x,x), \quad x_i(0) = x_{oi},\ 1 \le i \le m, \tag{2.4.8}$$

where $f_i \equiv 0$ for $1 \le i \le m-1$ and $f_m = g$, and

$$a_{ij} \equiv \begin{cases} 0, & j \ne i+1 \\ 1, & j = i+1 \end{cases} \quad \text{when } i = 1, \dots, m-1 \text{ and } a_{mj} = -a_j.$$

These choices and the hypotheses given for $g$ and $a_j$, $j = 1, \dots, m$, imply that all the hypotheses of proposition 2.4.4 are satisfied. Thus the system (2.4.8) has the extremal solutions $x_*$ and $x^*$, and they are nondecreasing with respect to $f_m$ and $x_{oi}$, $i = 1, \dots, m$. From the above discussion it then follows that $x_* = \underline{y}_* = (y_*, y_*', \dots, y_*^{(m-1)})$, where $y_*$ is the minimal solution of the IVP (2.4.13), and $\underline{y}_*$ is nondecreasing with respect to to the initial values $x_{oi}$ and $g$. Similarly, the first component of $x^* = \underline{y}^*$ is the maximal solution of (2.4.13) and $\underline{y}^*$ is nondecreasing with respect to the initial values $x_{oi}$ and $g$. □

By the similar reasoning it follows from proposition 2.4.5.

**Proposition 2.4.7:** *Given $q$, $\frac{1}{q} \in L^\infty(\mathbb{R}, \mathbb{R}_+)$, $q$ continuous, and $g: J \times \mathbb{R}^m \to \mathbb{R}$, assume that $g(t, y(t))$ is measurable in $t$ for each $y \in AC(J, \mathbb{R}^m)$, that $g(t,y)$ is nondecreasing in $y$ for a.a. $t \in J$, and that there exist $p \in L^1(J, \mathbb{R}_+)$ and a nondecreasing function $\psi: \mathbb{R}_+ \to (0, \infty)$ with and $\int_o^\infty \frac{dv}{\psi(v)} = \infty$ such that*

$$|g(t,x)| \le p(t)\psi(\|x\|) \text{ for all } x \in \mathbb{R}^m \text{ and for a.a. } t \in J.$$

*If $a_j \in L^1(J, \mathbb{R})$, $j = 1, \dots, m$, and $a_j(t) \le 0$ for a.a. $t \in J$ when $j < m$, then for each $x_o = (x_{o1}, \dots, x_{om}) \in \mathbb{R}^m$ the IVP*

$$\begin{aligned} &y^{(m)} + a_m(t)y^{(m-1)} + \cdots + a_1(t)y = q(y^{(m-1)})g(t, \underline{y}), \\ &y(0) = x_{o1},\ y'(0) = x_{o2},\ \dots,\ y^{(m-1)} = x_{om} \end{aligned} \tag{2.4.14}$$

*has the extremal solutions, which are nondecreasing up to their $m-1$:th derivatives with respect to $g$ and $x_{oi}$, $i = 1, \ldots, m$.*

When $a_{ij} \equiv 0$ in (2.4.13) we have the $m$:th order IVP

$$\begin{aligned} &y^{(m)} = g(t, (y, y', \ldots, y^{(m-1)}), (y, y', \ldots, y^{(m-1)})), \\ &y(0) = x_{o1},\ y'(0) = x_{o2},\ \ldots,\ y^{(m-1)} = x_{om}. \end{aligned} \tag{2.4.15}$$

Given $\beta \colon J \to I\!R$ such that $\beta^{(m-1)} \in AC(J, I\!R)$, we say that $\beta$ is an *upper solution* of the IVP (2.4.15) if

$$\begin{aligned} &\beta^{(m)} \geq g(t, (\beta, \beta', \ldots, \beta^{(m-1)}), (\beta, \beta', \ldots, \beta^{(m-1)})), \\ &\beta(0) \geq x_{o1},\ \beta'(0) \geq x_{o2},\ \ldots,\ \beta^{(m-1)} \geq x_{om}. \end{aligned}$$

If the reversed inequalities hold, we say that $\beta$ is a *lower solution* of (2.4.15).

As a consequence of theorem 2.4.1 we get by the reasoning used in the proof of theorem 2.4.4.

**Theorem 2.4.5:** *Assume there is a null set $Z$ in $J$ so that*

(ga) *(2.4.15) has a lower solution $\alpha$ and an upper solution $\beta$ so that $\underline{\alpha} \leq \underline{\beta}$.*

(gb) *$g(\cdot, x(\cdot), y(\cdot))$ is measurable for all $x$, $y \in [\underline{\alpha}, \underline{\beta}]$.*

(gc) *$g(t, (x_1, \ldots, x_m), y)$ is continuous in $x_m$ and nondecreasing in $x_i$, $i < m$, for all $y \in [\underline{\alpha}(t), \underline{\beta}(t)]$ and $t \in J \setminus Z$.*

(gd) *There is $M \in L^1(J, I\!R_+)$ and a nondecreasing function $\varphi \in C(I\!R, I\!R)$ such that the function $y \mapsto g(t, x, y) + M(t)\,\varphi(y_m)$ is nondecreasing in $[\underline{\alpha}(t), \underline{\beta}(t)]$ for all $x \in [\underline{\alpha}(t), \underline{\beta}(t)]$ and $t \in J \setminus Z$.*

(ge) *There is $N \in L^1(J, I\!R_+)$ such that $|g(t, x, y)| \leq N(t)$ for all $x, y \in [\underline{\alpha}(t), \underline{\beta}(t)]$ and $t \in J \setminus Z$.*

*Then the IVP (2.4.15) has the extremal solutions in $[\underline{\alpha}, \underline{\beta}]$.*

## 2.5. MIXED MONOTONE SYSTEMS

In this section we shall show that many of the results derived in section 2.2 for mixed monotone IVPs can be extended to corresponding mixed monotone systems

$$x_i' = f_i(t, x, x, x), \qquad x_i(0) = x_{oi}, \quad i = 1, \dots, m, \tag{2.5.1}$$

where $f = (f_1, \dots, f_m): J \times I\!R^{3m} \to I\!R^m$.

The functions $y = (y_1, \dots, y_m)$, $z = (z_1, \dots, z_m) \in AC(J, I\!R^m)$ are called *coupled quasisolutions* of (2.5.1) if for each $i = 1, \dots, m$

$$\begin{aligned} y_i'(t) &= f_i(t, y(t), y(t), z(t)) \text{ for a.a. } t \in J, \ \ y_i(0) = x_{oi}, \\ z_i'(t) &= f_i(t, z(t), z(t), y(t)) \text{ for a.a. } t \in J, \ \ z_i(0) = x_{oi}. \end{aligned} \tag{2.5.2}$$

We shall first consider an existence of coupled quasisolutions of the system (2.5.1) in the case when there is a null set $Z$ in $J$ such that $f: J \times I\!R^{3m} \to I\!R^m$ satisfies some of the following conditions.

(B0) $\alpha, \beta \in AC(J, I\!R^m)$, $\alpha \le \beta$ and $\alpha' \le f(t, \alpha, \alpha, \beta)$, $\beta' \ge f(t, \beta, \beta, \alpha)$ on $J \setminus Z$.

(B1) There is $N \in L^1(J, I\!R_+)$ such that $\|f(t, x, y, z)\| \le N(t)$ for all $x$, $y$, $z \in [\alpha(t), \beta(t)]$ and $t \in J \setminus Z$.

(B2) $f(\cdot, x(\cdot), y(\cdot), z(\cdot))$ is measurable for all $x$, $y$, $z \in AC(J, I\!R^m)$.

(B3) for each $i = 1, \dots, m$ the function $f_i(t, x, y, z)$ is continuous with respect to the $i$:th coordinate of $x$ for all $x$, $y$, $z \in I\!R^m$ and $t \in J \setminus Z$.

(B4) $f(t, \cdot, y, z)$ is quasimonotone nondecreasing, $f(t, y, \cdot, z)$ is nondecreasing and $f(t, y, z, \cdot)$ is nonincreasing for all $y$, $z \in I\!R^m$ and $t \in J \setminus Z$.

(B5) $f_i(t, (x_1, \dots, x_i + k, \dots, x_m), y, z) - f_i(t, x, y, z) \le g_i(t, k)$ for all $i = 1, \dots, m$ and for all $x = (x_1, \dots, x_m)$, $y$, $z \in I\!R^m$, $k \ge 0$, and $t \in J \setminus Z$, where $g_i: J \times I\!R_+ \to I\!R_+$

and $u(t) \equiv 0$ is the only lower solution of the IVP $u' = g_i(t,u), \ u(0) = 0$.

(B6) $\|f(t,x,y,z)\| \le p(t)\, h(\|x\|,\|y\|,\|z\|)$ for all $t \in J \setminus Z$ and $x, y, z \in I\!R^m$, where $p \in L^1(J, I\!R_+)$, $h\colon I\!R^3_+ \to (0,\infty)$ is nondecreasing in all its arguments and $\int_o^\infty \frac{dv}{h(v,v,v)} = \infty$.

(B7) $\|f(t,x,y,z)\| \le H(t,\|x\|,\|y\|,\|z\|)$ for all $x, y\ z \in I\!R^m$ and $t \in J \setminus Z$, where $H\colon J \times I\!R^3_+ \to I\!R_+$, $H(t,u,v,w)$ is nondecreasing in $u$, $v$ and $w$ the IVP
$w' = H(t,w,w,w)$, $w(0) = \|(x_{o1},\dots,x_{om})\|$
has an upper solution $w$ on $J$.

We shall see that conditions (B0)–(B4) imply the existence of coupled quasisolutions of (2.5.1) in the order interval $[\alpha,\beta]$ of $AC(J,I\!R^m)$ if $x_o = (x_{o1},\dots,x_{om}) \in [\alpha(0),\beta(0)]$. If also (B5) holds, there exist *extremal* coupled quasisolutions $y$, $z$ in $[\alpha,\beta]$. By condition (B7) one can construct $\alpha$, $\beta$ and $N$ so that (B0) and (B1) are valid. If (B2)–(B6) hold, it turns out that (2.5.1) has for each $x_o = (x_{o1},\dots,x_{om}) \in I\!R^m$ the extremals among all its coupled quasisolutions.

The only continuity hypothesis is that functions $f_i(t,x,y,z)$ are continuous with respect to the $i$:th coordinate of $x$. In the special case $f_i(t,x,y,z) = q_i(x_i)\, g_i(t,y,z)$, $i = 1,\dots,m$, no continuity assumptions are imposed on $f$.

### 2.5.1. Existence of coupled quasisolutions

Our first result for the system (2.5.1) is a generalization to theorem 2.2.1.

**Theorem 2.5.1:** *If there is a null set $Z$ in $J$ so that the hypotheses (B0)–(B4) hold, then the system (2.5.1) has coupled quasisolutions in $[\alpha,\beta]$ for each $x_o = (x_{o1},\dots,x_{om}) \in [\alpha(0),\beta(0)]$.*

*Proof.* Let $x_o = (x_{o1},\dots,x_{om}) \in [\alpha(0),\beta(0)]$ and $y$, $z \in [\alpha,\beta]$ be given. Consider the system

$$x_i' = F_i(t, x_i; y(t), z(t)), \qquad x_i(0) = x_{oi}, \quad i = 1,\dots,m, \quad (2.5.3)$$

where

$$F_i(t,x;y(t),z(t)) = \\ f_i(t,(y_1(t),\dots,y_{i-1}(t),x,y_{i+1}(t),\dots,y_m(t)),y(t),z(t)). \tag{2.5.4}$$

By (B2) and (B3) each function $(t,x) \mapsto F_i(t,x,y(t),z(t))$ is a Carathéodory function in $\Omega_i = \{(t,x) \mid x \in [\alpha_i(t),\beta_i(t)],\ t \in J\}$ for all choices of $y$, $z \in [\alpha,\beta]$. In view of (B4) we get

$$F_i(t,\alpha_i(t);\alpha(t),\beta(t)) \le F_i(t,\alpha_i(t);y(t),z(t))$$

and

$$F_i(t,\beta_i(t);\beta(t),\alpha(t)) \ge F_i(t,\beta_i(t);y(t),z(t))$$

on $J \setminus Z$. This and (B0) imply that

$$\alpha_i' \le F_i(t,\alpha_i;y(t),z(t)) \text{ and } \beta_i' \ge F_i(t,\beta_i;y(t),z(t)) \text{ on } J\setminus Z.$$

Condition (B1) implies that

$$|F_i(t,x;y(t),z(t))| \le N(t) \text{ for } x \in [\alpha_i(t),\beta_i(t)],\ t \in J\setminus Z. \tag{a}$$

Hence, by theorem 1.5.1 each IVP of the system (2.5.3) has for all choices of $y$, $z \in [\alpha,\beta]$ the maximal solution $x_i$ in $[\alpha_i,\beta_i]$. Since the system (2.5.3) is uncoupled, then $x = (x_1,\dots,x_m)$ is the maximal solution of (2.5.3) in $[\alpha,\beta]$.

We now define a map $A = (A_1,\dots,A_m)\colon [\alpha,\beta]\times[\alpha,\beta] \to [\alpha,\beta]$ by

$$A_i(y,z) = x_i, \quad y,\ z \in [\alpha,\beta],\ i = 1,\dots,m, \tag{2.5.5}$$

where $x = (x_1,\dots,x_m)$ is the maximal solution of (2.5.3) in $[\alpha,\beta]$. We shall show that $A$ is mixed monotone. Let $y$, $\bar{y}$, $z \in [\alpha,\beta]$, $\bar{y} \le$

$y$ be given, and denote $x = A(y,z)$ and $\bar{x} = A(\bar{y},z)$. Since $\bar{y} \le y$, we have by (B4) and (2.5.4) for each $i = 1, \ldots, m$

$$\bar{x}_i' \le F_i(t, \bar{x}_i; y(t), z(t)) \quad \text{on } J \setminus Z,$$

which implies by lemma 1.5.1 that $\bar{x}_i \le x_i$, $1 \le i \le m$. Thus $A(\bar{y},z) \le A(y,z)$, so that $A(y,z)$ is nondecreasing in $y$. Similarly, it can be shown that $A(y,z)$ is nonincreasing in $z$, whence $A$ is mixed monotone.

Because

$$(A_i(y,z))'(t) = F_i(t, A_i(y,z)(t); y(t), z(t))$$

for all $y$, $z \in [\alpha,\beta]$ and for a.a. $t \in J$, it follows by (a) that

$$\|(A(y,z))'(t)\| \le N(t) \quad \text{for all } y,\ z \in [\alpha,\beta] \text{ and for a.a. } \ t \in J.$$

The above proof implies that $A$ satisfies the hypotheses of proposition 1.4.6, whence $A$ has the extremal coupled fixed points $y$, $z$. In particular,

$$y = A(y,z), \qquad z = A(z,y). \tag{2.5.6}$$

From (2.5.2)–(2.5.6) it then follows that these functions $y = (y_1, \ldots, y_m)$, $z = (z_1, \ldots, z_m)$ are coupled quasisolutions of the (2.5.1) in $[\alpha,\beta]$. □

The coupled quasisolutions of the system (2.5.1) constructed in the proof of theorem 2.5.1 are extremal in $[\alpha,\beta]$ if condition (B5) holds.

**Proposition 2.5.1:** *Let $f\colon J \times \mathbb{R}^{3m} \to \mathbb{R}^m$ satisfy conditions (B0)–(B5). Then the system (2.5.1) has the extremal coupled quasisolutions in $[\alpha,\beta]$ for each $x_o = (x_{o1}, \ldots, x_{om}) \in [\alpha(0), \beta(0)]$.*

*Proof.* Cf. the proof of proposition 2.2.1. □

As consequence of theorem 2.5.1, proposition 2.5.1 and corollary 1.5.1 we obtain

**Corollary 2.5.1:** *Let $f$ in theorem 2.5.1 be defined by*

$$f(t,x,y,z) = f^1(t,x) + f^2(t,y) + f^3(t,z), \ t \in J, \ x, \ y, \ z \in \mathbb{R}^m,$$

*where $f^1$ is a Carathéodory function, $f^2$, $f^3$ are standard functions, $f^1(t,x)$ is quasimonotone nondecreasing in $x$, $f^2(t,y)$ is nondecreasing in $y$ and $f^3(t,z)$ is nonincreasing in $z$. If (B0) holds, and if there exists $N \in L^1(J\mathbb{R}_+)$ such that*

$$\sup\{\|f^j(t,y)\| \mid y \in [\alpha(t), \beta(t)]\} \leq N(t)$$

*for a.a. $t \in J$, $j = 1,2,3$, then the system (2.5.1) has for each $x_o = (x_{o1}, \cdots, x_{om}) \in [\alpha(0), \ \beta(0)]$ coupled quasisolutions in $[\alpha, \beta]$. Moreover, if there exist $p_i \in L^1(J, \mathbb{R}_+)$, $i = 1, \ldots, m$, such that*

$$f_i^1(t,(x_1,\ldots,x_i+k,\ldots,x_m) - f_i^1(t,(x_1,\ldots,x_i,\ldots,x_m) \leq p_i(t)\,k$$

*for all $x = (x_1, \ldots, x_m \in \mathbb{R}^m$, $k > 0$, and for a.a. $t \in J$, then (2.5.1) has extremal coupled quasisolutions in $[\alpha, \beta]$.*

The results of theorem 2.5.1, proposition 2.5.1 and corollary 2.5.1 hold also when conditions (B2)–(B4) are restricted to be valid in the set $D = \{(t,x,y,z) \mid t \in J, \ x, \ y, \ z \in [\alpha(t), \beta(t)]\}$.

The next result gives a method to find $\alpha, \beta \in AC(J, \mathbb{R}^m)$ such that (B0) and (B1) hold.

**Proposition 2.5.2:** *Let $f\colon J \times \mathbb{R}^{3m} \to \mathbb{R}^m$ satisfy condition (B7). Then conditions (B0) and (B1) hold with $\alpha = (-w, \dots, -w)$, $\beta = (w, \dots, w)$ and $N = w'$, where $w$ is any upper solution of the IVP*

$$w' = H(t, w, w, w), \qquad w(0) = \|(x_{o1}, \dots, x_{om})\|. \tag{2.5.7}$$

*Proof.* The assumed condition ensures that

$$\sup\{\|f(t,x,y,z)\| \mid \|x\| \le w(t),\ \|y\| \le w(t)\ \|z\| \le w(t)\}$$
$$\le H(t, w(t), w(t), w(t)) \le w'(t) \text{ for a.a. } t \in J,$$

which implies the assertions. □

As a consequence of theorem 2.5.1 and proposition 2.5.2 we obtain

**Corollary 2.5.2:** *If conditions (B2)-(B5) and (B7) hold, then the system (2.5.1) has extremal coupled quasisolutions in $[-\beta, \beta]$, where $\beta = (w. \dots, w)$ and $w$ is any upper solution of (2.5.7). The conclusion holds also when $f(t, \cdot, y, z)$ is quasimonotone nonincreasing.*

*Proof.* The first assertion is a direct consequence of theorem 2.5.1 and proposition 2.5.2.

Assume next that $f(t, \cdot, y, z)$ is quasimonotone nonincreasing for all $t \in J \setminus Z$ and $y$, $z \in \mathbb{R}^m$. Condition (B7) implies also that

$$-w' \le f(t, \beta, -\beta, \beta), \text{ and } w' \ge f(t, -\beta, \beta, -\beta) \text{ for a.a. } t \in J.$$

Thus the proof of theorem 2.5.1 can be applied to construct coupled quasisolutions $y$, $z$ of the system (2.5.1), when we define the functions $F_i$ in (2.5.3) by

$$F_i(t, x; y(t), z(t)) =$$
$$f_i(t, (z_1(t), \dots, z_{i-1}(t), x, z_{i+1}(t), \dots, z_m(t)), y(t), z(t)),$$

for $t \in J$, $x \in [-w(t), w(t)]$. □

### 2.5.2. Extremal coupled quasisolutions

As for the existence of the extremal coupled quasisolutions among all the coupled quasisolutions of the system (2.5.1) for any fixed initial values $x_{o1}, \ldots, x_{om}$, we have the following generalization to theorem 2.2.2.

**Theorem 2.5.2:** *Given $f = (f_1, \ldots, f_m)\colon J \times \mathbb{R}^{3m} \to \mathbb{R}^m$, assume there is a null set $Z$ of $J$ such that (B2)–(B6) hold. Then for each fixed $x_o = (x_{o1}, \ldots, x_{om}) \in \mathbb{R}^m$ the system (2.5.1) has the extremal coupled quasisolutions, and all the coupled quasisolutions of (2.5.1) belong to the order interval $[\alpha, \beta]$, given by*

$$\alpha_i(t) = x_{oi} + \|x_o\| - w(t),\ \beta(t) = x_{oi} - \|x_o\| + w(t),\ t \in J, \quad (2.5.8)$$

*where $w \in AC(J, \mathbb{R})$ is the solution of the IVP*

$$w' = p(t)\, h(w, w, w), \qquad w(0) = \|x_o\|. \qquad (2.5.9)$$

*Proof.* Let $x_o = (x_{o1}, \ldots, x_{om}) \in \mathbb{R}^m$ be given. Since $[\alpha_i, \beta_i] \subseteq [-w, w]$ for each $i = 1, \ldots, m$, it follows from (B6) that for all $x, y, z \in [\alpha, \beta]$ and for a.a. $t \in J$,

$$\|f(t, x(t), y(t), z(t))\| \le p(t)\, h(w(t), w(t), w(t)) = w'(t)$$

This ensures that the hypotheses (B0) and (B1) of theorem 2.5.1 hold for $\alpha$ and $\beta$ given by (2.5.8) and for $N = \beta'$. Since $f$ satisfies also the hypotheses (B2)–(B5) the system (2.5.1) has by proposition 2.5.1 the extremal coupled quasisolutions $y$, $z$ in $[\alpha, \beta]$.

If $u = (u_1, \ldots, u_m)$, $v = (v_1, \ldots, v_m)$ are coupled quasisolutions of (2.5.1), then they satisfy also the integral equations

$$u_i(t) = x_{oi} + \int_o^t f_i(s, u(s), u(s), v(s))ds, \quad t \in J, \qquad (a)$$

and

$$v_i(t) = x_{oi} + \int_o^t f_i(s, v(s), v(s), u(s))ds, \quad t \in J, \qquad \text{(b)}$$

for each $i = 1, \ldots, m$. Denoting $x(t) = \max\{\|u(t)\|, \|v(t)\|\}$, $t \in J$, it follows from (a), (b) and (B6) by the reasoning used in the proof of theorem 2.2.2 that $x(t) \leq w(t)$ on $J$, and that $u, v \in [\alpha, \beta]$. Since $y$ and $z$ are the extremal coupled quasisolutions of (2.5.1) in $[\alpha, \beta]$, then $u, v \in [y, z]$. Thus $y$ and $z$ are extremal coupled quasisolutions of the system (2.5.1), and all the coupled quasisolutions of (2.5.1) are contained in $[\alpha, \beta]$. □

Since condition (B6) is a special case of (B7), it follows from corollary 2.5.2 that $f(t, \cdot, y, z)$ can be also quasimonotone nonincreasing in theorem 2.5.2.

### 2.5.3. Allowing discontinuity in all arguments

The following result generalizes the result of theorem 2.2.3 for the system

$$x_i' = q_i(x_i)\, g_i(t, x, x), \qquad x_i(0) = x_{oi}, \quad i = 1, \ldots, m. \qquad (2.5.10)$$

**Theorem 2.5.3:** *Given $q_i$, $\frac{1}{q_i} \in L^\infty_{loc}(\mathbb{R}, \mathbb{R}_+)$ and a standard function $g = (g_1, \ldots, g_m)\colon J \times \mathbb{R}^{2m} \to \mathbb{R}^m$, assume that $g(t, \cdot, y)$ is nondecreasing and $g(t, y, \cdot)$ is nonincreasing for a.a. $t \in J$ and for all $y \in \mathbb{R}^m$, and denote*

$$\begin{aligned} &f(t, (x_1, \ldots, x_m), y, z) \\ &= (q_1(x_1)\, g_1(t, y, z), \ldots, q_m(x_m)\, g_m(t, y, z)). \end{aligned} \qquad (2.5.11)$$

a) *If (B0) holds, then (2.5.10) has extremal coupled quasisolutions in $[\alpha, \beta]$ when $x_o = (x_{o1}, \ldots, x_{om}) \in [\alpha(0), \beta(0)]$.*
b) *If condition (B6) holds, then (2.5.10) has extremal coupled quasisolutions for each $x_o = (x_{o1}, \ldots, x_{om}) \in \mathbb{R}^m$.*

*Proof.* a) The hypotheses given for $g$ imply that for all $i = 1, \ldots, m$ and $y$, $z \in [\alpha, \beta]$

$$\int_o^t g_i(s, \alpha(s), \beta(s))\, ds \leq \int_o^t g_i(s, y(s), z(s))\, ds$$
$$\leq \int_o^t g_i(s, \beta(s), \alpha(s))\, ds, \quad t \in J.$$

Thus the equations

$$\int_{x_o}^{A_i(y,z)(t)} \frac{dv}{q_i(v)} = \int_o^t g_i(s, y(s), z(s)) ds, \; t \in J, \tag{a}$$

define a mapping $A = (A_1, \ldots, A_m)$: $[\alpha, \beta] \times [\alpha, \beta] \to [\alpha, \beta]$. Moreover, $A(y, z)$ is nondecreasing in $y$ and nonincreasing in $z$ because $g$ is. Denoting $M_i =$ess $\sup\{q_i(x) \mid \min \alpha_i \leq x \leq \max \beta_i\}$, the reasoning used in the proof of theorem 2.2.3 shows that

$$|A_i(y, z)(t) - A_i(y, z)(\bar{t})|$$
$$\leq M_i \int_{\bar{t}}^t (|g_i(s, \alpha(s), \beta(s))| + |g_i(s, \beta(s), \alpha(s))|)\, ds \tag{b}$$

whenever $y$, $z \in [\alpha, \beta]$ and $0 \leq \bar{t} \leq t \leq T$. Thus $A$ satisfies the hypotheses of theorem 1.4.8, whence $A$ has the extremal coupled fixed points $y$, $z$.

From theorem 1.4.5 it follows that $u$, $v \in [\alpha, \beta]$ are coupled quasisolutions of (2.5.10) if and only if

$$\int_{x_o}^{u_i(t)} \frac{dx}{q_i(x)} = \int_o^t g_i(s, u(s), v(s)) ds, \quad t \in J,$$

and

$$\int_{x_o}^{v_i(t)} \frac{dx}{q_i(x)} = \int_o^t g_i(s, v(s), u(s)) ds, \quad t \in J,$$

which is by (a) equivalent that $u$, $v$ are coupled fixed points of $A$. Since $y$, $z$ are the extremal coupled fixed points of $A$, then they are the extremal coupled quasisolutions of the system (2.5.1) in $[\alpha, \beta]$.

b) If $f$, given by (2.5.11), satisfies condition (B6), then the proof of theorem 2.5.2 implies the existence of $\alpha$, $\beta \in AC(J, \mathbb{R}^m)$ such that (B0) holds, and that all the coupled quasisolutions of (2.5.10) are contained in $[\alpha, \beta]$. Moreover, (2.5.10) has by the above proof extremal coupled quasisolutions on $[\alpha, \beta]$, which concludes the proof. □

### 2.5.4. Special cases

As a consequence of theorem 2.5.2 we obtain

**Proposition 2.5.3:** *Given for each $i = 1, \ldots, m$ and $j = 1, \ldots, n$ a bounded and Lipschitz continuous function $q_i^j : \mathbb{R} \to \mathbb{R}_+$ and a bounded and standard function $g_i^j : J \times \mathbb{R}^{2m} \to \mathbb{R}$, such that $g_i^j(t, \cdot, z)$ is nondecreasing and $g_i^j(t, y, \cdot)$ is nonincreasing for a.a. $t \in J$, then the system*

$$x_i' = \sum_{j=1}^{n} q_i^j(x_i) g_i^j(t, x, x), \quad x_i(0) = x_{oi}, \ i = 1, \ldots, m, \quad (2.5.12)$$

*has for each $x_o = (x_{o1}, \ldots, x_{om}) \in \mathbb{R}^m$ the extremal coupled quasisolutions.*

*Proof.* Define a function $f = (f_1, \ldots, f_m) : J \times \mathbb{R}^{3m} \to \mathbb{R}^m$ by

$$f_i(t, x, y, z) = \sum_{j=1}^{n} q_i^j(x_i) g_i^j(t, y, z), \ t \in J, \ x, y, z \in \mathbb{R}^m$$

As in the proof of proposition 2.2.3 it can be shown that $f$ satisfies the hypotheses of theorem 2.5.2, which implies the assertion. □

From theorem 2.5.3 it follows

**Proposition 2.5.4:** *If the functions $q_i$, $g$ satisfy the hypotheses theorem 2.5.3 a), and if $g$ is a Carathéodory function, then for each $(x_{o1}, \ldots, x_{om}) \in [\alpha(0), \beta(0)]$ the sequences $(y_n)_{n=o}^{\infty}$ and $(z_n)_{n=o}^{\infty}$ of functions $y_n, z_n \colon J \to \mathbb{R}^m$, given by $y_o = \alpha$, $z_o = \beta$,*

$$\int_{x_o}^{(y_{n+1})_i(t)} \frac{dv}{q_i(v)} = \int_o^t g_i(s, y_n(s), z_n(s))ds, \tag{2.5.13}$$

$$\int_{x_o}^{(z_{n+1})_i(t)} \frac{dv}{q_i(v)} = \int_o^t g_i(s, z_n(s), y_n(s))ds, \tag{2.5.14}$$

*for $t \in J$, $i = 1, \ldots, m$, $n \in \mathbb{N}$, converge uniformly on $J$ to the extremal coupled quasisolutions of the system (2.5.10) in $[\alpha, \beta]$.*

*Proof.* Similar to that of proposition 2.2.4. □

When $q_i(x) = 1$ a.e. on $\mathbb{R}$, $i = 1, \ldots, m$, in (2.5.13) and in (2.5.14), they are reduced to the sequences of successive approximations

$$y_{n+1}(t) = x_o + \int_o^t g(s, y_n(s), z_n(s))\, ds, \quad t \in J,\ n \in \mathbb{N}, \tag{2.5.15}$$

$$z_{n+1}(t) = x_o + \int_o^t g(s, z_n(s), y_n(s))\, ds, \quad t \in J,\ n \in \mathbb{N}. \tag{2.5.16}$$

Thus we obtain

**Corollary 2.5.3:** *Assume that $q_i \colon \mathbb{R} \to (0, \infty)$, that $q_i(x) = 1$ a.e. on $\mathbb{R}$, $i = 1, \ldots, m$, that $g \colon J \times \mathbb{R}^{2m} \to \mathbb{R}^m$ is a Carathéodory function, and that $g(t, \cdot, y)$ is nondecreasing and $g(t, y, \cdot)$ is nonincreasing for all $y \in \mathbb{R}^m$ and for a.a. $t \in J$. If there exist $\alpha$, $\beta \in AC(J, \mathbb{R}^m)$ such that*

$$\alpha \le \beta,\ \alpha' \le g(t, \alpha, \beta), \quad \text{and } \beta' \ge g(t, \beta, \alpha) \text{ a.e. on } J,$$

*then the sequences* $(y_n)_{n=o}^{\infty}$ *and* $(z_n)_{n=o}^{\infty}$ *of the successive approximations defined by (2.5.15) and (2.5.16) converge uniformly on* $J$ *to the extremal coupled quasisolutions* $y$, $z$ *of the system*

$$x_i' = q_i(x_i)\,g_i(t,x,x), \qquad x_i(0) = x_{oi}, \quad i = 1,\dots,m \quad (2.5.17)$$

*in* $[\alpha,\beta]$ *if* $y_o = \alpha$, $z_o = \beta$ *and* $\alpha(0) \le (x_{o1},\dots,x_{om}) \le \beta(0)$.

As an immediate consequence of proposition 2.5.1 we have

**Corollary 2.5.4:** *Assume that for each* $i = 1,\dots,m$ *and* $j = 1,\dots,n$, $p_i^j : J \to I\!R_+$ *is Lebesgue integrable,* $q_i^j : I\!R \to I\!R_+$ *is bounded and Lipschitz continuous,* $g_i^j : I\!R^m \to I\!R_+$ *is bounded and nondecreasing, and* $h_i^j : I\!R^m \to I\!R_+$ *is bounded and nonincreasing. Then the system*

$$x_i' = \sum_{j=1}^{n} p_i^j(t) q_i^j(x_i) g_i^j(x) h_i^j(x), \qquad x_i(0) = x_{oi},\ i = 1,\dots,m,$$

*has for each* $x_o = (x_{o1},\dots,x_{om}) \in I\!R^m$ *extremal coupled quasisolutions.*

As a consequence of theorem 2.5.3 we obtain

**Corollary 2.5.5:** *If for each* $i = 1,\dots,m$, $q_i$, $\frac{1}{q_i} \in L^\infty(I\!R, I\!R_+)$, $p_i \in L^1(J, I\!R_+)$, *if* $g_i : I\!R^m \to R_+$ *is bounded and nondecreasing and if* $h_i : I\!R^m \to I\!R_+$ *is bounded and nonincreasing, then the system*

$$x_i' = p_i(t) q_i(x_i) g_i(x) h_i(x), \qquad x_i(0) = x_{oi},\ i = 1,\dots,m,$$

*has for each* $x_o = (x_{o1},\dots,x_{om}) \in I\!R^m$ *the extremal coupled quasisolutions.*

## 2.6. PERIODIC BOUNDARY VALUE SYSTEMS

Most of the results of section 2.3 have their counterparts within finite systems of periodic boundary value problems, as we shall show in this section.

### 2.6.1. Existence of extremal solutions

Consider the periodic boundary value system (PBVS for short)

$$x_i' = f_i(t,x,x), \qquad x_i(0) = x_i(T), \quad i = 1,\dots,m, \tag{2.6.1}$$

where $f = (f_1,\dots,f_m)\colon J \times \mathbb{R}^{2m} \to \mathbb{R}^m$, $J = [0,T]$. We shall base our study on the following assumptions.

(C0) $\alpha, \beta \in AC(J,\mathbb{R}^m)$, $\alpha \le \beta$, $\alpha(0) \le \alpha(T)$, $\beta(0) \ge \beta(T)$, $\alpha' \le f(t,\alpha,\alpha)$ and $\beta' \ge f(t,\beta,\beta)$ on $J \setminus Z$.

(C1) $\|f(t,x,y)\| \le N(t)$ for all $t \in J \setminus Z$ and $x, y \in [\alpha(t),\beta(t)]$.

(C2) $f(\cdot,x(\cdot),y(\cdot))$ is measurable for all $x, y \in [\alpha,\beta]$.

(C3) $f(t,\cdot,y)$ and $f(t,y,\cdot)$ are quasimonotone nondecreasing for all $t \in J \setminus Z$ and $y \in \mathbb{R}^m$.

(C4) $f_i(t,x,y)+M_i(t)\,y_i$ is nondecreasing in $y_i \in [\alpha_i(t),\beta_i(t)]$ for all $t \in J \setminus Z$, $x, y \in [\alpha(t),\beta(t)]$ and $i = 1,\dots,m$.

(C5) $f_i(t,x,y) - p_i(t)\,x_i$ is continuous and nonincreasing in $x_i \in [\alpha_i(t),\beta_i(t)]$ for all $t \in J \setminus Z$, $x, y \in [\alpha(t),\beta(t)]$ and $i = 1,\dots,m$.

(C6) $f_i(t,x,y)+p_i(t)\,x$ is nondecreasing in $x \in [\alpha(t),\beta(t)]$ for all $t \in J \setminus Z$, $x, y \in [\alpha(t),\beta(t)]$ and $i = 1,\dots,m$.

(C7) $N_1(t) + q_1(t)x + q_2(t)y \le f(t,x,y) \le N_2(t) + q_1(t)x + q_2(t)y$ for all $t \in J \setminus Z$ and $x, y \in \mathbb{R}^m$.

We shall show that the PBVS (2.6.1) has the extremal solutions in the order interval $[\alpha,\beta]$ of $AC(J,\mathbb{R}^m)$ if there is a null set $Z$ in $J$ such that either conditions (C0)–(C5) hold for some $N$, $M_i$, $p_i \in L^2(J,\mathbb{R})$ or conditions (C0)–(C4) and (C6) hold for some $N$, $M_i$, $p_i \in L^1(J,\mathbb{R})$, $i = 1,\dots,m$. If (C0) and (C1) are replaced by (C7) with $N_j \in L^2(J,\mathbb{R}^m)$, $q_j \in$

$L^2(J,\mathbb{R})$ (resp. $N_j \in L^1(J,\mathbb{R}^m)$, $q_j \in L^1(J,\mathbb{R})$), $j = 1,2$, with $\int_o^T (q_1(s)+q_2(s))\,ds < 0$, then these extremal solutions are the least and the greatest of all the solutions of (2.6.1). Moreover, in all the above cases the extremal solutions in question are nondecreasing with respect to $f$.

**Theorem 2.6.1:** *Assume there exist $N$, $M_i$, $p_i \in L^2(J,\mathbb{R})$, $i = 1,\dots,m$, and a null set $Z$ in $J$ such that conditions (C0)–(C5) hold. Then the PBVS (2.6.1) has extremal solutions in the order interval $[\alpha,\beta]$.*

*Proof.* We may choose the functions $M_i$ and $p_i$ in (C4) and in (C5) so that $\int_o^T (M_i(s)-p_i(s))\,ds > 0$. Let $y \in [\alpha,\beta]$ be given. Consider the PBVS

$$x_i' = F_i(t,x_i;y(t)), \qquad x_i(0) = x_i(T), \quad i = 1,\dots,m, \tag{2.6.2}$$

where

$$\begin{aligned} F_i(t,x;y(t)) &= M_i(t)(y_i(t)-x)) \\ &+ f_i(t,(y_1(t),\dots,y_{i-1}(t),x,y_{i+1}(t),\dots,y_m(t)),y(t)). \end{aligned} \tag{2.6.3}$$

In view of (C2) and (C5) $F_i$ is a Carathéodory function with respect to $(t,x)$ in $\Omega_i = \{(t,x) \mid t \in J,\ x \in [\alpha_i(t),\beta_i(t)]\}$. Applying (C0), (C3), (C4) and (2.6.3) one obtains for each $i = 1,\dots,m$,

$$\alpha_i' \le F_i(t,\alpha_i;y(t)) \ \text{ and } \ \beta_i' \ge F_i(t,\beta_i;y(t)) \ \text{ on } \ J\setminus Z.$$

Condition (C1) implies that for all $x \in [\alpha_i(t),\beta_i(t)]$ and $t \in J\setminus Z$,

$$|F_i(t,x;y(t))| \le K_i\, M_i(t) + N(t), \tag{2.6.4}$$

where $K_i = \max\beta_i - \min\alpha_i$, $i = 1,\dots,m$. Hence, by theorem 1.5.2 each PBVP of (2.6.2) has a solution $x_i$ in $[\alpha_i,\beta_i]$. In view of condition (C5) we see that $F_i(t,x;y(t)) + (M_i(t) - p_i(t))\,x$ is nonincreasing in $x \in [\alpha_i(t), \beta_i(t)]$ for all $t \in J\setminus Z$, whence $x_i$ is by corollary 1.5.3 the only solution of the $i$:th PBVP of (2.6.2). Since (2.6.2) is uncoupled, then $x = (x_1, \ldots, x_m)$ is a unique solution of the PBVS (2.6.2).

Define a map $G = (G_1, \ldots, G_m)\colon [\alpha, \beta] \to [\alpha, \beta]$ by

$$G_i y = x_i, \quad y \in [\alpha, \beta], \quad i = 1, \ldots, m, \tag{2.6.5}$$

where $x = (x_1, \ldots, x_m)$ is the solution of (2.6.2) in $[\alpha, \beta]$. To prove that $G$ is nondecreasing, let $y$, $\bar{y} \in [\alpha, \beta]$, $\bar{y} \le y$, be given, and suppose that $x = (x_1, \ldots, x_m)$ and $\bar{x} = (\bar{x}_1, \ldots, \bar{x}_m)$ are the corresponding solutions of (2.6.2) and

$$\bar{x}_i' = F_i(t, \bar{x}_i; \bar{y}(t)), \qquad \bar{x}_i(0) = \bar{x}_i(T), \quad i = 1, \ldots, m,$$

in $[\alpha, \beta]$, respectively. Since $\bar{y} \le y$, we have by (C4) and (2.6.3)

$$\bar{x}_i' \le F_i(t, \bar{x}_i; y(t)) \ \text{ a.e. on } J, \qquad \bar{x}_i(0) = \bar{x}_i(T).$$

Thus the hypotheses of lemma 1.5.6 hold when $y = \bar{x}_i$ and $z = x_i$, whence $\bar{x}_i \le x_i$, thus proving that $G_i\bar{y} \le G_i y$. This holds for each $i = 1, \ldots, m$, whence $G\bar{y} \le Gy$.

From (2.6.3), (2.6.4) and (2.6.5) it follows that for all $y \in [\alpha, \beta]$ and for a.a. $t \in J$,

$$\|(Gy)'(t)\| \le N(t) + K\,M(t),$$

where $K = \max\{K_1, \ldots, K_m\}$ and $M = \max\{M_1, \ldots, M_m\}$. Thus $G$ satisfies the hypotheses of proposition 1.4.4, whence $G$ has the least fixed point $x_*$ and the greatest fixed point $x^*$. From the definition of $G$ it follows that $x_* = (x_{*1}, \ldots, x_{*m})$ and $x^* = (x_1^*, \ldots, x_m^*)$ are also solutions of the PBVS (2.6.1) in $[\alpha, \beta]$.

If $x = (x_1, \ldots, x_m)$ is any solution of (2.6.1) in $[\alpha, \beta]$, then it satisfies also the PBVS (2.6.2) with $y = x$. But this means that $x$ is a fixed point of $G$, whence $x_* \leq x \leq x^*$. Thus $x_*$ is the minimal solution and $x^*$ is the maximal solution of the PBVS (2.6.1) in $[\alpha, \beta]$. □

If condition (C5) is replaced in theorem 2.6.1 by condition (C6), we obtain

**Theorem 2.6.2:** *If there exist $N$, $M_i$, $p_i \in L^1(J, \mathbb{R})$, $i = 1, \ldots, m$, and a null set $Z$ in $J$ such that conditions (C0)–(C4) and (C6) hold, then the PBVS (2.6.1) has the extremal solutions in $[\alpha, \beta]$.*

*Proof.* We can choose the functions $M_i$, $p_i \in L^1(J, \mathbb{R})$ in conditions (C4) and (C6) so that $\int_o^T (M_i(s) + p_i(s))\, ds > 0$, $i = 1, \ldots, m$. Denote $q = \max\{M_i + p_i \mid i = 1, \ldots, m\}$, and define an operator $G$ in $[\alpha, \beta]$ by

$$
\begin{aligned}
Gx(t) &= e^{-Q(t)} \int_o^t e^{Q(s)} g(s, x(s))\, ds \\
&\quad + \frac{e^{-Q(t)}}{e^{Q(T)} - 1} \int_o^T e^{Q(s)} g(s, x(s))\, ds, \quad t \in J,
\end{aligned}
\tag{2.6.6}
$$

where

$$
g(t, x) = f(t, x, x) + q(t)\, x, \quad \text{and} \quad Q(t) = \int_o^t q(s)\, ds. \tag{2.6.7}
$$

From (C0) and (2.6.7) it follows that

$$
e^{Q(t)} g(t, \alpha(t)) \geq e^{Q(t)} (\alpha'(t) + q(t)\alpha(t)) \quad t \in J \setminus Z. \tag{a}
$$

In view of (a), (C0) and (2.6.6) we then obtain

$$\begin{aligned}
e^{Q(t)}G\alpha(t) &\geq \int_o^t e^{Q(s)}(\alpha'(s)) + q(s)\alpha(s))\,ds \\
&+ \frac{1}{e^{Q(T)}-1}\int_o^T e^{Q(s)}(\alpha'(s) + q(s)\alpha(s))\,ds \\
&= e^{Q(t)}\alpha(t)) - \alpha(0) + \frac{1}{e^{Q(T)}-1}(e^{Q(T)}(\alpha(T)) - \alpha(0)) \\
&\geq e^{Q(t)}\alpha(t), \quad t \in J.
\end{aligned}$$

Thus $\alpha \leq G\alpha$. Similarly, it can be shown that $G\beta \leq \beta$. From (C3), (C4), (C6) and (2.6.7) it follows that $g(t,\cdot)$ is nondecreasing in $[\alpha,\beta]$ for all $t \in J \setminus Z$. This and the definition (2.6.6) of $G$ imply that $G$ is nondecreasing. Obviously, $Gx \in AC(J, I\!R^m)$ for each $x \in [\alpha,\beta]$, whence (2.6.6) defines a nondecreasing operator $G\colon [\alpha,\beta] \to [\alpha,\beta]$. From (2.6.6) and (2.6.7) it follows that

$$(Gy)'(t) = f(t, y(t), y(t)) + q(t)(y(t) - Gy(t))$$

for all $y \in [\alpha,\beta]$ and for a.a. $t \in J$. This and (C1) imply that for all $y \in [\alpha,\beta]$ and for a.a $t \in J$,

$$\|(Gy)'(t)\| \leq N(t) + Kq(t),$$

where $K = 2\max\{\|\alpha\|_o, \|\beta\|_o\}$. Thus $G$ satisfies the hypotheses of proposition 1.4.4, whence $G$ has the least fixed point $x_*$ and the greatest fixed point $x^*$. From the definitions of $g$ and $G$ and lemma 1.5.7 it follows that $x_* = (x_{*1},\ldots,x_{*m})$ and $x^* = (x_1^*,\ldots,x_m^*)$ are also solutions of the PBVS (2.6.1) in $[\alpha,\beta]$.

If $x = (x_1,\ldots,x_m)$ is any solution of (2.6.1) in $[\alpha,\beta]$, then it is by lemma 1.5.6 a fixed point of $G$, whence $x_* \leq x \leq x^*$. Thus $x_* = (x_{*1},\ldots,x_{*m})$ and $x^* = (x_1^*,\ldots,x_m^*)$ are the extremal solutions of the PBVS (2.6.1) in $[\alpha,\beta]$. □

Consider next the existence of the extremal solutions of (2.6.1) among all its solutions.

**Proposition 2.6.1:** *Assume there exist* $N_j \in L^2(J, I\!R^m)$ *and* $q_j \in L^2(J, I\!R)$, $j = 1, 2$, *with* $\int_o^T (q_1(s) + q_2(s))\, ds < 0$, *and* $M_i, p_i, \in L^2(J, I\!R)$ *and a null set* $Z$ *in* $J$ *such that condition (C7) holds, and that (C2)–(C5) hold when* $\alpha$, $\beta$ *are given by*

$$\alpha(t) = e^{Q(t)}[\int_o^t e^{-Q(s)} N_1(s)\, ds + \int_o^T \frac{e^{-Q(s)} N_1(s)}{e^{-Q(T)} - 1}]\, ds, \quad (2.6.8)$$

$$\beta(t) = e^{Q(t)}[\int_o^t e^{-Q(s)} N_2(s)\, ds + \int_o^T \frac{e^{-Q(s)} N_2(s)}{e^{-Q(T)} - 1}]\, ds, \quad (2.6.9)$$

*where* $Q(t) = \int_o^t (q_1(s) + q_2(s))\, ds$, $t \in J$. *Then there exist the extremal solutions to PBVS (2.6.1), and all the solutions of (2.6.1) lie within the order interval* $[\alpha, \beta]$.

*Proof:* From corollary 1.5.4 it follows that $\alpha$ and $\beta$, given by (2.6.8) and (2.6.9), are the unique solutions of the PBVP's

$$\alpha' = N_1(t) + (q_1(t) + q_2(t))\alpha(t), \quad \alpha(0) = \alpha(T), \qquad \text{(a)}$$

and

$$\beta' = N_2(t) + (q_1(t) + q_2(t))\beta(t), \quad \beta(0) = \beta(T). \qquad \text{(b)}$$

In view of (a), (b) and (C7) we see that condition (C0) holds, and that condition (C1) holds with

$$N(t) = \|N_1(t)\| + \|N_2(t)\| + (|q_1(t)| + |q_2(t)|)(\|\alpha\|_o + \|\beta\|_o).$$

Thus the PBVS (2.6.1) has the extremal solutions $x_*$ and $x^*$ in $[\alpha, \beta]$.

If $x = (x_1, \ldots, x_m)$ is a solution of (2.6.1), it follows from (2.6.1) and (C7) that

$$\begin{aligned} &x_i' \geq (N_1)_i(t) + (q_1(t) + q_2(t))x_i(t) \ \text{ a.e. on } \ J, \\ &x_i(0) = x_i(T), \ i = 1, \ldots, m, \end{aligned} \tag{c}$$

and

$$\begin{aligned} &x_i' \leq (N_2)_i(t) + (q_1(t) + q_2(t))x_i(t) \ \text{ a.e. on } \ J, \\ &x_i(0) = x_i(T), \ i = 1, \ldots, m. \end{aligned} \tag{d}$$

Hence, (a), (b), (c) and (d) imply by applying lemma 1.5.6 to the components of $x$, $\alpha$ and $\beta$ that $x \in [\alpha, \beta]$. In particular, $x \in [x_*, x^*]$, whence $x_*$ and $x^*$ are the extremals among all the solutions of (2.6.1). □

By the similar reasoning it follows from theorem 2.6.2 and lemmas 1.5.5 and 1.5.6.

**Proposition 2.6.2:** *Assume there exist $N_j \in L^1(J, \mathbb{R}^m)$ and $q_j \in L^1(J, \mathbb{R})$, $j = 1, 2$ with $\int_0^T (q_1(s) + q_2(s))\, ds < 0$, and $M_i$, $p_i, \in L^1(J, \mathbb{R})$ and a null set $Z$ in $J$ such that condition (C7) holds, and that conditions (C2)–(C4) and (C6) are valid with $\alpha$, $\beta$ defined by (2.6.8) and (2.6.9). Then the conclusions of proposition 2.6.1 hold.*

As for the dependence of the extremal solutions of (2.6.1) on $f$ we have

**Proposition 2.6.3:** *If the hypotheses of theorem 2.6.1 or theorem 2.6.2 hold, then the extremal solutions of the PBVS (2.6.1) in $[\alpha, \beta]$ are nondecreasing with respect to $f$.*

*Proof.* Let $f, \hat{f} \colon J \times \mathbb{R}^2 \to \mathbb{R}^m$ satisfy

$$f(t, x, y) \leq \hat{f}(t, x, y) \ \text{ for a.a. } t \in J \text{ and for all } \ x, y \in \mathbb{R}^m. \tag{a}$$

Assume first that the hypotheses of theorem 2.6.1 hold for $f$ and $\hat{f}$. Let $x_*$ be the minimal solution of (2.6.1) in $[\alpha,\beta]$, and let $\hat{x}_*$ be the minimal solution of the PBVS

$$x_i' = \hat{f}_i(t,x,x), \qquad x_i(0) = x_i(T), \quad i = 1,\dots,m. \tag{b}$$

If $F_i(t,x;y(t))$ are defined by (2.6.3), it follows from (a), (b) and (2.6.3) that

$$\begin{aligned} &\hat{x}_{*i}'(t) \geq F_i(t,\hat{x}_{*i}(t);\hat{x}_*(t)) \ \text{ a.e. on } \ J, \\ &\hat{x}_{*i}(0) = \hat{x}_{*i}(T), \ i = 1,\dots,m. \end{aligned} \tag{c}$$

From the definition (2.6.5) of the operator $G$ it follows that

$$\begin{aligned} &(G_i\hat{x}_*)'(t) = F_i(t,G_i\hat{x}_*(t);\hat{x}_*(t)) \text{ a.e. on } \ J, \\ &G_i\hat{x}_*(0) = G_i\hat{x}_*(T). \end{aligned} \tag{d}$$

Because the functions $(t,x) \mapsto F_i(t,x;\hat{x}_*(t))$ satisfy the hypotheses of lemma 1.5.6, then (c) and (d) imply that $G\hat{x}_* \leq \hat{x}_*$. Since $G$ satisfies the hypotheses of proposition 1.4.4, then $x_* \leq \hat{x}_*$ by (1.4.11).

The proof that $x^* \leq \hat{x}^*$, where $x^*$ denotes the maximal solution of (2.6.1) in $[\alpha,\beta]$ and $\hat{x}^*$ is the maximal solution of (b) in $[\alpha,\beta]$, is similar.

Assume next that (a) holds for the functions $f$, $\hat{f}$ which satisfy the hypotheses of theorem 2.6.2, and let $G\colon [\alpha,\beta] \to [\alpha,\beta]$ be defined by (2.6.6), where $g$ is given by (2.6.7). If $x_*$ and $\hat{x}_*$ are as above, it follows from lemma 1.5.7 that

$$\begin{aligned} \hat{x}_*(t) &= e^{-Q(t)} \int_o^t e^{Q(s)}\hat{g}(s,\hat{x}_*(s))\,ds \\ &\quad + \frac{e^{-Q(t)}}{e^{Q(T)}-1} \int_o^T e^{Q(s)}\hat{g}(s,\hat{x}_*(s))\,ds, \end{aligned} \tag{e}$$

where

$$\hat{g}(t,x) = \hat{f}(t,x,x) + q(t)\,x, \quad \text{and} \quad Q(t) = \int_0^t q(s)\,ds. \tag{f}$$

From (a), (f) and (2.6.7) it follows that

$$g(t,x) \le \hat{g}(t,x), \quad \text{for a.a. } t \in J \text{ and for all } x \in [\alpha(t), \beta(t)],$$

whence (2.6.6) and (e) imply that $G\hat{x}_* \le \hat{x}_*$. Since $G$ satisfies the hypotheses of proposition 1.4.4 and $x_*$ is the least fixed point of $G$, it follows from (1.4.11) that $x_* \le \hat{x}_*$.

Similarly, it can be shown that $x^* \le \hat{x}^*$, where $x^*$ denotes the maximal solution of (2.6.1) in $[\alpha, \beta]$ and $\hat{x}^*$ is the maximal solution of (b) in $[\alpha, \beta]$. □

The following corollaries are direct consequences of theorem 2.6.1 and proposition 2.6.3.

**Corollary 2.6.1:** *Let $f\colon J \times I\!R^2 \to I\!R^m$ satisfy conditions (C0) and (C1). If $f(t,x,y)$ is in $D = \{(t,x,y) \mid t \in J,\ x, y \in [\alpha(t), \beta(t)]\}$ Borel measurable, continuous and nonincreasing in $x$ and nondecreasing in $y$, then the PBVS (2.6.1) has the extremal solutions in $[\alpha, \beta]$, and they are nondecreasing with respect to $f$.*

**Corollary 2.6.2:** *If $f\colon J \times I\!R^2 \to I\!R^m$ satisfies condition (C0), if $f(t,x,y)$ is in $D = \{(t,x,y) \mid t \in J,\ x, y \in [\alpha(t), \beta(t)]\}$ continuous, nonincreasing in $x$ and nondecreasing in $y$, then the PBVS (2.6.1) has the extremal solutions in $[\alpha, \beta]$, and they are nondecreasing with respect to $f$.*

The proof of proposition 2.6.3 can also be used to verify the following result.

**Proposition 2.6.4:** *If the hypotheses of proposition 2.6.1 or proposition 2.6.2 hold, then the extremal solutions of the PBVS (2.6.1) are nondecreasing with respect to $f$.*

## 2.6.2. Existence of extremal coupled quasisolutions

The functions $y = (y_1, \ldots, y_m)$ and $z = (z_1, \ldots, z_m)$ of $AC(J, \mathbb{R}^m)$ are said to be *coupled quasisolutions* of the PBVS

$$x_i' = f_i(t, x, x, x), \qquad x_i(0) = x_i(T), \quad i = 1, \ldots, m, \quad (2.6.10)$$

if

$$\begin{aligned} &y_i'(t) = f_i(t, y(t), y(t), z(t)) \text{for a.a. } t \in J, \\ &z_i'(t) = f_i(t, z(t), z(t), y(t)) \text{ for a.a. } t \in J, \\ &y_i(0) = y_i(T), \text{ and } z_i(0) = z_i(T), \ i = 1, \ldots, m. \end{aligned} \quad (2.6.11)$$

Consider now the existence of the coupled quasisolutions of the PBVS (2.6.10) in the case when $f : J \times \mathbb{R}^{3m} \to \mathbb{R}^m$ satisfies some of the following hypotheses.

(D0) $\alpha$, $\beta \in AC(J, \mathbb{R}^m)$, $\alpha \leq \beta$, $\alpha(0) \leq \alpha(T)$, $\beta(0) \geq \beta(T)$, and $\alpha' \leq f(t, \alpha, \alpha, \beta)$ and $\beta' \geq f(t, \beta, \beta, \alpha)$ on $J \setminus Z$.

(D1) $\sup\{\|f(t, x, y, z)\| \mid x, y, z \in [\alpha(t), \beta(t)]\} \leq N(t)$ for all $t \in J \setminus Z$.

(D2) $f(\cdot, x(\cdot), y(\cdot), z(\cdot))$ is measurable for all $x$, $y$, $z \in [\alpha, \beta]$.

(D3) $f_i(t, x, y, z) - p_i(t)x$ is continuous and nonincreasing in $x_i \in [\alpha_i(t), \beta_i(t)]$ for all $t \in J \setminus Z$ and $x$, $y$, $z \in [\alpha(t), \beta(t)]$.

(D4) $f(t, \cdot, y, z)$ is quasimonotone nondecreasing, $f(t, y, \cdot, z)$ is nondecreasing and $f(t, y, z, \cdot)$ is nonincreasing for all $y$, $z \in \mathbb{R}^m$ and $t \in J \setminus Z$.

**Theorem 2.6.3:** *If there is a null set $Z$ in $J$ and functions $N$, $p_i \in L^2(J, \mathbb{R})$ such that the hypotheses (D0)–(D4) hold, then the PBVS (2.6.10) has coupled quasisolutions in the order interval $[\alpha, \beta]$ of $AC(J, \mathbb{R}^m)$.*

*Proof.* Denote $M_i(t) = |p_i(t)| + 1$, $t \in J$. Given $y$, $z \in [\alpha, \beta]$, consider the PBVS

$$x_i' = F_i(t, x_i; y(t), z(t)), \quad x_i(0) = x_i(T), \ i = 1, \ldots, m, \quad (2.6.12)$$

where

$$F_i(t,x;y(t),z(t)) = M_i(t)(y_i(t)-x)) \\ + f_i(t,(y_1(t),\dots,y_{i-1}(t),x,y_{i+1}(t),\dots,y_m(t)),y(t),z(t)). \tag{2.6.13}$$

From (D2) and (D3) it follows that $(t,x)\mapsto F_i(t,x;y(t),z(t))$ is a Carathéodory function in $\Omega_i = \{(t,x)\mid t\in J,\ x\in[\alpha_i(t),\beta_i(t)]\}$ for each $i=1,\dots,m$. Applying (D0), (D3), (D4) and (2.6.13) it is easy to see that

$$\alpha_i' \le F_i(t,\alpha_i;y(t),z(t)) \quad\text{and}\quad \beta_i' \ge F_i(t,\beta_i;y(t),z(t)) \text{ on } J\setminus Z.$$

Condition (D1) implies that for all $x\in[\alpha_i(t),\beta_i(t)]$ and $t\in J\setminus Z$,

$$|F_i(t,x;y(t),z(t))| \le K_i\, M_i(t) + N(t), \tag{a}$$

where $K_i = \max\beta_i - \min\alpha_i$. Hence, by theorem 1.5.2 each PBVP of (2.6.12) has a solution $x_i$ in $[\alpha_i,\beta_i]$. In view of (2.6.13) and condition (D3) we see that each of the functions $F_i(t,x;y(t),z(t)) + (M_i(t)-p_i(t))\,x$ is nonincreasing in $x\in[\alpha_i(t),\beta_i(t)]$ for all $t\in J\setminus Z$, whence $x_i$ is by corollary 1.5.3 the only solution of the $i$:th PBVP of (2.6.12). Since the system (2.6.12) is uncoupled, then $x=(x_1,\dots,x_m)$ is the only solution of the PBVS (2.6.12).

Define a map $A=(A_1,\dots,A_m)\colon [\alpha,\beta]\times[\alpha,\beta]\to[\alpha,\beta]$ by

$$A_i(y,z) = x_i, \quad y,\ z\in[\alpha,\beta], \tag{2.6.14}$$

where $x=(x_1,\dots,x_m)$ is the solution of (2.6.12) in $[\alpha,\beta]$. Applying the proof of theorem 2.3.4 to each component of $A$ it follows that $A$ is mixed monotone, and that $\|A'(y,z)\|$ is for all $y,\ z\in[\alpha,\beta]$ bounded a.e. on $J$ by a function of $L^1(J,I\!R_+)$. Thus $A$ satisfies the hypotheses of proposition 1.4.6, whence $A$ has the extremal coupled fixed points $y,\ z$. From (2.6.11)–(2.6.14) it follows that $u,\ v\in[\alpha,\beta]$ are coupled quasisolutions of (2.6.10) if and only if they are coupled fixed points of $A$. Thus $y \leq u$, $v \leq z$ for all coupled quasisolutions $u$, $v$ of (2.6.10) in $[\alpha, \beta]$. □

The reasoning similar to that used in the proof of proposition 2.6.1 implies.

**Proposition 2.6.5:** *Given $f = (f_1, \dots, f_m)\colon J \times \mathbb{R}^{3m} \to \mathbb{R}^m$ and a null set $Z$ of $J$, assume there exist $N_1$, $N_2 \in L^2(J, \mathbb{R}^m)$ and $q_1$, $q_2$, $q_3 \in L^2(J, \mathbb{R})$ with $\int_o^T (q_1(s) + q_2(s) + q_3(s))\,ds < 0$ such that*

$$N_1(t) + q_1(t)x + q_2(t)y + q_3(t)z \leq f(t, x, y, z)$$
$$\leq N_2(t) + q_1(t)x + q_2(t)y + q_3(t)z$$

*for all $t \in J \backslash Z$ and $x$, $y$, $z \in \mathbb{R}$. If $\alpha$, $\beta$ are defined by (2.6.8) and (2.6.9), where $Q(t) = \int_o^t (q_1(s) + q_2(s) + q_3(s))\,ds$, $t \in J$, and if conditions (D2)–(D4) hold for some $p_i \in L^2(J, \mathbb{R})$, $i = 1, \dots, m$, then the PBVS (2.6.10) has the extremal coupled quasisolutions, and all the coupled quasisolutions of (2.6.10) belong to the order interval $[\alpha, \beta]$.*

## 2.7. INFINITE DIFFERENTIAL SYSTEMS

In this section we shall study which of the results derived for finite systems in section 2.4 can be extended to the infinite differential systems

$$x_i' = f_i(t, x, x), \qquad x_i(0) = x_{oi}, \quad i = 1, 2, \dots, \tag{2.7.1}$$

in $l^p$-spaces, $1 \leq p \leq \infty$, equipped with $p$-norm and componentwise ordering. The quasimonotonicity concepts given in section 2.4 have obvious extensions to functions $g = (g_1, g_2, \dots)\colon V \to l^p$, $V \subseteq l^p$.

Consider the case when $f = (f_1, f_2, \dots): J \times l^p \times l^p \to l^p$, $J = [0, T]$, $1 \le p \le \infty$, satisfies some of the following hypotheses:

(A0) $\alpha, \beta \in AC(J, l^p)$, $\alpha \le \beta$, $\alpha_i' \le f_i(t, \alpha, \alpha)$, $\beta_i' \ge f_i(t, \beta, \beta)$ on $J \setminus Z$ for each $i = 1, 2, \dots$.

(A1) There is $N = (N_1, N_2, \dots) \in L^1(J, l^p_+)$ such that $|f_i(t, x, y)| \le N_i(t)$ for all $t \in J \setminus Z$, $x, y \in [\alpha(t), \beta(t)]$ and $i = 1, 2, \dots$.

(A2) $f_i(\cdot, x(\cdot), y(\cdot))$ is measurable for all $x, y \in AC(J, l^p)$ and $i = 1, 2, \dots$.

(A3) $f(t, \cdot, y)$ and $f(t, y, \cdot)$ are quasimonotone nondecreasing for all $t \in J \setminus Z$ and $y \in l^p$.

(A4) There is $M = (M_1, M_2, \dots) \in L^1(J, l^p_+)$ and a nondecreasing function $\varphi \in C(I\!R, I\!R)$ such that $f_i(t, (x_1, \dots, x_i, \dots), (y_1, \dots, y_i, \dots)) + M_i(t)\, \varphi(y_i)$ is continuous in $x_i$ and nondecreasing in $y_i$ for all $x = (x_1, x_2, \dots)$, $y = (y_1, y_2, \dots) \in l^p$, $t \in J \setminus Z$ and $i = 1, 2, \dots$.

(A5) $|f_i(t, x, y)| \le \mu_1(t)\, |x_i| + \mu_2(t)\, |y_i| + \nu_i(t)$ for all $t \in J \setminus Z$, $x, y \in l^p$ and $i = 1, 2, \dots$, where $\mu_1, \mu_2 \in L^1(J, I\!R_+)$ and $\nu = (\nu_1, \nu_2, \dots) \in L^1(J, l^p_+)$.

(A6) $|f_i(t, x, y)| \le H_i(t, (|x_1|, |x_2|, \dots), (|y_1|, |y_2|, \dots)$ for all $x = (x_1, x_2, \dots)$ and $y = (y_1, y_2, \dots) \in l^p$, $t \in J \setminus Z$ and $i = 1, 2, \dots$, where $H = (H_1, H_2, \dots): J \times l^p_+ \times l^p_+ \to l^p_+$, $H(t, u, v)$ is nondecreasing in $u$ and in $v$, and the system

$$w_i' = H_i(t, w, w), \qquad w_i(0) = |x_{oi}|, \quad i = 1, 2, \dots$$

has for given $(x_{o1}, x_{o2}, \dots) \in l^p$ an upper solution $w = (w_1, w_2, \dots) \in AC(J, l^p_+)$.

If there is a null set $Z$ in $J$ such that (A0)–(A4) hold, then the system (2.7.1) has for each $x_o = (x_{o1}, x_{o2}, \dots) \in [\alpha(0), \beta(0)]$ the extremal solutions between $\alpha$ and $\beta$ given by (A0). If (A0) and (A1) are replaced by (A6), the above conclusion holds for $\alpha = -w$ and $\beta = w$, where $w$ is any upper solution of the system

$$w_i' = H_i(t, w, w), \qquad w_i(0) = |x_{oi}|, \quad i = 1, 2, \dots$$

in $AC(J, l^p_+)$. Conditions (A2)–(A5) imply that the system (2.7.1) has for each $x_o = (x_{o1}, x_{o2}, \dots) \in l^p$ extremal solutions, which are nondecreasing with respect to $x_{oi}$ and $f_i$, $i = 1, 2, \dots$. The only continuity assumption is that each $f_i(t, x, y)$ is continuous with respect to the $i$:th coordinate of $x$. Also this continuity hypothesis can be dropped in the case when $f_i(t, x, y) = q_i(x_i)\, g_i(t, y)$, $i = 1, 2, \dots$.

### 2.7.1. Existence of extremal solutions

The main result of this subsection is the following theorem which extends theorem 2.4.1 to infinite systems.

**Theorem 2.7.1:** *Assume there is a null set $Z$ in $J$ and $p \in [1, \infty]$ so that conditions (A0)–(A4) hold. Then the system (2.7.1) has the extremal solutions in the order interval $[\alpha, \beta]$ of $AC(J, l^p)$ for each $x_o = (x_{o1}, x_{o2}, \dots) \in [\alpha(0), \beta(0)]$.*

*Proof.* Let $x_o = (x_{o1}, x_{o2}, \dots) \in [\alpha(0), \beta(0)]$ and $y = (y_1, y_2, \dots) \in [\alpha, \beta]$ be given. The given hypotheses imply that the equation

$$\begin{aligned} F_i(t, x; y(t)) =& f_i(t, (y_1(t), \dots, y_{i-1}(t), x, y_{i+1}(t), \dots), y(t)) \\ &+ M_i(t)(\varphi(y_i(t)) - \varphi(x)) \end{aligned} \tag{a}$$

defines for each $i = 1, 2, \dots$ a Carathéodory function $(t, x) \mapsto F_i(t, x; y(t))$ in $\Omega_i = \{(t, x) \in J \times I\!R \mid t \in J,\ x \in [\alpha_i(t), \beta_i(t)]\}$. From (A0), (A3) and (A4) it follows that for each $i = 1, 2, \dots,$

$$\alpha_i' \le F_i(t, \alpha_i; y(t)) \quad \text{and} \quad \beta_i' \ge F_i(t, \beta_i; y(t)) \quad \text{on} \quad J \setminus Z.$$

Condition (A1) implies that for all $x \in [\alpha_i(t), \beta_i(t)]$ and $t \in J \setminus Z$,

$$|F_i(t, x; y(t))| \le K_i\, M_i(t) + N_i(t), \tag{b}$$

where $K_i = \varphi(\max \beta_i) - \varphi(\min \alpha_i)$. Hence, by theorem 1.5.1, each IVP of the system

$$x_i' = F_i(t, x_i; y(t)), \qquad x_i(0) = x_{oi}, \quad i = 1, 2, \dots, \qquad (2.7.2)$$

has for each $y \in [\alpha, \beta]$ the maximal solution $x_i$ in $[\alpha_i, \beta_i]$. Since the system (2.7.2) is uncoupled, then $x = (x_1, x_2, \dots)$ is its maximal solution in $[\alpha, \beta]$.

We now define a map $G = (G_1, G_2, \dots)\colon [\alpha, \beta] \to [\alpha, \beta]$ by

$$G_i y = x_i, \quad y \in [\alpha, \beta], \quad i = 1, 2, \dots, \qquad (2.7.3)$$

where $x = (x_1, x_2, \dots)$ is the maximal solution of (2.7.2) in $[\alpha, \beta]$. To prove that $G$ is nondecreasing, let $y, \bar{y} \in [\alpha, \beta]$, $\bar{y} \le y$, be given, and suppose that $x = (x_1, x_1, \dots)$ and $\bar{x} = (\bar{x}_1, \bar{x}_2, \dots)$ are the corresponding maximal solutions of (2.7.2) and

$$\bar{x}_i' = F_i(t, \bar{x}_i; \bar{y}(t)), \qquad \bar{x}_i(0) = x_{oi}, \quad i = 1, 2, \dots,$$

in $[\alpha, \beta]$, respectively. Since $\bar{y} \le y$, we have by (A3), (A4), (a) and lemma 1.5.1 that $\bar{x}_i(t) \le x_i(t)$ on $J$, thus proving that $G_i\bar{y} \le G_i y$. This holds for each $i = 1, 2, \dots$, whence $G\bar{y} \le Gy$.

Because

$$(G_i y)'(t) = F_i(t, G_i y(t); y(t))$$

for all $y \in [\alpha, \beta]$ and for a.a. $t \in J$, $i = 1, 2, \dots$, it follows by (b) that for all $y \in [\alpha, \beta]$

$$|G_i y(t) - G_i y(\bar{t})| \le \int_{\bar{t}}^{t} (N_i(s) + K\, M_i(s))\, ds, \; 0 \le \bar{t} \le t \le T.$$

Since $M = (M_1, M_2, \dots)$, $N = (N_1, N_2, \dots) \in L^1(J, l^p)$, then

$$\|Gy_n(t) - Gy_n(\bar{t})\|_p \le \int_{\bar{t}}^{t} (\|N(s)\|_p + K\, \|M(s)\|_p)\, ds$$

for $y \in [\alpha, \beta]$, $0 \leq \bar{t} \leq t \leq T$.

The above proof implies that the hypotheses of proposition 1.4.5 hold. Thus $G$ has the greatest fixed point $x^* = (x_1^*, x_2^*, \dots)$, which by definitions of $F_i$ and $G_i$ is also a solution of the system (2.7.1) in $[\alpha, \beta]$.

If $x = (x_1, x_2, \dots)$ is any solution of (2.7.1) in $[\alpha, \beta]$, then it satisfies also the system (2.7.2) with $y = x$. But $Gx = (G_1 x, G_2 x, \dots)$ is the maximal solution of (2.7.2) with $y = x$, whence $x \leq Gx$. This implies by (1.4.11) that $x \leq x^*$. Thus $x^* = (x_1^*, x_2^*, \dots)$ is the maximal solution of the system (2.7.1) in $[\alpha, \beta]$.

The similar reasoning shows that the rule which assigns to each $y \in [\alpha, \beta]$ the minimal solution $x$ of the system (2.7.2) in $[\alpha, \beta]$, defines an operator $G\colon [\alpha, \beta] \to [\alpha, \beta]$ which satisfies the hypotheses of proposition 1.4.5. Thus $G$ has the least fixed point $x_* = (x_{*1}, x_{*2}, \dots)$, which is also the minimal solution of the system (2.7.1) in $[\alpha, \beta]$. □

As a generalization to corollary 2.4.1 we obtain

**Corollary 2.7.1:** *Given $p \in [1, \infty)$, let $f$ in theorem 2.7.1 be defined by*

$$f(t, x, y) = f^1(t, x) + f^2(t, y), \quad t \in J, \ x, \ y \in l^p,$$

*where $f^1(t, x)$ is a Carathéodory function and quasimonotone nondecreasing in $x$, and $f^2(t, y)$ is a standard function and nondecreasing in $y$. If (A0) holds, and if there is $N = (N_1, N_2, \dots) \in L^1(J, l_+^p)$ such that*

$$\sup\{|f_i^j(t, y)| \mid x, \ y \in [\alpha(t), \beta(t)]\} \leq N_i(t) \ \textit{for a.a.} \ t \in J,$$

*$j = 1, 2$, $i = 1, 2, \dots$, then the conclusion of theorem 2.7.1 is valid.*

The results of theorem 2.7.1 and corollary 2.7.1 are valid also when the conditions (A2)–(A4) are restricted to hold in the set $D = \{(t, x, y) \mid t \in J, \ x, \ y \in [\alpha(t), \beta(t)]\}$.

The next result is an extension of proposition 2.4.1 to infinite systems.

**Proposition 2.7.1:** *If $f = (f_1, f_2, \dots): J \times l^p \times l^p \to l^p$ satisfies condition (A6), then (A0) and (A1) hold with $\alpha = -w$, $\beta = w$ and $N = (w_1', w_2', \dots)$, where $w$ is any upper solution of the system*

$$w_i' = H_i(t, w, w), \qquad w_i(0) = |x_{oi}|, \quad i = 1, 2, \dots.$$

*Proof.* If $x,\ y \in [-w, w]$, then the given hypotheses imply that

$$|f_i(t, x(t), y(t))| \le H_i(t, w(t), w(t)) \le w_i'(t) \text{ for a.a. } t \in J. \quad \text{(a)}$$

In view of (a) and the choices of $\alpha$ and $\beta$ we obtain

$$\alpha_i' \le f_i(t, \alpha, \alpha) \text{ and } \beta_i' \ge f(t, \beta, \beta) \text{ a.e. on } J,\ i = 1, 2, \dots. \quad \text{(b)}$$

The conclusions follow from (a) and (b). □

Theorem 2.4.2 can also be extended to the infinite differential systems if the growth conditions for the functions $f_i$ are sufficiently strong.

**Theorem 2.7.2:** *Given $f = (f_1, f_2 \dots): J \times l^p \times l^p \to l^p$ and a null set $Z$ of $J$, assume that conditions (A2)–(A5) hold. Then for each fixed $x_o = (x_{o1}, x_{o2}, \dots) \in l^p$ the system (2.7.1) has the extremal solutions, and all the solutions of (2.7.1) belong to the order interval $[\alpha, \beta]$, given by*

$$\alpha_i(t) = x_{oi} + |x_{oi}| - w_i(t),\ \beta_i(t) = x_{oi} - |x_{oi}| + w_i(t), \quad (2.7.4)$$

*where*

$$w_i(t) = e^{P(t)}[|x_{oi}| + \int_o^t e^{-P(s)} \nu_i(s)\, ds], \qquad (2.7.5)$$

*with* $P(t) = \int_o^t (\mu_1(s) + \mu_2(s))\,ds$.

*Proof.* Let $x_o = (x_{o1}, x_{o2}, \dots) \in l^p$ be given. The function $w = (w_1, w_2, \dots)$ given by (2.7.5) is the solution of the system

$$w_i' = (\mu_1(t) + \mu_2(t))w_i + \nu_i(t), \quad w_i(0) = |x_{oi}|,\ i = 1, 2, \dots. \tag{a}$$

The assumptions given for $\mu_j$ and $\nu_i$ imply by (2.7.5) that $w = (w_1, w_2, \dots) \in AC(J, l_+^p)$. Since $[\alpha_i, \beta_i] \subseteq [-w_i, w_i]$ for each $i = 1, 2, \dots$, it follows from (2.7.4) that $\alpha$, $\beta \in AC(J, l^p)$. By condition (A5) and (a) we have for all $x$, $y \in [\alpha, \beta]$ and $i = 1, 2, \dots$, and for a.a. $t \in J$,

$$|f_i(t, x(t), y(t))| \le \mu_1(t)w_i(t) + \mu_2(t)w_i(t) + \nu_i(t) = w_i'(t). \tag{b}$$

This ensures that the hypotheses (A0) and (A1) of theorem 2.7.1 hold for $\alpha$ and $\beta$ given by (2.7.4) and for $N_i = w_i'$, $i = 1, 2, \dots$. Since $f$ satisfies also the hypotheses (A2)–(A4), then the system (2.7.1) has by theorem 2.7.1 the extremal solutions $x_*$ and $x^*$ in $[\alpha, \beta]$.

If $x = (x_1, x_2, \dots) \in AC(J, l^p)$ is a solution of (2.7.1), then for each $i = 1, 2, \dots$

$$x_i(t) = x_{oi} + \int_o^t f_i(s, x(s), x(s))ds, \quad t \in J. \tag{c}$$

The equations

$$u_i(t) = |x_{oi}| + \int_o^t |f_i(s, x(s), x(s))|\,ds, \quad t \in J, \tag{d}$$

define functions $u_i \in AC(J, I\!R_+)$, $i = 1, 2, \dots$. Because (c) and (d) imply that $|x_i(t)| \le u_i(t)$ on $J$, it follows from (A5) that for a.a. $t \in J$ and for all $i = 1, 2, \dots$

$$u_i'(t) = |f_i(t, x(t), x(t))| \le (\mu_1(t) + \mu_2(t))u_i(t) + \nu_i(t).$$

Thus $u_i$ is a lower solution of the $i$:th IVP of the system (a). Since $w_i$ is the solution of this IVP, it follows from lemma 1.5.3 that $u_i(t) \leq w_i(t)$ on $J$. This implies by (b), (c) and (d) that

$$|x_i(t) - x_{oi}| \leq u_i(t) - |x_{oi}| \leq w_i(t) - |x_{oi}| \qquad \text{(e)}$$

for each $i = 1, 2, \ldots$. From (e) and (2.7.4) we see that $x \in [\alpha, \beta]$. Since $x_*$ and $x^*$ are the extremal solutions of (2.7.1) in $[\alpha, \beta]$, then $x \in [x_*, x^*]$. But $x$ was an arbitrary solution of (2.7.1), whence $x_*$ and $x^*$ are the least and the greatest of all the solutions of the IVP (2.7.1), and all the solutions of (2.7.1) are contained in $[\alpha, \beta]$. □

**Remark 2.7.1:** If $f$ is defined in $J \times V \times V$, where $V \subset l^p$, and if conditions (A2)–(A5) hold, then the extremal solutions of (2.1.1) exist in such a subinterval $J_o = [0, c]$ of $J$ that for each $t \in J_o$, $[\alpha(t), \beta(t)] \subseteq V$, where the components of $\alpha$, $\beta$ are given by (2.7.4).

By theorem 1.4.4 and corollary 1.4.5 condition (A2) holds if $f$ is a standard function or a Borel function and $1 \leq p < \infty$. In the case when the values of $f$ are in $l^p_+$ and $f(t, x, y)$ is nondecreasing in $y$, it is sufficient that condition (A1) is valid for nondecreasing functions $x$, $y$ of the order interval $[x_o, \beta]$. In such a case theorems 1.4.1 and 1.4.2 imply that the so restricted condition (A1) holds if $1 \leq p < \infty$, and if one of the following hypotheses is valid:

(a) $f(\cdot, x, y)$ is right continuous or left continuous in $J \setminus Z$ for all $x$, $y \in l^p$.

(b) $f(t + h, x + k, y + l)$ has the limit in $l^p$ for all $t \in [0, T)$ and $x$, $y \in l^p$, as $h \to 0+$, $k \to 0$ in $l^p_+$ and $l \to 0$ in $l^p_+$.

(c) $f(t - h, x - k, y - l)$ has the limit in $l^p$ for all $t \in (0, T]$ and $x$, $y \in l^p$, as $h \to 0+$, $k \to 0$ in $l^p_+$ and $l \to 0$ in $l^p_+$.

Moreover, if either of conditions (b) or (c) holds, then theorem 1.4.2 ensures that each solution of the system (2.7.1) satisfy the differential equations of that system in the complement of a countable set.

**Example 2.7.1:** Let $g = (g_1, g_2, \dots)\colon J \times l^p \to l^p$ be a bounded Carathéodory function, and assume that $g(t,\cdot)$ is quasimonotone nondecreasing for a.a. $t \in J$. Let $C_i$, $i = 1, 2, \dots,$ be nonempty, bounded and well-ordered sets in $I\!R_+$, for instance, $C_i = \{\frac{i+1}{i} c(n_o, \dots, n_k) \mid (n_o, \dots, n_k) \in D\}$, where $c(n_o, \dots, n_k)$ is defined by (1.1.2). Denote for each $j = 1, 2, \dots$

$$z_j(v) = \begin{cases} v & \text{if } v \le \min C_j \text{ or } v \ge \sup C_j, \\ \min\{z \in C_j \mid v < z\} & \text{if } \min C_j < v < \sup C_j. \end{cases}$$

Define functions $f_i\colon J \times (l^p)^2 \to I\!R$, $i = 1, 2, \dots,$ by

$$f_i(t, x, (y_1, y_2, \dots)) = g_i(t, x) \sum_{j=1}^{i} z_j(y_j).$$

It is easy to see that $f = (f_1, f_2, \dots)$ satisfies conditions (A2)–(A5), whence the system (2.7.1) has for each $x_o = (x_{o1}, x_{o2}, \dots)$ the extremal solutions.

### 2.7.2. Systems with no continuous variables

Consider now the infinite system

$$x_i' = q_i(x_i)\, g_i(t, x), \;\; x_i(0) = x_{oi}, \;\; i = 1, 2, \dots. \tag{2.7.6}$$

**Theorem 2.7.3:** *Given $q_i \in L^\infty(I\!R, I\!R_+)$ such that $\frac{1}{q_i} \in L^\infty_{loc}\,(I\!R, I\!R_+)$ and $M = \sup_i\{\, ess\ sup\ q_i\} < \infty$, and $g = (g_1, g_2, \dots)\colon\ J \times l^p \to l^p$ such that $g(t,\cdot)$ is nondecreasing for a.a. $t \in J$ and each $g_i(\cdot, y(\cdot))$ is measurable for each $y \in AC(J, l^p)$, assume that the system (2.7.6) has a lower solution $\alpha$ and an upper solution $\beta$ in $AC(J, l^p)$ such that $\alpha \le \beta$, and that*

$g(\cdot, \alpha(\cdot))$, $g(\cdot, \beta(\cdot)) \in L^1(J, l^p)$. *Then (2.7.6) has the extremal solutions in the order interval* $[\alpha, \beta]$.

*Proof.* The given hypotheses imply that the equations

$$\int_{x_{oi}}^{G_i x(t)} \frac{dv}{q_i(v)} = \int_o^t g_i(s, x(s))ds, \quad t \in J, \quad i = 1, 2, \ldots, \tag{a}$$

define a nondecreasing mapping $G = (G_1, G_2 \ldots) : [\alpha, \beta] \to [\alpha, \beta]$. If $x \in [\alpha, \beta]$ and $0 \le \bar{t} \le t \le T$, then for each $i = 1, 2, \ldots$

$$\begin{aligned} \frac{|G_i x(t) - G_i x(\bar{t})|}{M} &\le \left| \int_{G_i x(\bar{t})}^{G_i x(t)} \frac{dv}{q_i(v)} \right| = \left| \int_{\bar{t}}^t g_i(s, x(s))\, ds \right| \\ &\le \int_{\bar{t}}^t (|g_i(s, \alpha(s))| + |g_i(s, \beta(s))|)\, ds. \end{aligned}$$

Hence

$$|G_i x(t) - G_i x(\bar{t})| \le M \int_{\bar{t}}^t (|g_i(s, \alpha(s))| + |g_i(s, \beta(s))|)\, ds \tag{b}$$

whenever $0 \le \bar{t} \le t \le T$ and $i = 1, 2, \ldots$.

From (b) it follows that if $1 \le p \le \infty$, then

$$\|Gx(t) - Gx(\bar{t})\|_p \le |w(t) - w(\bar{t})|, \ x \in C, \ t, \bar{t} \in J, \tag{c}$$

where

$$w(t) = M \int_o^t (\|g(s, \alpha(s))\|_p + \|g(s, \beta(s))\|_p)\, ds, \quad t \in J.$$

Thus $G$ satisfies the hypotheses of proposition 1.4.5, whence $G$ has the least fixed point $x_*$ and the greatest fixed point $x^*$.

From (2.7.6) it follows by theorem 1.4.5 that $x = (x_1, x_2 \ldots) \in [\alpha, \beta]$ is a solution of the system (2.7.6) if and only if

$$\int_{x_{oi}}^{x_i(t)} \frac{dv}{q_i(v)} = \int_o^t g_i(s, x(s))ds, \quad t \in J, \ i = 1, 2, \ldots.$$

By (a) this is equivalent to the fact that $x = (x_1, x_2, \ldots)$ is a fixed point of $G = (G_1, G_2, \ldots)$. Since $x_*$ and $x^*$ are the extremal fixed points of $G$, then their components form also the least and the greatest of all the solutions of (2.7.6) in $[\alpha, \beta]$. □

### 2.7.3. Dependence on the data and special cases

Consider first the dependence of the solutions of the system (2.7.1) on the initial values $x_{oi}$ and on the functions $f_i$. The following generalization to lemma 2.4.1 is needed.

**Lemma 2.7.1:** *Let $f = (f_1, f_2, \ldots): J \times l^p \times l^p \to l^p$ satisfy conditions (A2)–(A5), let $x_o = (x_{o1}, x_{o2} \ldots) \in l^p$ be given, and let $x_*$ and $x^*$ be the minimal and the maximal solutions of the system (2.7.1) on $J$.*

a) *If $x \in AC(J, l^p)$ is a lower solution of (2.7.1), then $x \leq x^*$.*

b) *If $x \in AC(J, l^p)$ is an upper solution of (2.7.1), then $x_* \leq x$.*

*Proof.* a) Let $x = (x_1, x_2, \ldots) \in AC(J, l^p)$ be a lower solution of (2.7.1). Denote $w_{oi} = \max\{|x_{oi}|, |x_i(0)|\}$, and let $w_i$ be the solution of the IVP

$$w_i' = (\mu_1(t) + \mu_2(t))\, w_i + \nu_i(t), \qquad w_i(0) = w_{oi}. \tag{a}$$

By choosing $\alpha = x$, $\beta = (w_1, w_2, , \ldots)$ and $N_i = w_i'$ in the proof of theorem 2.7.1, it follows that the system (2.7.1) has a solution

$z$ in $[x, \beta]$. Since $x^*$ is the greatest of all the solutions of (2.7.1), then $z \leq x^*$. Thus $x \leq z \leq x^*$.

The proof of the assertion b) is similar. □

As a direct application of lemma 2.7.1 we obtain

**Proposition 2.7.2:** *Let $f$, $\hat{f}$: $J \times l^p \times l^p \to l^p$ satisfy conditions (A2)–(A5). Given $x_o = (x_{o1}, x_{o2}, \ldots)$, $\hat{x}_o = (\hat{x}_{o1}, \hat{x}_{o2}, \ldots) \in l^p$, let $x_*$ be the minimal solution of the system (2.7.1), and let $\hat{x}^*$ be the maximal solution of the system*

$$x_i' = \hat{f}_i(t, x, x), \qquad x_i(0) = \hat{x}_{oi},\ i = 1, 2, \ldots. \tag{2.7.7}$$

*If $x_o \leq \hat{x}_o$ and $f(t, y, z) \leq \hat{f}(t, y, z)$ for a.a. $t \in J$ and for all $y$, $z \in l^p$, then $x_* \leq x \leq \hat{x}^*$ for any solution $x$ of (2.7.1) or (2.7.7).*

In particular, we have

**Corollary 2.7.2:** *If conditions (A2)–(A5) hold, then the extremal solutions of the system (2.7.1) exist and are nondecreasing with respect to $x_{oi}$ and $f_i$, $i = 1, 2, \ldots$.*

Consider next the case when the functions $q_i : I\!R \to I\!R_+$ and $g_i : J \times l^p \to I\!R$, $i = 1, 2, \ldots$, satisfy the following conditions:

(i) $q_i \in L^\infty(I\!R, I\!R_+)$, $\frac{1}{q_i} \in L^\infty_{loc}(I\!R, I\!R_+)$ and $M = \sup_i\{\text{ ess sup } q_i\} < \infty$;

(ii) $g_i(t, \cdot)$ is nondecreasing for a.a. $t \in J$;

(iii) there exist $a \in L^1(J, I\!R_+)$ and $b = (b_1, b_2, \ldots) \in L^1(J, l^p)$ such that $|g_i(t, x)| \leq a(t)|x_i| + b_i(t)$, $i = 1, 2, \ldots$, for all $x \in l^p$ and for a.a. $t \in J$;

(iv) $g_i(\cdot, y(\cdot))$ is measurable whenever $y \in AC(J, l^p)$.

We are going to prove the following result.

**Theorem 2.7.4:** *If the hypotheses (i)–(iv) hold, then for each $(x_{o1}, x_{o2}, \ldots) \in l^p$ the system (2.7.6) has extremal solutions, and they are nondecreasing with respect to $x_{oi}$ and $g_i$, $i = 1, 2, \ldots$.*

*Proof.* Let $(x_{o1}, x_{o2}, \dots) \in l^p$ be given. Choose $(u_1, u_2, \dots) \in l^p$ so that $|x_{oi}| \le u_i$, $i = 1, 2, \dots$. Each equation of the linear system

$$w_i' = M\,(a(t)w_i + b_i(t)), \qquad w_i(0) = u_i,\ i = 1, 2, \dots \tag{a}$$

has a unique solution

$$w_i(t) = e^{M\int_o^t a(s)ds}u_i + M\int_o^t e^{M\int_s^t a(\tau)d\tau}b_i(s)\,ds, \quad t \in J. \tag{b}$$

Since $(u_1, u_2, \dots) \in l^p_+$ and $b = (b_1, b_2, \dots) \in L^1(J, l^p_+)$, it follows from (b) that $w(t) = (w_1(t), w_2(t), \dots) \in l^p_+$ for each $t \in J$. In view of (a) we have

$$0 \le w_i(t) - w_i(\bar{t}) \le M\,w_i(T)\int_{\bar{t}}^t a(s)\,ds + M\int_{\bar{t}}^t b_i(s)\,ds$$

whenever $0 \le \bar{t} \le t \le T$ and $i = 1, 2, \dots$, whence

$$\|w(t) - w(\bar{t})\|_p \le M\,\|w(T)\|_p\int_{\bar{t}}^t a(s)\,ds + M\int_{\bar{t}}^t \|b(s)\|_p\,ds \tag{c}$$

for all $t$, $\bar{t} \in J$, $\bar{t} \le t$. Thus $w \in AC(J, l^p_+)$.

Let $x = (x_1, x_2, \dots) \in [-w, w]$ be given. Each $g_i(\cdot, x(\cdot))$ is measurable by condition (iv). Condition (iii) implies that

$$|g_i(t, x(t))| \le a(t)|x_i(t)| + b_i(t) \le a(t)w_i(t) + b_i(t) \tag{d}$$

for a.a. $t \in J$ and for each $i = 1, 2, \dots$. Thus $g_i(\cdot, x(\cdot)) \in L^1(J, I\!R)$, whence the equations

$$\int_{x_{oi}}^{G_i x(t)} \frac{dv}{q_i(v)} = \int_o^t g_i(s, x(s))ds, \quad t \in J, \quad i = 1, 2, \dots, \tag{e}$$

define mappings $G_ix\colon J \to I\!R$. Because of (i), (a) and (d) it follows from (e) that

$$|G_ix(t) - G_ix(\bar{t})| \le \left| \int_{G_ix(\bar{t})}^{G_ix(t)} \frac{M\,dv}{q_i(v)} \right| \le M \int_{\bar{t}}^{t} |g_i(s, x(s)|\,ds$$
$$\le M \int_{\bar{t}}^{t} (a(s)w_i(s) + b_i(s))ds = w_i(t) - w_i(\bar{t}) \tag{f}$$

whenever $0 \le \bar{t} \le t \le T$ and $i = 1, 2, \dots$. By choosing $\bar{t} = 0$ in (f) we obtain

$$|G_ix(t) - x_{oi}| \le w_i(t) - |x_{oi}|, \quad t \in J,\ i = 1, 2, ...,$$

which implies that $G_ix(t) \in [-w_i(t), w_i(t)]$, $t \in J$, $i = 1, 2, \dots$. In particular, $Gx(t) = (G_1x(t), G_2x(t), \dots) \in l^p$ for each $t \in J$. Moreover, it follows from (c) and (f) that

$$\|Gx(t) - Gx(\bar{t})\|_p \le M\,\|w(T)\|_p \int_{\bar{t}}^{t} a(s)\,ds + M \int_{\bar{t}}^{t} \|b(s)\|_p\,ds$$

for all $t$, $\bar{t} \in J$, $\bar{t} \le t$. Thus $Gx \in [-w, w]$. Because each $g_i(t, \cdot)$ is by (ii) nondecreasing for a.a. $t \in J$, then (e) implies that $G_ix(t) \le G_iy(t)$ for all $t \in J$ and $i = 1, 2, \dots$, whenever $x$, $y \in [-w, w]$, $x \le y$, whence $G$ is nondecreasing.

The above proof shows that the equations (e) define a mapping $G = (G_1, G_2, \dots)$ from the order interval $[-w, w]$ into itself which satisfies the hypotheses of proposition 1.4.5. Thus $G$ has the least fixed point $x_* = (x_{*1}, x_{*2}, \dots)$ and the greatest fixed point $x^* = (x_1^*, x_2^*, \dots)$. From the definition (e) of $G = (G_1, G_2, \dots)$ it follows by theorem 1.4.5 that $x_* = (x_{*1}, x_{*2}, \dots)$ and $x^* = (x_1^*, x_2^*, \dots)$ are solutions of the IVP (2.7.6).

If $x = (x_1, x_2, \dots) \in AC(J, l^p)$ is a solution of (2.7.6), then

$$x_i(t) = x_{oi} + \int_o^t q_i(x_i(s))g_i(s, x(s))\,ds, \quad t \in J,$$

for each $i = 1, 2, \ldots$. This, the choice of $u_i$, $i = 1, 2, \ldots$, and conditions (i) and (iii) imply that each $x_i$ satisfies

$$|x_i(t)| \leq u_i + M \int_o^t (a(s)|x_i(s)| + b_i(s))\, ds, \quad t \in J.$$

This and lemma 1.5.3 ensure that $|x_i(t)| \leq w_i(t)$, $t \in J$, $i = 1, 2, \ldots$, where $w = (w_1, w_2, \ldots)$ is the solution of the system (a). Thus $x = (x_1, x_2, \ldots) \in [-w, w]$. Because $(x_1, x_2, \ldots)$ is a solution of (2.7.6), then

$$\frac{x_i'(t)}{q_i(x_i(t))} = g_i(t, x(t)) \text{ a.e. on } J, \; x_i(0) = x_{oi}. \tag{g}$$

By integration and application of theorem 1.4.5 we see that

$$\int_{x_{oi}}^{x_i(t)} \frac{dv}{q_i(v)} = \int_o^t g_i(s, x(s))ds, \quad t \in J, \; i = 1, 2, \ldots.$$

Thus $x = (x_1, x_2, \ldots)$ is a fixed point of $G = (G_1, G_2, \ldots)$, defined by (e). Because $x_* = (x_{*1}, x_{*2}, \ldots)$ and $x^* = (x_1^*, x_2^*, \ldots)$ are the least and the greatest fixed points of $G$, then they are also the minimal and the maximal solutions of (2.7.6).

Assume that $f_i \colon J \times l^p \to I\!R$, $i = 1, 2, \ldots$, satisfy conditions (ii)–(iv), and that $(y_1, y_2, \ldots)$ belongs to $l^p$. The above proof implies that the system

$$x_i' = q_i(x_i) f_i(t, (x_1, x_2, \ldots)), \qquad x_i(0) = y_i, \; i = 1, 2, \ldots, \tag{h}$$

has the minimal solution $z = (z_1, z_2, \ldots) \in AC(J, l^p)$, and that

$$\int_{y_i}^{z_i(t)} \frac{dv}{q_i(v)} = \int_o^t f_i(s, z(s))ds, \quad t \in J, \; i = 1, 2, \ldots. \tag{j}$$

Assume further that $x_{oi} \le y_i$, $i = 1, 2, \ldots$, and that

$$g_i(t, x) \le f_i(t, x) \quad \text{for } x \in l^p \text{ and a.e. on } J, \; i = 1, 2, \ldots. \qquad \text{(k)}$$

By choosing $u_i = \max\{|x_{oi}|, |y_i|\}$, $i = 1, 2, \ldots$, in (a), it is easy to see that $z \in [-w, w]$. Moreover, it follows from (e), (j) and (k) that

$$\int_{x_{oi}}^{G_i z(t)} \frac{dv}{q_i(v)} = \int_o^t g_i(s, z(s))ds$$

$$\le \int_o^t f_i(s, z(s))ds = \int_{y_i}^{z_i(t)} \frac{dv}{q_i(v)}$$

for all $t \in J$ and $i = 1, 2, \ldots$. Because $x_{oi} \le y_i$, $i = 1, 2, \ldots$, we then have

$$\int_{x_{oi}}^{G_i z(t)} \frac{dv}{q_i(v)} \le \int_{y_i}^{z_i(t)} \frac{dv}{q_i(v)}, \qquad t \in J, \; i = 1, 2, \ldots.$$

This implies that $G_i z(t) \le z_i(t)$ for all $t \in J$ and $i = 1, 2, \ldots$, whence $Gz \le z$. In view of (1.4.11) we then have $x_* \le z$. Thus the minimal solution of the IVP (2.7.6) is nondecreasing with respect to $c_i$ and $g_i$, $i = 1, 2, \ldots$. The proof of the corresponding results to the maximal solution of (2.7.6) is similar. □

**Example 2.7.2:** Let $C_i$, $i = 1, 2, \ldots$, be inversely well-ordered chains in $(0, 1)$, for instance, $C_i = \{\frac{i}{i+n+1} \mid n \in \mathbb{N}\}$, $i = 1, 2, \ldots$. Denote $J = [0, 1]$, and define $\psi_i \colon J \to \mathbb{R}$, and $h_i \colon l^\infty \to \mathbb{R}$, $i = 1, 2, \ldots$, by

$$\psi_i(x) = \max\{y \in C_i \cup \{0, 1\} \mid x \ge y\},$$

and

$$h_i(x_1, x_2, \ldots) = (x_i + \psi_i(x_i - [x_i])) \left(1 + \prod_{j=i+1}^{2i} \psi_j \left(\frac{x_j}{|x_j| + 1}\right)\right).$$

Let $g_i\colon J \times l^\infty \to \mathbb{R}$ and $q_i\colon \mathbb{R} \to \mathbb{R}_+$, $i = 1, 2, \ldots$, be given by

$$g_i(t,x) = \begin{cases} \cos(\psi_1(t))h_i(x) + \sin(1 + \psi_i(t)), & t \in J,\ x \in l_+^\infty, \\ 0, & t \in J,\ x \notin l_+^\infty, \end{cases}$$

and

$$q_i(x) = 1 + \psi_i\left(\frac{|x|}{|x|+1}\right).$$

It is easy to see that the hypotheses (i)-(iv) hold when $p = \infty$. Thus for each bounded sequence of real numbers $x_{o1}, x_{o2}, \ldots$, the IVP (2.7.6) has by theorem 2.7.4 the extremal solutions which are nondecreasing with respect to $x_{oi}$, $i = 1, 2, \ldots$.

As a consequence of theorem 2.7.4 we get.

**Proposition 2.7.3:** *Assume that $q_i$ and $g_i$ satisfy the hypotheses (i)–(iii), and that $g = (g_1, g_2, \ldots)\colon J \times l^p \to l^p$, $1 \le p < \infty$, is a Carathéodory function. Then the sequence $(y_n)_{n=o}^\infty$ of functions $y_n\colon J \to l^p$, defined by*

$$\int_{x_{oi}}^{(y_{n+1})_i(t)} \frac{dv}{q_i(v)} = \int_o^t g_i(s, y_n(s))ds, \tag{2.7.8}$$

*converges uniformly on $J$ to the minimal (resp. the maximal) solution of the system (2.7.6) if $y_o = -w$ (resp. $y_o = w$), where $w$ is the solution of the system*

$$w_i' = M\left(a(t)w_i + b_i(t)\right), \qquad w_i(0) = |x_{oi}|,\ i = 1, 2, \ldots. \tag{2.7.9}$$

*Proof.* Since $g$ is a Carathéodory function, then also condition (iv) holds. By the proof of theorem 2.7.4 the equation

$$\int_{x_{oi}}^{G_i y(t)} \frac{dv}{q_i(v)} = \int_o^t g_i(s, y(s))ds, \quad t \in J,\ i = 1, 2, \ldots \tag{a}$$

defines a nondecreasing operator $G = (G_1, G_2, \dots) \colon [-w, w] \to [-w, w]$. By choosing $y_o = -w$, it follows from (2.7.8) and (a) that $(y_n)_{n=o}^{\infty} = (G^n(-w))_{n=o}^{\infty}$. Thus the sequence $(y_n)_{n=o}^{\infty}$ is nondecreasing and belongs to $[-w, w]$. Because $G$ satisfies the hypotheses of theorem 1.4.7, the proof of this theorem implies that $(y_n)_{n=o}^{\infty}$ converges uniformly on $J$ to a function $x_* \in [-w, w]$. Since $g$ is a Carathéodory function, it follows from (2.7.8) as $n \to \infty$ that for each $i = 1, 2, \dots$ the $i$:th component $x_{*i}$ of $x_*$ satisfies the integral equation

$$\int_{x_{oi}}^{x_{*i}(t)} \frac{dv}{q_i(v)} = \int_o^t g_i(s, x_*(s))ds, \quad t \in J. \tag{b}$$

Thus $x_*$ is a fixed point of $G$, and hence, by the proof of theorem 2.7.4, a solution of (2.7.6).

If $x = (x_1, x_2, \dots)$ is a solution of the system (2.7.6), the proof of theorem 2.7.4 implies that it is also a fixed point of $G$. Since $y_o = -w \leq x$ and $G$ is nondecreasing, it follows by induction that $y_n = G^n(-w) \leq x$ for each $n \in I\!N$. Thus $x_* = \lim_n y_n \leq x$, whence $x_* = (x_{*1}, x_{*2}, \dots)$ is the least solution of the system (2.7.6).

The proof that $(y_n)_{n=o}^{\infty}$ converges to the greatest solution of (2.7.6) when $y_o = w$ is similar to the above one. □

In the case when $q_i(x) = 1$ a.e. on $I\!R$, we obtain

**Corollary 2.7.3:** *Assume that $g = (g_1, g_2, \dots) \colon J \times l^p \to l^p$, $1 \leq p < \infty$ is a Carathéodory function, that conditions (ii) and (iii) hold, and that $q_i \colon I\!R \to I\!R$ is positive-valued and equals to 1 a.e. on $I\!R$ for each $i = 1, 2, \dots$. Then the sequence $(y_n)_{n=o}^{\infty}$ of the successive approximations $y_n \colon J \to I\!R$, defined by*

$$y_{n+1}(t) = y_o + \int_{t_o}^t g(s, y_n(s))ds, \quad t \in J, \ n \in I\!N,$$

*converges uniformly on $J$ to the minimal (resp. the maximal) solution of the system (2.7.6) if $y_o = -w$ (resp. $y_o = w$), where $w = (w_1, w_2, \dots)$ is the solution of the system (2.7.9).*

## 2.8. INFINITE MIXED MONOTONE SYSTEMS

It turns out that many of the results derived in section 2.5 for finite mixed monotone systems have their counterparts within infinite mixed monotone systems

$$x_i' = f_i(t,x,x,x), \qquad x_i(0) = x_{oi}, \quad i = 1,2,\dots, \tag{2.8.1}$$

where $f = (f_1, f_2, \dots): J \times (l^p)^3 \to l^p$, $J = [0,T]$, $1 \le p \le \infty$.

The concepts of coupled quasisolutions can be defined for the system (2.8.1) similarly as for finite systems.

We shall first list conditions for $f$, which are used as the hypotheses in subsequent theorems. $Z$ denotes a null set in $J$.

(B0) $\alpha, \beta \in AC(J, l^p)$, $\alpha' \le f(t,\alpha,\alpha,\beta)$ and $\beta' \ge f(t,\beta,\beta,\alpha)$ on $J \setminus Z$, and $\alpha \le \beta$.

(B1) There is $N = (N_1, N_2, \dots) \in L^1(J, l^p_+)$ such that $|f_i(t,x,y,z)| \le N_i(t)$ for all $t \in J \setminus Z$, $x$, $y$, $z \in [\alpha(t), \beta(t)]$ and $i = 1,2,\dots$ .

(B2) $f(\cdot, x(\cdot), y(\cdot), z(\cdot))$ is measurable for all $x$, $y$, $z \in AC(J, l^p)$.

(B3) for each $i = 1,2,\dots$ the function $f_i(t,x,y,z)$ is continuous with respect to the $i$:th coordinate of $x$ for all $x$, $y$, $z \in l^p$ and $t \in J \setminus Z$.

(B4) $f(t,\cdot,y,z)$ is quasimonotone nondecreasing, $f(t,y,\cdot,z)$ is nondecreasing and $f(t,y,z,\cdot)$ is nonincreasing for all $y$, $z \in l^p$ and $t \in J \setminus Z$.

(B5) $f_i(t,(x_1,\dots,x_i+k,\dots),y,z) - f_i(t,x,y,z) \le g_i(t,k)$ for all $i = 1,2,\dots$ and for all $x = (x_1,x_2,\dots)$, $y$, $z \in l^p$, $k \ge 0$, and $t \in J \setminus Z$, where $g_i: J \times I\!R_+ \to I\!R_+$ and $u(t) \equiv 0$ is the only lower solution of the IVP $u' = g_i(t,u)$, $u(0) = 0$ on $J$.

(B6) $|f_i(t,x,y,z)| \le \mu_1(t)\, x_i + \mu_2(t)\, y_i + \mu_3(t)\, z_i + q_i(t)$ for all $t \in J \setminus Z$ and $x$, $y$, $z \in l^p$, where $\mu_j \in L^1(J, I\!R_+)$, $j = 1,2,3$, and $\nu = (\nu_1, \nu_2, \dots) \in L^1(J, l^p_+)$.

(B7) $|f_i(t,x,y,z)| \le H_i(t, (|x_1|, |x_2|, \dots), (|y_1|, \dots), (|z_1|, \dots))$ for all $x = (x_1, x_2, \dots)$, $y = (y_1, y_2, \dots)$ and $z = (z_1, z_2, \dots) \in l^p$, $t \in J \setminus Z$ and $i = 1,2,\dots$, where $H = (H_1,$

$H_2, \dots): J \times (l^p_+)^3 \to l^p_+$, $H(t, u, v, w)$ is nondecreasing in $u$, $v$, $w$, and the system
$w_i' = H_i(t, w, w, w)$, $w_i(0) = |x_{oi}|$, $i = 1, 2, \dots$
has for a given $(x_{o1}, x_{o2}, \dots) \in l^p$ an upper solution $w = (w_1, w_2, \dots)$ in $AC(J, l^p_+)$.

As we shall see, conditions (B0)–(B4) imply the existence of coupled quasisolutions of (2.8.1) in the order interval $[\alpha, \beta]$ of $AC(J, l^p)$ if $(x_{o1}, x_{o2}, \dots) \in [\alpha(0), \beta(0)]$. If also (B5) holds, there exist *extremal* coupled quasisolutions $y$, $z$ in $[\alpha, \beta]$. Condition (B7) implies (B0) and (B1), and if (B2)–(B6) hold, then (2.8.1) has for each $(x_{o1}, x_{o2}, \dots) \in l^p$ extremals among all its coupled quasisolutions. In the case when $f_i(t, x, y, z) = q_i(x_i)\, g_i(t, y, z)$ no continuity assumptions are imposed on $f_i$.

### 2.8.1. Existence of coupled quasisolutions

Our first result is an infinite version of theorem 2.5.1.

**Theorem 2.8.1:** *If there is a null set $Z$ in $J$ and $p \in [1, \infty]$ so that the hypotheses (B0)–(B4) hold, then the system (2.8.1) has coupled quasisolutions in the order interval $[\alpha, \beta]$ of $AC(J, l^p)$ for each $x_o = (x_{o1}, x_{o2}, \dots) \in [\alpha(0), \beta(0)]$.*

**Proof:** Given $x_o = (x_{o1}, x_{o2}, \dots) \in [\alpha(0), \beta(0)]$ and $y$, $z \in [\alpha, \beta]$, consider the system

$$x_i' = F_i(t, x_i; y(t), z(t)), \qquad x_i(0) = x_{oi}, \quad i = 1, 2, \dots, \tag{2.8.2}$$

where

$$\begin{aligned} &F_i(t, x; y(t), z(t)) = \\ &f_i(t, (y_1(t), \dots, y_{i-1}(t), x, y_{i+1}(t), \dots), y(t), z(t)), \end{aligned} \tag{2.8.3}$$

$t \in J$, $x \in [\alpha_i(t), \beta_i(t)]$. From (B2) and (B3) it follows that $(t, x) \mapsto F_i(t, x, y(t), z(t))$ is a Carathéodory function in the set

$\Omega_i = \{(t,x) \mid x \in [\alpha_i(t), \beta_i(t)],\ t \in J\}$. By (B4) we get

$$F_i(t, \alpha_i(t); \alpha(t), \beta(t)) \leq F_i(t, \alpha_i(t); y(t), z(t))$$

and

$$F_i(t, \beta_i(t); \beta(t), \alpha(t)) \geq F_i(t, \beta_i(t); y(t), z(t))$$

on $J \setminus Z$ for each $i = 1, 2, \ldots$. This and (B0) imply that

$$\alpha_i' \leq F_i(t, \alpha_i; y(t), z(t)) \quad \text{and} \quad \beta_i' \geq F_i(t, \beta_i; y(t), z(t))$$

on $J \setminus Z$, $i = 1, 2, \ldots$. Condition (B1) implies that for each $i = 1, 2, \ldots$

$$|F_i(t, x; y(t), z(t))| \leq N_i(t) \quad \text{for } x \in [\alpha_i(t), \beta_i(t)]\ ,\ t \in J \setminus Z. \quad \text{(a)}$$

Hence, by theorem 1.5.1 each IVP of the system (2.8.2) has for all choices of $y$, $z \in [\alpha, \beta]$ the maximal solution $x_i$ in $[\alpha_i, \beta_i]$. Since the system (2.8.2) is uncoupled, then $x = (x_1, x_2, \ldots)$ is the maximal solution of (2.8.2) in $[\alpha, \beta]$.

We now define a map $A = (A_1, A_2, \ldots)\colon [\alpha, \beta] \times [\alpha, \beta] \to [\alpha, \beta]$ by

$$A_i(y, z) = x_i, \quad y,\ z \in [\alpha, \beta],\ i = 1, 2, \ldots, \tag{2.8.4}$$

where $x = (x_1, x_2, \ldots)$ is the maximal solution of (2.8.2) in $[\alpha, \beta]$. To show that $A$ is mixed monotone, let $y$, $\bar{y}$, $z \in [\alpha, \beta]$, $\bar{y} \leq y$, be given, and denote $x = A(y, z)$ and $\bar{x} = A(\bar{y}, z)$. Since $\bar{y} \leq y$, we have by (B4) and (2.8.3) for each $i = 1, 2 \ldots$

$$\bar{x}_i' \leq F_i(t, \bar{x}_i; y(t), z(t)) \quad \text{on } J \setminus Z,$$

which implies by lemma 1.5.1 that $\bar{x}_i \leq x_i$, $i = 1, 2, \ldots$. Thus $A(\bar{y}, z) \leq A(y, z)$, so that $A(y, z)$ is nondecreasing in $y$. Similarly, it can be shown that $A(y,z)$ is nonincreasing in $z$, whence $A$ is mixed monotone.

Because

$$(A_i(y,z))'(t) = F_i(t, A_i(y,z)(t); y(t), z(t))$$

for all $y$, $z \in [\alpha, \beta]$ and for a.a. $t \in J$, it follows by (a) that

$$|A_i(y,z)(t) - A_i(y,z)(\bar{t})| \le \int_{\bar{t}}^{t} N_i(s)\, ds$$

for $y$, $z \in [\alpha, \beta]$ and $0 \le \bar{t} \le t \le T$. This implies in turn that

$$\|A(y,z)(t) - A(y,z)(\bar{t})\|_p \le \int_{\bar{t}}^{t} \|N(s)\|_p\, ds,$$

whenever $y$, $z \in [\alpha, \beta]$ and $0 \le \bar{t} \le t \le T$. The above proof ensures by proposition 1.4.7 that $A$ has the extremal coupled fixed points $y$, $z$. In particular,

$$y = A(y,z), \qquad z = A(z,y). \tag{2.8.5}$$

From (2.8.2)–(2.8.5) it then follows that these functions $y = (y_1, y_2, \dots)$, $z = (z_1, z_2, \dots)$ are coupled quasisolutions of the (2.8.1) in $[\alpha, \beta]$. □

By adding condition (B5) to the hypotheses we obtain

**Proposition 2.8.1:** *Let $f\colon J \times (l^p)^3 \to l^p$ satisfy conditions (B0)–(B5). Then the system (2.8.1) has the extremal coupled quasisolutions in $[\alpha, \beta]$.*

*Proof.* Cf. the proof of proposition 2.2.1. □

As consequence of theorem 2.8.1, proposition 2.8.1 and corollary 1.5.1 we obtain

**Corollary 2.8.1:** *Given $p \in [1, \infty)$, let $f$ be defined by*

$$f(t, x, y, z) = f^1(t, x) + f^2(t, y) + f^3(t, z), \tag{2.8.6}$$

*where $f^1 : J \times l^p \to l^p$ is a Carathéodory function, $f^2$, $f^3 : J \times l^p \to l^p$ are standard functions, $f^1(t, x)$ is quasimonotone nondecreasing in $x$, $f^2(t, y)$ is nondecreasing in $y$ and $f^3(t, z)$ is nonincreasing in $z$. If (B0) holds, and if there exists $N = (N_1, N_2, \dots) \in L^1(J, l^p_+)$ such that*

$$\sup\{|f_i^j(t, y)| \mid y \in [\alpha(t), \beta(t)]\} \le N_i(t)$$

*for a.a. $t \in J$, and for $j = 1, 2, 3$, $i = 1, 2, \dots$, then the system (2.8.1) has for each $x_o = (x_{o1}, x_{o2}, \dots) \in [\alpha(0),\ \beta(0)]$ coupled quasisolutions in $[\alpha, \beta]$. Moreover, if there exist $p_i \in L^1(J, \mathbb{R}_+)$, $i = 1, 2, \dots$, such that*

$$f_i^1(t, (x_1, \dots, x_i + k, \dots) - f_i^1(t, (x_1, \dots, x_i, \dots) \le p_i(t)\, k$$

*for all $x \in l^p$, $k \ge 0$, and for a.a. $t \in J$, then (2.8.1) has extremal coupled quasisolutions in $[\alpha, \beta]$.*

The results of theorem 2.8.1, proposition 2.8.1 and corollary 2.8.1 are valid also when the conditions (B2)–(B4) are restricted to hold in the set $D = \{(t, x, y, z) \mid t \in J,\ x,\ y,\ z \in [\alpha(t), \beta(t)]\}$.

The next result gives a method to find $\alpha$, $\beta \in AC(J, l^p)$ such that (B0) and (B1) hold.

**Proposition 2.8.2:** *If $f = (f_1, f_2, \dots)\colon J \times (l^p)^3 \to l^p$ satisfies condition (B7), then (B0) and (B1) hold with $\alpha = -w$, $\beta = w$ and $N = (w'_1, w'_2, \dots)$, where $w\colon J \to l^p$ is any upper solution of the system*

$$w'_i = H_i(t, w, w, w), \qquad w_i(0) = |x_{oi}|, \quad i = 1, 2, \dots. \tag{2.8.7}$$

*Proof.* If $x$, $y$, $z \in [-w, w]$, the given hypotheses imply that

$$|f_i(t, x, y, z)| \le H_i(t, w(t), w(t), w(t)) \le w'_i(t) \tag{a}$$

for a.a. $t \in J$ and $i = 1, 2, \dots$. According to (a) and the choices of $\alpha$ and $\beta$ we have for each $i = 1, 2, \dots,$

$$\alpha'_i \le f_i(t, \alpha, \alpha, \beta) \text{ and } \beta'_i \ge f_i(t, \beta, \beta, \alpha) \text{ a.e. on } J. \tag{b}$$

The conclusions follow from (a) and (b). □

As a consequence of theorem 2.8.1 and proposition 2.8.2 we obtain

**Corollary 2.8.2:** *If conditions (B2)–(B5) and (B7) hold, then the system (2.8.1) has extremal coupled quasisolutions in $[-w, w]$, where $w\colon J \to l^p$ is any upper solution of (2.8.7). The conclusion holds also when $f(t, \cdot, y, z)$ is quasimonotone nonincreasing.*

*Proof.* Cf. the proof of corollary 2.5.2. □

Theorem 2.5.2 can also be extended to the infinite differential systems, when the growth conditions (B6) hold for functions $f_i$.

**Theorem 2.8.2:** *Given $f = (f_1, f_2 \dots)\colon J \times (l^p)^3 \to l^p$ and a null set $Z$ of $J$, assume that conditions (B2)–(B6) hold. Then for each fixed $x_o = (x_{o1}, x_{o2}, \dots) \in l^p$ the system (2.8.1) has the*

*extremal coupled quasisolutions, and all the coupled quasisolutions of (2.8.1) belong to the order interval* $[\alpha, \beta]$, *given by*

$$\alpha_i(t) = x_{oi} + |x_{oi}| - w_i(t), \; \beta_i(t) = x_{oi} - |x_{oi}| + w_i(t), \quad (2.8.8)$$

*where*

$$w_i(t) = e^{P(t)}[|x_{oi}| + \int_o^t e^{-P(s)} \nu_i(s)\, ds], \quad (2.8.9)$$

*with* $P(t) = \int_o^t (\mu_1(s) + \mu_2(s) + \mu_3(s))\, ds$.

*Proof.* Let $x_o = (x_{o1}, x_{o2}, \dots) \in l^p$ be given. The function $w = (w_1, w_2, \dots)$ given by (2.8.9) is the solution of the system

$$w_i' = (\mu_1(t) + \mu_2(t) + \mu_3(t))w_i + \nu_i(t), \; w_i(0) = |x_{oi}|. \quad \text{(a)}$$

The assumptions given for $\mu_j$ and $\nu_i$ imply by (2.8.9) that $w = (w_1, w_2, \dots) \in AC(J, l_+^p)$. Since $[\alpha_i, \beta_i] \subseteq [-w_i, w_i]$ for each $i = 1, 2, \dots$, it follows from (2.8.8) that $\alpha$, $\beta \in AC(J, l^p)$. By condition (B6) and (a) we have for all $x$, $y$, $z \in [\alpha, \beta]$ and $i = 1, 2, \dots$

$$|f_i(t, x(t), y(t), z(t))| \leq (\mu_1(t)w_i(t) + \mu_2(t)w_i(t)) + \nu_i(t) = w_i'(t),$$

for a.a. $t \in J$. This ensures that the hypotheses (B0) and (B1) of theorem 2.8.1 hold for $\alpha$ and $\beta$ given by (2.8.8) and for $N_i = w_i'$, $i = 1, 2, \dots$. Since $f$ satisfies also the hypotheses (B2)–(B5), then the system (2.8.1) has by proposition 2.8.1 the extremal coupled quasisolutions $x$, $y$, $z$ in $[\alpha, \beta]$.

If $u = (u_1, u_2, \dots)$, $v = (v_1, v_2, \dots) \in AC(J, l^p)$ are coupled quasisolutions of (2.8.1), then for each $i = 1, 2, \dots$

$$u_i(t) = x_{oi} + \int_o^t f_i(s, u(s), u(s), v(s))ds, \quad t \in J, \quad \text{(b)}$$

and

$$v_i(t) = x_{oi} + \int_o^t f_i(s, v(s), v(s), u(s))ds, \quad t \in J. \qquad \text{(c)}$$

Denoting $x_i(t) = \max\{|u_i(t)|, |v_i(t)|\}$, $i = 1, 2, \dots$, it follows from (B6) that for each $i = 1, 2, \dots$

$$x_i(t) \le |x_{oi}| + \int_o^t ((\mu_1(s) + \mu_2(s) + \mu_3(s))x_i(s) + \nu_i(s))\, ds, \; t \in J.$$

This implies by from lemma 1.5.3 that $x_i(t) \le w_i(t)$ on $J$. Thus, by (b) and (c),

$$|u_i(t) - x_{oi}| \le x_i(t) - |x_{oi}| \le w_i(t) - |x_{oi}|$$

and

$$|v_i(t) - x_{oi}| \le x_i(t) - |x_{oi}| \le w_i(t) - |x_{oi}|,$$

so that $u$, $v \in [\alpha, \beta]$. Since $y$ and $z$ are the extremal coupled quasisolutions of (2.8.1) in $[\alpha, \beta]$, then $u$, $v \in [y, z]$. Thus $y$ and $z$ are extremal coupled quasisolutions of the system (2.8.1), and all the coupled quasisolutions of (2.8.1) are contained in $[\alpha, \beta]$. □

Since condition (B6) is a special case of (B7), it follows from corollary 2.8.2 that $f(t, \cdot, y, z)$ can be also quasimonotone nonincreasing in theorem 2.8.2.

### 2.8.2. A case when $f$ is discontinuous in all its arguments

The following result is a generalization of theorem 2.5.3 for the infinite system

$$x_i' = q_i(x_i)\, g_i(t, x, x), \qquad x_i(0) = x_{oi}, \quad i = 1, 2, \dots. \qquad (2.8.10)$$

**Theorem 2.8.3:** *Given $q_i$, $\frac{1}{q_i} \in L^\infty_{loc}(I\!R, I\!R_+)$ such that $M = \sup_i\{ess\ sup\ q_i\} < \infty$, and a standard function $g = (g_1, g_2, \dots)$: $J \times (l^p)^2 \to l^p$, assume that $g(t, \cdot, y)$ is nondecreasing and $g(t, y, \cdot)$ is nonincreasing for a.a. $t \in J$ and for all $y \in l^p$, and denote*

$$f(t, x, y, z) = (q_1(x_1)\, g_1(t, y, z), q_2(x_2)\, g_2(t, y, z), \dots).$$

a) *If (B0) holds, and if $g(\cdot, \alpha(\cdot), \beta(\cdot))$, $g(\cdot, \beta(\cdot), \alpha(\cdot)) \in L^1(J, l^p)$, then the system (2.8.10) has the extremal coupled quasisolutions in $[\alpha, \beta]$ for each $x_o = (x_{o1}, x_{o2}, \dots) \in [\alpha(0), \beta(0)]$.*
b) *If (B6) holds, then the system (2.8.10) has the extremal coupled quasisolutions for each $x_o = (x_{o1}, x_{o2}, \dots) \in l^p$.*

*Proof.* a) The given hypotheses imply that for all $i = 1, 2, \dots$ and $y$, $z \in [\alpha, \beta]$ the equations

$$\int_{x_o}^{A_i(y,z)(t)} \frac{dv}{q_i(v)} = \int_o^t g_i(s, y(s), z(s))ds, \tag{a}$$

define a mapping $A = (A_1, A_2 \dots)$: $[\alpha, \beta] \times [\alpha, \beta] \to [\alpha, \beta]$ which is mixed monotone. The reasoning used in the proof of theorem 2.2.3 shows that

$$|A_i(y, z)(t) - A_i(y, z)(\bar{t})| \leq \\ M \int_{\bar{t}}^t (|g_i(s, \alpha(s), \beta(s))| + |g_i(s, \beta(s), \alpha(s))|)\, ds \tag{b}$$

whenever $y$, $z \in [\alpha, \beta]$ and $0 \leq \bar{t} \leq t \leq T$.

From (b) it follows that for all $y$, $z \in [\alpha, \beta]$,

$$\|A(y, z)(t) - A(y, z)(\bar{t})\|_p \leq \\ M \int_{\bar{t}}^t (\|g(s, \alpha(s), \beta(s))\|_p + \|g(s, \beta(s), \alpha(s))\|_p)ds. \tag{c}$$

The above proof shows that $A$ satisfies the hypotheses of proposition 1.4.7. Thus $A$ has the extremal coupled fixed points $y$, $z$. From theorem 1.4.5 it follows that $u = (u_1, u_2, \dots)$, and $v = (v_1, v_2, \dots)$ are coupled quasisolutions of (2.8.10) in $[\alpha, \beta]$ if and only if for each $i = 1, 2, \dots$

$$\int_{x_o}^{u_i(t)} \frac{dx}{q_i(x)} = \int_o^t g_i(s, u(s), v(s))ds,$$

and

$$\int_{x_o}^{v_i(t)} \frac{dx}{q_i(x)} = \int_o^t g_i(s, v(s), u(s))ds,$$

which is by (a) equivalent that $u$, $v$ are coupled fixed points of $A$. Since $y$, $z$ are the extremal coupled fixed points of $A$, then their components form the extremal coupled quasisolutions of the system (2.8.10) in $[\alpha, \beta]$.

b) Cf. the proof of theorem 2.5.3 b). □

From theorem 2.8.3 it follows

**Proposition 2.8.3:** *If the functions $q_i$, $g$ satisfy the hypotheses theorem 2.8.3 a) with $1 \leq p < \infty$, and if $g$ is a Carathéodory function, then for each $x_o = (x_{o1}, x_{o2}, \dots) \in [\alpha(0), \beta(0)]$ the sequences $(y_n)_{n=o}^{\infty}$ and $(z_n)_{n=o}^{\infty}$ of functions $y_n$, $z_n : J \to l^p$, defined by $y_o = \alpha$, $z_o = \beta$,*

$$\int_{x_o}^{(y_{n+1})_i(t)} \frac{dv}{q_i(v)} = \int_o^t g_i(s, y_n(s), z_n(s))ds, \tag{2.8.11}$$

*and*

$$\int_{x_o}^{(z_{n+1})_i(t)} \frac{dv}{q_i(v)} = \int_o^t g_i(s, z_n(s), y_n(s))ds, \tag{2.8.12}$$

*converge uniformly on $J$ to the extremal coupled quasisolutions of the system (2.8.10) in $[\alpha, \beta]$.*

*Proof.* Similar to that of proposition 2.2.4. □

When $q_i(x) = 1$ a.e. on $\mathbb{R}_+$, then (2.8.11) and (2.8.12) are reduced to the sequences of successive approximations

$$y_{n+1}(t) = x_o + \int_o^t g(s, y_n(s), z_n(s))\, ds, \quad t \in J, \ n \in \mathbb{N}, \quad (2.8.13)$$

and

$$z_{n+1}(t) = x_o + \int_o^t g(s, z_n(s), y_n(s))\, ds, \quad t \in J, \ n \in \mathbb{N}. \quad (2.8.14)$$

Thus we obtain

**Corollary 2.8.3:** *Assume that $g\colon J \times (l^p)^2 \to l^p$ is a Carathéodory function, and that $g(t, \cdot, y)$ is nondecreasing and $g(t, y, \cdot)$ is nonincreasing for all $y \in l^p$ and for a.a. $t \in J$. If $q_i\colon \mathbb{R} \to \mathbb{R}$ is positive-valued and $q_i(x) = 1$ a.e. on $J$, $i = 1, 2, \ldots$, and if there exist $\alpha$, $\beta \in AC(J, l^p)$ such that*

$$\alpha \le \beta, \ \alpha' \le g(t, \alpha, \beta), \quad \text{and } \beta' \ge g(t, \beta, \alpha) \ \text{a.e. on } J,$$

*then the sequences $(y_n)_{n=o}^{\infty}$ and $(z_n)_{n=o}^{\infty}$ of the successive approximations defined by (2.8.13) and (2.8.14) converge uniformly on $J$ to the extremal coupled quasisolutions $y$, $z$ of the system (2.8.10) in $[\alpha, \beta]$ if $y_o = \alpha$, $z_o = \beta$ and $\alpha(0) \le (x_{o1}, x_{o2}, \ldots) \le \beta(0)$.*

## 2.9. INFINITE SYSTEMS OF PERIODIC BVP'S

In this section we shall show that many of the results derived in section 2.6 for finite systems of periodic boundary value problems can be generalized to the corresponding infinite systems in $l^p$-spaces, where $1 \le p \le \infty$.

### 2.9.1. Existence of extremal solutions

Consider the infinite periodic boundary value system

$$x_i' = f_i(t, x, x), \qquad x_i(0) = x_i(T), \quad i = 1, 2, \ldots, \tag{2.9.1}$$

where $f = (f_1, f_2, \ldots) \colon J \times (l^p)^2 \to l^p$, $J = [0, T]$. We shall use the following conditions as assumptions of our theorems.

(C0) $\alpha$, $\beta \in AC(J, l^p)$, $\alpha \le \beta$, $\alpha(0) \le \alpha(T)$, $\beta(0) \ge \beta(T)$, $\alpha' \le f(t, \alpha, \alpha)$ and $\beta' \ge f(t, \beta, \beta)$ on $J \setminus Z$.

(C1) $|f_i(t, x, y)| \le N_i(t)$ for all $t \in J \setminus Z$ and $x$, $y \in [\alpha(t), \beta(t)]$.

(C2) $f(\cdot, x(\cdot), y(\cdot))$ is measurable for all $x$, $y \in [\alpha, \beta]$.

(C3) $f(t, \cdot, y)$ and $f(t, y, \cdot)$ are quasimonotone nondecreasing for all $t \in J \setminus Z$ and $y \in l^p$.

(C4) $f_i(t, x, y) + M_i(t)\, y_i$ is nondecreasing in $y_i \in [\alpha_i(t), \beta_i(t)]$ for all $t \in J \setminus Z$, $x$, $y \in [\alpha(t), \beta(t)]$ and $i = 1, 2, \ldots$.

(C5) $f_i(t, x, y) - \mu_i(t)\, x_i$ is continuous and nonincreasing in $x_i \in [\alpha_i(t), \beta_i(t)]$ for all $t \in J \setminus Z$, $x$, $y \in [\alpha(t), \beta(t)]$ and $i = 1, 2, \ldots$.

(C6) $f_i(t, x, y) + \mu_i(t)\, x$ is nondecreasing in $x \in [\alpha(t), \beta(t)]$ for all $t \in J \setminus Z$, $x$, $y \in [\alpha(t), \beta(t)]$ and $i = 1, 2, \ldots$.

(C7) $a(t) + q_1(t)x + q_2(t)y \le f(t, x, y) \le b(t) + q_1(t)x + q_2(t)y$ for all $t \in J \setminus Z$ and $x$, $y \in l^p$.

We shall show that the PBVS (2.9.1) has the extremal solutions in the order interval $[\alpha, \beta]$ of $AC(J, l^p)$ if there is a null set $Z$ in $J$ such that either conditions (C0)–(C5) hold for some $N =$

$(N_1, N_2, \dots)$, $M = (M_1, M_2, \dots)$, $\mu = (\mu_1, \mu_2, \dots) \in L^2(J, l^p)$ or conditions (C0)–(C4) and (C6) hold for some $N = (N_1, N_2, \dots) \in L^1(J, l^p)$, $M_i \equiv M$, $\mu_i \equiv \mu \in L^1(J, I\!R)$. If (C0) and (C1) are replaced by (C7) with $a, b \in L^2(J, l^p)$, $q_j \in L^2(J, I\!R)$ (resp. $a, b \in L^1(J, l^p)$, $q_j \in L^1(J, I\!R)$), $j = 1, 2$, with $\int_o^T (q_1(s) + q_2(s))\, ds < 0$, then these extremal solutions are the least and the greatest of all the solutions of (2.9.1). Moreover, in all these cases the extremal solutions in question are nondecreasing with respect to $f$.

**Theorem 2.9.1:** *Assume there exist a null set $Z$ in $J$ and $N = (N_1, N_2, \dots)$, $M = (M_1, M_2, \dots)$, $\mu = (\mu_1, \mu_2, \dots) \in L^2(J, l^p)$ such that conditions (C0)–(C5) hold. Then the PBVS (2.9.1) has extremal solutions in the order interval $[\alpha, \beta]$ of $AC(J, l^p)$.*

*Proof.* We may choose the functions $M_i$ and $\mu_i$ in conditions (C4) and (C5) so that $\int_o^T (M_i(s) - \mu_i(s))\, ds > 0$. Let $y \in [\alpha, \beta]$ be given. Consider the PBVS

$$x_i' = F_i(t, x_i; y(t)), \qquad x_i(0) = x_i(T), \quad i = 1, 2, \dots, \tag{2.9.2}$$

where

$$\begin{aligned} F_i(t, x; y(t)) &= M_i(t)(y_i(t) - x)) \\ &+ f_i(t, (y_1(t), \dots, y_{i-1}(t), x, y_{i+1}(t), \dots), y(t)). \end{aligned} \tag{2.9.3}$$

By (C2) and (C5) each $F_i$ is a Carathéodory function in the set $\Omega_i = \{(t, x) \mid t \in J,\ x \in [\alpha_i(t), \beta_i(t)]\}$. Applying (C0), (C3), (C4) and (2.9.3) we obtain for each $i = 1, 2, \dots$,

$$\alpha_i' \le F_i(t, \alpha_i; y(t)) \ \text{ and } \ \beta_i' \ge F_i(t, \beta_i; y(t)) \ \text{on} \ \ J \setminus Z.$$

Condition (C1) implies that for all $x \in [\alpha_i(t), \beta_i(t)]$ and $t \in J \setminus Z$,

$$|F_i(t, x; y(t))| \le K_i\, M_i(t) + N_i(t), \tag{a}$$

where $K_i = \max \beta_i - \min \alpha_i$, $i = 1, 2, \dots$. Hence, by theorem 1.5.2 each PBVS of (2.9.2) has a solution $x_i$ in $[\alpha_i, \beta_i]$. In view of the choice of $M_i$ in condition (C4) we see that each function $F_i(t, x; y(t)) + (M_i(t) - \mu_i(t))\, x$ is nonincreasing in $x \in [\alpha_i(t), \beta_i(t)]$ for all $t \in J \setminus Z$, whence $x_i$ is by corollary 1.5.3 the only solution of the $i$:th PBVP of (2.9.2). Since (2.9.2) is uncoupled, then $x = (x_1, x_2, \dots)$ is a unique solution of the PBVS (2.9.2).

Define a map $G = (G_1, G_2 \dots)\colon [\alpha, \beta] \to [\alpha, \beta]$ by

$$G_i y = x_i, \quad y \in [\alpha, \beta], \quad i = 1, 2, \dots, \tag{2.9.4}$$

where $x = (x_1, x_2 \dots)$ is the solution of (2.9.2) in $[\alpha, \beta]$. To prove that $G$ is nondecreasing, let $y$, $\bar{y} \in [\alpha, \beta]$, $\bar{y} \le y$, be given, and suppose that $x = (x_1, x_2, \dots)$ and $\bar{x} = (\bar{x}_1, \bar{x}_2, \dots)$ are the corresponding solutions of (2.9.2) and

$$\bar{x}_i' = F(t, \bar{x}_i; \bar{y}_i(t)), \qquad \bar{x}_i(0) = \bar{x}_i(T), \quad i = 1, 2, \dots$$

in $[\alpha, \beta]$, respectively. Since $\bar{y} \le y$, we have by (C4) and (2.9.3)

$$\bar{x}_i' \le F_i(t, \bar{x}_i; y(t)) \text{ a.e. on } J, \quad \bar{x}_i(0) = \bar{x}_i(T),\ i = 1, 2, \dots.$$

Thus the hypotheses of lemma 1.5.6 hold when $y = \bar{x}_i$ and $z = x_i$, whence $\bar{x}_i \le x_i$, thus proving that $G_i \bar{y} \le G_i y$. This holds for each $i = 1, 2, \dots$, whence $G\bar{y} \le Gy$, so that $G$ is nondecreasing.

From (2.9.3), (2.9.4) and (a) it follows that for a.a. $t \in J$ and $y \in [\alpha, \beta]$,

$$\|(Gy)'(t)\|_p \le \|N(t)\|_p + K\|M(t)\|_p, \tag{2.9.5}$$

where $K = \sup\{K_1, K_2 \dots\}$. Thus

$$\|Gy(t) - Gy(\bar{t})\|_p \le \int_{\bar{t}}^{t} (\|N(s)\|_p + K\,\|M(s)\|_p)\, ds, \tag{b}$$

whenever $0 \leq \bar{t} \leq t \leq T$ and $y \in [\alpha, \beta]$.

The above proof implies that $G$ satisfies the hypotheses of proposition 1.4.5, whence $G$ has the least fixed point $x_*$ and the greatest fixed point $x^*$.

From the definition of $G$ it follows that $x_* = (x_{*1}, x_{*2}, \dots)$ and $x^* = (x_1^*, x_2^*, \dots)$ are also solutions of the PBVS (2.9.1) in $[\alpha, \beta]$.

If $x = (x_1, x_2, \dots)$ is any solution of (2.9.1) in $[\alpha, \beta]$, then it satisfies also the PBVS (2.9.2) with $y = x$. But this means that $x$ is a fixed point of $G$, whence $x_* \leq x \leq x^*$. Thus $x_*$ is the minimal solution and $x^*$ is the maximal solution of the PBVS (2.9.1) in $[\alpha, \beta]$. □

If condition (C5) is replaced in theorem 2.9.1 by condition (C6), we obtain

**Theorem 2.9.2:** *If there exist null set $Z$ in $J$ and functions $N = (N_1, N_2, \dots) \in L^1(J, l^p)$ and $M$, $\mu \in L^1(J, \mathbb{R})$ such that conditions (C0)–(C4) and (C6) hold when $M_i = M$, $\mu_i = \mu$, $i = 1, 2, \dots,$ then the PBVS (2.9.1) has the extremal solutions in $[\alpha, \beta]$.*

*Proof.* We can choose the functions $M$, $\mu$ so that $\int_o^T (M(s) + \mu(s))\, ds > 0$. Denote $q = M + \mu$, and define an operator $G$ in $[\alpha, \beta]$ by

$$Gx(t) = e^{-Q(t)} \int_o^t e^{Q(s)} g(s, x(s))\, ds + \frac{e^{-Q(t)}}{e^{Q(T)} - 1} \int_o^T e^{Q(s)} g(s, x(s))\, ds, \tag{2.9.6}$$

where

$$g(t, x) = f(t, x, x) + q(t)\, x, \quad \text{and} \quad Q(t) = \int_o^t q(s)\, ds, \tag{2.9.7}$$

for $t \in J$ and $x \in [\alpha(t), \beta(t)]$. From (C0) and (2.9.7) it follows that

$$e^{Q(t)}g(t, \alpha(t)) \geq e^{Q(t)}(\alpha'(t) + q(t)\alpha(t)) \quad t \in J \setminus Z. \qquad \text{(a)}$$

In view of (a), (C0) and (2.9.6) we then obtain

$$\begin{aligned} e^{Q(t)}G\alpha(t) &\geq \int_o^t e^{Q(s)}(\alpha'(s)) + q(s)\alpha(s))\,ds \\ &+ \frac{1}{e^{Q(T)}-1}\int_o^T e^{Q(s)}(\alpha'(s) + q(s)\alpha(s))\,ds \\ &= e^{Q(t)}\alpha(t)) - \alpha(0) + \frac{1}{e^{Q(T)}-1}(e^{Q(T)}(\alpha(T)) - \alpha(0)) \\ &\geq e^{Q(t)}\alpha(t), \quad t \in J. \end{aligned}$$

Thus $\alpha \leq G\alpha$. Similarly, it can be shown that $G\beta \leq \beta$. From (C3), (C4), (C6) and (2.9.7) it follows that $g(t, \cdot)$ is nondecreasing in $[\alpha, \beta]$ for all $t \in J \setminus Z$. This and the definition (2.9.6) of $G$ imply that $G$ is nondecreasing. Obviously, $Gx \in AC(J, l^p)$ for each $x \in [\alpha, \beta]$, whence (2.9.6) defines a nondecreasing operator $G\colon [\alpha, \beta] \to [\alpha, \beta]$.

From (2.9.6) and (2.9.7) it follows that

$$(Gy)'(t) = f(t, y(t), y(t)) + q(t)(y(t) - Gy(t))$$

for all $y \in [\alpha, \beta]$ and for a.a. $t \in J$. This and (C1) imply that for all $y \in [\alpha, \beta]$ and for a.a. $t \in J$,

$$\|(Gy)'(t)\|_p \leq \|N(t)\|_p + Kq(t),$$

where $K = 2\max\{\|\alpha\|_p, \|\beta\|_p\}$. Thus for all $y \in [\alpha, \beta]$

$$\|Gy(t) - Gy(\bar{t})\|_p \leq \int_{\bar{t}}^t (\|N(s)\|_p + K\,q(s))\,ds, \; 0 \leq \bar{t} \leq t \leq T. \qquad \text{(b)}$$

The above proof ensures by proposition 1.4.5 that $G$ has the least fixed point $x_*$ and the greatest fixed point $x^*$. From the definitions of $g$ and $G$ and lemma 1.5.7 it follows that $x_* = (x_{*1}, x_{*2}, \dots)$ and $x^* = (x_1^*, x_2^*, \dots)$ are also solutions of the PBVS (2.9.1) in $[\alpha, \beta]$.

If $x = (x_1, x_2, \dots)$ is any solution of (2.9.1) in $[\alpha, \beta]$, then it is by lemma 1.5.7 a fixed point of $G$, whence $x_* \le x \le x^*$. Thus $x_* = (x_{*1}, x_{*2}, \dots)$ and $x^* = (x_1^*, x_2^*, \dots)$ are the minimal and the maximal solutions of the PBVS (2.9.1) in $[\alpha, \beta]$. □

Consider next the existence of the extremal solutions of (2.9.1) among all its solutions.

**Proposition 2.9.1:** *Assume there exists a null set $Z$ in $J$, functions $q_j \in L^2(J, \mathbb{R})$, $j = 1, 2$, with $\int_0^T (q_1(s) + q_2(s))\, ds < 0$, and $a$, $b$, $M$, $\mu \in L^2(J, l^p)$ such that condition (C7) holds, and that (C2)-(C5) hold when $\alpha$, $\beta$ are given by*

$$\alpha(t) = e^{Q(t)}\Big[\int_0^t e^{-Q(s)} a(s)\, ds + \int_0^T \frac{e^{-Q(s)} a(s)}{e^{-Q(T)} - 1} ds\Big], \qquad (2.9.8)$$

$$\beta(t) = e^{Q(t)}\Big[\int_0^t e^{-Q(s)} b(s)\, ds + \int_0^T \frac{e^{-Q(s)} b(s)}{e^{-Q(T)} - 1} ds\Big], \qquad (2.9.9)$$

*where $Q(t) = \int_0^t (q_1(s) + q_2(s))\, ds$, $t \in J$. Then there exist the extremal solutions to PBVS (2.9.1), and all the solutions of (2.9.1) lie within the order interval $[\alpha, \beta]$.*

*Proof.* From corollary 1.5.4 it follows that $\alpha$ and $\beta$, given by (2.9.8) and (2.9.9), are the unique solutions of the PBVP's

$$\alpha' = a(t) + (q_1(t) + q_2(t))\alpha(t), \quad \alpha(0) = \alpha(T), \qquad \text{(a)}$$

and

$$\beta' = b(t) + (q_1(t) + q_2(t))\beta(t), \quad \beta(0) = \beta(T). \qquad \text{(b)}$$

In view of (a), (b) and (C7) we see that condition (C0) holds, and that condition (C1) holds with

$$N(t) = \|a(t)\|_p + \|b(t)\|_p + (|q_1(t)| + |q_2(t)|)(\|\alpha\|_o + \|\beta\|_o).$$

Thus the PBVS (2.9.1) has by theorem 2.9.1 the extremal solutions $x_* = (x_{*1}, x_{*2}, \ldots)$ and $x^* = (x_1^*, x_2^*, \ldots)$ in $[\alpha, \beta]$.

If $x = (x_1, x_2, \ldots)$ is a solution of (2.9.1), it follows from (2.9.1) and (C7) that

$$x_i' \geq a_i(t) + (q_1(t) + q_2(t))x_i(t) \quad \text{a.e. on } J, \; x_i(0) = x_i(T), \quad \text{(c)}$$

and

$$x_i' \leq b_i(t) + (q_1(t) + q_2(t))x_i(t) \quad \text{a.e. on } J, \; x_i(0) = x_i(T). \quad \text{(d)}$$

Hence, (a), (b), (c) and (d) imply by lemma 1.5.6 that $x \in [\alpha, \beta]$. In particular, $x \in [x_*, x^*]$, whence $x_*$ and $x^*$ are the extremals among all the solutions of (2.9.1). □

By the similar reasoning it follows from theorem 2.9.2 and lemmas 1.5.5 and 1.5.6.

**Proposition 2.9.2:** *Assume there exists a null set $Z$ in $J$, functions $q_j \in L^1(J, \mathbb{R})$, $j = 1, 2$, with $\int_o^T (q_1(s) + q_2(s))\, ds < 0$, and $a$, $b$, $M$, $\mu \in L^1(J, l^p)$ such that condition (C7) holds, and that conditions (C2)–(C4) and (C6) are valid with $\alpha$, $\beta$ defined by (2.9.8) and (2.9.9). Then the conclusions of proposition 2.9.1 hold.*

As for the dependence of the extremal solutions of (2.9.1) on $f$ we have

**Proposition 2.9.3:** *If the hypotheses of theorem 2.9.1 or theorem 2.9.2 hold, then the extremal solutions of the PBVS (2.9.1) in $[\alpha, \beta]$ are nondecreasing with respect to $f$.*

*Proof.* Let $f$, $\hat{f}\colon J \times l^p \to l^p$ satisfy

$$f(t,x,y) \le \hat{f}(t,x,y) \text{ for a.a. } t \in J \text{ and for all } x,\ y \in I\!R. \tag{a}$$

Assume first that the hypotheses of theorem 2.9.1 hold for $f$ and $\hat{f}$. Let $x_*$ be the minimal solution of (2.9.1) in $[\alpha,\beta]$, and let $\hat{x}_*$ be the minimal solution of the PBVS

$$x' = \hat{f}(t,x,x), \qquad x(0) = x(T). \tag{b}$$

If $F(t,x;y(t))$ is defined by (2.9.3), it follows from (a), (b) and (2.9.3) that

$$\begin{aligned} \hat{x}'_{*i}(t) \ge & F(t,\hat{x}_{*i}(t);\hat{x}_*(t)) \text{ a.e. on } J, \\ & \hat{x}_{*i}(0) = \hat{x}_{*i}(T). \end{aligned} \tag{c}$$

From the definition (2.9.4) of the operator $G$ it follows that

$$\begin{aligned} (G_i\hat{x}_*)'(t) &= F(t,G_i\hat{x}_*(t);\hat{x}_*(t)) \text{ a.e. on } J, \\ G\hat{x}_*(0) &= G\hat{x}_*(T). \end{aligned} \tag{d}$$

Because the functions $(t,x) \mapsto F_i(t,x;\hat{x}_*(t))$, $i = 1,2,\ldots$, satisfy the hypotheses of lemma 1.5.6, then (c) and (d) imply that $G\hat{x}_* \le \hat{x}_*$. Thus $x_* \le \hat{x}_*$ by (1.4.11).

The proof that $x^* \le \hat{x}^*$, where $x^* = (x_1^*, x_2^*, \ldots)$ denotes the maximal solution of (2.9.1) in $[\alpha,\beta]$ and $\hat{x}^* = (\hat{x}_1^*, \hat{x}_2^*, \ldots)$ is the maximal solution of (b) in $[\alpha,\beta]$, is similar.

Assume next that (a) holds for the functions $f$, $\hat{f}$ which satisfy the hypotheses of theorem 2.9.2, and let $G\colon [\alpha,\beta] \to [\alpha,\beta]$ be defined by (2.9.6), where $g$ is given by (2.9.7). If $x_*$ and $\hat{x}_*$ are as above, it follows from lemma 1.5.7 that

$$\begin{aligned} \hat{x}_*(t) = {} & e^{-Q(t)} \int_0^t e^{Q(s)} \hat{g}(s,\hat{x}_*(s))\, ds \\ & + \frac{e^{-Q(t)}}{e^{Q(T)} - 1} \int_0^T e^{Q(s)} \hat{g}(s,\hat{x}_*(s))\, ds, \quad t \in J, \end{aligned} \tag{e}$$

where

$$\hat{g}(t,x) = \hat{f}(t,x,x) + q(t)\,x, \quad \text{and} \quad Q(t) = \int_0^t q(s)\,ds, \tag{f}$$

for $t \in J$ and $x \in [\alpha(t), \beta(t)]$. From (a), (f) and (2.9.7) it follows that

$$g(t,x) \leq \hat{g}(t,x), \quad \text{for a.a. } t \in J \text{ and for all } \; x \in [\alpha(t), \beta(t)],$$

whence (2.9.6) and (e) imply that $G\hat{x}_* \leq \hat{x}_*$. Since $x_*$ is the least fixed point of $G$, it follows from (1.4.11) that $x_* \leq \hat{x}_*$.

Similarly, it can be shown that $x^* \leq \hat{x}^*$, where $x^*$ denotes the maximal solution of (2.9.1) in $[\alpha, \beta]$ and $\hat{x}^*$ is the maximal solution of (b) in $[\alpha, \beta]$. □

The following corollaries are direct consequences of theorem 2.9.1 and proposition 2.9.3.

**Corollary 2.9.1:** *Given $p \in [1, \infty)$, let $f: J \times (l^p)^2 \to l^p$ satisfy conditions (C0) and (C1). If $f(t,x,y)$ is in $D = \{(t,x,y) \mid t \in J,\ x,\ y \in [\alpha(t), \beta(t)]\}$ Borel measurable, continuous and nonincreasing in $x$ and nondecreasing in $y$, then the PBVS (2.9.1) has the extremal solutions in $[\alpha, \beta]$, and they are nondecreasing with respect to $f$.*

**Corollary 2.9.2:** *If $f: J \times (l^p)^2 \to l^p$ satisfies condition (C0), if $f(t,x,y)$ is in $D = \{(t,x,y) \mid t \in J,\ x,\ y \in [\alpha(t), \beta(t)]\}$ continuous, nonincreasing in $x$ and nondecreasing in $y$, then the PBVS (2.9.1) has the extremal solutions in $[\alpha, \beta]$, and they are nondecreasing with respect to $f$.*

The proof of proposition 2.9.3 can also be used to verify the following result.

**Proposition 2.9.4:** *If the hypotheses of proposition 2.9.1 or proposition 2.9.2 hold, then the extremal solutions of the PBVS (2.9.1) are nondecreasing with respect to $f$.*

## 2.9.2. Existence of extremal coupled quasisolutions

Consider next the existence of the coupled quasisolutions of the PBVS

$$x_i' = f_i(t, x, x, x), \qquad x(0) = x(T). \tag{2.9.10}$$

The functions $y = (y_1, y_2, \dots)$, $z = (z_1, z_2, \dots)$ in $AC(J, l^p)$ are said to be *coupled quasisolutions* of (2.9.10) if

$$\begin{aligned} &y_i'(t) = f_i(t, y(t), y(t), z(t)) \text{ for a.a. } t \in J, \\ &z_i'(t) = f_i(t, z(t), z(t), y(t)) \text{ for a.a. } t \in J, \\ &y_i(0) = y_i(T), \quad \text{and } z_i(0) = z_i(T), \ i = 1, 2, \dots . \end{aligned} \tag{2.9.11}$$

Assume that the function $f \colon J \times (l^p)^3 \to l^p$ satisfies some of the following hypotheses.

(D0) $\alpha$, $\beta \in AC(J, l^p)$, $\alpha \le \beta$, $\alpha(0) \le \alpha(T)$, $\beta(0) \ge \beta(T)$, and $\alpha' \le f(t, \alpha, \alpha, \beta)$ and $\beta' \ge f(t, \beta, \beta, \alpha)$ on $J \setminus Z$.

(D1) $|f(t, x, y, z)| \le N_i(t)$ for $t \in J \setminus Z$, $x$, $y$, $z \in [\alpha(t), \beta(t)]$.

(D2) $f(\cdot, x(\cdot), y(\cdot), z(\cdot))$ is measurable for all $x$, $y$, $z \in [\alpha, \beta]$.

(D3) $f_i(t, x, y, z) - p_i(t)x$ is continuous and nonincreasing in $x_i \in [\alpha_i(t), \beta_i(t)]$ for all $t \in J \setminus Z$ and $x$, $y$, $z \in [\alpha(t), \beta(t)]$.

(D4) $f(t, \cdot, y, z)$ is quasimonotone nondecreasing, $f(t, y, \cdot, z)$ is nondecreasing and $f(t, y, z, \cdot)$ is nonincreasing for all $y$, $z \in l^p$ and $t \in J \setminus Z$.

**Theorem 2.9.3:** *If there is a null set $Z$ in $J$ and $N = (N_1, N_2, \dots)$, $\mu = (\mu_1, \mu_2, \dots) \in L^2(J, l^p)$ such that the hypotheses (D0)–(D4) hold, then the PBVS (2.9.10) has coupled quasisolutions in the order interval $[\alpha, \beta]$.*

*Proof.* Choose $(a_1, a_2, \dots) \in l^p$ so that $a_i > 0$ for each $i = 1, 2, \dots$ . Denote $M_i(t) = |\mu_i(t)| + a_i$, $t \in J$, $i = 1, 2, \dots$ . Given

$y,\ z \in [\alpha, \beta]$, consider the PBVS

$$x_i' = F_i(t, x_i; y(t), z(t)), \qquad x_i(0) = x_i(T),\ i = 1, 2, \ldots \tag{2.9.12}$$

where

$$\begin{aligned} &F_i(t, x; y(t), z(t)) = M_i(t)(y_i(t) - x)) \\ &+ f_i(t, (y_1(t), \ldots, y_{i-1}(t), x, y_{i+1}(t), \ldots), y(t), z(t)). \end{aligned} \tag{2.9.13}$$

By (D2) and (D3) each $F_i$ is a Carathéodory function with respect to $(t, x)$ in $\Omega_i = \{(t, x) \mid t \in J,\ x \in [\alpha_i(t), \beta_i(t)]\}$. Applying (D0), (D3), (D4) and (2.9.13) it is easy to see that

$$\alpha_i' \leq F_i(t, \alpha_i; y(t), z(t)) \ \text{ and } \ \beta_i' \geq F_i(t, \beta_i; y(t), z(t)) \ \text{ on } \ J \setminus Z.$$

Condition (D1) implies that for all $x \in [\alpha_i(t), \beta_i(t)]$ $t \in J \setminus Z$,

$$|F_i(t, x; y(t), z(t))| \leq K_i\, M_i(t) + N_i(t), \tag{a}$$

where $K_i = \max \beta_i - \min \alpha_i$. Hence, by theorem 1.5.2 each PBVP of (2.9.12) has a solution $x_i$ in $[\alpha_i, \beta_i]$. In view of (2.9.12) and condition (D3) we see that each of the functions $F_i(t, x; y(t), z(t)) + (M_i(t) - p_i(t))\, x$ is nonincreasing in $x \in [\alpha_i(t), \beta_i(t)]$ for all $t \in J \setminus Z$, whence $x_i$ is by corollary 1.5.3 the only solution of the $i$:th PBVP of (2.9.12). Since the system (2.9.12) is uncoupled, then $x = (x_1, x_2, \ldots)$ is the only solution of the PBVS (2.9.12).

Define a map $A = (A_1, A_2, \ldots)\colon [\alpha, \beta] \times [\alpha, \beta] \to [\alpha, \beta]$ by

$$A_i(y, z) = x_i, \quad y,\ z \in [\alpha, \beta], \tag{2.9.14}$$

where $x = (x_1, x_2, \ldots)$ is the solution of (2.9.12) in $[\alpha, \beta]$. Applying the proof of theorem 2.3.4 to each component of $A$ it follows that $A$ is mixed monotone.

From (2.9.13), (2.9.14) and (a) it follows that for all $y$, $z \in [\alpha, \beta]$ and for a.a. $t \in J$,

$$\|A(y,z))'(t)\|_p \leq K\|M(t)\|_p + \|N(t)\|_p$$

where $K = \max\{K_1, K_2, \dots\}$ . Thus for all $y$, $z \in [\alpha, \beta]$

$$\|A(y,z)(t) - A(y,z)(\bar{t})\|_p \leq \int_{\bar{t}}^{t} (K\|M(s)\|_p + \|N(s)\|_p)\, ds,$$

whenever $0 \leq \bar{t} \leq t \leq T$.

By the above proof $A$ satisfies the hypotheses of proposition 1.4.7, whence $A$ has the extremal coupled fixed points $y$, $z$.

From (2.9.11)–(2.9.14) it follows that If $u$, $v \in [\alpha, \beta]$ are coupled quasisolutions of (2.9.10) if and only if they are coupled fixed points of $A$. Thus $y \leq u$, $v \leq z$ for all coupled quasisolutions $u$, $v$ of (2.9.10) in $[\alpha, \beta]$. □

The reasoning similar to that used in the proof of proposition 2.3.1 implies.

**Proposition 2.9.5:** *Given $f = (f_1, f_2, \dots)\colon J \times (l^p)^3 \to l^p$ and a null set $Z$ of $J$, assume there exist $a$, $b$, $\mu \in L^2(J, l^p)$, $q_1$, $q_2$, $q_3 \in L^2(J, I\!R)$ with $\int_o^T (q_1(s) + q_2(s) + q_3(s))\, ds < 0$ such that*

$$\begin{aligned} a(t) + q_1(t)x + q_2(t)y + q_3(t)z &\leq f(t,x,y,z) \\ &\leq b(t) + q_1(t)x + q_2(t)y + q_3(t)z \end{aligned}$$

*for all $t \in J \setminus Z$ and $x$, $y$, $z \in l^p$. If $\alpha$, $\beta$ are defined by (2.9.8) and (2.9.9), where $Q(t) = \int_o^t (q_1(s) + q_2(s) + q_3(s))\, ds$, $t \in J$, and if conditions (D2)–(D4) hold, then the PBVS (2.9.11) has the extremal coupled quasisolutions, and all the coupled quasisolutions of (2.9.11) belong to the order interval $[\alpha, \beta]$.*

## 2.10. NOTES AND COMMENTS

In section 2 existence and comparison results are derived for first order IVP's and PBVP's and their systems. Theorem 2.1.1 and its corollaries are taken from Heikkilä and Lakshmikantham (1994a). The results of propositions 2.1.1 and 2.1.2, as well as those of subsections 2.1.2-2.1.4 are new. Some of the special cases treated in subsection 2.1.5 are adapted from Carl and Heikkilä (1992c), and Heikkilä, Kumpulainen and Lakshmikantham (1992). Subsections 2.2.1 and 2.2.2 contain new results, and subsections 2.2.3 and 2.2.4 are based on Heikkilä, Kumpulainen and Lakshmikantham (1992). Theorem 2.3.1 is taken from Heikkilä and Lakshmikantham (1994a), whereas theorem 2.3.2 and proposition 2.3.1 are based on Heikkilä (1991). The results of subsection 2.3.2 are new. The results of sections 2.4–2.9 are new, except subsections 2.4.2 and 2.5.3 which are based on Heikkilä, Kumpulainen and Lakshmikantham (1992) (see also Howell and Lakshmikantham (1990)), and theorem 2.7.4 and example 2.7.2 which are taken from Heikkilä (1994b). Systems which are finite or infinite, nondecreasing or quasimonotone nondecreasing, and continuous or of Carathéodory type, are considered, for instance, in Ladde, Lakshmikantham and Vatsala (1985), Mlak (1958), Mlak and Olech (1963), Walter (1965, 1970, 1971), and Wazewski (1950).

In Bressan (1988) uniqueness results are derived for a class of discontinuous differential equations. As for the existence of more general type of solutions to discontinuous differential equations, see Hájek (1979), and Matrosov (1967).

# 3

# Second Order Differential Equations

## 3.0. INTRODUCTION

This chapter deals with second order discontinuous nonlinear boundary value problems (BVP's for short) in the framework of the method of upper and lower solutions coupled with a generalized monotone iterative technique.

In section 3.1 we shall first present basic results concerning the method of upper and lower solutions for equations of Carathéodory type, namely, a comparison result and an existence theorem. We then proceed to prove the existence of extremal solutions for a general scalar boundary value problem, where the dependent variable allows a decomposition into continuous and discontinuous parts. The case when the given problem is independent on the first order derivative of the dependent variable is discussed first before launching the investigation of the general case. Section 3.2 is devoted to the theory of second order boundary value problems involving both continuous type and mixed monotone type dependencies on the dependent variable. In the use of the generalized monotone iterative method we follow the theory of mixed monotone operators in the context of coupled quasisolutions.

In section 3.3 we present existence results for finite and infinite systems of discontinuous second order BVP's, while in section 3.4 we generalize the results of section 3.2 to mixed monotone systems in the setup of coupled quasisolutions, obtaining existence of extremal coupled quasisolutions.

## 3.1. SECOND ORDER BVP'S

In this section we shall study the existence of extremal solutions of second order scalar boundary value problems of Carathéodory type. Before this study we shall derive some results concerning the method of upper and lower solutions.

### 3.1.1. Preliminaries

Given $J = [t_o, t_1]$, $t_o < t_1$, $a_j, b_j \in I\!R_+$, $c_j \in I\!R$, $j = 0, 1$, and $f\colon J \times I\!R^2 \to I\!R$, consider the boundary value problem

$$-x'' = f(t, x, x'), \; B_j x(t_j) = a_j\, x(t_j) - (-1)^j b_j\, x'(t_j) = c_j. \tag{3.1.1}$$

Denote by $AC^1(J, I\!R)$ the set of those functions $y\colon J \to I\!R$ possessing an absolutely continuous first derivative on $J$. A function $y \in AC^1(J, I\!R)$ is said to be a *lower solution* of (3.1.1) if

$$-y''(t) \le f(t, y(t), y'(t)) \text{ a.e. on } J \text{ and } \quad B_j y(t_j) \le c_j, \; j = 0, 1,$$

and an *upper solution* of (3.1.1) if the reversed inequalities hold. If equalities hold, we say that $y$ is a *solution* of the BVP (3.1.1).

In the following we shall always assume that the constants in the boundary condition of (3.1.1) satisfy $a_o a_1 + a_o b_1 + a_1 b_o > 0$. The following result is useful later.

**Lemma 3.1.1:** *Let $f: J \times \mathbb{R}^2 \to \mathbb{R}$ be a Carathéodory function. Assume that $f(t,\cdot,v)$ is decreasing, and that $f(t,u,\cdot)$ satisfies a generalized Lipschitz condition*

$$|f(t,u,v+k) - f(t,u,v)| \le p(t)|k|$$

*for all $u$, $v$, $k \in \mathbb{R}$ and for a.a. $t \in J$, and for some $p \in L^1(J,\mathbb{R}_+)$. If $y$ and $z$ are lower and upper solutions of (3.1.1), then $y \le z$.*

*Proof.* If the conclusion is not true, then $y(t) - z(t)$ attains its positive maximum at a point $t_2 \in J$. Assume first that $t_o < t_2 < t_1$, and let $\delta > 0$ be so chosen that $y(t) > z(t)$ for each $t \in J_\delta = [t_2, t_2+\delta]$. Then, setting $w = y'$ and $v = z'$ we have $w(t_2) = v(t_2)$. In view of the descending nature of $f(t,u,v)$ in $u$ we get

$$\begin{aligned} &w' > -f(t,z,w), \ v' \le -f(t,z,v) \\ &\text{a.e. on } J_\delta \text{ and } \ w(t_2) = v(t_2) = x_o. \end{aligned} \tag{3.1.2}$$

Because $(t,x) \mapsto f(t,z(t),x)$ is a Carathéodory function and satisfies a generalized Lipschitz condition with respect to $x$, it follows that the IVP

$$x' = -f(t,z(t),x), \qquad x(t_2) = x_o$$

has a unique solution $x$ on $J_\delta$ (cf. corollary 5.1.1). This and (3.1.2) imply by lemma 1.5.1 that $v(t) \le x(t) \le w(t)$ on $J_\delta$. But this means that $z' \le y'$ on $J_\delta$, whence $y(t) - z(t) \ge y(t_2) - z(t_2)$ on $J_\delta$. Because $t_2$ was the maximum point of $y(t) - z(t)$, then there is a positive constant $c$ such that $y(t) \equiv z(t) + c$ on $J_\delta$. In particular, $y'(t) = z'(t)$, i.e. $w(t) = v(t)$ on $J_\delta$. But then, integrating (3.1.2) we get $w(t) - v(t) > w(t_2) - v(t_2) = 0$ on $J_\delta$, and thus $y(t_2+\delta) - z(t_2+\delta) > y(t_2) - z(t_2)$. This contradicts with $y(t) - z(t) \equiv c$ on $J_\delta$.

If $y(t) - z(t)$ attains its maximum at $t_o$, then
$y(t_o) - z(t_o) \ge y(t_o+h) - z(t_o+h) > 0$ for small $h > 0$,

which implies that $y'(t_o) \leq z'(t_o)$. But the boundary condition yields $a_o[z(t_o) - y(t_o)] \geq b_o[z'(t_o) - y'(t_o)] \geq 0$, so that $z'(t_o) = y'(t_o)$. Hence, proceeding as above we obtain a contradiction. In the case when $y(t) - z(t)$ is assumed to obtain its positive maximum at $t_1$ we can proceed similarly to get a contradiction.

The above proof shows that the maximum of $y(t) - z(t)$ on $J$ cannot be positive, whence $y \leq z$. □

As an immediate consequence of lemma 3.1.1 we obtain.

**Corollary 3.1.1:** *If the hypotheses of lemma 3.1.1 hold, then the BVP (3.1.1) can posses at most one solution.*

As another consequence of lemma 3.1.1 we have.

**Lemma 3.1.2:** *If $q \in L^1(J,(0,\infty))$ and $p, g \in L^1(J\, \mathbb{R})$, and if $g(t_o+) = g(t_o)$ and $g(t_1-) = g(t_1)$, then the BVP*

$$-x''+p(t)x'+q(t)x = g(t), \qquad B_j x(t_j) = c_j, \quad j = 0,1, \quad (3.1.3)$$

*has a unique solution*

$$\begin{aligned} x(t) &= c_o x_1(t) + c_1 x_o(t) + x_1(t) \int_{t_o}^{t} \frac{x_o(s)}{-W(s)} g(s)\, ds \\ &\quad + x_o(t) \int_t^{t_1} \frac{x_1(s)}{-W(s)} g(s)\, ds, \end{aligned} \qquad (3.1.4)$$

*whenever the solutions $x_j$, $j = 0,1$, of the BVP:s*

$$-x_j'' + p(t)x_j' + q(t)x_j = 0, \;\; B_j x_j(t_j) = 0, \;\; B_{1-j}x_j(t_{1-j}) = 1,$$

*exist and $W(t) = x_o(t)x_1'(t) - x_1(t)x_o'(t)$, $t \in J$.*

*Proof.* By choosing $f(t,u,v) = g(t) - q(t)u - p(t)v$ in lemma 3.1.1 it follows by corollary 3.1.1 that the BVP (3.1.3) can have at most one solution. Hence it remains to show that the function $x$ given by (3.1.4) is a solution of (3.1.3). From the theory of the second order linear differential equations, it follows that if the solutions $x_j$ exist, they are linearly independent, so that $W(t) \neq 0$ for all $t \in J$. Routine calculations then show that the function $x$ defined by (3.1.4) is a solution of (3.1.3). □

As for the existence of the solution of the BVP (3.1.1) between the assumed lower and upper solutions, we have.

**Theorem 3.1.1:** *Let $f \colon J \times \mathbb{R}^2 \to \mathbb{R}$ be a Carathéodory function, and let $y, z \in AC^1(J, \mathbb{R})$, $y \leq z$, be lower and upper solutions of (3.1.1). Assume that $f$ is continuous at $(t_j, u, v)$, $j = 0, 1$, $u \in [y(t_j), z(t_j)]$, $v \in \mathbb{R}$, and satisfies a generalized Lipschitz condition*

$$|f(t, x, v + k) - f(t, x, v)| \leq p(t)|k| \tag{3.1.5}$$

*for all $x \in [y(t), z(t)]$, $v, k \in \mathbb{R}$ and for a.a. $t \in J$, where $p \in L^1(J, \mathbb{R}_+)$, and a Nagumo condition*

$$|f(t, x, v)| \leq h(|v|) \quad \text{for } t \in J,\ x \in [y(t), z(t)],\ v \in \mathbb{R}, \tag{3.1.6}$$

*where $h \in C(\mathbb{R}_+, (0, \infty))$ such that $\int_\lambda^N \frac{s\,ds}{h(s)} > \max z - \min y$ holds for some $N > 0$ and $\lambda = \max\{|z(t_j) - y(t_{1-j})| \mid j = 0, 1\}$. Then the BVP (3.1.1) has a solution in the order interval $[y, z]$.*

*Proof.* Set $C = 1 + \max\{N, \|y\|_o, \|z\|_o\}$, and define a function $F \colon J \times \mathbb{R}^2 \to \mathbb{R}$ by

$$F(t, u, v) = f(t, p(t, u), g(v)) + \frac{u - p(t, u)}{1 + u^2}, \tag{a}$$

where

$$p(t, u) = \max\{y(t), \min\{u, z(t)\}\}, \quad g(v) = \max\{-C, \min\{v, C\}\}. \tag{b}$$

It is clear that $F$ is a Carathéodory function, and there exists $M \in L^1(J, \mathbb{R}_+)$ such that

$$|F(t,u,v)| \leq M(t) \text{ for all } u,\ v \in \mathbb{R} \text{ and for a.a. } t \in J.$$

These properties ensure (cf. Deimling, Ladde and Lakshmikantham (1985)) that the BVP

$$-x'' = F(t,x,x'), \qquad B_j x(t_j) = c_j, \quad j = 0,1, \tag{3.1.7}$$

has a solution $x$ on $J$.

Applying the method used in the proof of lemma 3.1.1 it can be shown that $y \leq x \leq z$. Since $C > N$, then repeating the arguments used in the proof of theorem 1.4.1 in Bernfeld and Lakshmikantham (1974) we find that $|x'(t)| \leq N$ on $J$. From (a) and (b) it then follows that $F(t,x(t),x'(t)) = f(t,x(t),x'(t))$ on $J$, which implies by (3.1.7) that $x$ is a solution of the BVP (3.1.1) in $[y,z]$. □

It can be shown, by using an approximation technique (cf. Deimling, Ladde and Lakshmikantham (1985)), that the result of theorem 3.1.1 holds also when $f$ does not satisfy the generalized Lipschitz condition (3.1.5). Because the Nagumo condition (3.1.6) implies the boundedness of $f$ in $t$, we shall present another existence result in the case when the right hand side of (3.1.1) does not depend on $x'$.

**Theorem 3.1.2:** *Given $f\colon J \times \mathbb{R} \to \mathbb{R}$, assume that $y$, $z \in AC(J,\mathbb{R})$, $y \leq z$, are lower and upper solutions of the BVP*

$$-x'' = f(t,x), \qquad B_j x(t_j) = c_j, \quad j = 0,1. \tag{3.1.8}$$

*If $f$ is continuous at $(t_j,u)$, $j = 0,1$, $u \in [y(t_j), z(t_j)]$, if $f$ is a Carathéodory function in $\Omega = \{(t,x) \mid t \in J,\ x \in [y(t), z(t)]\}$, and if there exists $N \in L^1(J,\mathbb{R}_+)$ such that*

$$|f(t,x)| \leq N(t) \text{ for a.a. } t \in J \text{ and for } x \in [y(t), z(t)], \tag{3.1.9}$$

*then the BVP (3.1.8) has a solution in the order interval* $[y, z]$.

*Proof.* Define a function $F: J \times \mathbb{R} \to \mathbb{R}$ by

$$F(t,u) = f(t, p(t,u)) + \frac{u - p(t,u)}{1+u^2}, \quad t \in J, \ u \in \mathbb{R},$$

where

$$p(t,u) = \max\{y(t), \min\{u, z(t)\}\}.$$

$F$ is a Carathéodory function, and there exists $M \in L^1(J, \mathbb{R}_+)$ such that

$$|F(t,u)| \le M(t) \text{ for all } u \in \mathbb{R} \text{ and for a.a. } t \in J.$$

Thus the BVP

$$-x'' = F(t,x), \qquad B_j x(t_j) = c_j, \ j = 0, 1,$$

has a solution $x$ on $J$. By the reasoning used in the the proof of lemma 3.1.1 it can be shown that $y \le x \le z$, whence $x$ is also a solution of (3.1.8). □

### 3.1.2. $f$ not dependent on first derivative

Now we shall consider the case when the dependence of $f$ on the unknown function $x$ can be split to a continuous part and to a discontinuous part. To clarify ideas we shall first restrict ourselves to the case when $f$ does not depend on $x'$. So we shall begin with the BVP

$$-x'' = f(t,x,x), \qquad B_j x(t_j) = c_j, \ j = 0, 1, \tag{3.1.10}$$

where $f: J \times \mathbb{R}^2 \to \mathbb{R}$, $J = [t_o, t_1]$. We shall impose the following hypotheses on $f$.

(A0) $\alpha,\ \beta \in AC^1(J, I\!R)$, $\alpha \le \beta$, $-\alpha'' \le f(t, \alpha, \alpha)$ and $-\beta'' \ge f(t, \beta, \beta)$ on $J \setminus Z$, $B_j\alpha(t_j) \le c_j$ and $B_j\beta(t_j) \ge c_j$, $j = 0, 1$.

(A1) $|f(t, x, y)| \le N(t)$ for all $t \in J \setminus Z$ and $x,\ y \in [\alpha(t), \beta(t)]$ for some $N \in L^1(J, I\!R_+)$.

(A2) $f(\cdot, x(\cdot), y(\cdot))$ is measurable whenever $x,\ y \in AC(J, I\!R)$ and $\alpha \le x,\ y \le \beta$.

(A3) $f$ is continuous at $(t_j, x, y)$ for all $x,\ y \in [\alpha(t_j), \beta(t_j)]$, $j = 0, 1$.

(A4) There is $M \in L^1(J, I\!R_+)$ and a nondecreasing function $\varphi \in C(I\!R, I\!R)$ such that $f(t, x, y) + M(t)\,(\varphi(y) - \varphi(x))$ is continuous and decreasing in $x$ for each $(t, y) \in (J \setminus Z) \times [\alpha(t), \beta(t)]$, and nondecreasing in $y$ for each $(t, x) \in (J \setminus Z) \times [\alpha(t), \beta(t)]$.

The hypotheses (A0)–(A3) allow us to convert the BVP (3.1.10) to a Fredholm integral equation.

**Lemma 3.1.3:** *Assume there is a null set $Z$ in $J$ such that conditions (A0)–(A3) are valid. Then $x \in AC^1(J, I\!R)$ is a solution of the BVP (3.1.10) in the order interval $[\alpha, \beta]$ if and only if $x$ satisfies the integral equation*

$$x(t) = z_o(t) + \int_{t_o}^{t_1} k(t, s) f(s, x(s), x(s))\, ds, \quad t \in J, \tag{3.1.11}$$

*where*

$$z_o(t) = \frac{c_o a_1(t_1 - t) + c_o b_1 + c_1 a_o(t - t_o) + c_1 b_o}{a_o a_1(t_1 - t_o) + a_o b_1 + a_1 b_o}, \tag{3.1.12}$$

*and*

$$k(t, s) = \begin{cases} \frac{(a_1(t_1 - t) + b_1)(a_o(s - t_o) + b_o)}{a_o a_1(t_1 - t_o) + a_o b_1 + a_1 b_o}, & t_o \le s \le t \le t_1, \\ \frac{(a_1(t_1 - t) + b_1)(a_o(t - t_o) + b_o)}{a_o a_1(t_1 - t_o) + a_o b_1 + a_1 b_o}, & t_o \le t \le s \le t_1. \end{cases} \tag{3.1.13}$$

*Proof.* Assume first that $x \in [\alpha, \beta]$ satisfies the integral equation (3.1.11) with $z_o$ and $k$ given by (3.1.12) and (3.1.13). By choosing $p(t) = q(t) = 0$, $g(t) = f(t, x(t), x(t))$, $t \in J$,

$$x_o(t) = \frac{a_o(t - t_o) + b_o}{a_o a_1(t_1 - t_o) + a_o b_1 + a_1 b_o},$$

and

$$x_1(t) = \frac{a_1(t_1 - t) + b_1}{a_o a_1(t_1 - t_o) + a_o b_1 + a_1 b_o},$$

in lemma 3.1.2 we see that (3.1.11) is equivalent to (3.1.4). Thus $x$ is a solution of the BVP (3.1.10). Conversely, if $x$ is a solution of (3.1.10) in $[\alpha, \beta]$, then replacing $f(s, x(s), x(s))$ in the right-hand side of (3.1.11) by $-x''(s)$, applying partial integration twice and using boundary conditions one can show that the right-hand side of (3.1.11) equals $x(t)$. Thus $x$ satisfies the integral equation (3.1.11). □

Now we are ready to prove our main result concerning the existence of extremal solutions of the BVP (3.1.10).

**Theorem 3.1.3:** *Assume there exists a null set $Z$ in $J$ such that conditions (A0)-(A4) are valid. Then the BVP (3.1.10) has the extremal solutions in the order interval $[\alpha, \beta]$ of $AC(J, \mathbb{R})$.*

*Proof.* Given $y \in [\alpha, \beta]$, consider the BVP

$$-x'' = F(t, x; y(t)), \qquad B_j x(t_j) = c_j, \ j = 0, 1, \tag{3.1.14}$$

where

$$F(t, x; y(t)) = f(t, x, y(t)) + M(t)\left(\varphi(y(t)) - \varphi(x)\right), \tag{3.1.15}$$

for $t \in J$, $x \in [\alpha(t), \beta(t)]$. $F$ is a Carathéodory function by (A2) and (A4) in $\Omega = \{(t, x) \mid t \in J, \ x \in [\alpha(t), \beta(t)]\}$. In view of (A0), (A4) and (3.1.15) it is easy to see that

$$-\alpha'' \le F(t, \alpha; y(t)), \ \text{ and } \ -\beta'' \ge F(t, \beta; y(t)) \ \text{ on } \ J \setminus Z.$$

From (A1) and (3.1.15) it follows that

$$|F(t,x;y(t))| \leq K\,M(t) + N(t), \quad t \in J \setminus Z,\ x \in [\alpha(t), \beta(t)], \quad \text{(a)}$$

where $K = \varphi(\max \beta) - \varphi(\min \alpha)$. Thus the BVP (3.1.14) has by theorem 3.1.2 a solution $x$ in $[\alpha, \beta]$. Noticing also that $F(t,x;y(t))$ is by (A4) decreasing in $x$ on $[\alpha(t), \beta(t)]$ for each $t \in J \setminus Z$, it follows from corollary 3.1.1 that $x$ is the only solution of (3.1.14) in $[\alpha, \beta]$.

We can now define a map $G\colon [\alpha, \beta] \to [\alpha, \beta]$ by $Gy = x$, where $x$ is the solution of the BVP (3.1.14) in $[\alpha, \beta]$. To prove that $G$ is nondecreasing, let $y$, $\bar{y} \in [\alpha, \beta]$ be given, and let $x$, $\bar{x}$ be the corresponding solutions of (3.1.14). If $\bar{y} \leq y$, it follows from (A4), (3.1.14) and (3.1.15) that $\bar{x}$ is a lower solution of the BVP (3.1.14). Since $x$, as a solution, is an upper solution of (3.1.14), it follows from lemma 3.1.1 that $\bar{x} \leq x$, which means that $G\bar{y} \leq Gy$.

Since $x = Gy$ is the solution of (3.1.14) in $[\alpha, \beta]$, it follows from lemma 3.1.3 that

$$Gy(t) = z_o(t) + \int_{t_o}^{t_1} k(t,s) F(s, Gy(s); y(s))\, ds, \ t \in J, \qquad \text{(b)}$$

for $y \in [\alpha, \beta]$. From the definitions (3.1.12) and (3.1.13) of $z_o$ and $k(t,s)$ it follows that $z_o$ is Lipschitz continuous and $k(t,s)$ is Lipschitz continuous in $t$, uniformly over $s$. This, (a) and (b) imply an existence of a Lipschitz continuous function $w\colon J \to I\!R$ such that

$$|Gy(t) - Gy(s)| \leq |w(t) - w(s)|, \quad y \in [\alpha, \beta],\ s,\, t \in J.$$

The foregoing arguments show that $G$ satisfies the assumptions of theorem 1.4.7, whence $G$ has the least fixed point $x_*$ and the greatest fixed point $x^*$. By the definitions of $F$ and $G$ both these fixed points are solutions of the BVP (3.1.10) in $[\alpha, \beta]$.

If $x \in [\alpha, \beta]$ is a solution of (3.1.10), then it is also a solution of (3.1.14) with $y = x$, and hence a fixed point of $G$. Since $x_*$ and $x^*$ were the extremal fixed points of $G$, it follows that $x_* \leq x \leq x^*$. Thus $x_*$ and $x^*$ are the extremal solutions of the BVP (3.1.10) in $[\alpha, \beta]$. □

In the case when $f$ is of the form

$$f(t, x, y) = f_1(t, x) + f_2(t, y), \qquad t \in J, \ x, y \in \mathbb{R}, \qquad (3.1.16)$$

we obtain.

**Corollary 3.1.2:** *Let $f_1(t, x)$ be a Carathéodory function and decreasing in $x$, and $f_2(t, y)$ a standard function and nondecreasing in $y$. If $f$ is defined by (3.1.16) and if (A0) and (A1) hold, then the BVP (3.1.10) has the extremal solutions in the order interval $[\alpha, \beta]$.*

**Corollary 3.1.3:** *Given $f_1, f_2 \in C(J \times \mathbb{R}, \mathbb{R})$, assume that $f_1(t, x)$ is decreasing in $x$, and $f_2(t, y)$ is nondecreasing in $y$. If $f$ is defined by (3.1.16) and if (A0) holds, then the BVP (3.1.10) has the extremal solutions in the order interval $[\alpha, \beta]$.*

The conditions (A0) and (A1) can be replaced by the following growth condition:

(A5) $|f(t, x, y)| \leq H(t, |x|, |y|)$ for all $t \in J \setminus Z$ and $x, y \in \mathbb{R}$, where $H \colon J \times \mathbb{R}_+^2 \to \mathbb{R}_+$, $H(t, u, v)$ is nondecreasing in $(u, v)$ for all $t \in J \setminus Z$, and the BVP

$$-x'' = H(t, x, x), \ a_j\, x(t_j) - (-1)^j b_j\, x'(t_j) = |c_j|, \qquad (3.1.17)$$

has an upper solution in $AC^1(J, \mathbb{R}_+)$.

**Proposition 3.1.1:** *If there is a null set $Z$ in $J$ such that $f\colon J \times I\!R^2 \to I\!R$ satisfies condition (A5), then conditions (A0) and (A1) hold for $\alpha = -w$, $\beta = w$ and $N = -w''$, where $w \in AC^1(J, I\!R_+)$ is an upper solution of (3.1.17). If conditions (A2)–(A4) hold with these $\alpha$, $\beta$, then the BVP (3.1.10) has the extremal solutions in the order interval $[-w, w]$.*

*Proof.* Let $w \in AC^1(J, I\!R_+)$ be an upper solution of the BVP (3.1.17). From (A5) it follows that

$$\sup\{|f(t,x,y)| \mid |x|, |y| \le w(t)\} \le H(t, w(t), w(t)) \le -w''(t) \tag{a}$$

for a.a. $t \in J$. This implies that

$$-(-w)''(t) \le f(t, -w(t), -w(t)), \quad f(t, w(t), w(t)) \le -w''(t)$$

a.e. on $J$. Because

$$a_j\, w(t_j) - (-1)^j b_j\, w'(t_j) \ge |c_j|, \quad j = 0, 1,$$

it follows that

$$a_j\, w(t_j) - (-1)^j b_j\, w'(t_j) \ge c_j,$$

and

$$a_j\,(-w)(t_j) - (-1)^j b_j\,(-w)'(t_j) \le c_j \quad j = 0, 1.$$

Thus (A0) holds with $\alpha = -w$ and $\beta = w$. With these $\alpha$, $\beta$ it follows from (a) that (A1) holds when $N = -w''$. If conditions (A2)–(A4) hold with these $\alpha$, $\beta$, then the BVP (3.1.10) has by theorem 3.1.3 the extremal solutions in the order interval $[-w, w]$. □

The existence of the least and the greatest of all the solutions of the BVP (3.1.10) is ensured if condition (A0) is replaced by

(A6) There exist $p_i$, $g_i \in L^1(J, I\!R)$ with $p_1 + p_2$ a.e. positive-valued, $g_1(t) \le g_2(t)$ a.e. on $J$, $g_i(t_o+) = g_i(t_o)$ and $g_i(t_1-) = g_i(t_1)$, $i = 1, 2$, such that
$g_1(t) - p_1(t)x - p_2(t)y \le f(t, x, y) \le g_2(t) - p_1(t)x - p_2(t)y$ for a.a. $t \in J$ and for all $x$, $y \in I\!R$, and that the BVP's

$$-w_i'' + (p_1(t) + p_2(t))w_i = g_i(t), \; B_j w_i(t_j) = c_j \quad (3.1.18)$$

have solutions $w_i$, $i = 1, 2$.

**Proposition 3.1.2:** *If $f\colon J \times I\!R^2 \to I\!R$ satisfies condition (A6), then condition (A0) holds for $\alpha = w_1$, $\beta = w_2$, where $w_i \in AC^1(J, I\!R_+)$ are the solutions of the BVP's (3.1.18). If conditions (A1)–(A4) hold with these $\alpha$, $\beta$, then the BVP (3.1.10) has the extremal solutions, and they belong to the order interval $[w_1, w_2]$.*

*Proof.* From lemma 3.1.2 it follows that the solutions $w_i \in AC^1(J, I\!R)$ of (3.1.18) are uniquely determined. Since $w_1$ is a lower solution of (3.1.18) with $i = 2$, then $w_1 \le w_2$ by lemma 3.1.1. From (A6) and (3.1.18) it follows that

$$-w_1''(t) \le f(t, w_1, w_1) \quad \text{and} \quad f(t, w_2, w_2) \le -w_2''(t) \qquad \text{(a)}$$

for a.a. $t \in J$. Thus $w_1$ and $w_2$ are lower and upper solutions of (3.1.10). If conditions (A1)–(A4) hold with $\alpha = w_1$, $\beta = w_2$, then the BVP (3.1.10) has by theorem 3.1.3 the extremal solutions $x_*$ and $x^*$ in the order interval $[w_1, w_2]$.

If $x$ is any solution of the BVP (3.1.10), it follows from (A6) that $x$ is an upper solution of (3.1.18) with $i = 1$ and a lower solution of (3.1.18) with $i = 2$. From lemma 3.1.1 it then follows that $w_1 \le x \le w_2$. Thus $x_*$ and $x^*$ are the least and the greatest of all the solutions of (3.1.10). □

The existence of the least and the greatest of all the solutions of the BVP (3.1.10) is ensured also when the conditions (A0) and (A1) are replaced in theorem 3.1.3 by

(A7) $|f(t,x,y)| \le p_1(t)|x| + p_2(t)|y|$ for a.a. $t \in J$ and for all $x, y \in I\!R$, where $p_i \in L^\infty(J, I\!R_+)$, $i = 1, 2$, and $c = \text{ess sup}(p_1 + p_2) < \|K^n\|^{\frac{-1}{n}}$ for some $n = 1, 2, \dots$, with $Ku(t) = \int_{t_0}^{t_1} k(t,s)u(s)\,ds$.

**Proposition 3.1.3:** *If $f\colon J \times I\!R^2 \to I\!R$ satisfies condition (A7), then conditions (A0) and (A1) hold for $\alpha = -w$, $\beta = w$, and $N = -w''$, where $w \in AC^1(J, I\!R_+)$ is the solution of the BVP*

$$-w'' = c\,w, \qquad B_j w(t_j) = |c_j|,\ j = 0, 1. \tag{3.1.19}$$

*If conditions (A2)–(A4) hold with these $\alpha$, $\beta$, then (3.1.10) has the extremal solutions, and they are in the order interval $[-w, w]$.*

*Proof.* From the theory of the linear BVP's it follows that the BVP (3.1.19) has a unique solution $w$ in $AC^1(J, I\!R_+)$. Thus condition (A5) holds with $H(t,u,v) = p_1(t)u + p_2(t)v$, which implies by proposition 3.1.1 that conditions (A0) and (A1) hold with $\alpha = -w$, $\beta = w$ and $N = w''$. Hence, if (A2)–(A4) hold, then the BVP (3.1.10) has by proposition 3.1.1 the extremal solutions in $[-w, w]$.

If $x$ is any solution of the BVP (3.1.10), it follows from (A7) that $x$ is a lower solution of (3.1.19). This implies by the maximum principle (cf. Protter and Weinberger (1967)) that $x \le w$. Similarly, it can be shown that $x \ge -w$. Thus the extremal solutions of (3.1.10) in $[-w, w]$ are the least and the greatest of all the solutions of (3.1.10). □

### 3.1.3. The general case and an example

Next we shall formulate an existence result for extremal solutions of the BVP

$$-x'' = f(t, x, x, x'),\ \ B_j x(t_j) = a_j\, x(t_j) - (-1)^j b_j\, x'(t_j) = c_j. \tag{3.1.20}$$

Assume that $f: J \times \mathbb{R}^3 \to \mathbb{R}$, $J = [t_o, t_1]$, satisfies the following conditions:

(B0) $\alpha, \beta \in AC^1(J, \mathbb{R})$, $\alpha \le \beta$, $-\alpha'' \le f(t, \alpha, \alpha, \alpha')$ and $-\beta'' \ge f(t, \beta, \beta, \beta')$ on $J \setminus Z$, $B_j\alpha(t_j) \le c_j$ and $B_j\beta(t_j) \ge c_j$, $j = 0, 1$.

(B1) $f(\cdot, x(\cdot), y(\cdot), z(\cdot))$ is measurable for all $x$, $y$, $z \in AC(J, \mathbb{R})$, $\alpha \le x$, $y \le \beta$.

(B2) $f$ is continuous at $(t_j, x, y, z)$ for all $x$, $y \in [\alpha(t_j), \beta(t_j)]$, $j = 0, 1$ and $z \in \mathbb{R}$.

(B3) $f(t, x, y, z)$ is continuous and decreasing in $x$ for each $(t, y, z) \in (J \setminus Z) \times [\alpha(t), \beta(t)] \times \mathbb{R}$, and nondecreasing in $y$ for each $(t, x, z) \in (J \setminus Z) \times [\alpha(t), \beta(t)] \times \mathbb{R}$.

(B4) $p \in L^1(J, \mathbb{R}_+)$ and $|f(t, x, y, z + k) - f(t, x, y, z)| \le p(t)|k|$ for all $t \in J \setminus Z$, $x$, $y \in [\alpha(t), \beta(t)]$ and $z$, $k \in \mathbb{R}$.

(B5) $|f(t, x, y, z)| \le h(|z|)$ for all $t \in J$, $x$, $y \in [\alpha(t), \beta(t)]$ and $z \in \mathbb{R}$, where $h \in C(\mathbb{R}_+, (0, \infty))$, and $\int_\lambda^N \frac{s\,ds}{h(s)} > \max\beta - \min\alpha$ for some $N > 0$ and $\lambda = \max\{|\beta(t_j) - \alpha(t_{1-j})| \mid j = 0, 1\}$.

**Theorem 3.1.4:** *If there exists a null set $Z$ in $J$ such that conditions (B0)–(B5) hold, then the BVP (3.1.20) has the extremal solutions in the order interval $[\alpha, \beta]$.*

*Proof.* Given $y \in [\alpha, \beta]$, consider the BVP

$$-x'' = f(t, x, y(t), x'), \qquad B_j x(t_j) = c_j,\ j = 0, 1. \tag{3.1.21}$$

The function $(t, x, z) \mapsto f(t, x, y(t), z)$ satisfies the hypotheses of theorem 3.1.1 with $y$, $z$ replaced by $\alpha$, $\beta$, whence the BVP (3.1.21) has a solution $x$ in $[\alpha, \beta]$. Since $(t, u, z) \mapsto f(t, u, y(t), z)$ is decreasing in $u$, it follows from corollary 3.1.1 that $x$ is the only solution of (3.1.21) in $[\alpha, \beta]$. Moreover, the proof of theorem 3.1.1 ensures that $|x'(t)| \le N$ on $J$. This and (B5) imply also that

$$|f(t, x(t), y(t), x'(t)| \le h(N), \qquad t \in J. \tag{a}$$

We can now define a map $G\colon [\alpha,\beta] \to [\alpha,\beta]$ by $Gy = x$, where $x$ is the solution of the BVP (3.1.21) in $[\alpha,\beta]$. To prove that $G$ is nondecreasing, let $y$, $\bar{y} \in [\alpha,\beta]$ be given, and let $x$, $\bar{x}$ be the corresponding solutions of (3.1.21). If $\bar{y} \le y$, then $\bar{x}$ is by (B3) a lower solution of the BVP (3.1.21). Since $x$, as the solution, is an upper solution of (3.1.21), it follows from lemma 3.1.1 that $\bar{x} \le x$, which means that $G\bar{y} \le Gy$.

Lemma 3.1.3 implies that $x = Gy$, as the solution of (3.1.21) in $[\alpha,\beta]$, satisfies the integral equation

$$Gy(t) = z_o(t) + \int_{t_o}^{t_1} k(t,s)F(s, Gy(s); y(s), (Gy)'(s))\, ds, \qquad \text{(b)}$$

for $y \in [\alpha,\beta]$, $t \in J$. Because $z_o$ is Lipschitz continuous and $k(t,s)$ is Lipschitz continuous in $t$, uniformly over $s$, then (a) and (b) imply an existence of a Lipschitz continuous function $w\colon J \to I\!R$ such that

$$|Gy(t) - Gy(s)| \le |w(t) - w(s)|, \quad y \in [\alpha,\beta],\ s,\ t \in J.$$

Thus $G$ satisfies the assumptions of theorem 1.4.7, whence $G$ has the least fixed point $x_*$ and the greatest fixed point $x^*$. By the definition of $G$ both these fixed points are solutions of the BVP (3.1.20) in $[\alpha,\beta]$.

If $x \in [\alpha,\beta]$ is a solution of (3.1.20), then it is also a solution of (3.1.21) with $y = x$, and hence a fixed point of $G$. Since $x_*$ and $x^*$ were the extremal fixed points of $G$, it follows that $x_* \le x \le x^*$. Thus $x_*$ and $x^*$ are the extremal solutions of the BVP (3.1.20) in $[\alpha,\beta]$. □

**Example 3.1.1:** Let $C_j$, $j = 1, 2$, be nonempty well-ordered sets in $[\frac{1}{2}, 1)$, for instance $C_j = \{1 - \frac{1}{n+j+1} \mid j \in I\!N\}$. Define $g_j\colon I\!R \to I\!R$, $j = 1, 2$, by

$$g_j(x) = \begin{cases} x, & x \in [0, \min C_j) \cup [1, \infty), \\ \min\{y \in C_j \cup \{1\} \mid x < y\}, & x \in [\min C_j, 1), \\ -g_j(-x), & x < 0. \end{cases}$$

It is easy to see that each $g_j$ is nondecreasing and discontinuous at each point of $C_j$. Moreover, $|p_j(x)| \leq 2|x|$ for each $x \in I\!R$.

Choose $J = [0,1]$, and let $p_j \in L^\infty_+(J, I\!R)$, $j = 1,2$, satisfy $c =$ ess $\sup(p_1 + p_2) < \frac{\pi^2}{4}$, and $p_j(t) \to 0$ as $t \to 0+$ or $t \to 1-$.

Consider the boundary value problem

$$-x''(t) = f(t, x(t), x(t)), \quad \text{a.e. on } J, \; x(0) = x(1) = 1, \quad (3.1.22)$$

where

$$f(t, x, y) = p_1(t)g_1(x) + p_2(t)g_2(y), \quad t \in J, \; x, \, y \in I\!R. \quad (3.1.23)$$

It is easy to show that the function $f$ satisfies conditions (A1)–(A4). Moreover,

$$|f(t, x, y)| \leq 2\,c\,(|x| + |y|), \text{ a.e. on } J.$$

Thus $f$ satisfies also the hypothesis (A7), whence the BVP (3.1.22) has by proposition 3.1.3 the least and the greatest solution, and that all the solutions of (3.1.22) belong to the order interval $[-w, w]$, where $w$ is the solution of the BVP

$$-w''(t) = 4\,c\,w(t), \; t \in J, \quad w(0) = w(1) = 1. \quad (3.1.24)$$

## 3.2. SECOND ORDER MIXED MONOTONE BVP'S

We shall now consider the existence of extremal coupled quasisolutions of second order boundary value problems involving continuous dependence and mixed monotone type of discontinuous dependence on the dependent variable.

### 3.2.1. The case with no first order derivatives

Let $J = [t_o, t_1]$, $a_j$, $b_j \in I\!R_+$, $c_j \in I\!R$, $j = 0, 1$, be as in the previous section. Given $f\colon J \times I\!R^3 \to I\!R$, consider the BVP

$$-x'' = f(t, x, x, x), \;\; B_j x(t_j) = a_j\, x(t_j) - (-1)^j b_j\, x'(t_j) = c_j. \tag{3.2.1}$$

The functions $y$, $z \in AC^1(J, I\!R)$ are said to be *coupled quasisolutions* of (3.2.1) if

$$\begin{aligned} &- y''(t) = f(t, y(t), y(t), z(t)) \;\text{ a.e. on } J,\\ &- z''(t) = f(t, z(t), z(t), y(t)) \;\text{ a.e. on } J,\\ &B_j y(t_j) = c_j, \;\text{and}\;\; B_j z(t_j) = c_j, \; j = 0, 1. \end{aligned} \tag{3.2.2}$$

As in section 3.1 we shall always assume that the constants in the boundary condition of (3.2.1) satisfy $a_o a_1 + a_o b_1 + a_1 b_o > 0$. The following hypotheses are used in our considerations:

(C0) $\alpha$, $\beta \in AC^1(J, I\!R)$, $\alpha \le \beta$, $-\alpha'' \le f(t, \alpha, \alpha, \beta)$ and $-\beta'' \ge f(t, \beta, \beta, \alpha)$ on $J \setminus Z$, $B_j\alpha(t_j) \le c_j$ and $B_j\beta(t_j) \ge c_j$, $j = 0, 1$.

(C1) $|f(t, x, y, z)| \le N(t)$ for all $t \in J \setminus Z$ and $x$, $y$, $z \in [\alpha(t), \beta(t)]$ for some $N \in L^1(J, I\!R_+)$.

(C2) $f(\cdot, x(\cdot), y(\cdot), z(\cdot))$ is measurable for all $x$, $y$, $z \in AC(J, I\!R)$, $\alpha \le x$, $y \le \beta$.

(C3) $f$ is continuous at $(t_j, x, y, z)$ for all $x$, $y$, $z \in [\alpha(t_j), \beta(t_j)]$, $j = 0, 1$.

(C4) $f(t, \cdot, y, z)$ is continuous and decreasing, $f(t, y, \cdot, z)$ is nondecreasing and $f(t, y, z, \cdot)$ is nonincreasing for all $t \in J \setminus Z$ and $y$, $z \in [\alpha(t), \beta(t)]$.

Our main result concerning the existence of extremal coupled quasisolutions of the BVP (3.2.1) is.

**Theorem 3.2.1:** *Assume there exists a null set $Z$ in $J$ such that conditions (C0)–(C4) are valid. Then the BVP (3.2.1) has the extremal coupled quasisolutions in the order interval $[\alpha, \beta]$.*

*Proof.* Given $y$, $z \in [\alpha, \beta]$, consider the BVP

$$-x'' = f(t, x, y(t), z(t)), \qquad B_j x(t_j) = c_j, \; j = 0, 1. \tag{3.2.3}$$

From (C2) and (C4) it follows that $(t, x) \mapsto f(t, x, y(t), z(t))$ is a Carathéodory function in $\Omega = \{(t, x) \mid t \in J, \, x \in [\alpha(t), \beta(t)]\}$. By (C0) and (C4) we have

$$-\alpha'' \leq f(t, \alpha, y(t), z(t)), \text{ and } -\beta'' \geq f(t, \beta, y(t), z(t)) \text{ on } J \setminus Z.$$

From (C1) it follows that

$$|f(t, x, y(t), z(t))| \leq N(t), \quad t \in J \setminus Z, \; x \in [\alpha(t), \beta(t)]. \tag{a}$$

Noticing also that $f(t, x, y(t), z(t))$ is by (C4) decreasing in $x$ on $[\alpha(t), \beta(t)]$ for each $t \in J \setminus Z$, then theorem 3.1.2 and corollary 3.1.1 imply that (3.2.3) has a unique solution $x$ in $[\alpha, \beta]$.

Define a map $A: [\alpha, \beta] \times [\alpha, \beta] \to [\alpha, \beta]$ by $A(y, z) = x$, where $x$ is the solution of the BVP (3.2.3) in $[\alpha, \beta]$. To prove that $A$ is mixed monotone, let $y$, $\bar{y}$, $z \in [\alpha, \beta]$ be given, and denote $x = A(y, z)$ and $\bar{x} = A(\bar{y}, z)$. If $\bar{y} \leq y$, it follows from (C4) that $\bar{x}$ is a lower solution of the BVP (3.2.3). Since $x$, as a solution, is an upper solution of (3.2.3), it follows from lemma 3.1.1 that $\bar{x} \leq x$, which means that $A(\bar{y}, z) \leq A(y, z)$. Similarly, it can be shown that $A(z, \bar{y}) \geq A(z, y)$.

Since $x = A(y, z)$, $y$, $z \in [\alpha, \beta]$, is the solution of (3.2.3) in $[\alpha, \beta]$, it follows from lemma 3.1.3 that

$$A(y, z)(t) = z_o(t) + \int_{t_o}^{t_1} k(t, s) f(s, A(y, z)(s), y(s), z(s)) \, ds. \tag{b}$$

The definitions (3.1.12) and (3.1.13) of $z_o$ and $k(t, s)$, (a) and (b) ensure an existence of a Lipschitz continuous function $w: J \to \mathbb{R}$ such that

$$|A(y, z)(t) - A(y, z)(s)| \leq |w(t) - w(s)|, \quad y, \, z \in [\alpha, \beta], \; s, \, t \in J.$$

The above arguments show that $A$ satisfies the assumptions of theorem 1.4.8, whence $A$ has the extremal coupled fixed points $y$, $z$. In view of the definition of $A$ these coupled fixed points are coupled quasisolutions of (3.2.1).

If $u, v \in [\alpha, \beta]$ are coupled quasisolutions of (3.2.1), then they are also coupled fixed points of $A$, whence $y \leq u, v \leq z$. Thus $y$, $z$ are the extremal coupled quasisolutions of the BVP (3.2.1) in $[\alpha, \beta]$. □

In the case when $f$ is of the form

$$f(t, x, y, z) = f_1(t, x) + f_2(t, y) + f_3(t, z), \tag{3.2.4}$$

we obtain the following consequences of theorem 3.2.1.

**Corollary 3.2.1:** *Let $f_1(t, x)$ be a Carathéodory function and decreasing in $x$, $f_2(t, y)$ a standard function and nondecreasing in $y$, and $f_3(t, z)$ a standard function and nonincreasing in $z$. If $f$ is defined by (3.2.4) and if (C0), (C1) and (C3) hold, then the BVP (3.2.1) has the extremal coupled quasisolutions in the order interval $[\alpha, \beta]$.*

**Corollary 3.2.2:** *Given $f_1, f_2, f_3 \in C(J \times I\!R, I\!R)$, assume that $f_1(t, x)$ is decreasing in $x$, and $f_2(t, y)$ is nondecreasing in $y$, and that $f_3(t, z)$ is nonincreasing in $z$. If $f$ is defined by (3.2.4) and if (C0) holds, then the BVP (3.2.1) has the extremal coupled quasisolutions in the order interval $[\alpha, \beta]$.*

The conditions (C0) and (C1) can be replaced by

(C5) $|f(t, x, y, z)| \leq H(t, |x|, |y|, |z|)$ for all $t \in J \setminus Z$ and $x, y, z \in I\!R$, where $H : J \times I\!R^3_+ \to I\!R_+$, $H(t, u, v, w)$ is nondecreasing in $(u, v, w)$ for all $t \in J \setminus Z$, and the BVP

$$-x'' = H(t, x, x, x), \quad a_j x(t_j) - (-1)^j b_j x'(t_j) = |c_j|, \tag{3.2.5}$$

has an upper solution in $AC^1(J, \mathbb{R}_+)$.

**Proposition 3.2.1:** *If there is a null set $Z$ in $J$ such that $f: J \times \mathbb{R}^3 \to \mathbb{R}$ satisfies condition (C5), then conditions (C0) and (C1) hold for $\alpha = -w$, $\beta = w$ and $N = -w''$, where $w \in AC^1(J, \mathbb{R}_+)$ is an upper solution of (3.2.5). If conditions (C2)–(C4) hold with these $\alpha$, $\beta$, then the BVP (3.2.1) has the extremal coupled quasisolutions in the order interval $[-w, w]$.*

*Proof.* Let $w \in AC^1(J, \mathbb{R}_+)$ be an upper solution of the BVP (3.2.5). From (C5) it follows that

$$\begin{aligned} &\sup\{|f(t,x,y,z)| \mid |x|, |y|, |z| \le w(t)\} \\ &\le H(t, w(t), w(t)) \le -w''(t) \quad \text{for a.a.} \quad t \in J. \end{aligned} \tag{a}$$

From (a) it follows that

$$-(-w)''(t) \le f(t, -w(t), -w(t), w(t))$$

and

$$f(t, w(t), w(t), -w(t)) \le -w''(t)$$

a.e. in $J$. Because

$$a_j\, w(t_j) - (-1)^j b_j\, w'(t_j) \ge |c_j|, \quad j = 0, 1,$$

then

$$a_j\, w(t_j) - (-1)^j b_j\, w'(t_j) \ge c_j,$$

and

$$a_j\, (-w)(t_j) - (-1)^j b_j\, (-w)'(t_j) \le c_j, \ j = 0, 1.$$

Thus (C0) holds with $\alpha = -w$ and $\beta = w$. With these $\alpha$, $\beta$ it follows from (a) that (C1) holds when $N = -w''$. If conditions (C2)–(C4) hold with these $\alpha$, $\beta$, then the BVP (3.2.1) has by theorem 3.2.1 the extremal coupled quasisolutions in the order interval $[-w, w]$. □

### 3.2.2. The case with first order derivatives

Next we shall consider an existence of the extremal coupled quasisolutions of the BVP

$$-x'' = f(t, x, x, x, x'), \quad B_j x(t_j) = c_j,\ j = 0, 1. \tag{3.2.6}$$

Assume that $f\colon J \times \mathbb{R}^4 \to \mathbb{R}$, $J = [t_o, t_1]$, satisfies the following conditions:

(D0) $\alpha, \beta \in AC^1(J, \mathbb{R})$, $\alpha \le \beta$, $-\alpha'' \le f(t, \alpha, \alpha, \beta, \alpha')$ and $-\beta'' \ge f(t, \beta, \beta, \alpha, \beta')$ on $J \setminus Z$, $B_j\alpha(t_j) \le c_j$ and $B_j\beta(t_j) \ge c_j$, $j = 0, 1$.

(D1) $f(\cdot, x(\cdot), y(\cdot), z(\cdot), w(\cdot))$ is measurable for all $x$, $y$, $z, w \in AC(J, \mathbb{R})$, $\alpha \le x$, $y$, $z \le \beta$.

(D2) $f$ is continuous at $(t_j, x, y, z, w)$ for $x, y, z \in [\alpha(t_j), \beta(t_j)]$, $j = 0, 1$ and $w \in \mathbb{R}$.

(D3) $f(t, \cdot, y, z, w)$ is continuous and decreasing, $f(t, y, \cdot, z, w)$ is nondecreasing and $f(t, y, z, \cdot, w)$ is nonincreasing for all $t \in J \setminus Z$, $y$, $z \in [\alpha(t), \beta(t)]$ and $w \in \mathbb{R}$.

(D4) There is $p \in L^1(J, \mathbb{R}_+)$ such that
$$|f(t, x, y, z, w + k) - f(t, x, y, z, w)| \le p(t)|k|$$
for all $t \in J \setminus Z$, $x$, $y$, $z \in [\alpha(t), \beta(t)]$ and $w$, $k \in \mathbb{R}$.

(D5) $|f(t, x, y, z, w)| \le h(|w|)$ for all $t \in J$, $x$, $y$, $z \in [\alpha(t), \beta(t)]$ and $w \in \mathbb{R}$, where $h \in C(\mathbb{R}_+, (0, \infty))$ such that condition $\int_\lambda^N \frac{s\,ds}{h(s)} > \max\beta - \min\alpha$ holds for some $N > 0$ and for $\lambda = \max\{|\beta(t_j), -\alpha(t_{1-j})| \mid j = 0, 1\}$.

**Theorem 3.2.2:** *If there exists a null set $Z$ in $J$ such that conditions (D0)–(D5) hold, then the BVP (3.2.6) has the extremal coupled quasisolutions in the order interval $[\alpha, \beta]$.*

*Proof.* Given $y$, $z \in [\alpha, \beta]$, consider the BVP

$$-x'' = f(t, x, y(t), z(t), x'), \qquad B_j x(t_j) = c_j, \ j = 0, 1. \quad (3.2.7)$$

The function $(t, x, w) \mapsto f(t, x, y(t), z(t), w)$ satisfies the hypotheses of theorem 3.1.1 with $y$, $z$ replaced by $\alpha$, $\beta$, whence (3.2.7) has a solution $x$ in $[\alpha, \beta]$. Since $(t, u, w) \mapsto f(t, u, y(t), z(t), w)$ is decreasing in $u$, it follows from corollary 3.1.1 that $x$ is the only solution of (3.2.7) in $[\alpha, \beta]$. Moreover, the proof of theorem 3.1.1 ensures that $x'(t) \leq N$ on $J$. This and (D5) imply also that

$$|f(t, x(t), y(t), z(t), x'(t)| \leq h(N), \qquad t \in J. \quad \text{(a)}$$

We can now define a map $A\colon [\alpha, \beta] \times [\alpha, \beta] \to [\alpha, \beta]$ by $A(y, z) = x$, where $x$ is the solution of the BVP (3.2.7) in $[\alpha, \beta]$. To prove that $A$ is mixed monotone, let $y$, $\bar{y} \in [\alpha, \beta]$ be given, and denote $x = A(y, z)$ and $\bar{x} = A(\bar{y}, z)$. If $\bar{y} \leq y$, then $\bar{x}$ is by (D3) a lower solution of the BVP (3.2.7). Since $x$, as a solution, is an upper solution of (3.2.7), it follows from lemma 3.1.1 that $\bar{x} \leq x$, which means that $A(\bar{y}, z) \leq A(y, z)$. Similarly, it can be shown that $A(y, z)$ is nonincreasing in $z$ on $[\alpha, \beta]$ for each $y \in [\alpha, \beta]$.

Since $x = A(y, z)$, $y$, $z \in [\alpha, \beta]$, is the solution of (3.2.7) in $[\alpha, \beta]$, it follows from lemma 3.1.3 that

$$\begin{aligned} A(y, z)(t) &= z_o(t) \\ &+ \int_{t_o}^{t_1} k(t, s) f(s, A(y, z)(s), y(s), z(s), A(y, z)'(s))\, ds. \end{aligned} \quad \text{(b)}$$

The definitions (3.1.12) and (3.1.13) of $z_o$ and $k(t, s)$, (a) and (b) ensure an existence of a Lipschitz continuous function $w\colon J \to \mathbb{R}$ such that

$$|A(y, z)(t) - A(y, z)(s)| \leq |w(t) - w(s)|, \quad y, z \in [\alpha, \beta], \ s, t \in J.$$

The above arguments imply that $A$ satisfies the assumptions of theorem 1.4.8, whence $A$ has the extremal coupled fixed points $y$, $z$. By the definition of $A$ these coupled fixed points are coupled quasisolutions of (3.2.6).

If $u, v \in [\alpha, \beta]$ are coupled quasisolutions of (3.2.6), then they are also coupled fixed points of $A$, whence $y \leq u, v \leq z$. Thus $y$, $z$ are the extremal coupled quasisolutions of the BVP (3.2.6) in $[\alpha, \beta]$. □

## 3.3. SYSTEMS OF SECOND ORDER BVP'S

In this section we shall study the existence of extremal solutions of finite and infinite systems of second order boundary value problems. We shall see that many of the results derived for one dimensional BVP's can be extended to the corresponding systems.

### 3.3.1. Finite systems of BVP's

Let $J = [t_o, t_1]$, $a_i^j, b_i^j \in I\!R_+$, $c_i^j \in I\!R$, $i = 1, \ldots, m$, $j = 0, 1$, and $f = (f_1, \ldots, f_m)\colon J \times I\!R^{2m} \to I\!R^m$ be given. Consider the following system of the second order boundary value problems (BVS for short)

$$\begin{aligned} &-x_i'' = f_i(t, x, x), \ i = 1, \ldots, m, \\ &B_i^j x_i(t_j) = a_i^j x_i(t_j) - (-1)^j b_i^j x_i'(t_j) = c_i^j, \ j = 0, 1. \end{aligned} \tag{3.3.1}$$

By assuming that $a_i^o a_i^1 + a_i^o b_i^1 + a_i^1 b_i^o > 0$ for each $i = 1, \ldots, m$, we shall show that the system (3.3.1) has the extremal solutions in an order interval of $AC(J, I\!R^m)$ if there is a null set $Z$ in $J$ such that the following conditions hold.

(A0) $\alpha, \beta \in AC^1(J, I\!R^m)$, $\alpha \leq \beta$, $-\alpha'' \leq f(t, \alpha, \alpha)$ and $-\beta'' \geq f(t, \beta, \beta)$ on $J \setminus Z$, $B_i^j \alpha_i(t_j) \leq c_i^j$ and $B_i^j \beta_i(t_j) \geq c_i^j$, $i = 1, \ldots, m$, $j = 0, 1$.

(A1) For each $i = 1, \dots, m$ there exists $N_i \in L^1(J, I\!R_+)$ such that $|f_i(t,x,y)| \le N_i(t)$ for all $t \in J \setminus Z$ and $x, y \in [\alpha(t), \beta(t)]$.

(A2) $f(\cdot, x(\cdot), y(\cdot))$ is measurable for all $x, y \in AC(J, I\!R^m)$ and
$\alpha \le x,\ y \le \beta$.

(A3) $f$ is continuous at $(t_j, x, y)$ for all $x, y \in [\alpha(t_j), \beta(t_j)]$ and $j = 0, 1$.

(A4) $f(t, \cdot, y)$ and $f(t, y, \cdot)$ are quasimonotone nondecreasing for all $t \in J \setminus Z$ and $y \in [\alpha, \beta]$.

(A5) For each $i = i, \dots, m$ there is $M_i \in L^1(J, I\!R_+)$ and a nondecreasing function $\varphi_i \in C(I\!R, I\!R)$ such that the function $f_i(t,x,y) + M_i(t)\,(\varphi_i(y_i) - \varphi_i(x_i))$ is continuous and decreasing in $x_i$ for each $(t, y) \in (J \setminus Z) \times [\alpha(t), \beta(t)]$, and nondecreasing in $y_i$ for each $(t, x) \in (J \setminus Z) \times [\alpha(t), \beta(t)]$.

**Theorem 3.3.1:** *Assume there exists a null set $Z$ in $J$ such that conditions (A0)–(A5) are valid. Then the BVS (3.3.1) has the extremal solutions in the order interval $[\alpha, \beta]$.*

*Proof.* Given $y \in [\alpha, \beta]$, consider the BVS

$$-x_i'' = F_i(t, x_i; y(t)), \quad B_i^j x_i(t_j) = c_i^j, \ i = 1, \dots, m, \tag{3.3.2}$$

where

$$\begin{aligned} F_i(t, x; y(t)) &= M_i(t)\,(\varphi_i(y_i(t)) - \varphi_i(x)) \\ &+ f_i(t, y_1(t), \dots, y_{i-1}(t), x, y_{i+1}(t), \dots, y_m(t), y(t)), \end{aligned} \tag{3.3.3}$$

$(t, x) \in \Omega_i = \{(t, x) \mid t \in J,\ x \in [\alpha_i(t), \beta_i(t)]\}$. Each function $(t, x) \mapsto F_i(t, x; y(t))$ is by (A2) and (A5) a Carathéodory function in $\Omega_i$. By (A0), (A4), (A5) and (3.3.3) it is easy to see that

$$-\alpha_i'' \le F_i(t, \alpha_i; y(t)), \quad \text{and} \quad -\beta_i'' \ge F_i(t, \beta_i; y(t)) \text{ on } J \setminus Z.$$

From (A1) and (3.3.3) it follows that for each $i = 1, \ldots, m$,

$$|F_i(t, x; y(t))| \le K_i\, M_i(t) + N_i(t), \quad t \in J \setminus Z,\ x \in [\alpha(t), \beta(t)],$$

where $K_i = \varphi_i(\max \beta_i) - \varphi_i(\min \alpha_i)$. Thus each BVP of the BVS (3.3.2) has by theorem 3.1.2 a solution $x_i$ in $[\alpha_i, \beta_i]$. Noticing also that each $F_i(t, x; y(t))$ is by (A5) decreasing in $x$ on $[\alpha_i(t), \beta_i(t)]$ for each $t \in J \setminus Z$, it follows from corollary 3.1.1 that $x_i$ is the only solution of the $i$:th BVP of (3.3.2) in $[\alpha_i, \beta_i]$. Since the system (3.3.2) is uncoupled, then $x = (x_1, \ldots, x_m)$ is the only solution of the BVS (3.3.2) in $[\alpha, \beta]$.

We now define a map $G = (G_1, \ldots, G_m) \colon [\alpha, \beta] \to [\alpha, \beta]$ by

$$G_i y = x_i, \qquad y \in [\alpha, \beta],\ i = 1, \ldots, m, \tag{3.3.4}$$

where $x = (x_1, \ldots, x_m)$ is the solution of the BVS (3.3.2) in $[\alpha, \beta]$. To prove that $G$ is nondecreasing, let $y$, $\bar{y} \in [\alpha, \beta]$ be given, and let $x = (x_1, \ldots, x_m)$, $\bar{x} = (\bar{x}_1, \ldots, \bar{x}_m)$ be the corresponding solutions of (3.3.2). If $\bar{y} \le y$, it follows from (3.3.2)–(3.3.4), (A4) and (A5) that for each $i = 1, \ldots, m$, $\bar{x}_i$ is a lower solution of the $i$:th BVP of (3.3.2). Since $x_i$, as the solution, is an upper solution of the same BVP, it follows from lemma 3.1.1 that $\bar{x}_i \le x_i$, which means that $G_i \bar{y} \le G_i y$. This holds for each $i = 1, \ldots, m$, whence $G$ is nondecreasing.

Since $x_i = G_i y$ is the solution of the $i$:th BVP of (3.3.2) in $[\alpha_i, \beta_i]$, it follows from lemma 3.1.3 that for each $i = 1, \ldots, m$,

$$G_i y(t) = z_i^o(t) + \int_{t_o}^{t_1} k_i(t, s) F_i(s, G_i y(s); y(s))\, ds, \tag{3.3.5}$$

for $y \in [\alpha, \beta]$, $t \in J$, where

$$z_i^o(t) = \frac{c_i^o a_i^1 (t_1 - t) + c_i^o b_i^1 + c_i^1 a_i^o (t - t_o) + c_i^1 b_i^o}{a_i^o a_i^1 (t_1 - t_o) + a_i^o b_i^1 + a_i^1 b_i^o}, \tag{3.3.6}$$

and

$$k_i(t,s) = \begin{cases} \frac{(a_i^1(t_1-t)+b_i^1)(a_i^o(s-t_o)+b_i^o)}{a_i^o a_i^1(t_1-t_o)+a_i^o b_i^1+a_i^1 b_i^o}, & t_o \le s \le t \le t_1, \\ \frac{(a_i^1(t_1-s)+b_i^1)(a_i^o(t-t_o)+b_i^o)}{a_i^o a_i^1(t_1-t_o)+a_i^o b_i^1+a_i^1 b_i^o}, & t_o \le t \le s \le t_1. \end{cases} \tag{3.3.7}$$

Each $z_i^o$ is Lipschitz continuous and each $k_i(t,s)$ is Lipschitz continuous in $t$, uniformly over $s$. This, (a) and (3.3.5) imply an existence of a Lipschitz continuous function $w\colon J \to \mathbb{R}$ such that

$$\|Gy(t) - Gy(s)\| \le |w(t) - w(s)|, \quad y \in [\alpha,\beta],\ s,\ t \in J.$$

The above proof shows that the hypotheses of theorem 1.4.7 hold, so that $G$ has the least fixed point $x_* = (x_{*1},\ldots,x_{*m})$ and the greatest fixed point $x^* = (x_1^*,\ldots,x_m^*)$. By the definitions of $F_i$ and $G_i$ both these fixed points are solutions of the BVS (3.3.1) in $[\alpha,\beta]$.

If $x = (x_1,\ldots,x_m) \in [\alpha,\beta]$ is a solution of (3.3.1), then it is also a solution of (3.3.2) with $y = x$, and hence a fixed point of $G$. Since $x_*$ and $x^*$ were the extremal fixed points of $G$, it follows that $x_* \le x \le x^*$. Thus $x_* = (x_{*1},\ldots,x_{*m})$ and $x^* = (x_1^*,\ldots,x_m^*)$ are the extremal solutions of the BVS (3.3.1) in $[\alpha,\beta]$. □

In the case when $f$ is of the form

$$f(t,x,y) = f_1(t,x) + f_2(t,y), \qquad t \in J,\ x,\ y \in \mathbb{R}^m, \tag{3.3.8}$$

we have the following consequences to theorem 3.3.1.

**Corollary 3.3.1:** *Given $f_1$, $f_2\colon J \times \mathbb{R}^m \to \mathbb{R}^m$, assume that $f_1(t,x)$ is a Carathéodory function and decreasing in $x$, and that $f_2(t,y)$ is a standard function and nondecreasing in $y$. If $f$ is defined by (3.3.8) and if (A0), (A1) and (A3) hold, then the BVS (3.3.1) has the extremal solutions in the order interval $[\alpha,\beta]$.*

**Corollary 3.3.2:** *Given $f_1, f_2 \in C(J \times \mathbb{R}^m, \mathbb{R}^m)$, assume that $f_1(t,x)$ is decreasing in $x$, and $f_2(t,y)$ is nondecreasing in $y$. If $f$ is defined by (3.3.8) and if (A0) holds, then the BVS (3.3.1) has the extremal solutions in the order interval $[\alpha, \beta]$.*

Sufficient for the validity of conditions (A0) and (A1) is that the following growth condition holds.

(A6) $|f_i(t,x,y)| \le H_i(t, |x_1|, \ldots, |x_m|, |y_1|, \ldots, |y_m|)$ for all $t \in J \setminus Z$, $x, = (x_1, \ldots, x_m)$ $y = (y_1, \ldots, y_m) \in \mathbb{R}^m$ and $i = 1, \ldots, m$, where $H = (H_1, \ldots, H_m) \colon J \times \mathbb{R}_+^{2m} \to \mathbb{R}_+^m$, $H(t,u,v)$ is nondecreasing in $(u,v)$ for $t \in J \setminus Z$, and the BVS

$$-x_i'' = H_i(t,x,x), \; a_i^j x_i(t_j) - (-1)^j b_i^j x_i'(t_j) = |c_i^j|, \tag{3.3.9}$$

$i = 1, \ldots, m$, $j = 0, 1$, has an upper solution in $AC^1(J, \mathbb{R}_+^m)$.

**Proposition 3.3.1:** *If there is a null set $Z$ in $J$ such that $f = (f_1, \ldots, f_m) \colon J \times \mathbb{R}^{2m} \to \mathbb{R}^m$ satisfies condition (A6), then conditions (A0) and (A1) hold for $\alpha = -w$, $\beta = w$ and $N_i = -w_i''$, where $w = (w_1, \ldots, w_m) \in AC^1(J, \mathbb{R}_+^m)$ is an upper solution of (3.3.9). If conditions (A2)–(A5) hold with these $\alpha$, $\beta$, then the BVS (3.3.1) has the extremal solutions in the order interval $[-w, w]$.*

*Proof.* Let $w = (w_1, \ldots, w_m) \in AC^1(J, \mathbb{R}_+^m)$ be an upper solution of the BVS (3.3.9). From (A6) it follows that for each $i = 1, \ldots, m$

$$\begin{aligned} &\sup\{|f_i(t,x,y)| \mid |x_j| \le w_j(t),\; |y_j| \le w_j(t),\; 1 \le j \le m\} \\ &\le H_i(t, w(t), w(t)) \le -w_i''(t) \quad \text{for a.a. } t \in J. \end{aligned} \tag{a}$$

Thus we have

$$-(-w)''(t) \le f(t, -w(t), -w(t)), \; f(t, w(t), w(t)) \le -w''(t)$$

a.e. on $J$. Because

$$a_i^j\, w_i(t_j) - (-1)^j b_i^j\, w_i'(t_j) \ge |c_i^j|, \quad i = 1,\ldots,m,\ j = 0,1,$$

it follows that

$$a_i^j\, w_i(t_j) - (-1)^j b_i^j\, w_i'(t_j) \ge c_i^j,$$

and

$$a_i^j\,(-w_i)(t_j) - (-1)^j b_i^j\,(-w_i)'(t_j) \le c_i^j$$

for $i = 1,\ldots,m$, $j = 0,1$. Thus (A0) holds with $\alpha = -w$ and $\beta = w$. With these $\alpha$, $\beta$ it follows from (a) that (A1) holds when $N_i = -w_i''$. If conditions (A2)–(A5) hold with these $\alpha$, $\beta$, then the BVS (3.3.1) has by theorem 3.3.1 the extremal solutions in the order interval $[-w, w]$. □

The existence of the least and the greatest of all the solutions of the BVS (3.3.1) is ensured if the condition (A0) is replaced by

(A7) For all $i = 1,\ldots,m$ and $n = 1,2$ there exist $p_i^n$, $g_i^n \in L^1(J,\mathbb{R})$ with $p_i^1 + p_i^2$ a.e. positive-valued, $g_i^1(t) \le g_i^2(t)$ a.e. on $J$, $g_i^n(t_o+) = g_i^n(t_o)$ and $g_i^n(t_1-) = g_i^n(t_1)$, such that $g_i^1(t) - p_i^1(t)x_i - p_i^2(t)y_i \le f_i(t,x,y) \le g_i^2(t) - p_i^1(t)x_i - p_i^2(t)y_i$ for all $t \in J\backslash Z$ and $x = (x_1,\ldots,x_m)$, $y = (y_1,\ldots,y_m) \in \mathbb{R}^m$, and the boundary value systems

$$-(w_i^n)'' + (p_i^1(t) + p_i^2(t))w_i^n = g_i^n(t),\ B_j w_i^n(t_j) = c_i^j, \tag{3.3.10}$$

have solutions $w_i^n$, $i = 1,\ldots,m$, $n = 1,2$.

**Proposition 3.3.2:** *If $f = (f_1,\ldots,f_m)\colon J \times \mathbb{R}^{2m} \to \mathbb{R}^m$ satisfies condition (A7), then condition (A0) holds for $\alpha = w^1$ and $\beta = w^2$, where $w^n = (w_1^n,\ldots,w_m^n) \in AC^1(J,\mathbb{R}_+^m)$, $n = 1,2$, are the solutions of the boundary value systems (3.3.10). If conditions (A1)–(A5) hold with these $\alpha$, $\beta$, then the BVS (3.3.1)* *has the extremal solutions, and they belong to the order interval* $[w^1, w^2]$.

*Proof.* From lemma 3.1.2 it follows that the solutions $w^n = (w_1^n, \dots, w_m^n) \in AC^1(J, I\!R_+^m)$ of (3.3.10) are uniquely determined. Since the $i$:th component of $w^1$ is a lower solution of the $i$:th BVP of (3.3.10) with $n = 2$, then $w^1 \le w^2$ by lemma 3.1.1. From (A7) and (3.3.10) it follows that

$$-(w^1)''(t) \le f(t, w^1, w^1) \quad \text{and} \quad f(t, w^2, w^2) \le -(w^2)''(t) \qquad \text{(a)}$$

for a.a. $t \in J$. Thus $w^1$ and $w^2$ are lower and upper solutions of (3.3.1). If conditions (A1)–(A5) hold with $\alpha = w^1$, $\beta = w^2$, then the BVS (3.3.1) has by theorem 3.3.1 the extremal solutions $x_* = (x_{*1}, \dots, x_{*m})$ and $x^* = (x_1^*, \dots, x_m^*)$ in the order interval $[w^1, w^2]$.

If $x = (x_1, \dots, x_m)$ is any solution of the BVS (3.3.1), it follows from (A7) that for each $i = 1, \dots, m$, $x_i$ is an upper solution of the $i$:th BVP of (3.3.10) with $n = 1$ and a lower solution of the $i$:th BVP of (3.3.10) with $n = 2$. From lemma 3.1.1 it then follows that $w_i^1 \le x_i \le w_i^2$, so that $x \in [w^1, w^2]$. Thus $x_*$ and $x^*$ are the least and the greatest of all the solutions of (3.3.1). □

The existence of the least and the greatest of all the solutions of the BVS (3.3.1) is ensured also when the conditions (A0) and (A1) are replaced in theorem 3.3.1 by

(A8) $|f_i(t, x, y)| \le p_i^1(t)|x_i| + p_i^2(t)|y_i|$ for a.a. $t \in J$ and for all $x = (x_1, \dots, x_m)$, $y = (y_1, \dots, y_m) \in I\!R^m$, where $p_i^n \in L^\infty(J, I\!R_+)$, $n = 1, 2$, and $c_i = \text{ess sup}(p_i^1 + p_i^2) < \|K_i^{n_i}\|^{\frac{-1}{n_i}}$ for some $n_i = 1, 2, \dots$, with $K_i u(t) = \int_{t_0}^{t_1} k_i(t, s) u(s)\, ds$, $i = 1, \dots, m$.

**Proposition 3.3.3:** *If* $f = (f_1, \dots, f_m): J \times I\!R^{2m} \to I\!R^m$ *satisfies condition (A8), then (A0)–(A1) hold for* $\alpha = (-w_1, \dots, -w_m)$,

*$\beta = (w_1, \ldots, w_m)$, and $N_i = -w_i''$, where $w = (w_1, \ldots, w_m) \in AC^1(J, \mathbb{R}_+^m)$ is the solution of the BVS*

$$-w_i'' = c_i\, w_i, \qquad B_i^j w_i(t_j) = |c_j^i|,\ i = 1, \ldots, m. \tag{3.3.11}$$

*If conditions (A2)–(A5) hold with these $\alpha$, $\beta$, then the BVS (3.3.1) has the extremal solutions, and they belong to $[\alpha, \beta]$.*

*Proof.* From the theory of the linear BVP's it follows that each BVP of the BVS (3.3.11) has a unique solution $w_i$ in $AC^1(J, \mathbb{R}_+)$. Thus condition (A6) holds with $H_i(t, u, v) = p_i^1(t)u_i + p_i^2(t)v_i$, which implies by proposition 3.3.1 that (A0) and (A1) hold with $\alpha = (-w_1, \ldots, -w_m)$, $\beta = (w_1, \ldots, w_m)$ and $N_i = w_i''$. Hence, if (A2)–(A5) hold, then the BVS (3.3.1) has by proposition 3.3.1 the extremal solutions in $[\alpha, \beta]$.

If $x = (x_1, \ldots, x_m)$ is any solution of the BVS (3.3.1), it follows from (A8) that $x$ is a lower solution of (3.3.11). This implies by the maximum principle that $x_i \le w_i$, $i = 1, \ldots, m$. Similarly, it can be shown that $x_i \ge -w_i$, $i = 1, \ldots, m$. Thus the extremal solutions of (3.3.1) in $[\alpha, \beta]$ are the least and the greatest of all the solutions of (3.3.1). □

### 3.3.2. Infinite boundary value systems

Let $J = [t_o, t_1]$, $a^j = (a_1^j, a_2^j, \ldots)$, $b^j = (b_1^j, b_2^j, \ldots) \in l_+^\infty$, $c^j = (c_1^j, c_2^j, \ldots) \in l^\infty$, $j = 0, 1$, and $f = (f_1, f_2 \ldots) \colon J \times (l^\infty)^2 \to l^\infty$ be given. Consider the following infinite system of the second order boundary value problems (BVS for short):

$$-x_i'' = f_i(t, x, x),\ B_i^j x_i(t_j) = a_i^j\, x_i(t_j) - (-1)^j b_i^j\, x_i'(t_j) = c_i^j, \tag{3.3.12}$$

$i = 1, 2, \ldots,\ j = 0, 1$. We shall assume that

$$a_i^o a_i^1 (t_1 - t_o) + a_i^o b_i^1 + a_i^1 b_i^o \ge c > 0 \text{ for each } i = 1, 2, \ldots.$$

It turns out that the system (3.3.12) has the extremal solutions in an order interval of $AC(J, l^\infty)$ if there is a null set $Z$ in $J$ such that the following conditions hold.

(B0) $\alpha, \beta \in AC^1(J, l^\infty)$, $\alpha \le \beta$, $-\alpha'' \le f(t, \alpha, \alpha)$ and $-\beta'' \ge f(t, \beta, \beta)$ on $J \setminus Z$, $B_i^j \alpha_i(t_j) \le c_i^j$ and $B_i^j \beta_i(t_j) \ge c_i^j$, $i = 1, 2, \ldots$, $j = 0, 1$.

(B1) There exists $N = (N_1, N_2, \ldots) \in L^1(J, l^\infty)$ such that $|f_i(t, x, y)| \le N_i(t)$ for all $i = 1, 2, \ldots$, $t \in J \setminus Z$ and $x$, $y \in [\alpha(t), \beta(t)]$.

(B2) $f(\cdot, x(\cdot), y(\cdot))$ is measurable whenever $x$, $y \in AC(J, l^\infty)$ and
$\alpha \le x$, $y \le \beta$.

(B3) $f$ is continuous at $(t_j, x, y)$ for all $x$, $y \in [\alpha(t_j), \beta(t_j)]$, $j = 0, 1$.

(B4) $f(t, \cdot, y)$ and $f(t, y, \cdot)$ are quasimonotone nondecreasing for all $t \in J \setminus Z$ and $y \in [\alpha, \beta]$.

(B5) There is $M = (M_1, M_2, \ldots) \in L^1(J, l^\infty)$ and a nondecreasing function $\varphi \in C(I\!R, I\!R)$ such that the function $f_i(t, x, y) + M_i(t)\,(\varphi(y_i) - \varphi_i(x_i))$ is continuous and decreasing in $x_i$ for each $(t, y) \in (J \setminus Z) \times [\alpha(t), \beta(t)]$, and nondecreasing in $y_i$ for each $(t, x) \in (J \setminus Z) \times [\alpha(t), \beta(t)]$.

**Theorem 3.3.2:** *Assume there exists a null set $Z$ in $J$ such that conditions (B0)–(B5) are valid. Then the BVS (3.3.11) has the extremal solutions in the order interval $[\alpha, \beta]$ of $AC(J, l^\infty)$.*

*Proof.* Given $y \in [\alpha, \beta]$, consider the BVS

$$-x_i'' = F_i(t, x_i; y(t)), \quad B_i^j x_i(t_j) = c_i^j, \; i = 1, 2, \ldots, \tag{3.3.13}$$

where

$$\begin{aligned} F_i(t, x; y(t)) &= M_i(t)\,(\varphi(y_i(t)) - \varphi(x)) \\ &+ f_i(t, (y_1(t), \ldots, y_{i-1}(t), x, y_{i+1}(t), \ldots), y(t)), \end{aligned} \tag{3.3.14}$$

$(t, x) \in \Omega_i = \{(t, x) \mid t \in J,\, x \in [\alpha_i(t), \beta_i(t)]\}$. Each function $(t, x) \mapsto F_i(t, x; y(t))$ is by (B2) and (B5) a Carathéodory function in $\Omega_i$. By applying (B0), (B4), (B5) and (3.3.14) it can be shown that for each $i = 1, 2, \ldots,$

$$-\alpha_i'' \leq F_i(t, \alpha_i; y(t)), \quad \text{and} \quad -\beta_i'' \geq F_i(t, \beta_i; y(t)) \text{ on } J \setminus Z.$$

From (B1) and (3.3.14) it follows that for all $i = 1, 2, \ldots, t \in J \setminus Z$ and $x \in [\alpha(t), \beta(t)]$,

$$|F_i(t, x; y(t))| \leq K_i M_i(t) + N_i(t), \tag{a}$$

where $K_i = \varphi(\max \beta_i) - \varphi(\min \alpha_i)$. Thus each BVP of the BVS (3.3.13) has by theorem 3.1.2 a solution $x_i$ in $[\alpha_i, \beta_i]$. Since each $F_i(t, x; y(t))$ is by (B5) decreasing in $x$ on $[\alpha_i(t), \beta_i(t)]$ for each $t \in J \setminus Z$, it follows from corollary 3.1.1 that $x_i$ is the only solution of the $i$:th BVP of (3.3.13) in $[\alpha_i, \beta_i]$. Because the system (3.3.13) is uncoupled, then $x = (x_1, x_2, \ldots)$ is the only solution of the BVS (3.3.13) in $[\alpha, \beta]$.

Define a map $G = (G_1, G_2 \ldots)\colon [\alpha, \beta] \to [\alpha, \beta]$ by

$$G_i y = x_i, \qquad y \in [\alpha, \beta], \ i = 1, 2, \ldots, \tag{3.3.15}$$

where $x = (x_1, x_2, \ldots)$ is the solution of the BVS (3.3.13) in $[\alpha, \beta]$. To prove that $G$ is nondecreasing, let $y, \bar{y} \subset [\alpha, \beta]$ be given, and let $x = (x_1, x_2, \ldots)$, $\bar{x} = (\bar{x}_1, \bar{x}_2, \ldots)$ be the corresponding solutions of (3.3.13). If $\bar{y} \leq y$, it follows from (3.3.13)–(3.3.15), (B4) and (B5) that for each $i = 1, 2, \ldots$, $\bar{x}_i$ is a lower solution of the $i$:th BVP of (3.3.13). Since $x_i$, as the solution, is an upper solution of the same BVP, it follows from lemma 3.1.1 that $\bar{x}_i \leq x_i$, which means that $G_i \bar{y} \leq G_i y$. This holds for each $i = 1, 2, \ldots$, whence $G$ is nondecreasing.

Since $x_i = G_i y$ is the solution of the $i$:th BVP of (3.3.13) in $[\alpha_i, \beta_i]$, it follows from lemma 3.1.3 that for all $i = 1, 2, \ldots$ $y \in [\alpha, \beta]$ and $t \in J$,

$$G_i y(t) = z_i^o(t) + \int_{t_o}^{t_1} k_i(t, s) F_i(s, G_i y(s); y(s))\, ds, \tag{b}$$

where

$$z_i^o(t) = \frac{c_i^o a_i^1(t_1 - t) + c_i^o b_i^1 + c_i^1 a_i^o(t - t_o) + c_i^1 b_i^o}{a_i^o a_i^1(t_1 - t_o) + a_i^o b_i^1 + a_i^1 b_i^o}, \tag{3.3.16}$$

and

$$k_i(t,s) = \begin{cases} \frac{(a_i^1(t_1-t)+b_i^1)(a_i^o(s-t_o)+b_i^o)}{a_i^o a_i^1(t_1-t_o)+a_i^o b_i^1+a_i^1 b_i^o}, & t_o \le s \le t \le t_1, \\ \frac{(a_i^1(t_1-s)+b_i^1)(a_i^o(t-t_o)+b_i^o)}{a_i^o a_i^1(t_1-t_o)+a_i^o b_i^1+a_i^1 b_i^o}, & t_o \le t \le s \le t_1. \end{cases} \tag{3.3.17}$$

Since $(a_1^j, a_2^j, \dots)$, $(b_1^j, b_2^j, \dots) \in l_+^\infty$, $c^j = (c_1^j, c_2^j, \dots) \in l^\infty$, $j = 0, 1$, and $a_i^o a_i^1(t_1 - t_o) + a_i^o b_i^1 + a_i^1 b_i^o \ge c > 0$ for each $i = 1, 2, \dots$, it follows that each $z_i^o$ is Lipschitz continuous, uniformly over $i$ and each $k_i(t,s)$ is Lipschitz continuous in $t$, uniformly over $s$ and $i$. This, (a) and (b) imply an existence of a Lipschitz continuous function $w\colon J \to \mathbb{R}$ such that

$$\|Gy(t) - Gy(s)\|_\infty \le |w(t) - w(s)|, \quad y \in [\alpha, \beta], \ s, t \in J.$$

The above proof shows that the hypotheses of proposition 1.4.5 hold whence $G$ has the least fixed point $x_* = (x_{*1}, x_{*2}, \dots)$ and the greatest fixed point $x^* = (x_1^*, x_2^*, \dots)$. By the definitions of $F_i$ and $G_i$ both these fixed points are solutions of the BVS (3.3.12) in $[\alpha, \beta]$.

If $x = (x_1, x_2, \dots) \in [\alpha, \beta]$ is a solution of (3.3.12), then it is also a solution of (3.3.13) with $y = x$, and hence a fixed point of $G$. Since $x_*$ and $x^*$ were the extremal fixed points of $G$, it follows that $x_* \le x \le x^*$. Thus $x_* = (x_{*1}, x_{*2}, \dots)$ and $x^* = (x_1^*, x_2^*, \dots)$ are the extremal solutions of the BVS (3.3.12) in $[\alpha, \beta]$. □

In the case when $f$ is of the form

$$f(t, x, y) = f_1(t, x) + f_2(t, y), \qquad t \in J, \ x, y \in l^\infty, \tag{3.3.18}$$

we have the following corollary to theorem 3.3.2.

**Corollary 3.3.3:** *Given $f_j \colon J \times l^\infty \to l^\infty$, $j = 1,2$, assume that $f_1(t,x)$ is a Carathéodory function and decreasing in $x$, that $f_2(t,\cdot)$ is nondecreasing, and that $f_2(\cdot, y(\cdot))$ is measurable for each $y \in AC(J, l^\infty)$. If $f$ is defined by (3.3.18) and if (B0), (B1) and (B3) hold, then the BVS (3.3.12) has the extremal solutions in the order interval $[\alpha, \beta]$.*

*Conditions (B0) and (B1) hold if we assume that*

(B6) $|f_i(t,x,y)| \le H_i(t, (|x_1|, |x_2|, \dots), (|y_1|, |y_2|, \dots))$ *for all* $t \in J \setminus Z$, $x = (x_1, x_2, \dots)$, $y = (y_1, y_2, \dots) \in l^\infty$ *and* $i = 1, 2, \dots$, *where* $H = (H_1, H_2, \dots) \colon J \times (l_+^\infty)^2 \to l_+^\infty$, $H(t,u,v)$ *is nondecreasing in* $(u,v)$ *for all* $t \in J \setminus Z$, *and the BVS*

$$-x_i'' = H_i(t,x,x), \; a_i^j\, x_i(t_j) - (-1)^j b_i^j\, x_i'(t_j) = |c_i^j|, \tag{3.3.19}$$

$i = 1, 2, \dots$, $j = 0, 1$, *has an upper solution in* $AC^1(J, l_+^\infty)$.

**Proposition 3.3.4:** *If there is a null set $Z$ in $J$ such that $f = (f_1, f_2, \dots) \colon J \times (l^\infty)^2 \to l^\infty$ satisfies condition (B6), then (B0) and (B1) hold for $\alpha = -w$, $\beta = w$ and $N_i = -w_i''$, where $w = (w_1, w_2, \dots) \in AC^1(J, l_+^\infty)$ is an upper solution of (3.3.19). If conditions (B2)–(B5) hold with these $\alpha$, $\beta$, then the BVS (3.3.12) has the extremal solutions in the order interval $[-w, w]$.*

*Proof.* Let $w = (w_1, w_2, \dots) \in AC^1(J, l_+^\infty)$ be an upper solution of the BVS (3.3.19). From (B6) it follows that for each $i = 1, 2, \dots$

$$\begin{aligned} &\sup\{|f_i(t,x,y)| \mid |x_j| \le w_j(t),\ |y_j| \le w_j(t),\ 1 \le j < \infty\} \\ &\le H_i(t, w(t), w(t)) \le -w_i''(t) \end{aligned} \tag{a}$$

for a.a. $t \in J$. Thus we have

$$-(-w)''(t) \le f(t, -w(t), -w(t)), \; f(t, w(t), w(t)) \le -w''(t)$$

a.e. on $J$. Because

$$a_i^j\, w_i(t_j) - (-1)^j b_i^j\, w_i'(t_j) \ge |c_i^j|, \quad i = 1, 2, \ldots, \ j = 0, 1,$$

it follows that

$$a_i^j\, w_i(t_j) - (-1)^j b_i^j\, w_i'(t_j) \ge c_i^j,$$

and

$$a_i^j\, (-w_i)(t_j) - (-1)^j b_i^j\, (-w_i)'(t_j) \le c_i^j$$

for $i = 1, 2, \ldots,\ j = 0, 1$. Thus (B0) holds with $\alpha = (-w_1, -w_2, \ldots)$ and $\beta = (w_1, w_2, \ldots)$. With these $\alpha$, $\beta$ it follows from (a) that (B1) holds when $N_i = -w_i''$. If conditions (B2)–(B5) hold with these $\alpha$, $\beta$, then the BVS (3.3.12) has by theorem 3.3.2 the extremal solutions in the order interval $[\alpha, \beta]$. □

The least and the greatest of all the solutions of the BVS (3.3.12) exist if the condition (B0) is replaced by

(B7) There exist $(p_1^n, p_2^n, \ldots)$, $(g_1^n, g_2^n, \ldots) \in L^1(J, l^\infty)$, $n = 1, 2$, with $p_i^1 + p_i^2$ a.e. positive-valued, $g_i^1(t) \le g_i^2(t)$ a.e. on $J$, $g_i^n(t_o+) = g_i^n(t_o)$ and $g_i^n(t_1-) = g_i^n(t_1)$, such that for each $i = 1, 2, \ldots,$ $g_i^1(t) - p_i^1(t)x_i - p_i^2(t)y_i \le f_i(t, x, y) \le g_i^2(t) - p_i^1(t)x_i - p_i^2(t)y_i$ for all $t \in J \setminus Z$ and $x = (x_1, x_2 \ldots)$, $y = (y_1, y_2, \ldots) \in l^\infty$, and the boundary value systems

$$-(w_i^n)'' + (p_i^1(t) + p_i^2(t))w_i^n = g_i^n(t), \ B_j w_i^n(t_j) = c_i^j, \quad (3.3.20)$$

have solutions $w_i^n$, $i = 1, 2, \ldots$, $n = 1, 2$.

**Proposition 3.3.5:** *If there exists a null set $Z$ in $J$ such that $f = (f_1, f_2, \ldots)\colon J \times (l^\infty)^2 \to l^\infty$ satisfies condition (B7),* *then condition (B0) holds for* $\alpha = w^1$, $\beta = w^2$, *where* $w^n = (w_1^n, w_2^n, \ldots) \in AC^1(J, l_+^\infty)$, $n = 1, 2$, *are the solutions of the boundary value systems (3.3.20). If conditions (B1)–(B5) hold with these* $\alpha$, $\beta$, *then the BVS (3.3.12) has the extremal solutions, and they belong to the order interval* $[w^1, w^2]$.

*Proof.* From lemma 3.1.2 it follows that the solutions $w^n = (w_1^n, w_2^n, \ldots)$ of (3.3.20) are uniquely determined. Moreover, the given hypotheses imply that $w^n \in AC^1(J, l^\infty)$. Since the $i$:th component of $w^1$ is a lower solution of the $i$:th BVP of (3.3.20) with $n = 2$, then $w_1 \leq w_2$ by lemma 3.1.1. From (B7) and (3.3.20) it follows that

$$-(w^1)''(t) \leq f(t, w^1, w^1), \; f(t, w^2, w^2) \leq -(w^2)''(t) \qquad \text{(a)}$$

for a.a. $t \in J$. Thus $w^1$ and $w^2$ are lower and upper solutions of (3.3.12). If conditions (B1)–(B6) hold with $\alpha = w^1$, $\beta = w^2$, then the BVS (3.3.12) has by theorem 3.3.2 the extremal solutions $x_* = (x_{*1}, x_{*2}, \ldots)$ and $x^* = (x_1^*, x_2^*, \ldots)$ in the order interval $[w^1, w^2]$.

If $x = (x_1, x_2, \ldots)$ is any solution of the BVS (3.3.12), it follows from (B7) that for each $i = 1, 2, \ldots$, $x_i$ is an upper solution of the $i$:th BVP of (3.3.20) with $n = 1$ and a lower solution of the $i$:th BVP of (3.3.20) with $n = 2$. From lemma 3.1.1 it then follows that $w_i^1 \leq x_i \leq w_i^2$, so that $x \in [w^1, w^2]$. Thus $x_*$ and $x^*$ are the least and the greatest of all the solutions of (3.3.12). □

The existence of the extremal solutions of the BVS (3.3.12) is ensured also when the conditions (B0) and (B1) are replaced in theorem 3.3.2 by

(B8) $|f_i(t, x, y)| \leq p_i^1(t)|x_i| + p_i^2(t)|y_i|$ for a.a. $t \in J$ and for all $x = (x_1, x_2, \ldots)$, $y = (y_1, y_2, \ldots) \in l^\infty$, where $p_i^n \in L^\infty(J, \mathbb{R}_+)$, $n = 1, 2$, and $c_i = \text{ess sup}(p_i^1 + p_i^2) < \|K_i^{n_i}\|^{\frac{-1}{n_i}}$ for some $n_i = 1, 2, \ldots$, with $K_i u(t) = \int_{t_0}^{t_1} k_i(t, s) u(s)\, ds$, $i = 1, 2, \ldots$ .

**Proposition 3.3.6:** *If $f = (f_1, f_2 \dots)\colon J \times (l^\infty)^2 \to l^\infty$ satisfies condition (B8), then (B0)–(B1) hold for $\alpha = (-w_1, -w_2, \dots)$, $\beta = (w_1, w_2, \dots)$, and $N_i = -w_i''$, where $w = (w_1, w_2, \dots) \in AC^1(J, l^\infty_+)$ is the solution of the BVS*

$$-w_i'' = c_i w_i, \quad B_i^j w_i(t_j) = |c_j^i|, \; i = 1, 2, \dots, \; j = 0, 1. \quad (3.3.21)$$

*If (B2)–(B5) hold with these $\alpha$, $\beta$, then the BVS (3.3.12) has the extremal solutions, and they belong to $[\alpha, \beta]$.*

*Proof.* Each BVP of the BVS (3.3.21) has a unique solution $w_i$ in $AC^1(J, I\!R_+)$. Thus condition (B6) holds with $H_i(t, u, v) = p_i^1(t)u_i + p_i^2(t)v_i$, which implies by proposition 3.3.4 that (B0) and (B1) hold with $\alpha = (-w_1, -w_2, \dots)$, $\beta = (w_1, w_2, \dots)$ and $N_i = w_i''$. Hence, if (B2)–(B5) hold, then the BVS (3.3.12) has by proposition 3.3.4 the extremal solutions in $[\alpha, \beta]$.

If $x = (x_1, x_2, \dots)$ is any solution of the BVS (3.3.12), it follows from (B8) that each $x_i$ is a lower solution of the $i$:th BVP of (3.3.21). This implies by the maximum principle that $x_i \le w_i$ for each $i = 1, 2, \dots$. Similarly, it can be shown that $x_i \ge -w_i$, $i = 1, 2, \dots$. Thus the extremal solutions of (3.3.12) in $[\alpha, \beta]$ are the least and the greatest of all the solutions of (3.3.12). □

**Example 3.3.1:** Let $C_i^n$, $i = 1, 2, \dots$, $n = 1, 2$, be nonempty inversely well-ordered sets in $(0, 1)$, for instance $C_i^n = \{\frac{1}{n+j+i} \mid j \in I\!N\}$. Denote $J = [0, 1]$, and define functions $\psi_i^n\colon J \to I\!R$, and $g_i^n : l^\infty_+ \to I\!R$, $i = 1, 2, \dots$, $n = 1, 2$, by

$$\psi_i^n(x) = \max\{y \in C_i^n \cup \{0, 1\} \mid x \ge y\},$$

and

$g_i^n(x_1, x_2, \dots) = (x_i + \psi_i^n(x_i - [x_i]))\left(1 + \prod_{j=i+1}^{2i} \psi_i^n\left(\frac{x_j}{|x_j|+1}\right)\right)$.

It is easy to see that each $g_i^n$ is nondecreasing and $g_i^n(x) \le 4x_i$ for each $x \in l^\infty_+$. Let $p_i^n \in L^\infty_+(J, I\!R)$, $i = 1, 2, \dots$, $n = 1, 2$, satisfy $c = \text{ess sup}(p_i^1 + p_i^2) < \frac{\pi^2}{8}$, and $p_i^n(t) \to 0$ as $t \to 0+$ or $t \to 1-$.

Consider the boundary value system

$$-x_i''(t) = f_i(t, x, x), \quad x_i(0) = x_i(1) = 1, \tag{3.3.22}$$

where

$$f_i(t, x, y) = \begin{cases} p_i^1(t) g_i^1(x) + p_i^2(t) g_i^2(y), & t \in J, x, y \in l_+^\infty, \\ 0, & \text{otherwise.} \end{cases}$$

It is easy to show that the function $f = (f_1, f_2, \dots)$ satisfies conditions (B1)–(B5). Moreover,

$$|f_i(t, x, y)| \le 4\,c\,(|x_i| + |y_i|), \text{ a.e. on } J.$$

Thus $f$ satisfies also the hypothesis (B8), whence the BVP (3.3.22) has by proposition 3.3.5 the least and the greatest solution, and that all the solutions of (3.3.22) belong to the order interval $[\alpha, \beta]$, where $\alpha = (-w, -w, \dots)$, $\beta = (w, w, \dots)$, $w$ being the solution of the BVP

$$-w''(t) = 8\,c\,w(t),\ t \in J,\ w(0) = w(1) = 1. \tag{3.3.23}$$

## 3.4. SECOND ORDER MIXED MONOTONE BVPS

In this section we shall generalize some results of section 3.2 to finite and infinite systems of the mixed monotone second order boundary value problems.

### 3.4.1. Finite mixed monotone boundary value systems

Let $J = [t_o, t_1]$, $a_i^j, b_i^j \in I\!R_+$, $c_i^j \in I\!R$, $i = 1, \dots, m$, $j = 0, 1$, and $f = (f_1, \dots, f_m)\colon J \times I\!R^{3m} \to I\!R^m$ be given. Consider the following system of the second order boundary value problems

$$-x_i'' = f_i(t,x,x,x),\ B_i^j x_i(t_j) = a_i^j\, x_i(t_j) - (-1)^j b_i^j\, x_i'(t_j) = c_i^j, \tag{3.4.1}$$

$i = 1, \dots, m,\ j = 0, 1.$

The functions $y = (y_1, \dots, y_m)$, $z = (z_1, \dots, z_m)$ of $AC^1(J, I\!R^m)$ are said to be *coupled quasisolutions* of (3.4.1) if

$$\begin{aligned} &-y_i''(t) = f_i(t, y(t), y(t), z(t)) \text{ a.e. on } J, \\ &-z_i''(t) = f_i(t, z(t), z(t), y(t)) \text{ a.e. on } J, \\ &B_i^j y_i(t_j) = c_i^j,\ B_i^j z_i(t_j) = c_i^j,\ j = 0, 1,\ i = 1, \dots, m. \end{aligned} \tag{3.4.2}$$

By assuming that $a_i^o a_i^1 + a_i^o b_i^1 + a_i^1 b_i^o > 0$ for each $i = 1, \dots, m$, we shall prove that the system (3.4.1) has the extremal coupled quasisolutions in an order interval of $AC(J, I\!R^m)$ if there is a null set $Z$ in $J$ such that the following conditions hold.

(C0) $\alpha, \beta \in AC^1(J, I\!R^m)$, $\alpha \le \beta$, $-\alpha'' \le f(t, \alpha, \alpha, \beta)$ and $-\beta'' \ge f(t, \beta, \beta, \alpha)$ on $J \backslash Z$, $B_i^j \alpha_i(t_j) \le c_i^j$ and $B_i^j \beta_i(t_j) \ge c_i^j$, $i = 1, \dots, m$, $j = 0, 1$.

(C1) For each $i = 1, \dots, m$ there exists $N_i \in L^1(J, I\!R_+)$ such that $|f_i(t, x, y, z)| \le N_i(t)$ for all $t \in J \setminus Z$ and $x$, $y$, $z \in [\alpha(t), \beta(t)]$.

(C2) $f(\cdot, x(\cdot), y(\cdot), z(\cdot))$ is measurable for all $x$, $y$, $z \in AC(J, I\!R^m)$, $\alpha \le x, y, z \le \beta$.

(C3) $f$ is continuous at $(t_j, x, y, z)$ for all $x$, $y$, $z \in [\alpha(t_j), \beta(t_j)]$, $j = 0, 1$.

(C4) $f(t, \cdot, y, z)$ is quasimonotone nondecreasing, $f(t, y, \cdot, z)$ is nondecreasing and $f(t, y, z, \cdot)$ is nonincreasing for all $t \in J \setminus Z$ and $y$, $z \in [\alpha(t), \beta(t)]$.

(C5) The functions $x_i \mapsto f_i(t, (x_1, \dots, x_m), y, z)$ are continuous and decreasing in $[\alpha_i(t), \beta_i(t)]$ for all $t \in J \setminus Z$ and $x = (x_1, \dots, x_m)$, $y$, $z \in [\alpha(t), \beta(t)]$.

**Theorem 3.4.1:** *Assume there exists a null set $Z$ in $J$ such that conditions (C0)–(C5) hold. Then the BVS (3.4.1) has the extremal coupled quasisolutions in the order interval* $[\alpha, \beta]$.

*Proof.* Given $y$, $z \in [\alpha, \beta]$, consider the BVS

$$\begin{aligned} -x_i'' &= F_i(t, x_i; y(t), z(t)), \\ B_i^j x_i(t_j) &= c_i^j, \ i = 1, \dots, m, \ j = 0, 1, \end{aligned} \tag{3.4.3}$$

where

$$\begin{aligned} &F_i(t, x; y(t), z(t)) = \\ &f_i(t, (y_1(t), \dots, y_{i-1}(t), x, y_{i+1}(t), \dots, y_m(t)), y(t), z(t)), \end{aligned} \tag{3.4.4}$$

for $(t, x) \in \Omega_i = \{(t, x) \mid t \in J,\ x \in [\alpha_i(t), \beta_i(t)]\}$. Each function $(t, x) \mapsto F_i(t, x; y(t), z(t))$ is by (C2) and (C5) a Carathéodory function in $\Omega_i$.

From (C0), (C4) and (3.4.4) it follows that

$$-\alpha_i'' \le F_i(t, \alpha_i; y(t), z(t)), \quad \text{and} \quad -\beta_i'' \ge F_i(t, \beta_i; y(t), z(t))$$

on $J \setminus Z$. From (C1) and (3.4.4) we see that that for each $i = 1, \dots, m$

$$|F_i(t, x; y(t), z(t))| \le N_i(t), \quad t \in J \setminus Z,\ x \in [\alpha(t), \beta(t)].$$

Thus each BVP of the BVS (3.4.3) has by theorem 3.1.2 a solution $x_i$ in $[\alpha_i, \beta_i]$. Because each $F_i(t, x; y(t), z(t))$ is by (C5) decreasing in $x$ on $[\alpha_i(t), \beta_i(t)]$ for each $t \in J \setminus Z$, it follows from corollary 3.1.1 that $x_i$ is the only solution of the $i$:th BVP of (3.4.3) in $[\alpha_i, \beta_i]$. Since the system (3.4.3) is uncoupled, then $x = (x_1, \dots, x_m)$ is the only solution of the BVS (3.4.3) in $[\alpha, \beta]$.

We now define a map $A = (A_1, \ldots, A_m)\colon [\alpha, \beta] \times [\alpha, \beta] \to [\alpha, \beta]$ by

$$A_i(y, z) = x_i, \qquad y,\ z \in [\alpha, \beta],\ i = 1, \ldots, m, \tag{3.4.5}$$

where $x = (x_1, \ldots, x_m)$ is the solution of the BVS (3.4.3) in $[\alpha, \beta]$. To prove that $A$ is mixed monotone, let $y$, $\bar{y}$, $z \in [\alpha, \beta]$ be given, and let $x = (x_1, \ldots, x_m)$, $\bar{x} = (\bar{x}_1, \ldots, \bar{x}_m)$ be the corresponding solutions of (3.4.3). If $\bar{y} \leq y$, it follows from (3.4.3)–(3.4.5) and (C4) that for each $i = 1, \ldots, m$, $\bar{x}_i$ is a lower solution of the the $i$:th BVP of (3.4.3). Since $x_i$, as the solution, is an upper solution of this BVP, it follows from lemma 3.1.1 that $\bar{x}_i \leq x_i$, which means that $A_i(\bar{y}, z) \leq A_i(y, z)$. Similarly, it can be shown that $A_i(z, \bar{y}) \geq A_i(z, y)$. These results hold for each $i = 1, \ldots, m$, whence $A$ is mixed monotone.

Since $x_i = A_i(y, z)$ is the solution of the $i$:th BVP of (3.4.3) in $[\alpha_i, \beta_i]$, it follows from lemma 3.1.3 that for all $i = 1, \ldots, m$, $y \in [\alpha, \beta]$ and $t \in J$,

$$\begin{aligned} A_i(y, z)(t) &= z_i^o(t) \\ &+ \int_{t_o}^{t_1} k_i(t, s) F_i(s, A_i(y, z)(s); y(s), z(s))\, ds, \end{aligned} \tag{3.4.6}$$

where

$$z_i^o(t) = \frac{c_i^o a_i^1 (t_1 - t) + c_i^o b_i^1 + c_i^1 a_i^o (t - t_o) + c_i^1 b_i^o}{a_i^o a_i^1 (t_1 - t_o) + a_i^o b_i^1 + a_i^1 b_i^o}, \tag{3.4.7}$$

and

$$k_i(t, s) = \begin{cases} \dfrac{(a_i^1(t_1 - t) + b_i^1)(a_i^o(s - t_o) + b_i^o)}{a_i^o a_i^1 (t_1 - t_o) + a_i^o b_i^1 + a_i^1 b_i^o}, & t_o \leq s \leq t \leq t_1, \\[2ex] \dfrac{(a_i^1(t_1 - s) + b_i^1)(a_i^o(t - t_o) + b_i^o)}{a_i^o a_i^1 (t_1 - t_o) + a_i^o b_i^1 + a_i^1 b_i^o}, & t_o \leq t \leq s \leq t_1. \end{cases} \tag{3.4.8}$$

Each $z_i^0$ is Lipschitz continuous and each $k_i(t,s)$ is Lipschitz continuous in $t$, uniformly over $s$. This, together with (a) and (3.4.6), imply an existence of a Lipschitz continuous function $w\colon J \to I\!R$ such that

$$\|A(y,z)(t) - A(y,z)(s)\| \le |w(t) - w(s)|, \quad y \in [\alpha,\beta], \ s,\, t \in J.$$

The foregoing arguments show that $A$ satisfies the assumptions of theorem 1.4.8, whence $A$ has the extremal coupled fixed points $y$, $z$. By (3.4.2)–(3.4.5) these coupled fixed points are coupled quasisolutions of (3.4.1).

If $u$, $v \in [\alpha,\beta]$ are coupled quasisolutions of (3.4.1), then they are also coupled fixed points of $A$, whence $y \le u$, $v \le z$. Thus $y$, $z$ are the extremal coupled quasisolutions of the BVS (3.4.1) in $[\alpha,\beta]$. □

In the case when $f$ is of the form

$$f(t,x,y,z) = f_1(t,x) + f_2(t,y) + f_3(t,z), \tag{3.4.9}$$

we obtain the following consequences of theorem 3.4.1.

**Corollary 3.4.1:** *Let $f_1(t,x)$ be a Carathéodory function and decreasing in $x$, $f_2(t,y)$ a standard function and nondecreasing in $y$, and $f_3(t,z)$ a standard function and nonincreasing in $z$. If $f$ is defined by (3.4.9) and if (C0) and (C1) hold, then the BVS (3.4.1) has the extremal coupled quasisolutions in the order interval $[\alpha,\beta]$.*

**Corollary 3.4.2:** *Given $f_1$, $f_2$, $f_3 \in C(J \times I\!R^m, I\!R^m)$, assume that $f_1(t,x)$ is decreasing in $x$, and $f_2(t,y)$ is nondecreasing in $y$, and that $f_3(t,z)$ is nonincreasing in $z$. If $f$ is defined by (3.4.9) and if (C0) holds, then the BVS (3.4.1) has the extremal coupled quasisolutions in the order interval $[\alpha,\beta]$.*

Sufficient for the validity of conditions (C0) and (C1) is that the following growth condition holds.

(C6) $|f_i(t,x,y,z)| \le H_i(t,|x_1|,\ldots,|x_m|,|y_1|,\ldots,|y_m|,|z_1|,\ldots |z_m|)$ for all $t \in J \setminus Z$, $x$, $y$, $z \in I\!R^m$ and $i = 1,\ldots,m$, where $H = (H_1,\ldots,H_m)\colon J \times I\!R_+^{3m} \to I\!R_+^m$, $H(t,u,v,w)$ is nondecreasing in $(u,v,w)$ for all $t \in J \setminus Z$, and the BVS

$$-x_i'' = H_i(t,x,x,x),\ a_i^j\, x_i(t_j) - (-1)^j b_i^j\, x_i'(t_j) = |c_i^j|, \tag{3.4.10}$$

has an upper solution in $AC^1(J, I\!R_+^m)$.

**Proposition 3.4.1:** *If $f = (f_1,\ldots,f_m)\colon J \times I\!R^{3m} \to I\!R^m$ satisfies condition (C6), then conditions (C0) and (C1) hold for $\alpha = -w$, $\beta = w$ and $N_i = -w_i''$, where $w = (w_1,\ldots,w_m) \in AC^1(J, I\!R_+^m)$ is an upper solution of (3.4.10). If conditions (C2)–(C5) hold with these $\alpha$, $\beta$, then the BVS (3.4.1) has the extremal coupled quasisolutions in the order interval $[-w, w]$.*

*Proof.* Let $w = (w_1,\ldots,w_m) \in AC^1(J, I\!R_+^m)$ be an upper solution of the BVS (3.4.10). From (C6) it follows that for each $i = 1,\ldots,m$

$$\begin{aligned}&\sup\{|f_i(t,x,y,z)| \mid |x_j|,\ |y_j|,\ |z_j| \le w_j(t),\ 1 \le j \le m\}\\ &\le H_i(t,w(t),w(t),w(t)) \le -w_i''(t)\end{aligned} \tag{a}$$

for a.a. $t \in J$. Thus we have

$$-(-w)''(t) \le f(t,-w(t),-w(t),w(t))$$

and

$$f(t,w(t),w(t),-w(t)) \le -w''(t)$$

a.e. in $J$. Because

$$a_i^j\, w_i(t_j) - (-1)^j b_i^j\, w_i'(t_j) \ge |c_i^j|,$$

for $i = 1, \ldots, m,\ j = 0, 1$, it follows that for these $i,\ j$,

$$a_i^j\, w_i(t_j) - (-1)^j b_i^j\, w_i'(t_j) \geq c_i^j,$$

and

$$a_i^j\,(-w_i)(t_j) - (-1)^j b_i^j\,(-w_i)'(t_j) \leq c_i^j.$$

Thus (C0) holds with $\alpha = -w$ and $\beta = w$. With these $\alpha$, $\beta$ it follows from (a) that (C1) holds when $N_i = -w_i''$. If conditions (C2)–(C5) hold with these $\alpha$, $\beta$, then the BVS (3.4.1) has by theorem 3.4.1 the extremal coupled quasisolutions in the order interval $[-w, w]$. □

### 3.4.2. Infinite mixed monotone boundary value systems

Let $J = [t_o, t_1]$, $a^j = (a_1^j, a_2^j, \ldots)$, $b^j = (b_1^j, b_2^j, \ldots) \in l_+^\infty$, $c^j = (c_1^j, c_2^j, \ldots) \in l^\infty$, $j = 0, 1$ and $f = (f_1, f_2 \ldots)\colon J \times (l^\infty)^3 \to l^\infty$ be given. Consider the following infinite system of the second order boundary value problems

$$-x_i'' = f_i(t, x, x, x),\ B_i^j x(t_j) = a_i^j\, x(t_j) - (-1)^j b_i^j\, x'(t_j) = c_i^j, \tag{3.4.11}$$

$i = 1, 2, \ldots,\ j = 0, 1$. We shall assume that

$a_i^o a_i^1(t_1 - t_o) + a_i^o b_i^1 + a_i^1 b_i^o \geq c > 0$ for each $i = 1, 2, \ldots$.

It turns out that the system (3.4.11) has the extremal coupled quasisolutions in an order interval of $AC(J, l^\infty)$ if there is a null set $Z$ in $J$ such that the following conditions hold.

(D0) $\alpha, \beta \in AC^1(J, l^\infty)$, $\alpha \leq \beta$, $-\alpha'' \leq f(t, \alpha, \alpha, \beta)$ and $-\beta'' \geq f(t, \beta, \beta, \alpha)$ on $J\setminus Z$, $B_i^j \alpha_i(t_j) \leq c_i^j$ and $B_i^j \beta_i(t_j) \geq c_i^j$, $i = 1, 2, \ldots$, $j = 0, 1$.

(D1) $N = (N_1, N_2, \ldots) \in L^1(J, l^\infty)$ and $|f_i(t, x, y, z)| \leq N_i(t)$ for all $i = 1, 2, \ldots$, $t \in J \setminus Z$ and $x,\ y,\ z \in [\alpha(t), \beta(t)]$.

(D2) $f(\cdot, x(\cdot), y(\cdot), z(\cdot))$ is measurable for $x, y, z \in AC(J, l^\infty)$, $\alpha \leq x,\ y,\ z \leq \beta$.

(D3) $f$ is continuous at $(t_j, x, y, x)$ for all $x,\ y,\ z \in [\alpha(t_j), \beta(t_j)]$, $j = 1, 2$.

(D4) $f(t,\cdot,y,z)$ is quasimonotone nondecreasing, $f(t,y,\cdot,z)$ is nondecreasing and $f(t,y,z,\cdot)$ is nonincreasing for all $t \in J \setminus Z$ and $y$, $z \in [\alpha(t),\beta(t)]$.

(D5) For each $i = 1,2,\ldots$, $x_i \mapsto f_i(t,(x_1,x_2,\ldots),y,z)$ is continuous and decreasing in $[\alpha_i(t),\beta_i(t)]$ for all $t \in J \setminus Z$ and $x = (x_1,x_2,\ldots)$, $y$, $z \in [\alpha(t),\beta(t)]$.

**Theorem 3.4.2:** *Assume there exists a null set $Z$ in $J$ such that conditions (D0)–(D5) are valid. Then the BVS (3.4.11) has the extremal coupled quasisolutions in the order interval $[\alpha,\beta]$ of $AC(J,l^\infty)$.*

*Proof.* Given $y, z \in [\alpha,\beta]$, consider the BVS

$$-x_i'' = F_i(t,x_i;y(t),z(t)), \quad B_i^j x_i(t_j) = c_i^j, \tag{3.4.12}$$

where

$$F_i(t,x;y(t),z(t)) = f_i(t,(y_1(t),\ldots,y_{i-1}(t),x,y_{i+1}(t),\ldots),y(t)), \tag{3.4.13}$$

$(t,x) \in \Omega_i = \{(t,x) \mid t \in J,\ x \in [\alpha_i(t),\beta_i(t)]\}$. As in the proof of theorem 3.4.1 it can be shown that each BVP of the BVS (3.4.12) has a unique solution $x_i$ in $[\alpha_i,\beta_i]$. Since the system (3.4.12) is uncoupled, then $x = (x_1,x_2,\ldots)$ is the only solution of the BVS (3.4.12) in $[\alpha,\beta]$.

Define a map $A = (A_1,A_2\ldots)\colon [\alpha,\beta] \to [\alpha,\beta]$ by

$$A_i(y,z) = x_i, \qquad y,\ z \in [\alpha,\beta],\ i = 1,2,\ldots, \tag{3.4.14}$$

where $x = (x_1,x_2,\ldots)$ is the solution of the BVS (3.4.12) in $[\alpha,\beta]$. Referring to the proof of theorem 3.4.1 it is easy to show that $A$ is mixed monotone.

Since $x_i = A_i(y,z)$ is the solution of the $i$:th BVP of (3.4.12) in $[\alpha_i,\beta_i]$, it follows from lemma 3.1.3 that for all $i = 1,2,\ldots$

$y \in [\alpha, \beta]$ and $t \in J$,

$$A_i(y,z)(t) = z_i^o(t) + \int_{t_o}^{t_1} k_i(t,s)F_i(s, G_i y(s); y(s), z(s))\, ds, \tag{3.4.15}$$

where $z_i^o$ and $k_i$ are defined by (3.4.7) and (3.4.8). Since $a^j = (a_1^j, a_2^j, \dots)$, $b^j = (b_1^j, b_2^j, \dots) \in l_+^\infty$, $c^j = (c_1^j, c_2^j, \dots) \in l^\infty$, $j = 0, 1$ and $a_i^o a_i^1 (t_1 - t_o) + a_i^o b_i^1 + a_i^1 b_i^o \geq c > 0$ for each $i = 1, 2, \dots$, it follows that each $z_i^o$ is Lipschitz continuous, uniformly over $i$ and each $k_i(t,s)$ is Lipschitz continuous in $t$, uniformly over $s$ and $i$. This, (a) and (3.4.15) imply an existence of a Lipschitz continuous function $w\colon J \to I\!R$ such that

$$\|A(y,z)(t) - A(y,z)(s)\|_\infty \leq |w(t) - w(s)|, \quad y \in [\alpha, \beta],\ s,\, t \in J.$$

The above proof shows that $A$ satisfies the assumptions of proposition 1.4.7, whence $A$ has the extremal coupled fixed points $y = (y_1, y_2, \dots)$, $z = (z_1, z_2, \dots)$. By the definitions of $F_i$ and $A_i$ these coupled fixed points are coupled quasisolutions of the BVS (3.4.11) in $[\alpha, \beta]$.

If $u = (u_1, u_2, \dots)$, $v = (v_1, v_2, \dots) \in [\alpha, \beta]$ are coupled quasisolutions of (3.4.11), then they are also coupled quasisolutions of (3.4.12), and hence coupled fixed points of $A$. Since $y$, $z$ were the extremal coupled fixed points of $A$, it follows that $y \leq u$, $v \leq z$. Thus $y = (y_1, y_2, \dots)$, $z = (z_1, z_2, \dots)$ are the extremal coupled quasisolutions of the BVS (3.4.11) in $[\alpha, \beta]$. □

Corollary 3.4.1 can also be extended to $l^\infty$-valued case.

## 3.5. NOTES AND COMMENTS

Chapter 3 deals with existence and extremality results of second order BVP's involving discontinuous nonlinearities. Section 3.1 is taken from Heikkilä and Lakshmikantham (1994b). The results of section 3.2 are new. Sections 3.3 and 3.4 extend most of the results of sections 3.1 and 3.2 to finite and infinite systems of BVP's, and the results are new. As for second order BVP's with deviating arguments and discontinuous right-hand side see Dhage and Heikkilä (1993).

# 4

# Second Order Partial Differential Equations

## 4.0. INTRODUCTION

In this chapter we shall study the existence of weak extremal solutions of elliptic and parabolic partial differential equations and systems involving discontinuous nonlinearities, by combined application of the method of upper and lower solutions and the method of generalized monotone iterations.

Section 4.1 deals with a general quasilinear elliptic PDE with discontinuous lower order terms and discontinuous boundary conditions. Weak formulations to the solution, as well as upper and lower solutions of the considered problem, and the hypotheses are given in subsection 4.1.1. We shall first consider in subsection 4.1.2 an associated boundary value problem of Carathéodory type. The main result proved in subsection 4.1.3 gives an existence result for extremal solutions to a quasilinear elliptic boundary value problem, by allowing both continuous and discontinuous dependence on the dependent variable both in the PDE and in the boundary condition.

Dirichlet boundary value problem is studied in section 4.2 in the case when the differential operator in the PDE is uniformly elliptic, the dependent variable being decomposed in the nonlinear term to continuous and to discontinuous part. The so obtained existence result for extremal solutions is then extended in section 4.3 to the corresponding systems. The case when the discontinuous dependence of the dependent variable is of mixed monotone type is also included and existence results for extremal coupled quasisolutions are derived.

Section 4.4 presents existence results for extremal solutions to a general class of parabolic initial-boundary value problems (IBVP) involving discontinuous nonlinearities, when the partial differential operator of the PDE is a second order quasilinear elliptic operator, and when the initial- boundary conditions are Dirichlet type. Again we shall first derive existence results for extremal solutions of the associated Carathéodory type IBVP.

The finite systems of parabolic IBVP's are considered in section 4.5 when the partial derivatives in each equation of the system form an uniformly elliptic differential operator of the corresponding component of the unknown function. Both the cases when the discontinuous parts of the dependent variables in nonlinear terms are of nondecreasing and of mixed monotone type are studied.

Examples are presented to illustrate the obtained results and their possible applicability.

In the presentation of this chapter we assume that the reader is familiar with the basic notations and concepts of the theory of second order elliptic and parabolic boundary value problems (cf. Gilbarg and Trudinger (1983), Zeidler (1990a,b), Kufner, John and Fučic (1977)).

## 4.1. QUASILINEAR ELLIPTIC BVP'S

Let $\Omega \subset \mathbb{R}^N$ be a bounded domain with Lipschitz boundary $\partial\Omega$ (cf. Kufner, John and Fučic (1977)). In the present section we are concerned with a quasilinear elliptic boundary value problem (BVP) of the following type

$$\begin{aligned} A\,u(x) &= f(x, u(x), u(x), \nabla u(x)), \quad x \in \Omega, \\ \frac{\partial u(x)}{\partial \nu} &= g(x, u(x), u(x)), \quad x \in \partial\Omega, \end{aligned} \tag{4.1.1}$$

where

$$A\,u(x) = -\sum_{i=1}^{N} \frac{\partial}{\partial x_i} a_i(x, \nabla u(x)), \quad x \in \Omega,$$

$$\frac{\partial u(x)}{\partial \nu} = \sum_{i=1}^{N} a_i(x, \nabla u(x)) n_i(x), \quad x \in \partial\Omega,$$

$n_i(x)$ being the components of the outward unit normal vector of $\partial\Omega$ at $x$, and $\nabla u = (\frac{\partial u}{\partial x_1}, \ldots, \frac{\partial u}{\partial x_N})$.

First we shall prove extremal solution results for an associated BVP of Carathéodory type. Applying these results and a generalized monotone iteration method we shall then consider the existence of extremal solutions of the BVP (4.1.1) between assumed upper and lower solutions in the case when $f$ and $g$ are discontinuous with respect to first two variables. Examples are given to illustrate the results.

### 4.1.1. Hypotheses

Let $W^{1,p}(\Omega)$ denote the usual Sobolev space of those functions $u\colon \Omega \to \mathbb{R}$ which, together with their generalized derivatives $\frac{\partial u}{\partial x_i}$, $i = 1, \ldots, N$, belong to $L^p(\Omega) = L^p(\Omega, \mathbb{R})$ (cf. subsection 5.8.5). Denote by $\langle \cdot, \cdot \rangle$ the duality pairing between the elements of the dual space $(W^{1,p}(\Omega))^*$ and $W^{1,p}(\Omega)$. A partial ordering is defined in $L^p(\Omega)$ and in $W^{1,p}(\Omega)$ by $u \le v$ if and only if $v - u$ belongs to the set $L^p_+(\Omega)$ of all nonnegative elements of $L^p(\Omega)$. If $u,\, v \in W^{1,p}(\Omega)$, denote $[u,v] = \{w \in W^{1,p}(\Omega) \mid u \le w \le v\}$.

Let $p$ and $q$ be fixed constants satisfying $1 < p < \infty$ and $\frac{1}{p} + \frac{1}{q} = 1$. On the functions $a_i \colon \Omega \times \mathbb{R}^N \to \mathbb{R}$, $i = 1, \dots, N$, we impose the following standard conditions of Leray–Lions type (cf. e.g., Lions (1969), Ch. 2).

(A0) Each $a_i$ is a Carathéodory function.

(A1) There exists a constant $c_o \ge 0$ and $k_o \in L^q_+(\Omega)$ such that $|a_i(x,\xi)| \le k_o(x) + c_o\,|\xi|^{p-1}$, $i = 1, \dots, N$, for a.a. $x \in \Omega$ and for all $\xi \in \mathbb{R}^N$.

(A2) $\sum_{i=1}^N (a_i(x,\xi) - a_i(x,\xi'))(\xi_i - \xi_i') > 0$ for a.a. $x \in \Omega$ and for all $\xi,\, \xi' \in \mathbb{R}^N$ with $\xi \ne \xi'$.

(A3) $\sum_{i=1}^N a_i(x,\xi)\,\xi_i \ge c_1 |\xi|^p - k_1(x)$ for a.a. $x \in \Omega$ and for all $\xi \in \mathbb{R}^N$ with some positive constant $c_1$ and a function $k_1 \in L^1_+(\Omega)$.

As a consequence of (A1) the semilinear form $\ell$ associated with the operator $A$ by

$$\ell(u,\varphi) = \sum_{i=1}^N \int_\Omega a_i(x, \nabla u) \frac{\partial \varphi}{\partial x_i}\, dx$$

is well-defined on $W^{1,p}(\Omega) \times W^{1,p}(\Omega)$.

Notice that the partial ordering for elements of $W^{1,p}(\Omega)$ introduced above induces a corresponding partial ordering for their traces which belong to $L^p(\partial\Omega)$. Thus $u,\, v \in W^{1,p}(\Omega)$ and $u \le v$ in $\Omega$ imply that $u \le v$ on $\partial\Omega$ in the sense of traces and defined by the order cone $L^p_+(\partial\Omega)$. The function $g(x,u,u)$ in the boundary condition of (4.1.1) is defined for the traces of $u$.

A function $u \in W^{1,p}(\Omega)$ is called a *solution* of (4.1.1) if

$$\ell(u,\varphi) = \int_\Omega f(x,u,u,\nabla u)\, \varphi\, dx + \int_{\partial\Omega} g(x,u,u)\, \varphi\, d\omega$$

for all $\varphi \in W^{1,p}(\Omega)$.

A function $\bar{u} \in W^{1,p}(\Omega)$ is an *upper solution* of (4.1.1) if

$$\ell(\bar{u},\varphi) \ge \int_\Omega f(x,\bar{u},\bar{u},\nabla \bar{u})\, \varphi\, dx + \int_{\partial\Omega} g(x,\bar{u},\bar{u})\, \varphi\, d\omega$$

for all $\varphi \in W^{1,p}(\Omega) \cap L^p_+(\Omega)$.

A lower solution is defined similarly, by reversing the inequality.

We shall impose the following hypotheses on the functions $f: \Omega \times \mathbb{R} \times \mathbb{R} \times \mathbb{R}^N \to \mathbb{R}$ and $g: \partial\Omega \times \mathbb{R} \times \mathbb{R} \to \mathbb{R}$.

(H1) The BVP (4.1.1) has a lower solution $\underline{u}$ and an upper solution $\bar{u}$ such that $\underline{u} \leq \bar{u}$.

(H2) There exist functions $k_2 \in L_+^q(\Omega)$ and $k_3 \in L_+^q(\partial\Omega)$ and a constant $c_2 \geq 0$ such that

$$|f(x,r,s,\xi)| \leq k_2(x) + c_2|\xi|^{p-1} \tag{4.1.2}$$

for a.a. $x \in \Omega$ and for all $\xi \in \mathbb{R}^N$, $r, s \in [\underline{u}(x), \bar{u}(x)]$, and

$$|g(x,r,s)| \leq k_3(x) \tag{4.1.3}$$

for all $r, s \in [\underline{u}(x), \bar{u}(x)]$ and for a.a. $x \in \partial\Omega$.

(H3) The functions $(s,\xi) \mapsto f(x,r,s,\xi)$, $s \mapsto g(x,r,s)$ are continuous for all $r \in \mathbb{R}$ and for a.a. $x \in \Omega$, and there exist continuous functions $a$, $b$: $\mathbb{R} \to \mathbb{R}$ and positive constants $c_3$, $c_4$ such that

$$\begin{aligned} &|a(r)| \leq c_3\,(1 + |r|^{p-1}), \\ &(a(r_1) - a(r_2))(r_1 - r_2) \geq c_4|r_1 - r_2|^p, \end{aligned} \tag{4.1.4}$$

$$|b(r)| \leq c_3\,(1 + |r|^{p-1}), \quad (b(r_1) - b(r_2))(r_1 - r_2) \geq 0, \tag{4.1.5}$$

and that the functions $r \mapsto f(x,r,s,\xi) + a(r)$ and $r \mapsto g(x,r,s) + b(r)$ are nondecreasing for a.a. $x \in \Omega$ and for each $(s,\xi) \in \mathbb{R} \times \mathbb{R}^N$ and $s \in \mathbb{R}$, respectively.

(H4) The functions $(x,r) \mapsto f(x,r,s,\xi)$ and $(x,r) \mapsto g(x,r,s)$ are standard functions for each $(s,\xi) \in \mathbb{R} \times \mathbb{R}^N$ and $s \in \mathbb{R}$, respectively.

Our purpose is to show in this section that if the hypotheses (A0)–(A3) and (H1)–(H4) are satisfied, then the BVP (4.1.1) has the extremal solutions $u_*$, $u^*$ in $[\underline{u}, \bar{u}]$.

### 4.1.2. Preliminaries

We shall first derive preliminary results for a Carathéodory type BVP associated with the BVP (4.1.1). Throughout this section we shall assume that the hypotheses (A0)–(A3) and (H1)–(H4) are satisfied. For the sake of completeness we recall the following lemma which will be needed later.

**Lemma 4.1.1:** (Troianello (1987), lemma 4.22) *Let* $(u_n)_{n=0}^{\infty}$ *be a sequence which converges weakly to* $u$ *in* $W^{1,p}(\Omega)$. *If*

$$\ell(u_n, u_n - u) - \ell(u, u_n - u) \to 0 \text{ as } n \to \infty,$$

*then* $(u_n)$ *converges to* $u$ *strongly in* $W^{1,p}(\Omega)$.

As a consequence of proposition 1.3.7 we obtain

**Lemma 4.1.2:** *A bounded and well-ordered chain* $C$ *of* $W^{1,p}(\Omega)$ *has a nondecreasing sequence which converges to* $\sup C$ *weakly in* $W^{1,p}(\Omega)$ *and strongly both in* $L^p(\Omega)$ *and in* $L^p(\partial\Omega)$.

*Proof.* Let $C$ be a bounded and well-ordered chain in $W^{1,p}(\Omega)$. Since $W^{1,p}(\Omega)$ is continuously embedded in $L^p(\Omega)$, then $C$ is bounded also in $L^p(\Omega)$. Because $W^{1,p}(\Omega)$ is reflexive and the order cone $L_+^p(\Omega)$ of $L^p(\Omega)$ is fully regular, then proposition 1.3.7 implies the existence of a nondecreasing sequence $(v_n)_{n=1}^{\infty}$ which converges strongly in $L^p(\Omega)$ and weakly in $W^{1,p}(\Omega)$ to $w = \sup C$. This and the compactness of the trace operator $W^{1,p}(\Omega) \subset L^p(\partial\Omega)$ (cf. Kufner, John and Fučic (1977), p. 344) implies that $(v_n)_{n=1}^{\infty}$ converges to $w$ strongly in $L^p(\partial\Omega)$. □

Now, let $\bar{u}$, $\underline{u} \in W^{1,p}(\Omega)$ be upper and lower solutions of the BVP (4.1.1), as assumed in (H1). We assign to any upper solution $v \in [\underline{u}, \bar{u}]$ of (4.1.1) the following BVP:

$$\begin{aligned} Au + a(u) &= f(x, v, u, \nabla u) + a(v) \quad \text{in } \Omega, \\ \frac{\partial u}{\partial \nu} + b(u) &= g(x, v, u) + b(v) \quad \text{on } \partial\Omega. \end{aligned} \tag{4.1.6}$$

**Lemma 4.1.3:** *Let $v \in [\underline{u}, \bar{u}]$ be any upper solution of the BVP (4.1.1). If $u_i$, $i = 1, \dots, m$, are lower solutions of the BVP (4.1.6) in $[\underline{u}, v]$, then (4.1.6) has at least one solution $u$ in $[u_o, v]$, where $u_o = \max\{u_i \mid i = 1, \dots, m\}$.*

*Proof.* It can readily be seen that $v$ is also an upper solution of the BVP (4.1.6). For $u \in W^{1,p}(\Omega)$ we define the truncation mappings $T_i$, $i = 0, 1, \dots, m$, by

$$(T_i u)(x) = \begin{cases} u_i(x) & \text{if } u(x) < u_i(x), \\ u(x) & \text{if } u_i(x) \le u(x) \le v(x), \\ v(x) & \text{if } v(x) < u(x). \end{cases}$$

In Deuel and Hess (1974/75) it has been proved that these truncation mappings are bounded and continuous mappings from $W^{1,p}(\Omega)$ into itself. Analogously one defines the truncations of the trace $u \in L^p(\partial\Omega)$ which are bounded and continuous self-mappings of $L^p(\partial\Omega)$. Conditions (H3) and (H4) ensure that the functions $(x, s, \xi) \mapsto f(x, v(x), s, \xi) + a(v(x))$ and $(x, s) \mapsto g(x, v(x), s) + b(v(x))$ satisfy the Carathéodory conditions. If $F$ and $G$ denote the superposition operators associated with these functions, respectively, defined by

$$\begin{aligned} Fu(x) &= f(x, v(x), u(x), \nabla u(x)) + a(v(x)), \\ Gu(x) &= g(x, v(x), u(x)) + b(v(x)), \end{aligned} \tag{4.1.7}$$

then the BVP (4.1.6) can be written in the form

$$Au + a(u) = Fu \ \text{ in } \Omega, \quad \frac{\partial u}{\partial \nu} + b(u) = Gu \ \text{ on } \partial\Omega. \tag{4.1.8}$$

As a consequence of (H2) and (H3) the mappings $F$ and $G$ are bounded and continuous from $[\underline{u}, \bar{u}] \subset W^{1,p}(\Omega)$ into $L^q(\Omega)$ and from $[\underline{u}, \bar{u}] \subset L^p(\partial\Omega)$ into $L^q(\partial\Omega)$, respectively. Thus the composed operators $F \circ T_i \colon W^{1,p}(\Omega) \to L^q(\Omega)$ and $G \circ T_i \colon L^p(\partial\Omega) \to L^q(\partial\Omega)$, $i = 0, 1, \dots, m$, are bounded and continuous.

Now we associate with (4.1.8) (resp. (4.1.6)) the following auxiliary BVP.

$$
\begin{aligned}
Au + a(u) =& F \circ T_o u + \sum_{i=1}^{m} |F \circ T_i u - F \circ T_o u| \text{ in } \Omega, \\
\frac{\partial u}{\partial \nu} + b(u) =& G \circ T_o u + \sum_{i=1}^{m} |G \circ T_i u - G \circ T_o u| \text{ on } \partial\Omega.
\end{aligned}
\tag{4.1.9}
$$

The proof of the lemma is accomplished if we show that there exist solutions of the BVP (4.1.9), and if each solution of (4.1.9) belongs to $[u_o, v]$, since then we have $T_i u = u$ for each $i = 0, 1, \ldots, m$, and thus $u$ must be also a solution of the BVP (4.1.8).

Step 1: Existence of a solution of (4.1.9).
We define operators $L$, $L_1$, $L_2$: $W^{1,p}(\Omega) \to (W^{1,p}(\Omega))^*$ as follows.

$$
\begin{aligned}
\langle Lu, \varphi \rangle =& \ell(u, \varphi) + \int_{\Omega} a(u)\, \varphi \, dx, \\
\langle L_1 u, \varphi \rangle =& -\int_{\Omega} \{F \circ T_o u + \sum_{i=1}^{m} |F \circ T_i u - F \circ T_o u|\} \varphi \, dx, \\
\langle L_2 u, \varphi \rangle =& -\int_{\partial\Omega} \{G \circ T_o u + \sum_{i=1}^{m} |G \circ T_i u - G \circ T_o u|\} \varphi \, d\omega \\
& + \int_{\partial\Omega} b(u)\, \varphi \, d\omega,
\end{aligned}
$$

for $u$, $\varphi \in W^{1,p}(\Omega)$. Thus the corresponding weak formulation of the BVP (4.1.9) reads as

$$
\langle (L + L_1 + L_2)u, \varphi \rangle = 0 \quad \text{for all } \varphi \in W^{1,p}(\Omega). \tag{4.1.10}
$$

We shall show that the operator $L + L_1 + L_2$ is bounded, pseudomonotone and coercive in order to apply the main theorem on pseudomonotone operators.

It can easily be shown that $L$ and $L_1$ are bounded and continuous operators from $W^{1,p}(\Omega)$ into $(W^{1,p}(\Omega))^*$. The assumptions (A1), (A2), (4.1.2) and (4.1.4) imply that the operator $S = L+L_1$ is a Leray–Lions operator, and hence *pseudomonotone*, i.e. if $(u_n)$ converges weakly to $u$ in $W^{1,p}(\Omega)$ and $\lim\sup\langle Su_n, u_n - u\rangle \leq 0$, then $\lim\inf\langle Su_n, u_n - v\rangle \geq \langle Su, u - v\rangle$ for each $v \in W^{1,p}(\Omega)$ (cf. Gossez and Mustonen (1993), Troianello (1987)). The operator $L_2\colon W^{1,p}(\Omega) \to (W^{1,p}(\Omega))^*$ is strongly continuous. This is a consequence of the compactness of the trace operator acting from $W^{1,p}(\Omega)$ into $L^p(\partial\Omega)$, and of the continuity and boundedness of the mappings $b(\cdot)$, $G \circ T_i\colon L^p(\partial\Omega) \to L^q(\partial\Omega)$, due to (4.1.3) and (4.1.5). Hence, it follows that also the sum of $L + L_1$ and $L_2$ is a pseudomonotone operator (cf. Zeidler (1990b), Prop. 27.7).

In the following, let $d_i$ denote some positive constants. By means of (A3) and (4.1.4) we obtain

$$\langle Lu, u\rangle \geq c_1\|\nabla u\|_p^p + c_4\|u\|_p^p - d_0(1 + \|u\|_p). \tag{4.1.11}$$

From (4.1.2) and (4.1.4) we get

$$\left|\int_\Omega (F \circ T_i u)\, u\, dx\right| \leq d_1(1 + \|\nabla u\|_p^{p-1})\|u\|_p. \tag{4.1.12}$$

Applying Young's inequality to (4.1.12) we obtain for any $\epsilon > 0$

$$\left|\int_\Omega (F \circ T_i u)\, u\, dx\right| \leq d_1\|u\|_p + d_2(\epsilon)\|u\|_p^p + \epsilon\,\|\nabla u\|_p^p \tag{4.1.13}$$

for all $i = 0, 1, \ldots, m$. This implies an estimate of the form

$$\langle L_1 u, u\rangle \geq -\epsilon\,\|\nabla u\|_p^p - d_3\|u\|_p - d_4(\epsilon)\|u\|_p^p. \tag{4.1.14}$$

Further, (4.1.3) and (4.1.5) yield the estimate

$$\left|\int_{\partial\Omega} (G \circ T_i u)\, u\, d\omega\right| \leq d_5\|u\|_{L^p(\partial\Omega)} \leq d_6\|u\|_{W^{1,p}(\Omega)},$$

and thus by (4.1.5) we have

$$\langle L_2 u, u\rangle \geq -d_7 \|u\|_{W^{1,p}(\Omega)}. \tag{4.1.15}$$

Since any multiple $ka$ of the function $a$ also satisfies the hypothesis (H3) for $k \geq 1$, we may always choose the constant $c_4$ sufficiently large. Thus (4.1.11), (4.1.14) and (4.1.15) imply, by choosing $\epsilon < c_1$ and $c_4 > d_4(\epsilon)$, that $L + L_1 + L_2$ is *coercive*, i.e.

$$\frac{\langle (L + L_1 + L_2)u, u\rangle}{\|u\|_{W^{1,p}(\Omega)}} \to \infty \quad \text{for } \|u\|_{W^{1,p}(\Omega)} \to \infty.$$

Now, the main theorem on pseudomonotone operators (cf. Zeidler (1990b), Thm. 27.A) can be applied to ensure the existence of a solution of (4.1.10), and hence of (4.1.9).

Step 2: Each solution of (4.1.9) belongs to $[u_o, v]$.
Let $u_k$, $k \in \{1, 2, \ldots, m\}$, be any of the assumed lower solutions, and let $u$ be a solution of (4.1.9). Denoting $w^+ = \max\{w, 0\}$, then $(u_k - u)^+ \in W^{1,p}(\Omega) \cap L^p_+(\Omega)$, so that

$$\langle Lu_k - Lu, (u_k - u)^+\rangle + \langle Lu, (u_k - u)^+\rangle \leq \int_\Omega Fu_k (u_k - u)^+ \, dx + \int_{\partial\Omega} (Gu_k - b(u_k))(u_k - u)^+ \, d\omega.$$

We replace $\langle Lu, (u_k - u)^+\rangle$ by means of (4.1.10) and obtain

$$\begin{aligned}
\langle Lu_k - Lu, (u_k - u)^+\rangle &\leq \int_{\partial\Omega} (b(u) - b(u_k))(u_k - u)^+ \, d\omega \\
&+ \int_\Omega \{Fu_k - F\circ T_o u - \sum_{i=1}^m |F\circ T_i u - F\circ T_o u|\}(u_k - u)^+ \, dx \\
&+ \int_{\partial\Omega} \{Gu_k - G\circ T_o u - \sum_{i=1}^m |G\circ T_i u - G\circ T_o u|\}(u_k - u)^+ \, d\omega \\
&\leq \int_{\partial\Omega} (b(u) - b(u_k))(u_k - u)^+ \, d\omega \leq 0.
\end{aligned} \tag{4.1.16}$$

It is well-known that for $w \in W^{1,p}(\Omega)$ the following rule holds (cf. e.g., Gilbarg and Trudinger (1983), Lemma 7.6).

$$\frac{\partial w^+}{\partial x_i} = \begin{cases} \frac{\partial w}{\partial x_i} & \text{if } w > 0, \\ 0 & \text{if } w \le 0. \end{cases}$$

By means of (A2) and (4.1.4) we then get

$$\begin{aligned} &\langle Lu_k - Lu, (u_k - u)^+ \rangle \\ &= \sum_{i=1}^{N} \int_\Omega (a_i(\cdot, \nabla u_k) - a_i(\cdot, \nabla u)) \frac{\partial (u_k - u)^+}{\partial x_i}\, dx \\ &+ \int_\Omega (a(u_k) - a(u))(u_k - u)^+\, dx \ge c_4 \int_\Omega |(u_k - u)^+|^p\, dx \ge 0. \end{aligned} \tag{4.1.17}$$

Finally, from (4.1.16) and (4.1.17) it follows that $(u_k - u)^+ = 0$, i.e. $u_k \le u$ for each $k = 1, \dots, m$, and thus $u_o \le u$. In a similar way it can be shown that $u \le v$. This completes the proof of lemma 4.1.3. □

**Lemma 4.1.4:** *If $v \in [\underline{u}, \bar{u}]$ is any upper solution of (4.1.1), then the BVP (4.1.6) has at least one solution in $[\underline{u}, v]$, and all such solutions are uniformly bounded in $W^{1,p}(\Omega)$.*

*Proof.* The first part of lemma 4.1.4 follows from the fact that the lower solution $\underline{u}$ of the BVP (4.1.1) is in particular also a lower solution of the BVP (4.1.6). Thus lemma 4.1.3 can be applied. The uniform boundedness of the solutions of (4.1.6) can be derived from the weak formulation (4.1.10) by taking $\varphi = u$ as a special test function and by using the estimates (4.1.11), (4.1.14) and (4.1.15), which yields

$$\begin{aligned} &(c_1 - \epsilon)\|\nabla u\|_p^p + (c_4 - d_4(\epsilon))\|u\|_p^p \\ &\le d_0(1 + \|u\|_p) + d_3\|u\|_p + d_7\|u\|_{W^{1,p}(\Omega)}, \end{aligned} \tag{4.1.18}$$

where the coefficients $d_i$ still depend on $v$ in the form of $\|a(v)\|_{L^q(\Omega)}$ and $\|b(v)\|_{L^q(\partial\Omega)}$. However, due to the monotonicity of the functions $a$ and $b$ and since $v \in [\underline{u}, \bar{u}]$ we obtain the following uniform bounds

$$\|a(v)\|_q \leq \|a(\underline{u})\|_q + \|a(\bar{u})\|_q,$$

$$\|b(v)\|_{L^q(\partial\Omega)} \leq \|b(\underline{u})\|_{L^q(\partial\Omega)} + \|b(\bar{u})\|_{L^q(\partial\Omega)},$$

and therefore also bounds for the coefficients $d_i$ independent of $v$. If we apply Young's inequality to $d_7\|u\|_{W^{1,p}(\Omega)}$ of (4.1.18) we obtain for any $\delta > 0$

$$d_7\|u\|_{W^{1,p}(\Omega)} \leq d_8(\delta) + \delta\|u\|^p_{W^{1,p}(\Omega)}. \tag{4.1.19}$$

Since $\|u\|_{L^p(\Omega)} \leq \|\underline{u}\|_{L^p(\Omega)} + \|\bar{u}\|_{L^p(\Omega)}$, we finally obtain, by choosing $\epsilon$ and $\delta$ sufficiently small and $c_4$ sufficiently large,

$$\|u\|_{W^{1,p}(\Omega)} \leq \text{ const.} \tag{4.1.20}$$

for any solution of (4.1.6) which is in $[\underline{u}, \bar{u}]$. □

**Lemma 4.1.5:** *Let $v \in [\underline{u}, \bar{u}]$ be any upper solution of the BVP (4.1.1), and denote $S_v = \{u \in [\underline{u}, v] \mid u$ is a solution of the BVP (4.1.6). If $C$ is a well-ordered chain in the set $S_v$, then $w = \sup C$ exists and belongs to $S_v$.*

*Proof.* By lemma 4.1.4 the set $S_v$ is nonempty and possesses the uniform $W^{1,p}(\Omega)$-bound (4.1.20). Thus $C$ contains by lemma 4.1.2 a nondecreasing sequence $(w_n)_{n=0}^{\infty}$ which converges to $w = \sup C$ weakly in $W^{1,p}(\Omega)$ and strongly both in $L^p(\Omega)$ and in $L^p(\partial\Omega)$. By means of lemma 4.1.1 we are going to show that $w_n \to w$ strongly in $W^{1,p}(\Omega)$. By the weak convergence of $(w_n)$ to $w$ in $W^{1,p}(\Omega)$ we get $\ell(w, w_n - w) \to 0$ as $n \to \infty$. Hence it is enough to prove that

$$\ell(w_n, w_n - w) \to 0 \quad \text{as} \quad n \to \infty. \tag{4.1.21}$$

Since $w_n \in S_v$, then

$$\ell(w_n, \varphi) + \int_\Omega a(w_n)\,\varphi\,dx = \int_\Omega F w_n\,\varphi\,dx + \int_{\partial\Omega} (G w_n - b(w_n))\,\varphi\,d\omega$$

for all $\varphi \in W^{1,p}(\Omega)$. Taking $\varphi = w_n - w$ we get

$$\ell(w_n, w_n - w) = \int_{\partial\Omega} (G w_n - b(w_n))\,(w_n - w)\,d\omega - \int_\Omega a(w_n)\,(w_n - w)\,dx + \int_\Omega F w_n\,(w_n - w)\,dx. \tag{4.1.22}$$

The strong convergence of $(w_n)$ in $L^p(\Omega)$ and in $L^p(\partial\Omega)$ imply that the first two integrals on the right hand side of (4.1.22) tend to zero as $n \to \infty$. Moreover, we have the estimate

$$\left|\int_\Omega F w_n\,(w_n - w)\,dx\right| \leq c(1 + \|\nabla w_n\|_p^{p-1})\|w_n - w\|_p. \tag{4.1.23}$$

Taking into account the uniform boundedness of $w_n$ in $W^{1,p}(\Omega)$, due to (4.1.20), the right hand side of (4.1.23), and hence also its left hand side, tends to zero as $n \to \infty$. Thus (4.1.21) holds, which implies the strong convergence of the sequence $(w_n)$ in $W^{1,p}(\Omega)$. This allows us to pass to the limit in the weak formulation of (4.1.6) as $n \to \infty$, which proves that $w = \lim_{n\to\infty} w_n$ is a solution of (4.1.6), and thus also belongs to $S_v$. □

**Lemma 4.1.6:** *The solution set $S_v$ has the greatest element.*

*Proof.* Let $I$ denote the identity mapping of $S_v$. Since $I[S_v] = S_v$, it follows from lemma 4.1.5 that each well-ordered chain in $I[S_v]$ has an upper bound in $S_v$. Thus $I$ has by proposition 1.1.2

a maximal fixed point $z$, i.e. $z$ is a maximal element of $S_v$. To prove that $z$ is the greatest element of $S_v$, assume there is $w \in S_v$ such that $w \not\leq z$. By lemma 4.1.3 there is $u \in S_v$ for which $\max\{w, z\} \leq u \leq v$. In particular, this implies that $z < u$, which contradicts the fact that $z$ is a maximal element of $S_v$. Thus $z$ is the greatest element of $S_v$. □

### 4.1.3. Existence of extremal solutions of (4.1.1)

By means of the results proved in the previous section and proposition 1.2.1 we shall now prove our main result.

**Theorem 4.1.1:** *Let the hypotheses (A0)–(A3) and (H1)–(H3) be satisfied. Then the BVP (4.1.1) has the extremal solutions* $u_*$, $u^*$ *in the order interval* $[\underline{u}, \bar{u}]$ *of* $W^{1,p}(\Omega)$ *in the sense that if* $u$ *is any solution of (4.1.1) in* $[\underline{u}, \bar{u}]$, *then* $u_* \leq u \leq u^*$.

*Proof.* The proof will only be given for the existence of the maximal solution, since the existence of the minimal solution can be proved quite similarly by dual reasoning.

Let $V$ be the set of all upper solutions of the BVP (4.1.1) lying between $\underline{u}$ and $\bar{u}$. Define an operator $T: V \to V$ by assigning to each upper solution $v \in V$ the greatest element $Tv$ of $S_v$, i.e. the maximal solution of the BVP (4.1.6) within the order interval $[\underline{u}, v]$. By using the monotonicity conditions of (H3) and the fact that $Tv \leq v$ one readily verifies that $Tv \in V$. The result of lemma 4.1.6 ensures that the operator $T$ is well-defined. Let $C$ denote the i.w.o. chain of $T$-iterations of $\bar{u}$, defined in proposition 1.2.1. By the definition of $C$, $T[C]$ is also an inversely well-ordered chain in $W^{1,p}(\Omega)$. Moreover, it is bounded in $W^{1,p}(\Omega)$ by (4.1.20). Applying the result of lemma 4.1.2 to $-T[C]$ it follows that $T[C]$ contains a nonincreasing sequence $(Tw_n)_{n=o}^{\infty}$ which converges to $u^* = \inf T[C]$ weakly in $W^{1,p}(\Omega)$ and strongly both in $L^p(\Omega)$ and in $L^p(\partial\Omega)$. Obviously, $u^*$ belongs to $[\underline{u}, \bar{u}]$. Next we shall show that $(Tw_n)_{n=0}^{\infty}$ converges to $u^*$ strongly in $W^{1,p}(\Omega)$. Since $u^* \leq Tw_n \leq w_n$ for each $n \in \mathbb{N}$, we obtain by noticing that

$Tw_n \in S_{w_n}$ and using the monotonicity conditions of (H3) that

$$\begin{aligned}
&\ell(Tw_n, \varphi) + \int_\Omega a(Tw_n)\, \varphi\, dx + \int_{\partial\Omega} b(Tw_n)\, \varphi\, d\omega \\
&= \int_\Omega (f(x, w_n, Tw_n, \nabla Tw_n) + a(w_n))\varphi\, dx \\
&+ \int_{\partial\Omega} (g(x, w_n, Tw_n) + b(w_n))\varphi\, d\omega \\
&\geq \int_\Omega (f(x, u^*, Tw_n, \nabla Tw_n) + a(u^*))\varphi\, dx \\
&+ \int_{\partial\Omega} (g(x, u^*, Tw_n) + b(u^*))\varphi\, d\omega
\end{aligned} \tag{a}$$

for all $\varphi \in W^{1,p}(\Omega) \cap L^p_+(\Omega)$. If $F$ and $G$ denote the superposition operators defined by (4.1.7) with $v = u^*$, it follows from (a) that

$$\begin{aligned}
\ell(Tw_n, \varphi) \geq &\int_\Omega (F(Tw_n) - a(Tw_n))\, \varphi\, dx \\
&+ \int_{\partial\Omega} (G(Tw_n) - b(Tw_n))\varphi\, d\omega
\end{aligned} \tag{b}$$

for all $\varphi \in W^{1,p}(\Omega) \cap L^p_+(\Omega)$. Since $w_n \leq \bar{u}$, it follows by the similar reasoning as above that for all $\varphi \in W^{1,p}(\Omega) \cap L^p_+(\Omega)$

$$\begin{aligned}
\ell(Tw_n, \varphi) \leq &\int_\Omega (\bar{F}(Tw_n) - a(Tw_n))\, \varphi\, dx \\
&+ \int_{\partial\Omega} (\bar{G}(Tw_n) - b(Tw_n))\varphi\, d\omega,
\end{aligned} \tag{c}$$

where $\bar{F}$ and $\bar{G}$ denote the superposition operators defined by (4.1.7) with $v = \bar{u}$. Noticing that $Tw_n - u^* \in W^{1,p}(\Omega) \cap L^p_+(\Omega)$ for each $n \in \mathbb{N}$, and that $(Tw_n)_{n=0}^\infty$ converges to $u^*$ weakly in $W^{1,p}(\Omega)$, the right hand sides of (b) and (c) with $\varphi = Tw_n - u^*$

tend to zero as $n \to \infty$, by the reasoning used in the proof of lemma 4.1.5. Thus (b) and (c) imply that $\ell(Tw_n, Tw_n - u^*) \to 0$ as $n \to \infty$. Since also $\ell(u^*, Tw_n - u^*) \to 0$ as $n \to \infty$ by the weak convergence of $(Tw_n)_{n=0}^{\infty}$ to $u^*$ in $W^{1,p}(\Omega)$, it follows that $\lim_{n\to\infty}(\ell(Tw_n, Tw_n - u^*) - \ell(u^*, Tw_n - u^*)) = 0$. Thus, by lemma 4.1.1, the sequence $(Tw_n)_{n=0}^{\infty}$ converges to $u^*$ strongly in $W^{1,p}(\Omega)$, and also in $L^p(\Omega)$ and in $L^p(\partial\Omega)$, as stated above. This implies, as $n \to \infty$ in (b), that

$$\begin{aligned}\ell(u^*, \varphi) &\geq \int_\Omega (Fu^* - a(u^*))\, \varphi\, dx + \int_{\partial\Omega} (Gu^* - b(u^*))\varphi\, d\omega \\ &= \int_\Omega f(x, u^*, u^*, \nabla u^*)\varphi\, dx + \int_{\partial\Omega} g(x, u^*, u^*)\varphi\, d\omega\end{aligned}$$

for all $\varphi \in W^{1,p}(\Omega) \cap L_+^p(\Omega)$. Hence, $u^*$ is an upper solution of the BVP (4.1.1) in $[\underline{u}, \bar{u}]$, whence $Tu^*$ is defined. By definition, $Tu^* \leq u^*$.

The above proof shows that $T$ satisfies the hypotheses of proposition 1.2.1 a), whence $u^*$ is a fixed point of $T$. Since $u^* = Tu^* \in S_{u^*}$, then (4.1.6) holds with $u = v = u^*$, whence $u^*$ is a solution of the BVP (4.1.1) in $[\underline{u}, \bar{u}]$.

Let $\tilde{u}$ be any solution of (4.1.1) in $[\underline{u}, \bar{u}]$. To prove that $\tilde{u} \leq u^*$, we shall first show that $\tilde{u} \leq u$ for each $u \in C$, where $C$ is the i.w.o. chain of $G$-iterations of $\bar{u}$, defined in proposition 1.2.1. Make a counter-hypothesis, $\tilde{u}$ is not a lower bound of $C$. Then $C$ contains the greatest element $u$ such that $\tilde{u} \not\leq u$. Since $\tilde{u} \leq \bar{u}$, then $u < \bar{u}$. If $w \in C$ and $w > u$, then $\tilde{u} \leq w$, whence the BVP (4.1.6) has a solution in $[\tilde{u}, w]$ by lemma 4.1.3. Since $Tw$ is the maximal solution of (4.1.6) in $[\underline{u}, w]$, then $\tilde{u} \leq Tw$. This proves that $\tilde{u}$ is a lower bound of $T\{w \in C \mid w > u\}$. But then $\tilde{u} \leq \inf T\{w \in C \mid w > u\} = u$, which contradicts with $\tilde{u} \not\leq u$. Thus $\tilde{u} \leq w$ for all $w \in C$, whence $\tilde{u} \leq \min C = u^*$. This proves that $u^*$ is the maximal solution of (4.1.1) in $[\underline{u}, \bar{u}]$. □

**Remark 4.1.1:** In the BVP (4.1.1) considered above a more general quasilinear operator $A$ of the form

$$A\,u(x) = -\sum_{i=1}^{N} \frac{\partial}{\partial x_i} a_i(x, u(x), \nabla u(x)),$$

which depends also on $u$, can be chosen. However, in this case further restrictions have to be imposed on the coefficients $a_i(x, \eta, \xi)$ in order to obtain the similar conclusions. (cf. remarks 4.4.1).

### 4.1.4. Examples

In Ambrosetti and Turner (1988) the following discontinuous BVP arising in plasma physics has been studied:

$$-\Delta u = h(u-a)\,p(x,u) \ \text{ in } \Omega, \qquad u = 0 \ \text{ on } \partial\Omega. \tag{4.1.24}$$

Here $h$ is the Heaviside function defined by $h(t) = \begin{cases} 0, t \le 0, \\ 1, t > 0, \end{cases}$ $a$ is a positive constant and the function $p \in C(\Omega \times \mathbb{R}, \mathbb{R}_+)$ is supposed to be nondecreasing in the second argument and to satisfy a growth condition of the form $p(x,u) \le \alpha\, u + \beta$, where $\alpha$ and $\beta$ are some positive constants. By means of a dual variational principle the existence of nontrivial solutions of (4.1.24) has been proved in Ambrosetti and Turner (1988), provided that the parameters $a$ and $b := \inf\{p(x,a) \mid x \in \Omega\}$ satisfy a certain inequality.

Now, instead of the Dirichlet condition we are able to treat discontinuous flux conditions. As an example to the theory developed in the previous subsections, let us consider the BVP

$$\begin{aligned} -\,\Delta u &= h(u-a)\,p(x,u) \ \text{ in } \Omega, \\ \frac{\partial u}{\partial n} + q(u) &= h(u-a) \text{ on } \partial\Omega, \end{aligned} \tag{4.1.25}$$

where $\frac{\partial}{\partial n}$ denotes the outer normal derivative on $\partial\Omega$. We shall impose the following hypotheses on the functions $p$ and $q$:

(i) $p\colon \Omega \times \mathbb{R} \to \mathbb{R}_+$ is a Carathéodory function and

$$0 < b \leq p(x,s) \leq \alpha\, s + \beta \tag{4.1.26}$$

for a.a. $x \in \Omega$ and for all $s \geq a$, where $\alpha$ and $\beta$ are some positive constants.

(ii) $q\colon \mathbb{R} \to \mathbb{R}$ is nondecreasing and continuous, and $q(0) = 0$.

One readily verifies that $u = 0$ is a solution of the BVP (4.1.25), and that there are no other constant solutions. Due to hypothesis (ii) the BVP

$$-\Delta v = 0 \ \text{ in } \Omega, \quad \frac{\partial v}{\partial n} + q(v) = 0 \text{ on } \partial\Omega$$

possesses the uniquely defined solution $v = 0$. Thus by comparison argument any nontrivial solution $u$ of (4.1.25) must be nonnegative in $\Omega$ and, in fact, must exceed $a$ on a set of positive Lebesgue measure.

We shall show the existence of nontrivial solutions of (4.1.25) by the method developed in the previous chapters. To this end we prove the existence of a nontrivial lower solution $\underline{u}$ and an upper solution $\bar{u}$ satisfying $\underline{u} \leq \bar{u}$, where the lower solution must be positive at least on a set of a positive Lebesgue measure. For simplicity we consider in the following the one-dimensional case, i.e. $\Omega = (0,1)$, where the upper and lower solutions can be constructed explicitly. The one-dimensional version of the BVP (4.1.25) reads as

$$\begin{aligned} -u'' &= h(u-a)\, p(x,u) \ \text{ in } (0,1), \\ -u'(0) &= h(u(0)-a) - q(u(0)), \\ u'(1) &= h(u(1)-a) - q(u(1)). \end{aligned} \tag{4.1.27}$$

We are looking for an upper solution of the form

$$\bar{u}(x) = c\,x\,(1-x) + a, \tag{4.1.28}$$

where the constant $c$ has to be chosen in such a way that the following inequalities are satisfied.

$$-\bar{u}'' \geq \alpha\bar{u} + \beta \quad \text{in } (0,1), \tag{4.1.29}$$

$$-\bar{u}'(0) \geq -q(a), \quad \bar{u}'(1) \geq -q(a). \tag{4.1.30}$$

Since $\max\{\bar{u}(x) \mid 0 \leq x \leq 1\} = \frac{c}{4} + a$, we obtain from (4.1.29) the condition $2\,c \geq \alpha\,(\frac{c}{4} + a) + \beta$, or

$$c \geq \frac{4}{8-\alpha}\,(\alpha\,a + \beta), \tag{4.1.31}$$

which requires that $0 \leq \alpha < 8$.

Conditions (4.1.30) are fulfilled if

$$c \leq q(a). \tag{4.1.32}$$

A function $\underline{u}$ that satisfies the inequalities

$$-\underline{u}'' \leq b\,h(\underline{u} - a) \quad \text{a.e. in } (0,1), \tag{4.1.33}$$

$$-\underline{u}'(0) \leq h(\underline{u}(0) - a) - q(\underline{u}(0)), \tag{4.1.34}$$

$$\underline{u}'(1) \leq h(\underline{u}(1) - a) - q(\underline{u}(1)), \tag{4.1.35}$$

is a lower solution of the BVP (4.1.27). Assuming that $16\,a \leq b$, a lower solution $\underline{u}$ is given by

$$\underline{u}(x) = \begin{cases} \frac{a}{d}\,x, & x \in [0,d], \\ \frac{a}{d}\,(1-x), & x \in [1-d,1], \\ \frac{b}{2}\,(x - x^2 - d + d^2) + a, & x \in (d, 1-d), \end{cases} \tag{4.1.36}$$

where $d$ can be one of the two values $\frac{1}{4}(1 \pm \sqrt{1 - \frac{16a}{b}})$. Since $\underline{u}(0) = \underline{u}(1) = 0$, the inequalities (4.1.34) and (4.1.35) are readily seen to be satisfied. Furthermore, $\underline{u}$ given by (4.1.36) fulfills the equation

$$-\underline{u}'' = b\,h(\underline{u} - a) \text{ a.e. in } (0,1),$$

and hence (4.1.33) holds. Finally, the condition $c \geq \frac{b}{2}$, which holds due to hypothesis (i) (cf. (4.1.26)) and condition (4.1.31), ensures that $\bar{u}(x) \geq \underline{u}(x)$ for $x \in [0,1]$.

Summarizing the above reasoning we obtain the following result: The functions $\bar{u}$ and $\underline{u}$ given by (4.1.28) and (4.1.36), respectively, are upper and lower solutions of the BVP (4.1.27) satisfying $\bar{u} \geq \underline{u}$, provided that the parameters of the problem satisfy the following inequalities:

$$q(a) \geq c \geq \frac{4}{8-\alpha}(\alpha\, a + \beta) \quad \text{and} \quad b \geq 16\, a. \tag{4.1.37}$$

Thus the existence of a nontrivial solution of the BVP (4.1.27) follows by theorem 4.1.1.

The inequalities (4.1.37) hold in the special case when

$$q(s) = 16\, s, \quad \alpha = 4, \quad \beta = 3, \quad a = \tfrac{1}{4} \quad \text{and} \quad b = 4.$$

In this case $\alpha\, a + \beta = 4 = b$ and $q(a) = c = \frac{4}{8-\alpha}(\alpha\, a + \beta)$.

In a similar way one can treat also the slightly modified BVP

$$-\Delta u = h(u - a_1)\, p(x,u) \text{ in } \Omega, \quad \frac{\partial u}{\partial n} = h(u - a_2) - q(u) \text{ on } \partial\Omega,$$

where $a_1$, $a_2$ are unequal positive constants.

## 4.2. ELLIPTIC EQUATIONS

Let $\Omega \subset \mathbb{R}^N$ be a bounded domain with Lipschitz boundary $\partial\Omega$. In this section we shall study an elliptic boundary value problem with Dirichlet boundary conditions. Instead of the quasilinear equation considered in section 4.1 we shall now consider for simplicity the BVP

$$-Lu = f(x,u,u) \text{ in } \Omega, \qquad u = 0 \text{ on } \partial\Omega, \tag{4.2.1}$$

where $f: \Omega \times \mathbb{R} \times \mathbb{R} \to \mathbb{R}$, and $L$ is assumed to be a linear operator of the form

$$Lu = \sum_{i,j=1}^{N} \frac{\partial}{\partial x_i}\left(a_{ij}\frac{\partial u}{\partial x_j}\right) - \sum_{i=1}^{N} b_i \frac{\partial u}{\partial x_i}, \tag{4.2.2}$$

with coefficients $a_{ij}, b_i \in L^\infty(\Omega)$, $i,j = 1,\dots N$. We shall also assume that $L$ is *uniformly elliptic*, i.e. there exists $\nu > 0$ such that $\sum_{i,j=1}^{N} a_{ij}(x)\,\xi_i\xi_j \geq \nu\,\|\xi\|^2$ for a.a. $x \in \Omega$ and for all $\xi = (\xi_1,\dots,\xi_N) \in \mathbb{R}^N$.

### 4.2.1. Hypotheses and preliminaries

Denote by $H^1(\Omega)$ the Sobolev space of those functions which, together with their generalized derivatives, belong to $L^2(\Omega)$, and by $H_o^1(\Omega)$ the space of all the elements of $H^1(\Omega)$ which possess generalized homogeneous boundary values. Define a partial ordering in the spaces $L^2(\Omega)$, $H^1(\Omega)$ and $H_o^1(\Omega)$ by $u \leq v$ if and only if $v - u$ belongs to the set $L_+^2(\Omega)$ of all nonnegative elements of $L^2(\Omega)$, and denote $[u,v] = \{w \in H^1(\Omega) \mid u \leq w \leq v\}$.

Let $\ell$ be the bilinear form associated with the elliptic differential operator $L$ by

$$\ell(u,v) = \int_\Omega \left(\sum_{i,j=1}^{N} a_{ij}\frac{\partial u}{\partial x_i}\frac{\partial v}{\partial x_j} + \sum_{i=1}^{N} b_i\frac{\partial u}{\partial x_i} v\right) dx$$

for $u$, $v \in H^1(\Omega)$. A function $u \in H_o^1(\Omega)$ is called a *solution* of the BVP (4.2.1) if

$$\ell(u,\varphi) = \int_\Omega f(x,u,u)\varphi\,dx \quad \text{for all } \varphi \in H_o^1(\Omega).$$

A function $\bar{u} \in H^1(\Omega)$ is called an *upper solution* of (4.2.1) if

$$\ell(\bar{u},\varphi) \geq \int_\Omega f(x,\bar{u},\bar{u})\varphi dx, \ \ \varphi \in H_o^1(\Omega)\cap L_+^2(\Omega), \ \ \bar{u} \geq 0 \text{ on } \partial\Omega.$$

A lower solution is defined similarly, by reversing the inequality signs.

We shall impose the following hypotheses on the function $f\colon \Omega \times \mathbb{R} \times \mathbb{R} \to \mathbb{R}$.

(f0) The BVP (4.2.1) has a lower solution $\underline{u}$ and an upper solution $\bar{u}$ such that $\underline{u} \leq \bar{u}$.

(f1) $f(\cdot,\cdot,s)$ is a Carathéodory function for each $s \in \mathbb{R}$.

(f2) There exists $M \in C(\mathbb{R},\mathbb{R})$ and $c > 0$ for which $M(\underline{u})$, $M(\bar{u}) \in L^2(\Omega)$ and $(M(r)-M(s))(r-s) \geq c(r-s)^2$ for all $r$, $s \in \mathbb{R}$, such that the function $s \mapsto f(x,r,s)+M(s)$ is nondecreasing in $[\underline{u}(x),\bar{u}(x)]$ for a.a. $x \in \Omega$ and for all $r \in [\underline{u}(x),\bar{u}(x)]$.

(f3) $f(\cdot,u(\cdot),v(\cdot))$ is measurable for all $u$, $v \in [\underline{u},\bar{u}]$.

(f4) There is a $p \in L^2(\Omega)$ such that $|f(x,r,s)| \leq p(x)$ for a.a. $x \in \Omega$ and for all $r,s \in [\underline{u}(x),\bar{u}(x)]$.

We are going to show that under hypotheses (f0)–(f4) the BVP (4.2.1) has the extremal solutions in the order interval $[\underline{u},\bar{u}]$. In the proof of this result we shall apply theorem 1.1.1 to the operator $F$ defined in the following lemma.

**Lemma 4.2.1:** *Let (f0)–(f4) hold and let $P$ be the set of all lower solutions of (4.2.1) lying between $\underline{u}$, $\bar{u}$, given by (f0). Then for each $w \in P$, the BVP*

$$-Lu = f(x,u,w)+M(w)-M(u) \ \text{in } \Omega, \qquad u=0 \ \text{on } \partial\Omega, \quad (4.2.3)$$

*has the minimal solution* $u = Fw$ *in* $[w, \bar{u}]$. *Moreover,* $Fw \in P$, *and there is* $c > 0$ *such that* $\|Fw\|_{H^1(\Omega)} \leq c$ *for all* $w \in P$.

*Proof.* Let $w \in P$ be given. Since $w$ is a lower solution of (4.2.3) and $\bar{u}$ is its upper solution, then the reasoning similar to that used in subsection 4.1.2 ensures that (4.2.3) has the minimal solution $u$ in $[w, \bar{u}]$. Noticing that $w \leq u$, the definition of the solution and conditions (f2)–(f4) imply that

$$\begin{aligned} \ell(u, \varphi) &= \int_\Omega (f(x, u, w) + M(w) - M(u))\, \varphi\, dx \\ &\leq \int_\Omega f(x, u, u)\varphi\, dx \end{aligned}$$

for all $\varphi \in H_o^1(\Omega) \cap L_+^2(\Omega)$. Denoting $u = Fw$ we then have $Fw \in P$. The last assertion of the lemma follows from the a-priori estimates for the solutions of (4.2.3) (cf. Gilbarg and Trudinger (1983)), and from the fact that order intervals of $L^2(\Omega)$ are norm bounded. □

### 4.2.2. On extremal solutions of (4.2.1)

The main result of this section is

**Theorem 4.2.1:** *Under the hypotheses (f0)–(f4) the BVP (4.2.1) has the extremal solutions* $u_*$, $u^*$ *in the order interval* $[\underline{u}, \bar{u}]$ *of* $H^1(\Omega)$ *in the sense that if* $u \in H^1(\Omega)$ *is any solution of (4.2.1) in* $[\underline{u}, \bar{u}]$, *then* $u_* \leq u \leq u^*$.

*Proof.* We shall first show that the mapping $F$ defined in lemma 4.2.1 satisfies the hypotheses of theorem 1.1.1. Let $C$ be the w.o. chain of $F$-iterations of $\underline{u}$. Since $F[C]$ is a well-ordered and, by lemma 4.2.1, norm bounded chain in $H^1(\Omega)$, then by lemma 4.1.2 there is a nondecreasing sequence $(v_j)_{j=0}^\infty$ in $F[C]$ which converges strongly in $L^2(\Omega)$ and weakly in $H^1(\Omega)$ and in $L^2(\partial\Omega)$ to $u_* = \sup F[C]$. Obviously, $u_* \in [\underline{u}, \bar{u}]$. Each $v_j$ is by lemma 4.2.1 a lower solution of (4.2.1), and $v_j \le u_*$, whence the definition of a lower solution and conditions (f1)–(f3) imply that

$$\ell(v_j,\varphi) + \int_\Omega M(v_j)\varphi\,dx \le \int_\Omega (f(x,v_j,v_j) + M(v_j))\,\varphi\,dx$$
$$\le \int_\Omega (f(x,v_j,u_*) + M(u_*))\,\varphi\,dx$$

for each $\varphi \in H_o^1(\Omega) \cap L_+^2(\Omega)$. Because $(v_j)_{j=o}^\infty$ converges strongly in $L^2(\Omega)$ and weakly in $H^1(\Omega)$ to $u_*$, it follows from the above inequality by (f3) and (f4), as $j \to \infty$, that

$$\ell_k(u_*,\varphi) \le \int_\Omega f(x,u_*,u_*)\varphi\,dx \quad \text{for all } \varphi \in H_o^1(\Omega) \cap L_+^2(\Omega)\,.$$

Since $v_j \in H_o^1(\Omega)$ and $v_j \rightharpoonup u_*$ in $H_o^1(\Omega)$, then $u_* = 0$ on $\partial\Omega$. Thus $u_*$ is by definition a lower solution of (4.2.1). Since $u_* \in [\underline{u}, \bar{u}]$, then $u_* = \sup F[C] \in P$. By definition, $Fu_* \in [u_*, \bar{u}]$ so that $\underline{u} \le u_* \le Fu_*$. Thus all the hypotheses of theorem 1.1.1 hold, whence $u_*$ is a fixed point of $F$. Since $Fu_* = u_*$ is the solution of the BVP (4.2.3) with $w = u_*$, then $u_*$ is a solution of (4.2.1) in $[\underline{u}, \bar{u}]$.

If $u$ is any upper solution of (4.2.1) in $[\underline{u}, \bar{u}]$, then $\underline{u} \le u$, which allows us to replace $\bar{u}$ by $u$ in the above reasonings, so that $u_* \le u$. Thus $u_*$ is the minimal upper solution, and hence the minimal solution of (4.2.1) in $[\underline{u}, \bar{u}]$.

The assertions concerning the existence of the maximal solution $u^*$ can be proved by dual argumentation, replacing lower solutions by upper solutions and applying proposition 1.2.1. □

**Corollary 4.2.1:** *If the function $f$ is replaced in (4.2.1) by another function $\tilde{f}$ satisfying (f0)–(f4) with the same $\underline{u}$, $\bar{u}$, and if $\tilde{u}_*$ and $\tilde{u}^*$ denote the corresponding minimal and maximal solutions of the BVP (4.2.1) in $[\underline{u}, \bar{u}]$, then $f \le \tilde{f}$ pointwise implies that $u_* \le \tilde{u}_*$ and $u^* \le \tilde{u}^*$.*

*Proof.* The proof of theorem 4.2.1 ensures that the revised system (4.2.1) has the minimal solution $\tilde{u}_*$ in $[\underline{u}, \bar{u}]$. If $f \leq \tilde{f}$, then $\tilde{u}_*$ is an upper solution of the original system (4.2.1) in $[\underline{u}, \bar{u}]$. Since $u_*$ is by the proof of theorem 4.2.1 the smallest of such upper solutions, then $u_* \leq \tilde{u}_*$.

The assertion concerning the maximal solutions follows similarly. □

### 4.2.3. An example

Consider the boundary value problem

$$-u'' = h(u-a)p(u,u) \ \text{ a.e. in } \ (0,1), \quad u(0) = u(1) = 0, \tag{4.2.4}$$

where $h$ is the Heaviside function. Given positive constants $a$, $b_1$ and $b_2$ which satisfy $16\,a \leq b_1 \leq b_2$ and a function $p\colon \mathbb{R} \times \mathbb{R} \to \mathbb{R}_+$, assume that $p(\cdot, s)$ is continuous for each $s$, that $p(r, \cdot)$ is nondecreasing for all $r$, and that

(p1) $b_1 \leq p(r,s) \leq b_2$ for $a \leq r$, $a \leq s$.

Since $u \equiv 0$ is obviously a solution of (4.2.4) we are interested in the existence of nontrivial nonnegative solutions of the BVP (4.2.4). Denote $d_i = \frac{1}{4} - \sqrt{\frac{1}{16} - \frac{a}{b_i}}$, $i = 1, 2$. It is easy to see that, for each $i = 1, 2$ the function

$$u_i(x) = \begin{cases} \frac{a\,x}{d_i}, & 0 \leq x \leq d_i, \\ \frac{a(1-x)}{d_i}, & 1 - d_i \leq x \leq 1, \\ \frac{b_i}{2}(x(1-x) - d_i(1-d_i)) + a, & d_i < x < 1 - d_i \end{cases}$$

is a solution of the BVP

$$-u'' = h(u-a)b \ \text{ in } \ (0,1), \quad u(0) = u(1) = 0, \tag{4.2.5}$$

when $b = b_i$. The hypotheses given for $p$ imply that $\underline{u} = u_1$ and $\bar{u} = u_2$ are lower and upper solutions of the BVP (4.2.4), respectively. Moreover, it can be shown that $\underline{u} \leq \bar{u}$, and that the function

$$f(x,r,s) = h(s-a)p(r,s), \quad x \in [0,1], \ r,s \in \mathbb{R}$$

satisfies the conditions (f1)–(f4) with $\underline{u}$, $\bar{u}$ as above. Thus the BVP (4.2.4) has by theorem 4.2.1 the minimal solution $u_*$ and the maximal solution $u^*$ lying between $\underline{u}$ and $\bar{u}$.

To obtain numerical results, choose $a = \frac{1}{4}$, $b_1 = 4$ and $b_2 = 7$. In this case the BVP (4.2.4) has under the given hypotheses the minimal solution $u_*$ and the maximal solution $u^*$ in the order interval $[\underline{u}, \bar{u}]$, where

$$\underline{u}(x) = \begin{cases} x, & 0 \leq x \leq \frac{1}{4}, \\ 1-x, & \frac{3}{4} \leq x \leq 1, \\ 2x - 2x^2 - \frac{1}{8}, & \frac{1}{4} < x < \frac{3}{4}, \end{cases}$$

and

$$\bar{u}(x) = \begin{cases} \frac{(7+\sqrt{21})x}{4}, & 0 \leq x \leq \frac{7-\sqrt{21}}{28}, \\ \frac{(7+\sqrt{21})(1-x)}{4}, & \frac{21+\sqrt{21}}{28} \leq x \leq 1, \\ \frac{7}{2}x - \frac{7}{2}x^2 + \frac{\sqrt{21}-5}{16}, & \frac{7-\sqrt{21}}{28} < x < \frac{21+\sqrt{21}}{28}. \end{cases}$$

The greatest difference between $\bar{u}(x)$ and $\underline{u}(x)$ is obtained when $x = \frac{1}{2}$. For instance, in our numerical example $\epsilon \leq \frac{3+\sqrt{21}}{16} \simeq 0.47391$.

Since the greatest value of $\bar{u} = u_2$ is $u_2(\frac{1}{2}) = \frac{a}{2} + \frac{b_2}{16}(1 + \sqrt{1 - \frac{16a}{b_2}})$, then condition (p1) can be replaced by

(p2) $b_1 \leq p(r,s) \leq b_2$ for $a \leq r, s \leq \frac{a}{2} + \frac{b_2}{8}$.

If $p$ depends only on $s$ and if

(p3) $b_1 \leq p(s) \leq c^2 s + k$ for $a \leq s$,

then (p2) holds, if $c^2 < 8$ and $b_2 \geq \frac{8c^2}{8-c^2}(\frac{a}{2} + \frac{k}{c^2})$. Now, instead of calculating the upper solution $\bar{u}$ by means of (4.2.5) with $b = b_2 = \frac{8c^2}{8-c^2}(\frac{a}{2} + \frac{k}{c^2})$, we can find an improved upper solution as follows.

$$\bar{u}(x) = \begin{cases} \frac{ax}{d}, & 0 \leq x \leq d, \\ \frac{a(1-x)}{d}, & d \leq x \leq 1-d, \\ (a + \frac{k}{c^2})\frac{\cos(\frac{c}{2} - cx)}{\cos(\frac{c}{2} - cd)} - \frac{k}{c^2}, & d < x < 1-d, \end{cases}$$

is a solution of the BVP

$$-u'' = h(u-a)(c^2 u + k) \text{ in } (0,1), \quad u(0) = u(1) = 0, \quad (4.2.6)$$

if $d > 0$ is chosen in such a way that $\bar{u}$ and its derivative are continuous at $x = d$ and at $x = 1-d$. Such a $d$ exists if $c < \pi$. In fact, $x = d$ is the least positive solution of the equation

$$(ac + \tfrac{k}{c})\tan(\tfrac{c}{2} - cx) = \tfrac{a}{x}.$$

The so obtained function $\bar{u}$ can be chosen as the upper solution in the previous considerations, if $p\colon \mathbb{R} \to \mathbb{R}_+$ is continuous *or* nondecreasing and satisfies (p3).

If condition (p3) holds, the above defined solution of (4.2.6) yields a better upper estimates for the solutions of (4.2.4) as $\bar{u} = u_2$, defined above by (4.2.5) with $b = b_2$. For instance, if $b_1 = 4$, $a = \frac{1}{4}$, $c = 2$ and $k = 3$ in (p3), then the maximum difference between the values of the above defined upper solution and the $\underline{u}$ defined by (4.2.6) with $b = 4$ is less than 0.19128.

## 4.3. ELLIPTIC SYSTEMS

Let $\Omega \subset I\!R^N$ be a bounded domain with Lipschitz boundary $\partial\Omega$. In this section we shall study the following system of boundary value problems

$$-L_k u_k = f_k(x, u, u) \text{ in } \Omega, \; u_k = 0 \text{ on } \partial\Omega, \; k = 1, \ldots, n, \quad (4.3.1)$$

where $L_k$'s are differential operators of the form

$$L_k u_k = \sum_{i,j=1}^{N} \frac{\partial}{\partial x_i}\left(a_{ij}^k \frac{\partial u_k}{\partial x_j}\right) - \sum_{i=1}^{N} b_i^k \frac{\partial u_k}{\partial x_i}, \quad k = 1, \ldots, n,$$

with coefficients $a_{ij}^k$, $b_i^k \in L^\infty(\Omega)$, $i, j = 1, \ldots, N$. We shall also assume that each $L_k$ is uniformly elliptic.

### 4.3.1 Hypotheses and preliminaries

Let $H^1(\Omega)$ and $H_o^1(\Omega)$ be as in section 4.2. The corresponding $n$-dimensional Cartesian products are denoted by $H^1(\Omega, I\!R^n)$ and $H_o^1(\Omega, I\!R^n)$, respectively. Assume that $H^1(\Omega, I\!R^n)$, $H_o^1(\Omega, I\!R^n)$ and $I\!R^n$ are ordered componentwise.

Let $\ell_k$ be the bilinear form associated with the differential operator $L_k$ by

$$\ell_k(u, \varphi) = \int_\Omega \left( \sum_{i,j=1}^{N} a_{ij}^k \frac{\partial u}{\partial x_i} \frac{\partial \varphi}{\partial x_j} + \sum_{i=1}^{N} b_i^k \frac{\partial u}{\partial x_i} \varphi \right) dx, \quad (4.3.2)$$

for $u$, $\varphi \in H^1(\Omega)$. A function $u = (u_1, \ldots, u_n) \in H_o^1(\Omega, I\!R^n)$ is called a *solution* of the system (4.3.1) if

$\ell_k(u_k, \varphi) = \int_\Omega f_k(x, u, u)\varphi \, dx$ for all $\varphi \in H_o^1(\Omega)$ and $k = 1, \ldots, n$.

A function $\underline{u} = (\underline{u}_1, \ldots, \underline{u}_n) \in H^1(\Omega, \mathbb{R}^n)$ is called a *lower solution* of (4.3.1) if

(i) $\underline{u} \le 0$ on $\partial\Omega$,
(ii) $\ell_k(\underline{u}_k, \varphi) \le \int_\Omega f_k(x, \underline{u}, \underline{u})\varphi\, dx$ for all $\varphi \in H_o^1(\Omega) \cap L_+^2(\Omega)$,
$k = 1, \ldots, n$.

An upper solution $\bar{u}$ is defined similarly, by reversing inequality signs in (i), (ii).

We shall impose the following hypotheses on the functions $f_k \colon \Omega \times \mathbb{R}^{2n} \to \mathbb{R}$.

(f0) The system (4.3.1) has a lower solution $\underline{u} = (\underline{u}_1, \ldots, \underline{u}_n)$ and an upper solution $\bar{u} = (\bar{u}_1, \ldots, \bar{u}_n)$ such that $\underline{u} \le \bar{u}$.
(f1) Each $f_k(\cdot, u(\cdot), v(\cdot))$ is measurable when $u, v \in [\underline{u}, \bar{u}]$, and there exist $p_k \in L_+^2(\Omega)$ such that
$$|f_k(x, s, \underline{u}(x))| + |f_k(x, s, \bar{u}(x))| \le p_k(x)$$
for a.a. $x \in \Omega$ and for all $s \in [\underline{u}(x), \bar{u}(x)]$, $k = 1, \ldots, n$.
(f2) Each $f_k(x, \cdot, s)$, and $f_k(x, s, \cdot)$ is quasimonotone nondecreasing in $[\underline{u}(x), \bar{u}(x)]$ for a.a. $x \in \Omega$ and for all $s \in [\underline{u}(x), \bar{u}(x)]$.
(f3) There are $M_k \in C(\mathbb{R}, \mathbb{R})$ and $c_k > 0$, $k = 1, \ldots, n$, such that $(M_k(s) - M_k(t))(s - t) \ge c_k(s - t)^2$ for all $s, t \in \mathbb{R}$, that $M_k(\underline{u}_k)$, $M_k(\bar{u}_k) \in L^2(\Omega)$, and that the functions $s_k \mapsto f_k(x, r, s) + M_k(s_k)$ are nondecreasing in $[\underline{u}_k(x), \bar{u}_k(x)]$ for a.a. $x \in \Omega$ and for all $r, s \in [\underline{u}(x), \bar{u}(x)]$.
(f4) The functions $r_k \mapsto f_k(x, r, s)$, $k = 1, \ldots, n$ are continuous in $[\underline{u}_k(x), \bar{u}_k(x)]$ for a.a. $x \in \Omega$ and for all $r, s \in [\underline{u}(x), \bar{u}(x)]$.

We are going to show that under hypotheses (f0)–(f4) the system (4.3.1) has the extremal solutions in the order interval $[\underline{u}, \bar{u}]$ of $H^1(\Omega, \mathbb{R}^n)$, by applying theorem 1.1.1 to the operator $F$ defined in the following lemma.

**Lemma 4.3.1:** *Let (f0)–(f4) hold and let $P$ be the set of all lower solutions of (4.3.1) in $[\underline{u}, \bar{u}]$, given by (f0). Then for each $w = (w_1, \ldots, w_n) \in P$, the system*

$$\begin{aligned} -L_k u_k = f_k(x, (w_1, \ldots, w_{k-1}, u_k, w_{k+1}, \ldots, w_n), w) \\ + M_k(w_k) - M_k(u_k) \text{ in } \Omega, \quad u_k = 0 \text{ on } \partial\Omega \end{aligned} \tag{4.3.3}$$

*has the minimal solution $u = Fw$ in $[w, \bar{u}]$. Moreover, $Fw \in P$, and there is $c > 0$ such that $\|(Fw)_k\|_{H^1(\Omega)} \leq c$ for all $w \in P$ and $k = 1, \ldots, n$.*

**Proof:** Let $w = (w_1, \ldots, w_n) \in P$ be given. For each fixed $k = 1, \ldots, n$, $w_k$ is a lower solution of (4.3.3) and $\bar{u}_k$ is its upper solution. The hypotheses (f1)–(f4) ensure then by lemma 4.2.1 that (4.3.3) has for each fixed $k = 1, \ldots, n$ the minimal solution $u_k$ in $[w_k, \bar{u}_k]$. Obviously, $u = (u_1, \ldots, u_n)$ is the minimal solution of the system (4.3.3) in $[w, \bar{u}]$. Since $w \leq u$, the definition of the solution and conditions (f2)–(f4) imply that

$$\ell_k(u_k, \varphi) \leq \int_\Omega f_k(x, (w_1, \ldots, w_{k-1}, u_k, w_{k+1}, \ldots, w_n), u)\varphi \, dx$$

for all $\varphi \in H_o^1(\Omega) \cap L_+^2(\Omega)$ and $k = 1, \ldots, n$. Denoting $u = Fw$ we then have $Fw \in P$. The last assertion of the lemma follows from the a-priori estimates for the solutions of (4.3.3) for fixed $k$ (cf. Gilbarg and Trudinger (1983)) and from the fact that order intervals of $L^2(\Omega)$ are norm bounded. □

### 4.3.2. Extremal solutions of (4.3.1)

The main result of this section is

**Theorem 4.3.1:** *Under the hypotheses (f0)–(f4) the system (4.3.1) has the extremal solutions in the order interval $[\underline{u}, \bar{u}]$ of $H^1(\Omega, \mathbb{R}^n)$.*

*Proof.* We shall first show that the mapping $F$ defined in lemma 4.3.1 satisfies the hypotheses of theorem 1.1.1. Let $C$ be the w.o. chain of $F$-iterations of $\underline{u}$. Since also $F[C]$ is a well-ordered chain, the $k$:th components $(Fw)_k$, $w \in C$ form a well-ordered and, by lemma 4.3.1, norm bounded chain in $H^1(\Omega)$. By lemma 4.1.2 there is for each $k = 1, \ldots, n$ a nondecreasing sequence $(v(k)^j)_{j=0}^{\infty}$ in $F[C]$ such that the sequence $(v(k)_k^j)_{j=0}^{\infty}$ of $k$:th components of $v(k)^j$'s converges strongly in $L^2(\Omega)$ and weakly in $H^1(\Omega)$ to $u_k = \sup\{(Fw)_k \mid w \in C\}$. Obviously, $u_* = (u_1, \ldots, u_n)$ is the supremum of $F[C]$ in $[\underline{u}, \bar{u}]$. Each $v(k)^j$ is by lemma 4.3.1 a lower solution of (4.3.1), and $v(k)^j \leq u_*$, whence the definition of a lower solution and conditions (f1)–(f3) imply that

$$\begin{aligned}
&\ell_k(v(k)_k^j, \varphi) + \int_\Omega M_k(v(k)_k^j)\varphi\, dx \\
&\leq \int_\Omega \Big( f_k(x, v(k)^j, v(k)^j) + M_k(v(k)_k^j) \Big)\, \varphi\, dx \\
&\leq \int_\Omega \left( f_k(x, v(k)^j, u_*) + M_k(u_k) \right) \varphi\, dx
\end{aligned}$$

for each $\varphi \in H_o^1(\Omega) \cap L_+^2(\Omega)$. Since $(v(k)_k^j)_{j=0}^{\infty}$ converges to $u_k$ according to lemma 4.1.2, it follows from the above inequality by (f3) and (f4), as $j \to \infty$, that

$$\ell_k(u_k, \varphi) \leq \int_\Omega f_k(x, u_*, u_*)\varphi\, dx, \ \varphi \in H_o^1(\Omega) \cap L_+^2(\Omega), \ k = 1, \ldots, n,$$

and $u_k = 0$ on $\partial\Omega$. Thus $u_*$ is by definition a lower solution of (4.3.1). Since $u_* \in [\underline{u}, \bar{u}]$, then $u_* = \sup F[C] \in P$. By definition, $Fu_* \in [u_*, \bar{u}]$ so that $\underline{u} \leq u_* \leq Fu_*$. Thus all the hypotheses of theorem 1.1.1 hold, whence $u_*$ is a fixed point of $F$. Since $Fu_*$ is the solution of the system (4.3.3) with $w = u_*$, we have

$$\begin{aligned}
&-L_k(Fu_*)_k = f_k(x, Fu_*, u_*) + M_k(u_k) - M_k((Fu_*)_k) \ \text{ in } \Omega, \\
&(Fu_*)_k = 0 \text{ on } \partial\Omega, \ k = 1, \ldots, n.
\end{aligned}$$

Since $(Fu_*)_k = (u_*)_k = u_k$, this implies that $u_*$ is a solution of (4.3.1) in $[\underline{u}, \bar{u}]$.

If $u$ is any upper solution of (4.3.1) in $[\underline{u}, \bar{u}]$, then $\underline{u} \leq u$, which allows us to replace $\bar{u}$ by $u$ in the above reasonings, so that $u_* \leq u$. Thus $u_*$ is the minimal upper solution, and hence the minimal solution of (4.3.1) in $[\underline{u}, \bar{u}]$.

The assertions concerning the existence of the maximal solution $u^*$ can be proved by dual argumentation, replacing lower solutions by upper solutions and vice versa and applying proposition 1.2.1. □

**Corollary 4.3.1:** *If the functions $f_1, \dots, f_n$ are replaced in (4.3.1) by other functions $\tilde{f}_1, \dots, \tilde{f}_n$ satisfying (f0)–(f4) with the same $\underline{u}$, $\bar{u}$, and if $\tilde{u}_*$ and $\tilde{u}^*$ denote the corresponding minimal and maximal solutions of the system (4.3.1) in $[\underline{u}, \bar{u}]$, then $f_k \leq \tilde{f}_k$, $k = 1, \dots, n$ pointwise implies that $u_* \leq \tilde{u}_*$ and $u^* \leq \tilde{u}^*$.*

*Proof.* The proof of theorem 4.3.1 ensures that the revised system (4.3.1) has the minimal solution $\tilde{u}_*$ in $[\underline{u}, \bar{u}]$. If $f_k \leq \tilde{f}_k$ pointwise for each $k = 1, \dots, n$, then $\tilde{u}_*$ is an upper solution of the original system (4.3.1) in $[\underline{u}, \bar{u}]$. Since $u_*$ is by the proof of theorem 4.3.1 the smallest of such upper solutions, then $u_* \leq \tilde{u}_*$.

The conclusion concerning the maximal solutions follows similarly. □

**Remark 4.3.1:** The hypotheses (f2)–(f4) hold if

$$f_k(x, r, s) = f_k^1(r) + f_k^2(s), \quad x \in \Omega, \ r, \ s \in \mathbb{R}^N,$$

where $f_k^1$ and $f_k^2$ are quasimonotone nondecreasing, the functions $r_k \to f_k^1(r) + M_k \cdot r_k$ are nondecreasing for some $M_k \geq 0$, and the functions $s_k \to f_k^2(s_1, \dots, s_k, \dots, s_n)$ are continuous.

The result of theorem 4.3.1 can be extended to the more general situation by replacing the linear differential operators $L_k$ in (4.3.1) by quasilinear elliptic differential operators of the divergence form which satisfy the standard Leray-Lions type conditions (cf. (A0)–(A3) in section 4.1).

### 4.3.3 An example

The results of theorem 4.3.1 and corollary 4.3.1 are now applied to a system of two-point boundary value problems. Consider the system

$$\begin{aligned} -u_k'' &= f_k(u_k)h(u_1 - a_k^1)h(u_2 - a_k^2) \ \text{ in } \ (0,1), \\ u_k(0) &= u_k(1) = 0, \ k = 1,2, \end{aligned} \tag{4.3.4}$$

where $a_k^i > 0$, $i$, $k = 1,2$, and the functions $f_1$, $f_2$ and $h$ are assumed to satisfy

(i) The functions $f_1, f_2\colon I\!R \to I\!R_+$ are continuous or non-decreasing;
(ii) there exists $c \geq \max\{a_k^i \mid i, k = 1,2\}$ such that $f_k(s) \geq 16\,c$ whenever $s \geq c$ and $k = 1,2$;
(iii) there exist $\alpha \in [0,\pi^2)$ and $\beta > 0$ such that $f_k(s) \leq \alpha\,s + \beta$ for $s \geq \frac{3}{2}\,c$;
(iv) the function $h$ is the Heaviside step function.

We shall show by means of theorem 4.3.1 that system (4.3.4) possesses a non-trivial positive solution. It is easy to see that a nontrivial lower solution of (4.3.4) is $\underline{u} = (\underline{u}_1, \underline{u}_1)$, where

$$\underline{u}_1(x) = \begin{cases} 4\,c\,x, & x \in [0, \frac{1}{4}], \\ 8\,c\,(x - \frac{1}{4})(\frac{3}{4} - x) + c, & x \in (\frac{1}{4}, \frac{3}{4}), \\ -4\,c\,(x-1), & x \in [\frac{3}{4}, 1]. \end{cases} \tag{4.3.5}$$

An upper solution of the system (4.3.4) is $\bar{u} = (\bar{u}_1, \bar{u}_1)$, where

$$\bar{u}_1(x) = (\frac{3}{2}\,c + \frac{\beta}{\alpha})\,(\cos\sqrt{\alpha}x + \frac{1-\cos\sqrt{\alpha}}{\sin\sqrt{\alpha}}\sin\sqrt{\alpha}x) - \frac{\beta}{\alpha}. \tag{4.3.6}$$

$u = \bar{u}_1$ is a solution of the boundary value problem

$$-u'' = \alpha\,u + \beta, \qquad u(0) = u(1) = \frac{3}{2}\,c. \tag{4.3.7}$$

Since $0 \leq \alpha < \pi^2$, then $u = \bar{u}_1$ is the unique solution of (4.3.7). Moreover, $\underline{u}_1(x) \leq \frac{3}{2}c \leq \bar{u}_1(x)$ for each $x \in [0,1]$, whence $\underline{u} \leq \bar{u}$. It is easy to see that with these lower and upper solutions the hypotheses of theorem 4.3.1 hold, whence the system (4.3.4) has the minimal solution $u_*$ and the maximal solution $u^*$ in the order interval $[\underline{u}, \bar{u}]$.

If the functions $f_1$, $f_2$ are replaced in (4.3.4) by functions $\tilde{f}_1$, $\tilde{f}_2$ which satisfy conditions (i)–(iii) with the same constants $c$, $\alpha$, $\beta$, we obtain a new system which has the minimal solution $\tilde{u}_*$ and the maximal solution $\tilde{u}^*$ in $[\underline{u}, \bar{u}]$. Moreover, if $f_k(s) \leq \tilde{f}_k(s)$ for all $s \in I\!R$ and $k = 1, 2$, it follows from corollary 4.3.1 that $u_* \leq \tilde{u}_*$ and $u^* \leq \tilde{u}^*$.

### 4.3.4. Mixed monotone elliptic systems

In this subsection we shall consider the system (4.3.1) in the case when the function $f = (f_1, \ldots, f_n)$ is mixed monotone in its last two arguments. As we shall show by an example, the system (4.3.1) does not necessarily have solutions in the ordinary sense in this case. Therefore we shall give the following definitions:

The functions $v$, $w \in H_o^1(\Omega, I\!R^n)$ are said to be *coupled quasisolutions* of (4.3.1) if for each $k = 1, \ldots, n$,

(i) $\ell_k(v_k, \varphi) = \int_\Omega f_k(x, v, w)\varphi\, dx, \quad \varphi \in H_o^1(\Omega)$,
(ii) $\ell_k(w_k, \varphi) = \int_\Omega f_k(x, w, v)\varphi\, dx, \quad \varphi \in H_o^1(\Omega)$.

The functions $v$, $w \in H^1(\Omega, I\!R^n)$ are called *coupled lower and upper quasisolutions* of (4.3.1), respectively if

(i) $v \leq 0 \leq w$ on $\partial\Omega$,
(ii) $\ell_k(v_k, \varphi) \leq \int_\Omega f_k(x, v, w)\varphi\, dx$, for all $\varphi \in H_o^1(\Omega) \cap L_+^2(\Omega)$,
(iii) $\ell_k(w_k, \varphi) \geq \int_\Omega f_k(x, w, v)\varphi\, dx$ for all $\varphi \in H_o^1(\Omega) \cap L_+^2(\Omega)$.

We shall impose the following hypotheses on the functions $f_k : \Omega \times I\!R^{2n} \to I\!R$.

(H1) The system (4.3.1) has coupled lower and upper quasisolutions $\underline{v}$, $\bar{w}$ satisfying $\underline{v} \leq \bar{w}$.

(H2) $f_k(\cdot, u(\cdot), v(\cdot))$ is measurable whenever $u$, $v \in [\underline{v}, \bar{w}]$, and there exist $h_k \in L_+^2(\Omega)$ such that $|f_k(x, r, s)| \leq h_k(x)$ for a.a. $x \in \Omega$ and for all $r$, $s \in [\underline{v}(x), \bar{w}(x)]$.

(H3) For each $k = 1, \dots, n$ there is $M_k \in C(\mathbb{R}, \mathbb{R})$ and $c_k > 0$, for which $M_k(\underline{v}_k)$, $M_k(\bar{w}_k) \in L^2(\Omega)$ and
$$(M_k(s) - M_k(t))(s - t) \geq c_k(s - t)^2 \text{ for all } s,\, t \in \mathbb{R},$$
such that $f_k(x, r, s) + M_k(r_k)$ is nondecreasing in $r$ and nonincreasing in $s$ on $[\underline{v}(x), \bar{w}(x)]$ for a.a. $x \in \Omega$.

The main result of this subsection is

**Theorem 4.3.2:** *Under the hypotheses (H1)–(H3) the system (4.3.1) possesses the extremal coupled quasisolutions $u_*, u^*$ in the order interval $[\underline{v}, \bar{w}]$ of $H^1(\Omega, \mathbb{R}^n)$.*

*Proof.* The hypothesis (H2) implies that for all $v$, $w \in [\underline{v}, \bar{w}]$ the functions

$$F_k(v, w)(x) = f_k(x, v(x), w(x)) \tag{4.3.8}$$

are well-defined and belong to $L^2(\Omega)$. Given any $v$, $w \in [\underline{v}, \bar{w}]$ consider the following decoupled system of elliptic equations.

$$\begin{aligned} &-L_k u_k = F_k(v, w) + M_k(v_k) - M_k(u_k) \text{ in } \Omega, \\ &u_k = 0 \text{ on } \partial\Omega, \quad k = 1, \dots, n. \end{aligned} \tag{4.3.9}$$

One readily verifies that $\bar{w}_k$ and $\underline{v}_k$ are upper and lower solutions, respectively, to the boundary value problem (4.3.9) satisfying $\underline{v}_k \leq \bar{w}_k$ by (H1). Thus the system (4.3.9) has a unique solution $u \in H_o^1(\Omega, \mathbb{R}^n)$ with $\underline{v} \leq u \leq \bar{w}$, by (H3).

Define an operator $A\colon [\underline{v}, \bar{w}] \times [\underline{v}, \bar{w}] \to [\underline{v}, \bar{w}]$ by

$$A(v, w) = u, \tag{4.3.10}$$

where $u \in H_o^1(\Omega, \mathbb{R}^n)$ is the unique solution of the system (4.3.9) in $[\underline{v}, \bar{w}]$. By means of (H3) and comparison techniques we see that the operator $A$ is of mixed monotone type. From the weak formulation of (4.3.9) we get by taking $\varphi = u_k$ as a special test function

$$\ell_k(u_k, u_k) = \int_\Omega (F_k(v,w) + M_k(v_k) - M_k(u_k))\, u_k\, dx. \quad (4.3.11)$$

By hypothesis (H2), and since $u$, $v$, $w \in [\underline{v}, \bar{w}]$, the right-hand side of (4.3.11) is uniformly bounded. The left hand side of (4.3.11) satisfies for $c_k$ large enough an estimate

$$\ell_k(u_k, u_k) + \int_\Omega M_k(u_k) u_k\, dx \geq \mu_k |u_k|^2_{H^1(\Omega)} - m_k \|u_k\|_{L^2(\Omega)},$$

where $\mu_k$ and $m_k$ are some positive constants. Since $u \in [\underline{v}, \bar{w}]$, the last inequality implies the existence of a uniform $H^1(\Omega, \mathbb{R}^n)$-bound $c$ to the solutions of the system (4.3.9) for all $v$, $w \in [\underline{v}, \bar{w}]$. Thus, in view of the definition of $A$, we have

$$\|A(v,w)\|_{H^1(\Omega, \mathbb{R}^n)} \leq c \quad \text{for all } v,\ w \in [\underline{v}, \bar{w}].$$

Noticing also that $H^1(\Omega, \mathbb{R}^n)$ is a reflexive Banach space, it follows that $A$ satisfies the hypotheses of proposition 1.2.4. This ensures that $A$ has the extremal coupled fixed points, i.e. there exist $u_*, u^* \in [\underline{v}, \bar{w}]$ such that $u_* \leq u^*$, $u_* = A(u_*, u^*)$ and $u^* = A(u^*, u_*)$, and if $v$, $w$ are any coupled fixed points of $A$ in $[\underline{v}, \bar{w}]$ with $v \leq w$ then $u_* \leq v \leq w \leq u^*$. This and the definition of $A$ imply that $u_*$, $u^*$ are extremal coupled quasisolutions of the system (4.3.1) in $[\underline{v}, \bar{w}]$. □

**Remark 4.3.2:** As a consequence of theorem 4.3.2 any solution $u$ of the system (4.3.1) which belongs to $[\underline{v}, \bar{w}]$ must also belong to $[u_*, u^*]$, since $\{u, u\}$ is in particular a pair of coupled quasisolutions of (4.3.1) in $[\underline{v}, \bar{w}]$. However, unlike to the case of continuous right-hand sides, the existence of solutions between coupled lower and upper quasisolutions of (4.3.1) can no longer be ensured in the case of discontinuous right-hand sides considered above.

**Example 4.3.2:** Consider the system

$$\begin{aligned} &-\Delta u = f_1(u,v) \ \text{ in } \ B_\rho, \ -\Delta v = f_2(u,v) \ \text{ in } \ B_\rho, \\ &u = v = 0 \ \text{ on } \ \partial B_\rho, \end{aligned} \tag{4.3.12}$$

where $B_\rho = \{x \in I\!R^N \mid |x|^2 < \rho^2\}$, and the functions $f_1$ and $f_2$ are defined by $f_1(r,s) = \begin{cases} 1 & \text{for } \ s \le 0, \\ 0 & \text{for } \ s > 0, \end{cases}$ and $f_2(r,s) \equiv r$. A pair of coupled upper and lower quasisolutions of (4.3.12) is given by

$$\underline{v} = (0,0) \ \text{ and } \ \bar{w} = \left(\frac{1}{2N}(\rho^2 - |x|^2), \frac{\rho^2}{4N^2}(\rho^2 - |x|^2)\right). \tag{4.3.13}$$

It is also easy to see that the right hand sides of (4.3.12) satisfy conditions (H2)–(H3). Thus the system (4.3.12) has by theorem 4.3.2 the extremal coupled quasisolutions $u_*$, $u^*$ in $[\underline{v}, \bar{w}]$. But, the system (4.3.12) does not possess any solutions in $[\underline{v}, \bar{w}]$. To prove this, make a counter-hypothesis. There exist $p > 1$ and $(u,v) \in L^p(B_\rho, I\!R^2)$ such that $(u,v)$ satisfies (4.3.12) at least in a distributional sense. Since $f_1(u,v) \in L^\infty(B_\rho)$, it follows by standard regularity results for elliptic equations (cf. e.g., Gilbarg and Trudinger (1983), Thm. 9.15) that $u \in W^{2,q}(B_\rho) \cap W_o^{1,q}(B_\rho)$ for $1 < q < \infty$. From the second equation of (4.3.12) it then follows that $v$ belongs at least to $W^{2,q}(B_\rho) \cap W_o^{1,q}(B_\rho)$ for each $q \in (1,\infty)$. Hence, $(u,v)$ is a weak solution of (4.3.12) in $H_o^1(B_\rho, I\!R^2)$ satisfying the following relations

$$\int_{B_\rho} \nabla u \, \nabla \varphi \, dx = \int_{B_\rho} f_1(u,v)\, \varphi \, dx \ \text{ for } \ \varphi \in H_o^1(B_\rho), \tag{4.3.14}$$

$$\int_{B_\rho} \nabla v \, \nabla \varphi \, dx = \int_{B_\rho} u \, \varphi \, dx \quad \text{for } \ \varphi \in H_o^1(B_\rho). \tag{4.3.15}$$

By choosing $\varphi = v$ in (4.3.14) and $\varphi = u$ in (4.3.15) we get

$$\int_{B_\rho} \nabla u \, \nabla v \, dx = \int_{B_\rho} f_1(u,v)\, v \, dx, \quad \int_{B_\rho} \nabla v \, \nabla u \, dx = \int_{B_\rho} u^2 \, dx.$$

Since

$$\int_{B_\rho} f_1(u,v)\, v\, dx = \int_{v\le 0} v\, dx + \int_{v>0} 0 \cdot v \, dx \le 0,$$

we obtain

$$0 \le \int_{B_\rho} u^2 \, dx \le 0, \quad \text{i.e. } u = 0 \text{ in } B_\rho.$$

This, the second equation of (4.3.12) and the homogeneous boundary conditions imply that $v = 0$, too. Thus $(u,v) = (0,0)$ would be a solution of (4.3.12), which contradicts the fact that $u = v = 0$ does not satisfy the first equation of (4.3.12).

**Remark 4.3.3:** If in (4.3.12) the discontinuous function $f_1$ is approximated by the following continuous function

$$f_{1,\epsilon}(r,s) = \begin{cases} 1, & s \le 0, \\ -\frac{1}{\epsilon}(s-\epsilon), & 0 < s \le \epsilon, \\ 0, & s > \epsilon, \end{cases}$$

we obtain a new system $(4.3.12_\epsilon)$ of boundary value problems. The pair of vector functions given in (4.3.13) is a pair of coupled upper and lower quasisolutions of $(4.3.12_\epsilon)$ and the existence of solutions lying between them can be ensured (cf. e.g., Khavanin and Lakshmikantham (1986), Ladde, Lakshmikantham and Vatsala (1984)). This holds for each $\epsilon > 0$, whereas the limiting system (4.3.12) of $(4.3.12_\epsilon)$ does not posses any solution, as shown above.

## 4.4. PARABOLIC IBVP'S

Let $\Omega \subset I\!R^N$ be a bounded domain with Lipschitz boundary $\partial\Omega$, $Q = \Omega \times (0, \tau)$ and $\Gamma = \partial\Omega \times (0, \tau)$, $\tau > 0$. In this section we consider the existence of weak solutions of the following initial-boundary value problem (IBVP for short):

$$\begin{aligned}&\frac{\partial u(x,t)}{\partial t} + A\,u(x,t) = f(x,t,u(x,t),u(x,t),\nabla u(x,t)),\\&(x,t) \in Q,\ u(x,t) = 0,\ (x,t) \in \Gamma, \quad u(x,0) = 0,\ x \in \Omega,\end{aligned} \tag{4.4.1}$$

where $A$ is a second order quasilinear elliptic differential operator of divergence form, i.e.

$$A\,u(x,t) = -\sum_{i=1}^{N} \frac{\partial}{\partial x_i} a_i(x,t,\nabla u(x,t)).$$

First we shall study an associated boundary value problem of Carathéodory type. The derived results and a generalized monotone iteration method are then applied to prove the existence of extremal solutions of the BVP (4.4.1) between assumed upper and lower solutions in the case when $f$ is discontinuous with respect to the first three variables.

### 4.4.1. Hypotheses

Let $W^{1,p}(\Omega)$ denote the usual Sobolev space and $W^{1,p}(\Omega)^*$ its dual space. For the sake of simplicity we shall assume that $p \geq 2$, and let $q \in I\!R$ satisfy $\frac{1}{p} + \frac{1}{q} = 1$. It is well-known that $W^{1,p}(\Omega) \subset L^2(\Omega) \subset W^{1,p}(\Omega)^*$ forms an evolution triple with all the embeddings being continuous, dense and compact (cf. Zeidler (1990a)).

We set $\mathcal{V} = L^p(0,\tau; W^{1,p}(\Omega))$, denote its dual space by $\mathcal{V}^* = L^q(0,\tau; (W^{1,p}(\Omega))^*)$, and define a function space $\mathcal{W}$ by

$$\mathcal{W} = \{w \in \mathcal{V} \mid \frac{\partial w}{\partial t} \in \mathcal{V}^*\},$$

where the derivative $\frac{\partial}{\partial t}$ is understood in the sense of vector valued distributions (cf. Carl (1988b), Lions (1969), Zeidler (1990a)). The space $\mathcal{W}$ endowed with the norm

$$\|w\|_{\mathcal{W}} = \|w\|_{\mathcal{V}} + \|\frac{\partial w}{\partial t}\|_{\mathcal{V}^*}$$

is a Banach space, which is separable and reflexive due to the separability and reflexivity of $\mathcal{V}$ and $\mathcal{V}^*$, respectively. Furthermore it is well known that the embedding $\mathcal{W} \subset C([0,\tau], L^2(\Omega))$ is continuous (cf. Lions (1969), Zeidler (1990a)). Because the embedding $W^{1,p}(\Omega) \subset L^p(\Omega)$ is compact, then also $\mathcal{W} \subset L^p(Q)$ is compact (cf. Zeidler (1990a)).

By $W_o^{1,p}(\Omega)$ we denote the subspace of $W^{1,p}(\Omega)$ whose elements have generalized homogeneous boundary values. We set $\mathcal{V}_o = L^p(0,\tau; W_o^{1,p}(\Omega))$ and define analogously $\mathcal{V}_o^*$ and $\mathcal{W}_o$.

On the functions $a_i\colon Q \times I\!R^N \to I\!R$, $i = 1, \ldots, N$, we impose the following standard conditions of Leray–Lions type (cf. Deuel and Hess (1978), Mustonen (1990)).

(A1) Each $a_i(x,t,\xi)$ is a Carathéodory function, i.e. measurable in $(x,t) \in Q$ for all $\xi \in I\!R^N$ and continuous in $\xi \in I\!R^N$ for a.a. $(x,t) \in Q$.

(A2) There exists a constant $c_o \geq 0$ and a function $k_o \in L_+^q(Q)$ such that $|a_i(x,t,\xi)| \leq k_o(x,t) + c_o\,|\xi|^{p-1}$, $i = 1, \ldots, N$ for a.a. $(x,t) \in Q$ and for all $\xi \in I\!R^N$.

(A3) $\sum_{i=1}^N (a_i(x,t,\xi) - a_i(x,t,\xi'))(\xi_i - \xi_i') > 0$ for a.a. $(x,t) \in Q$ and for all $\xi$, $\xi' \in I\!R^N$ with $\xi \neq \xi'$.

(A4) There exists a positive constant $c_1$ and $k_1 \in L_+^1(Q)$ such that $\sum_{i=1}^N a_i(x,t,\xi)\,\xi_i \geq c_1\,|\xi|^p - k_1(x,t)$ for a.a. $(x,t) \in Q$ and for all $\xi \in I\!R^N$.

As a consequence of (A1) and (A2) the semilinear form $\ell$ associated with the operator $A$ by

$$\ell(u, \varphi) = \sum_{i=1}^{N} \int_Q a_i(x, t, \nabla u) \frac{\partial \varphi}{\partial x_i} \, dxdt \tag{4.4.2}$$

is well-defined in $\mathcal{V} \times \mathcal{V}$.

A partial ordering in $L^p(Q)$, induced by the order cone $L_+^p(Q)$ of the nonnegative elements of $L^p(Q)$, induces a corresponding partial ordering also in the subset $\mathcal{W}$ of $L^p(Q)$. If $\underline{u}$, $\bar{u} \in \mathcal{W}$ and $\underline{u} \leq \bar{u}$, denote $[\underline{u}, \bar{u}] = \{u \in \mathcal{W} \mid \underline{u} \leq u \leq \bar{u}\}$.

Denote by $\langle \cdot, \cdot \rangle$ the duality pairing between the elements of $\mathcal{V}^*$ and $\mathcal{V}$. The norm (strong) convergence is denoted by $\rightarrow$, and the weak convergence by $\rightharpoonup$.

A function $u \in \mathcal{W}_o$ is called a *solution* of the IBVP (4.4.1) if

(i) $\langle \frac{\partial u}{\partial t}, \varphi \rangle + \ell(u, \varphi) = \int_Q f(x, t, u, u, \nabla u) \, \varphi \, dxdt$ for all $\varphi \in \mathcal{V}_o$;

(ii) $u(x, 0) = 0$ in $\Omega$.

A function $u \in \mathcal{W}$ is an *upper solution* of (4.4.1) if

(i) $\langle \frac{\partial u}{\partial t}, \varphi \rangle + \ell(u, \varphi) \geq \int_Q f(x, t, u, u, \nabla u) \, \varphi \, dxdt$ for all $\varphi \in \mathcal{V}_o \cap L_+^p(Q)$;

(ii) $u(x, 0) \geq 0$ in $\Omega$ and $u(x, t) \geq 0$ on $\Gamma$.

A lower solution is defined similarly, by reversing the inequality signs in (i), (ii).

We shall make the following hypotheses on the function $f\colon Q \times I\!R \times I\!R \times I\!R^N \rightarrow I\!R$.

(B1) The IBVP (4.4.1) has a lower solution $\underline{u}$ and an upper solution $\bar{u}$ such that $\underline{u} \leq \bar{u}$.

(B2) There exist a function $k_2 \in L_+^q(Q)$ and a constant $c_2 \geq 0$ such that

$$|f(x, t, r, s, \xi)| \leq k_2(x, t) + c_2 |\xi|^{p-1} \tag{4.4.3}$$

for a.a. $(x, t) \in Q$ and for all $\xi \in I\!R^N$, $r, s \in [\underline{u}(x, t), \bar{u}(x, t)]$.

(B3) The function $(s,\xi) \mapsto f(x,t,r,s,\xi)$ is continuous for a.a. $(x,t) \in Q$ and for all $r \in \mathbb{R}$, and $r \mapsto f(x,t,r,s,\xi)$ is nondecreasing for a.a. $(x,t) \in Q$ and for $(s,\xi) \in \mathbb{R} \times \mathbb{R}^N$.

(B4) The function $(x,t,r) \mapsto f(x,t,r,s,\xi)$ is a standard function in $Q \times \mathbb{R}$ for each $(s,\xi) \in \mathbb{R} \times \mathbb{R}^N$.

As the main result of the present section we shall prove that if the hypotheses (A1)–(A4) and (B1)–(B4) are satisfied, then the IBVP (4.4.1) has the extremal solutions $u_*$, $u^*$ in $[\underline{u}, \bar{u}]$.

### 4.4.2. Preliminaries

We shall first derive results concerning a Carathéodory type IBVP associated with the IBVP (4.4.1). Throughout this subsection we shall assume that the hypotheses (A1)–(A4) and (B1)–(B4) are satisfied.

The following lemma is a consequence of proposition 1.3.7.

**Lemma 4.4.1:** *A bounded and well-ordered chain $C$ of $\mathcal{W}$ contains a nondecreasing sequence which converges to* $\sup C$ *weakly in $\mathcal{W}$ and strongly in $L^p(Q)$.*

*Proof.* The conclusion follows from proposition 1.3.7, since $\mathcal{W}$ is continuously and compactly embedded in $L^p(Q)$. □

Let us introduce the function $b\colon Q \times \mathbb{R} \to \mathbb{R}$ defined by

$$b(x,t,s) = \begin{cases} (s - \bar{w}(x,t))^{p-1}, & s > \bar{w}(x,t), \\ 0, & \underline{w}(x,t) \le s \le \bar{w}(x,t), \\ -(\underline{w}(x,t) - s)^{p-1}, & s < \underline{w}(x,t), \end{cases} \tag{4.4.4}$$

where $\underline{w}$, $\bar{w}$ are fixed functions in $L^p(Q)$ satisfying $\underline{w} \le \bar{w}$. We recall the following lemma proved in Deuel (1976), and in Deuel and Hess (1974/75).

**Lemma 4.4.2:** *The function $b$ satisfies Carathéodory conditions, and there exist positive constants $c_3$, $c_4$, $c_5$ and a function $k_3 \in L^q(Q)$ such that*

$$|b(x,t,s)| \leq k_3(x,t) + c_3|s|^{p-1}, \tag{4.4.5}$$

*and*

$$\int_Q b(x,t,u(x,t))u(x,t)\,dxdt \geq c_4\|u\|_p^p - c_5 \tag{4.4.6}$$

*for each $u \in L^p(Q)$.*

Let $\bar{u}$, $\underline{u} \in \mathcal{W}$ be upper and lower solutions of the IBVP (4.4.1), as assumed in (B1). We assign to any upper solution $v \in [\underline{u}, \bar{u}]$ of the IBVP (4.4.1) the following IBVP:

$$\begin{aligned} &\frac{\partial u}{\partial t} + Au = f(x,t,v,u,\nabla u) \;\text{ in } Q, \\ &u = 0 \;\text{ on } \Gamma, \; u = 0 \;\text{ in } \Omega. \end{aligned} \tag{4.4.7}$$

**Lemma 4.4.3:** *Let $v \in [\underline{u}, \bar{u}]$ be any upper solution of the IBVP (4.4.1). If $u_i$, $i = 1,\ldots,m$, are lower solutions of the IBVP (4.4.7) in $[\underline{u}, v]$, then (4.4.7) has at least one solution $u$ in $[u_o, v]$, where $u_o = \max\{u_i \mid i = 1,\ldots,m\}$.*

*Proof.* It is easy to see that $v$ is also an upper solution of the IBVP (4.4.7). For $u \in \mathcal{W}$ we define the following truncation mappings $T_i$, $i = 0,1,\ldots,m$, by

$$(T_i u)(x,t) = \begin{cases} u_i(x,t) & \text{if } u(x,t) < u_i(x,t), \\ u(x,t) & \text{if } u_i(x,t) \leq u(x,t) \leq v(x,t), \\ v(x,t) & \text{if } v(x,t) < u(x,t). \end{cases}$$

By results of Deuel (1976) and Deuel and Hess (1974/75) one can show that these mappings $T_i$ are bounded and continuous from $\mathcal{V}$ into itself. The hypotheses (B3) and (B4) ensure that the function $(x, t, s, \xi) \mapsto f(x, t, v(x, t), s, \xi)$ satisfies the Carathéodory conditions. If $F$ denotes the superposition operator associated with this function, defined by

$$Fu(x,t) = f(x,t,v(x,t),u(x,t),\nabla u(x,t)), \qquad (4.4.8)$$

then the BVP (4.4.7) can be written in the form

$$\frac{\partial u}{\partial t} + Au = Fu \text{ in } Q, \quad u = 0 \text{ on } \Gamma, \; u = 0 \text{ in } \Omega. \qquad (4.4.9)$$

As a consequence of (B2) and (B3) the mapping $F$ is bounded and continuous from $[\underline{u}, \bar{u}] \subset \mathcal{V}$ into $L^q(Q)$. Thus the composed operator $F \circ T_i \colon \mathcal{V} \to L^q(Q)$ is bounded and continuous. From (4.4.5) it follows that the superposition operator $B$ associated with the function $b$ is a bounded and continuous function from $L^p(Q)$ to $L^q(Q)$.

We shall associate with (4.4.9) (resp. (4.4.7)) the following auxiliary IBVP

$$\frac{\partial u}{\partial t} + Au - F \circ T_o u - \sum_{i=1}^{m} |F \circ T_i u - F \circ T_o u| + \gamma Bu = 0 \text{ in } Q,$$
$$u = 0 \text{ on } \Gamma, \; u = 0 \text{ in } \Omega, \qquad (4.4.10)$$

where $\gamma$ is a positive constant specified later, and where we have chosen $\underline{w} = u_o$ and $\bar{w} = v$ in the definition (4.4.4) of the function $b$. The assertion is proved if we show that there exist solutions of the IBVP (4.4.10), and if each solution $u$ of (4.4.10) belongs to $[u_o, v]$, since then we have $T_i u = u$ for each $i = 0, 1, \ldots, m$ and $Bu = 0$, so that $u$ must be also a solution of the IBVP (4.4.9). This proof will be done in several steps.

Step 1: Existence of a solution of (4.4.10).
We are looking for solutions $u \in \mathcal{W}_o$ having homogeneous initial values. The operator $\frac{\partial}{\partial t}$ induces a linear mapping $L$ from the subset

$$D(L) = \{v \in \mathcal{W}_o \mid v(\cdot, 0) = 0 \text{ in } \Omega\}$$

of $\mathcal{V}_o$ into $\mathcal{V}_o^*$. It can be shown that $L$ is closed, densely defined and maximal monotone operator (cf. Mustonen (1990), Zeidler (1990b)). The assumptions (A1) and (A2), lemma 4.4.2 and the properties of the composed mappings $F \circ T_i$ give rise to bounded and continuous mappings $L_1$ and $L_2$ from $\mathcal{V}_o$ to $\mathcal{V}_o^*$, defined by

$$\begin{aligned}\langle L_1 u, \varphi\rangle &= \ell(u, \varphi) + \gamma \int_Q (Bu)\, \varphi \, dx dt, \\ \langle L_2 u, \varphi\rangle &= -\int_Q \{F \circ T_o u + \sum_{i=1}^{m} |F \circ T_i u - F \circ T_o u|\} \varphi \, dx dt\end{aligned}$$

for $u$, $\varphi \in \mathcal{V}_o$. Thus the corresponding weak formulation of the BVP (4.4.10) reads as follows: Find $u \in D(L)$ such that

$$\langle (L + L_1 + L_2)u, \varphi\rangle = 0 \quad \text{for all } \varphi \in \mathcal{V}_o. \tag{4.4.11}$$

By hypothesis (A3) the operator $L_1 + L_2$ is pseudomonotone with respect to $D(L)$, in the sense defined in Lions (1965), or equivalently, in Mustonen (1990). Thus the existence of solutions of (4.4.11) follows if the operator $L_1 + L_2$ is coercive, i.e.

$$\frac{\langle (L_1 + L_2)u, u\rangle}{\|u\|_{\mathcal{V}_o}} \to \infty \quad \text{as } \|u\|_{\mathcal{V}_o} \to \infty. \tag{4.4.12}$$

Due to (A4) and the estimate (4.4.6) of lemma 4.4.2 we get

$$\begin{aligned}\langle L_1 u, u\rangle &= \ell(u, u) + \gamma \int_Q (Bu) u \, dx dt \\ &\geq c_1 \|\nabla u\|_p^p + c_4 \, \gamma \|u\|_p^p - c_5 \, \gamma - \|k_1\|_1.\end{aligned} \tag{4.4.13}$$

In view of the hypothesis (B2) and the relation

$$(\nabla T_i u)(x,t) = \begin{cases} \nabla u_i(x,t) & \text{if } u(x,t) < u_i(x,t), \\ \nabla u(x,t) & \text{if } u_i(x,t) \le u(x,t) \le v(x,t), \\ \nabla v(x,t) & \text{if } v(x,t) < u(x,t), \end{cases}$$

(cf. Gilbarg and Trudinger (1983)) we have the estimate

$$\left| \int_Q (F \circ T_i u)\, u\, dxdt \right| \le (\|k_2\|_q + c_2 \|\nabla u\|_p^{p-1}) \|u\|_p.$$

Applying Young's inequality we obtain for arbitrary $\epsilon > 0$ and some positive constants $d_1(\epsilon)$ and $d_2$

$$\left| \int_Q (F \circ T_i u)\, u\, dxdt \right| \le \epsilon\, \|\nabla u\|_p^p + d_1(\epsilon) \|u\|_p^p + d_2 \|u\|_p \quad (4.4.14)$$

for all $i = 0, 1, \ldots, m$. This implies an estimate

$$\langle L_2 u, u \rangle \ge -\epsilon\, \|\nabla u\|_p^p - d_3(\epsilon) \|u\|_p^p - d_4 \|u\|_p \quad (4.4.15)$$

with some positive constants $d_3(\epsilon)$ and $d_4$, which yields, together with (4.4.13) the estimate

$$\begin{aligned} \langle (L_1 + L_2)u, u \rangle &\ge (c_1 - \epsilon) \|\nabla u\|_p^p \\ &+ (c_4\, \gamma - d_3(\epsilon)) \|u\|_p^p - d_4\, \|u\|_p - c_5\, \gamma. \end{aligned} \quad (4.4.16)$$

Hence, the coercivity of the operator $L_1 + L_2$ follows from (4.4.16) by choosing $\epsilon < c_1$ and $\gamma$ so large that $c_4\, \gamma - d_3(\epsilon) > 0$ (Notice that $\|u\|_{\mathcal{V}_o}^p = \int_Q (|u(x,t)|^p + |\nabla u(x,t)|^p)\, dxdt$).

Step 2: Each solution of (4.4.11) belongs to $[u_o, v]$.
Let $u_k$, $k \in \{1, 2, \ldots, m\}$, be any of the assumed lower solutions,

and let $u$ be a solution of (4.4.11). Denoting $w^+ = \max\{w, 0\}$, we have $(u_k - u)^+ \in \mathcal{V}_o \cap L^p_+(Q)$ and $(u_k - u)^+(x, 0) = 0$ in $\Omega$. The definition of the lower solution implies that

$$\langle \frac{\partial u_k}{\partial t}, \varphi\rangle + \ell(u_k, \varphi) \leq \int_Q (Fu_k)\varphi\, dxdt, \;\; \varphi \in \mathcal{V}_o \cap L^p_+(Q),$$
$$u_k(x, 0) \leq 0 \;\text{ in } \Omega \text{ and } u_k(x, t) \leq 0 \text{ on } \;\Gamma. \tag{4.4.17}$$

Because $u$ is also a solution of (4.4.10), then

$$\langle \frac{\partial u}{\partial t}, \varphi\rangle + \ell(u, \varphi) + \gamma \int_Q (Bu)\varphi\, dxdt$$
$$= \int_Q \{F \circ T_o u + \sum_{k=1}^{m} |F \circ T_i u - F \circ T_o u|\}\varphi\, dxdt, \tag{4.4.18}$$
$$\varphi \in \mathcal{V}_o, \;\; u(x, 0) = 0 \;\text{ in } \Omega \text{ and } u(x, t) = 0 \text{ on } \;\Gamma.$$

By subtracting (4.4.18) from (4.4.17) and taking $\varphi = (u_k - u)^+$ we obtain

$$\langle \frac{\partial(u_k - u)}{\partial t}, (u_k - u)^+\rangle + \ell(u_k, (u_k - u)^+) - \ell(u, (u_k - u)^+)$$
$$\leq \int_Q \{Fu_k - F \circ T_o u - \sum_{k=1}^{m} |F \circ T_i u - F \circ T_o u|\}(u_k - u)^+\, dxdt$$
$$+ \gamma \int_Q Bu(u_k - u)^+\, dxdt \leq \gamma \int_Q Bu(u_k - u)^+\, dxdt. \tag{4.4.19}$$

By hypothesis (A3) and due to $\langle \frac{\partial w}{\partial t}, w^+\rangle \geq 0$ for any $w \in \mathcal{W}$ satisfying $w(x, 0) \leq 0$ (cf. Lions (1969)), the left hand side of (4.4.19) is nonnegative, and we get because of $u_k \leq u_o$

$$0 \leq \gamma \int_Q Bu(u_k - u)^+\, dxdt$$
$$= \gamma \int_{(u<u_k)} -(u_o - u)^{p-1}(u_k - u)^+\, dxdt \leq 0.$$

From the last inequality it follows that

$$0 \geq \int_{(u<u_k)} (u_o - u)^{p-1}(u_k - u)^+ \, dxdt$$
$$\geq \int_{(u<u_k)} (u_k - u)^p \, dxdt = \int_Q ((u_k - u)^+)^p \, dxdt,$$

which implies that $(u_k - u)^+ = 0$, i.e. $u_k \leq u$ for each $k = 1, \dots, m$. Thus $u_o = \max\{u_1, \dots, u_m\} \leq u$.

By a similar reasoning it can be shown that $u \leq v$. This completes the proof of lemma 4.4.3. □

**Lemma 4.4.4:** *If $v \in [\underline{u}, \bar{u}]$ is an upper solution of (4.4.1), then the IBVP (4.4.9) has at least one solution in $[\underline{u}, v]$, and each such a solution $u$ satisfies*

$$\|u\|_{\mathcal{W}_o} \leq c, \tag{4.4.20}$$

*where the constant $c$ is independent on $u$ and on $v$.*

*Proof.* For the convenience we recall that

$$\|u\|_{\mathcal{W}_o} = \|u\|_{\mathcal{V}_o} + \|\frac{\partial u}{\partial t}\|_{\mathcal{V}_o^*}, \quad \text{and} \quad \|u\|_{\mathcal{V}_o}^p = \int_Q (|u|^p + |\nabla u|^p) \, dxdt.$$

By definition, a solution of (4.4.9) is a function $u \in \mathcal{W}_o$ satisfying $u(x,0) = 0$ in $\Omega$ and

$$\langle \frac{\partial u}{\partial t}, \varphi \rangle + \ell(u, \varphi) = \int_Q (Fu)\varphi \, dxdt \quad \text{for all} \quad \varphi \in \mathcal{V}_o. \tag{4.4.21}$$

Since $u(x,0) = 0$ we have $\langle \frac{\partial u}{\partial t}, u \rangle \geq 0$, and by hypothesis (A4) it follows that

$$\langle \frac{\partial u}{\partial t}, u \rangle + \ell(u, u) \geq c_1 \|\nabla u\|_p^p - \|k_1\|_1. \tag{4.4.22}$$

By inspection of the proof of lemma 4.4.3 we have, due to condition (B2),

$$\begin{aligned}|\int_Q (Fu)u\,dxdt| &\le (\|k_2\|_q + c_2\|\nabla u\|_p^{p-1})\|u\|_p \\ &\le \|k_2\|_q\|u\|_p + \epsilon\|\nabla u\|_p^p + d_1(\epsilon)\|u\|_p^p.\end{aligned} \tag{4.4.23}$$

Taking $\varphi = u$ as a special test function in (4.4.21) we obtain by using the estimates (4.4.22), and (4.4.23)

$$(c_1 - \epsilon)\|\nabla u\|_p^p \le \|k_1\|_1 + \|k_2\|_q\|u\|_p) + d_1(\epsilon)\|u\|_p^p. \tag{4.4.24}$$

Adding $(c_1 - \epsilon)\|u\|_p^p$ to both sides of the last inequality and taking $c$ as a generic positive constant we get

$$(c_1 - \epsilon)\|u\|_{\mathcal{V}_o}^p \le c(1 + \|u\|_p^p + \|u\|_p). \tag{4.4.25}$$

Each $u \in [\underline{u}, \bar{u}]$ is uniformly bounded in $L^p(Q)$. Thus from (4.4.25) with $\epsilon < c_1$ we get

$$\|u\|_{\mathcal{V}_o}^p \le c. \tag{4.4.26}$$

To complete the proof we have to show that $\|\frac{\partial u}{\partial t}\|_{\mathcal{V}_o^*} \le c$. By (4.4.21) we get

$$|\langle \frac{\partial u}{\partial t}, \varphi\rangle| \le |\ell(u,\varphi)| + |\int_Q (Fu)\varphi\,dxdt|,$$

and due to (A2) and (B2) we have with the generic constant $c$

$$\begin{aligned}|\langle \frac{\partial u}{\partial t}, \varphi\rangle| &\le c(\|k_o\|_q + c_o\|\nabla u\|_p^{p-1})\|\nabla\varphi\|_p \\ &\quad + c(\|k_2\|_q + c_2\|\nabla u\|_p^{p-1})\|\varphi\|_p.\end{aligned} \tag{4.4.27}$$

This and (4.4.26) imply

$$\|\frac{\partial u}{\partial t}\|_{\mathcal{V}_o^*} = \sup\{|\langle\frac{\partial u}{\partial t}, \varphi\rangle| \mid \|\varphi\|_{\mathcal{V}_o} = 1\} \leq c,$$

which completes the proof. □

**Lemma 4.4.5:** *Let $\ell$ be the semilinear form defined in (4.4.2), and let $(u_n)$ be any sequence in $D(L)$ such that $u_n \rightharpoonup u$ in $\mathcal{V}_o$, $Lu_n \rightharpoonup Lu$ in $\mathcal{V}_o^*$ and* $\limsup \ell(u_n, u_n - u) \leq 0$. *Then $(u_n)$ converges strongly in $\mathcal{V}_o$.*

*Proof.* Let $A: \mathcal{V}_o \to \mathcal{V}_o^*$ be the operator associated with the semilinear form $\ell$ by

$$\langle Au, v\rangle = \ell(u, v) \quad \text{for all} \quad u,\ v \in \mathcal{V}_o.$$

Then the statement of lemma 4.4.5 is equivalent with the $(S_+)$ property with respect to $D(L)$ of the operator $A$, which has been proved in Berkovits and Mustonen (1992) (see also Mustonen (1990), Deuel and Hess (1978)). □

**Lemma 4.4.6:** *Let $v \in [\underline{u}, \bar{u}]$ be any upper solution of the IBVP (4.4.1), and denote*

$$S_v = \{u \in [\underline{u}, v] \mid u \ \ \textit{is a solution of the BVP (4.4.9)}\}. \tag{4.4.28}$$

*If $C$ is a well-ordered chain in the set $S_v$, then $w = \sup C$ exists and belongs to $S_v$.*

*Proof.* By lemma 4.4.4 the set $S_v$ is nonempty and is norm-bounded in $\mathcal{W}_o$. Thus, if $C$ is a well-ordered chain in $S_v$, then $C$ contains by lemma 4.4.1 a nondecreasing sequence $(w_n)_{n=o}^{\infty}$ which converges to $w = \sup C$ weakly in $\mathcal{W}_o$, i.e. $w_n \rightharpoonup w$ in $\mathcal{V}_o$ and

$Lw_n \rightharpoonup Lw$ in $\mathcal{V}_o^*$, and strongly in $L^p(Q)$. Each $w_n \in S_v$ satisfies $w_n \in \mathcal{W}_o$, $w_n(x,0) = 0$ in $\Omega$ and

$$\langle \frac{\partial w_n}{\partial t}, \varphi \rangle + \ell(w_n, \varphi) = \int_Q (Fw_n)\varphi \, dxdt \quad \text{for all} \quad \varphi \in \mathcal{V}_o. \tag{4.4.29}$$

Since $w_n \rightharpoonup w$ in $\mathcal{W}_o$, $w_n \in D(L)$ for each $n$, and since $D(L)$ is closed and convex we get $w \in D(L)$, i.e. $w \in \mathcal{W}_o$ and $w(x,0) = 0$ in $\Omega$. Hence the assertion follows if $w$ can be proved to satisfy relation (4.4.29). To this end we shall show that $(w_n)$ converges strongly in $\mathcal{V}_o$ which enables us to pass to the limit in (4.4.29). As special test function in (4.4.29) we take $\varphi = w_n - w$. Then due to

$$\begin{aligned} \langle \frac{\partial w_n}{\partial t}, w_n - w \rangle &= \langle \frac{\partial (w_n - w)}{\partial t}, w_n - w \rangle + \langle \frac{\partial w}{\partial t}, w_n - w \rangle \\ &\geq \langle \frac{\partial w}{\partial t}, w_n - w \rangle, \end{aligned}$$

we get from (4.4.29)

$$\ell(w_n, w_n - w) \leq \int_Q (Fw_n)(w_n - w)\, dxdt - \langle \frac{\partial w}{\partial t}, w_n - w \rangle. \tag{4.4.30}$$

The second term on the right-hand side of (4.4.30) tends to zero as $n \to \infty$ due to the weak convergence of $(w_n)$ to $w$ in $\mathcal{W}_o$. By hypothesis (B2) we deduce the estimate

$$\begin{aligned} &|\int_Q (Fw_n)(w_n - w)\, dxdt| \\ &\leq (\|k_2\|_q + c_2 \|\nabla w_n\|_p^{p-1}) \|w_n - w\|_p. \end{aligned} \tag{4.4.31}$$

The uniform norm-boundedness of $w_n$ in $\mathcal{W}_o$ and the strong convergence $w_n \to w$ in $L^p(Q)$ imply that the right hand side of

(4.4.31) tends to zero as $n \to \infty$. Thus we obtain from (4.4.30) that

$$\limsup \ell(w_n, w_n - w) \le 0.$$

The strong convergence $w_n \to w$ in $\mathcal{V}_o$ follows then from lemma 4.4.5, which completes the proof by passing to the limit as $n \to \infty$ in (4.4.29). □

**Lemma 4.4.7:** *The solution set $S_v$ has the greatest element.*

*Proof.* Let $I$ denote the identity mapping of $S_v$. Since $I[S_v] = S_v$, it follows from lemma 4.4.6 that each well-ordered chain in $I[S_v]$ has an upper bound in $S_v$. Thus $I$ has by proposition 1.1.2 a maximal fixed point $z$, i.e. $z$ is a maximal element of $S_v$. To prove that $z$ is the greatest element of $S_v$, assume there is $w \in S_v$ such that $w \not\le z$. Then, by lemma 4.4.3, there is $u \in S_v$ for which $\max\{w, z\} \le u \le v$. In particular, this implies that $z < u$, which contradicts the fact that $z$ is a maximal element of $S_v$. Thus $z$ is the greatest element of $S_v$. □

### 4.4.3. Existence of extremal solutions (4.4.1)

We are now ready to prove our main result concerning the existence of the extremal solutions of the IBVP (4.4.1).

**Theorem 4.4.1:** *Let the hypotheses (A1)–(A4) and (B1)–(B4) be satisfied. Then the IBVP (4.4.1) has the extremal solutions $u_*$, $u^*$ in the order interval $[\underline{u}, \bar{u}]$ of $\mathcal{W}$ in the sense that if $u$ is any solution of (4.4.1) in $[\underline{u}, \bar{u}]$, then $u_* \le u \le u^*$.*

*Proof.* The proof will only be given for the existence of the maximal solution $u^*$, since the existence of the minimal solution $u_*$ can be proved quite similarly by dual reasoning.

Let $P$ be the set of all upper solutions of the IBVP (4.4.1) lying between $\underline{u}$ and $\bar{u}$. Define an operator $G\colon P \to P$ by assigning to each upper solution $v \in P$ the greatest element $Gv$ of $S_v$, i.e. the maximal solution of the IBVP (4.4.9) within the order interval $[\underline{u}, v]$. By using the monotonicity conditions of (B3) and the fact that $Gv \leq v$ one readily verifies that $Gv \in P$. The result of lemma 4.4.7 ensures that the operator $G$ is well-defined. Let $C$ denote the i.w.o. chain of $G$-iterations of $\bar{u}$, defined in proposition 1.2.1. By the definition of $C$, $G[C]$ is also an inversely well-ordered chain in $\mathcal{W}_o$. Moreover, it is bounded in $\mathcal{W}_o$ by lemma 4.4.4. Applying the result of lemma 4.4.1 to $-G[C]$ it follows that $G[C]$ contains a nonincreasing sequence $(Gw_n)_{n=o}^{\infty}$ which converges to $u^* = \inf G[C]$ weakly in $\mathcal{W}_o$ and strongly in $L^p(Q)$. Obviously, $u^*$ belongs to $[\underline{u}, \bar{u}]$. Next we shall show that $(Gw_n)_{n=0}^{\infty}$ converges to $u^*$ strongly in $\mathcal{V}_o$. Since $u^* \leq Gw_n \leq w_n$ for each $n \in \mathbb{N}$, we obtain by noticing that $Gw_n \in S_{w_n}$ and using the monotonicity conditions of (B3) that

$$\langle \frac{\partial Gw_n}{\partial t}, \varphi \rangle + \ell(Gw_n, \varphi) = \int_Q f(x,t,w_n,Gw_n,\nabla Gw_n)\varphi dxdt$$
$$\geq \int_Q f(x,t,u^*,Gw_n,\nabla Gw_n)\,\varphi\, dxdt \tag{a}$$

for all $\varphi \in \mathcal{V}_o \cap L_+^p(Q)$. If $\bar{F}$ denotes the superposition operator defined by (4.4.8) with $v = \bar{u}$, it follows from (a) that

$$\langle \frac{\partial Gw_n}{\partial t}, \varphi \rangle + \ell(Gw_n, \varphi) \leq \int_Q (\bar{F}Gw_n)\varphi\, dxdt \tag{b}$$

for all $\varphi \in \mathcal{V}_o \cap L_+^p(Q)$. Because $\langle \frac{\partial(Gw_n - u^*)}{\partial t}, Gw_n - u^* \rangle \geq 0$ since $(Gw_n - u^*)(x,0) = 0$, it follows from (b) that

$$\ell(Gw_n, Gw_n - u^*) \leq \int_Q (\bar{F}Gw_n)(Gw_n - u^*)\, dxdt$$
$$- \langle \frac{\partial u^*}{\partial t}, Gw_n - u^* \rangle. \tag{c}$$

Due to condition (B2) we have

$$\begin{aligned} &|\int_Q (\bar{F}Gw_n)(Gw_n - u^*)\,dxdt| \\ &\le (\|k_2\|_q + c_2\|\nabla Gw_n\|_p^{p-1})\|Gw_n - u^*\|_p. \end{aligned} \tag{d}$$

Noticing that $Gw_n - u^* \in \mathcal{V}_o \cap L^p_+(Q)$ for each $n \in I\!N$, that the sequence $(Gw_n)$ is uniformly bounded in $\mathcal{W}_o$, and that $(Gw_n)_{n=o}^{\infty}$ converges to $u^*$ strongly in $L^p(Q)$ and weakly in $\mathcal{W}_o$, it follows from (d) that the right hand side of (c) goes to zero as $n \to \infty$. Thus $\limsup \ell(Gw_n, Gw_n - u^*) \le 0$, which implies by lemma 4.4.5 the strong convergence of the sequence $(Gw_n)_{n=0}^{\infty}$ to $u^*$ in $\mathcal{V}_o$. This ensures that when $n \to \infty$ in (a), we obtain

$$\langle \frac{\partial u^*}{\partial t}, \varphi\rangle + \ell(u^*, \varphi) \ge \int_Q f(x,t,u^*,u^*,\nabla u^*)\,\varphi\,dxdt.$$

Thus $u^*$ is an upper solution of the IBVP (4.4.1) in $[\underline{u}, \bar{u}]$, whence $Gu^*$ is defined. By definition, $Gu^* \le u^*$.

The above proof shows that $G$ satisfies the hypotheses of proposition 1.2.1 a), whence $u^*$ is a fixed point of $G$. Since $u^* = Gu^* \in S_{u^*}$, then (4.4.9) holds with $u = v = u^*$, so that $u^*$ is a solution of the IBVP (4.4.1) in $[\underline{u}, \bar{u}]$.

Let $\tilde{u}$ be any solution of (4.4.1) in $[\underline{u}, \bar{u}]$. To prove that $\tilde{u} \le u^*$, we shall first show that $\tilde{u} \le u$ for each $u \in C$, where $C$ is the i.w.o. chain of $G$-iterations of $\bar{u}$, defined in proposition 1.2.1. Make a counter-hypothesis, $\tilde{u}$ is not a lower bound of $C$. Then $C$ contains the greatest element $u$ such that $\tilde{u} \not\le u$. Since $\tilde{u} \le \bar{u}$, then $u < \bar{u}$. If $w \in C$ and $w > u$, then $\tilde{u} \le w$, whence the BVP (4.4.9) has a solution in $[\tilde{u}, w]$ by lemma 4.4.4. Since $Gw$ is the maximal solution of (4.4.9) in $[\underline{u}, w]$, then $\tilde{u} \le Gw$. This shows that $\tilde{u}$ is a lower bound of $G\{w \in C \mid w > u\}$. But then $\tilde{u} \le \inf G\{w \in C \mid w > u\} = u$, which contradicts with $\tilde{u} \not\le u$. Thus $\tilde{u} \le w$ for all $w \in C$, whence $\tilde{u} \le \min C = u^*$. This proves that $u^*$ is the maximal solution of (4.4.1) in $[\underline{u}, \bar{u}]$. □

**Remark 4.4.1:** (i) The method used above can be applied also to the IBVP with inhomogeneous boundary conditions:

$$\begin{aligned} &\frac{\partial u}{\partial t} + A\,u = f(x,t,u,u,\nabla u) + h, \;\; (x,t) \in Q, \\ &u(x,t) = g(x,t) \;\text{ on } \Gamma, \;\;\; u(x,0) = \psi(x) \text{ in } \;\Omega, \end{aligned} \tag{4.4.32}$$

where $A$ an $f$ are as above, $h \in \mathcal{V}^*$, $g \in L^p(0,T;W^{1-\frac{1}{p},p}(\Gamma))$ and $\psi \in L^2(\Omega)$. The so chosen $g$ and $\psi$ admit an extension $w \in \mathcal{W}$ such that $w(x,t) = g(x,t)$ on $\Gamma$ and $w(x,0) = \psi(x)$ in $\Omega$ (cf. Lions (1969)). Performing the change of variable $u \to u - w$, the IBVP (4.4.32) is reduced to the IBVP

$$\begin{aligned} &\frac{\partial u}{\partial t} + \bar{A}\,u = \bar{f}(x,t,u,u),\nabla u) + \bar{h}, \;\; (x,t) \in Q, \\ &u(x,t) = 0 \;\text{ on }\; \Gamma, \;\;\; u(x,0) = 0 \;\text{ in }\; \Omega, \end{aligned} \tag{4.4.33}$$

where

$$\begin{aligned} &\bar{A}\,u(x,t) = -\sum_{i=1}^{N} \frac{\partial}{\partial x_i} a_i(x,t,\nabla(u+w)(x,t)), \\ &\bar{f}(x,t,u,u,\nabla u) = f(x,t,u+w,u+w,\nabla(u+w)), \\ &\bar{h}(x,t) = h(x,t) - \frac{\partial w(x,t)}{\partial t}. \end{aligned} \tag{4.4.34}$$

One readily verifies that $\bar{a}_i(x,t,\xi) = a_i(x,t,\,\xi + \nabla w(x,t))$ satisfy (A1)–(A4), and that $\bar{h}$ belongs to $\mathcal{V}^*$. If $\bar{u}$ and $\underline{u}$ are upper and lower solutions of the IBVP (4.4.32), then $\bar{u} - w$ and $\underline{u} - w$ are, correspondingly, upper and lower solutions of the transformed problem (4.4.33). The nonlinearity $\bar{f}$ satisfies hypotheses (B2)–(B4) with respect to the shifted order interval $[\underline{u} - w, \bar{u} - w]$. Obviously, the existence of extremal solutions of (4.4.32) in $[\underline{u}, \bar{u}]$ is equivalent with the existence of extremal solutions of (4.4.33) in $[\underline{u} - w, \bar{u} - w]$. Thus the inhomogeneous IBVP (4.4.32) can be reduced to (4.4.33), which is of the form (4.4.1), with the exception of the term $\bar{h} \in \mathcal{V}^*$. This extra term causes only slight differences to the reasonings made above (cf. Carl (1994)).

(ii) A more general quasilinear operator $A$ of the form

$$A u(x,t) = -\sum_{i=1}^{N} \frac{\partial}{\partial x_i} a_i(x,t,u(x,t),\nabla u(x,t)),$$

which depends also on $u$, may be taken into account. However, in this case further restrictions have to be imposed on the coefficients $a_i(x,t,\eta,\xi)$ in order to obtain the same results. For instance, it suffices to assume that these coefficients possess an appropriate modulus of continuity with respect to $\eta$ which is related to a strong $p$-ellipticity with respect to $\xi$, i.e.

$$\begin{aligned}&|a_i(x,t,\eta,\xi) - a_i(x,t,\eta',\xi)|\\ &\leq (k_1(x,t) + |\eta|^{p-1} + |\eta'|^{p-1} + |\xi|^{p-1})\omega(|\eta - \eta'|),\end{aligned}$$

where $\omega$ is a modulus of continuity satisfying $\int_{0+} \frac{dr}{\omega(r)^q} = +\infty$, $(q = \frac{p}{p-1})$ and

$$\sum_{i=1}^{N}(a_i(x,t,\eta,\xi) - a_i(x,t,\eta,\xi'))(\xi_i - \xi_i') \geq \alpha|\xi - \xi'|^p, \ \alpha > 0.$$

(cf. Chipot and Rodrigues (1988)).

(iii) In Kuiper (1973) there has been considered elliptic differential equations with discontinuous nonlinearities which describe, among other things, the steady state situation of the Joule heating of a body subjected to an electric current given by

$$-\sum_{i=1}^{N} \frac{\partial}{\partial x_i}\left(K(x,u(x))\,\frac{\partial u(x)}{\partial x_i}\right) = \sigma(x,u(x)) \qquad (4.4.35)$$

under Dirichlet boundary conditions, where $K \in C^1(\bar{\Omega} \times I\!R)$ and the nonlinearity $\sigma(x, r)$ on the right-hand side of (4.4.35) may have countable jump discontinuities of monotone type with respect to $r$. However, in order to treat real situations one wants to require as little smoothness of $K$ as possible. In view of remark (ii) we are able to treat also the corresponding non-steady state situation under less regularity assumptions on $K$, i.e. under Carathéodory conditions, and without restristing the nonlinearity to possess only countable jump discontinuities.

## 4.5. PARABOLIC SYSTEMS

In this section we shall study the existence of extremal solutions or extremal coupled quasisolutions of the following system of initial-boundary value problems

$$\begin{aligned} &\frac{\partial u_k}{\partial t} - L_k u_k = f_k(x, t, u, u) \ \text{ in } Q \\ &u_k = 0 \ \text{ on } \Gamma, \quad u_k(x, 0) = 0 \ \text{ in } \Omega, \end{aligned} \tag{4.5.1}$$

$k = 1, \ldots, n$, where $f_k \colon Q \times I\!R^{2n} \to I\!R$, and where $L_k$'s are uniformly elliptic differential operators of the form

$$L_k u_k = \sum_{i,j=1}^{N} \frac{\partial}{\partial x_i}\left(a_{ij}^k \frac{\partial u_k}{\partial x_j}\right) - \sum_{i=1}^{N} b_i^k \frac{\partial u_k}{\partial x_i}, \quad k = 1, \ldots, n.$$

### 4.5.1. Existence of the extremal solutions

Appropriate function spaces where to consider the system (4.5.1) are the $n$-dimensional Cartesian products of the function spaces defined in section 4.4 with the index $p = 2$. The spaces $L^2(Q, I\!R^n)$

and $\mathcal{V}^n$ are ordered componentwise. Denoting

$$\ell_k(u,\varphi) = \sum_{i,j=1}^{N} \int_Q a_{ij}^k \frac{\partial u}{\partial x_i}\frac{\partial \varphi}{\partial x_j} dx\, dt + \sum_{i=1}^{N} \int_Q b_i^k \frac{\partial u}{\partial x_i} \varphi\, dx\, dt,$$

we define solutions of the system (4.5.1) as follows:

We say that the function $u \in \mathcal{W}_o^n$ is a *solution* of the system (4.5.1) if

(i) $\langle \frac{\partial u_k}{\partial t}, \varphi\rangle + \ell_k(u_k,\varphi) = \int_Q f_k(x,t,u(x,t),u(x,t))\varphi(x,t)dx\, dt$ for all $\varphi \in \mathcal{V}_o$;

(ii) $u_k(x,0) = 0$ in $\Omega$.

The function $v \in \mathcal{W}^n$ is said to be a *lower solution* of the system (4.5.1) if

(i) $\langle \frac{\partial v_k}{\partial t}, \varphi\rangle + \ell_k(v_k,\varphi) \le \int_Q f_k(x,t,v(x,t),v(x,t))\varphi(x,t)dx\, dt$, for all $\varphi \in \mathcal{V}_o \cap L_+^2(Q)$;

(ii) $v_k(x,0) \le 0$ for all $x \in \Omega$ and $v_k \le 0$ on $\Gamma$,

and an *upper solution* if the reversed inequalities hold.

The following hypotheses are imposed on the functions $f_k\colon Q \times \mathbb{R}^{2n} \to \mathbb{R}$.

(f0) The system (4.5.1) has a lower solution $\underline{u} = (\underline{u}_1,\dots,\underline{u}_n)$ and an upper solution $\bar{u} = (\bar{u}_1,\dots,\bar{u}_n)$ such that $\underline{u} \le \bar{u}$.

(f1) Each $f_k(\cdot,\cdot,u(\cdot,\cdot),v(\cdot,\cdot))$ is measurable when $u$, $v \in [\underline{u},\bar{u}]$, and there exist $p_k \in L_+^2(Q)$ such that for all $k = 1,\dots,n$,

$$|f_k(x,t,s,\underline{u}(x,t))| + |f_k(x,t,s,\bar{u}(x,t))| \le p_k(x,t)$$

for a.a. $(x,t) \in Q$ and for all $s \in [\underline{u}(x,t),\bar{u}(x,t)]$.

(f2) Each $f_k(x,t,\cdot,s)$, and $f_k(x,t,s,\cdot)$ is quasimonotone nondecreasing in $[\underline{u}(x,t),\bar{u}(x,t)]$ for a.a. $(x,t) \in Q$ and for all $s \in [\underline{u}(x,t),\bar{u}(x,t)]$.

(f3) The functions $s_k \mapsto f_k(x,t,r,s)$, $k = 1,\dots,n$ are nondecreasing in $[\underline{u}_k(x,t),\bar{u}_k(x,t)]$ for a.a. $(x,t) \in Q$ and for all $r$, $s \in [\underline{u}(x,t),\bar{u}(x,t)]$.

(f4) The functions $r_k \mapsto f_k(x,t,r,s)$, $k = 1,\dots,n$ are continuous in $[\underline{u}_k(x), \bar{u}_k(x)]$ for a.a. $(x,t) \in Q$ and for all $r$, $s \in [\underline{u}(x,t), \bar{u}(x,t)]$.

We shall show that under hypotheses (f0)–(f4) the system (4.5.1) has the extremal solutions in the order interval $[\underline{u}, \bar{u}]$. In the proof we shall apply theorem 1.1.1 and the following lemma.

**Lemma 4.5.1:** *Let (f0)–(f4) hold and let $P$ be the set of all lower solutions of (4.5.1) in $[\underline{u}, \bar{u}]$, given by (f0). Then for each $w = (w_1,\dots,w_n) \in P$, the system*

$$\frac{\partial u_k}{\partial t} - L_k u_k = f_k(x,t,(w_1,\dots,w_{k-1},u_k,w_{k+1},\dots,w_n),w)$$
$$\text{in } Q,\ u_k = 0 \quad \text{on } \Gamma,\ u_k(x,0) = 0 \quad \text{in } \Omega,\ k = 1,\dots,n \tag{4.5.2}$$

*has the minimal solution $u = Fw$ in $[w, \bar{u}]$. Moreover, $Fw \in P$, and there is $c > 0$ such that $\|(Fw)_k\|_{\mathcal{W}_o} \le c$ for all $w \in P$ and $k = 1,\dots,n$.*

*Proof.* Let $w = (w_1,\dots,w_n) \in P$ be given. For each fixed $k = 1,\dots,n$, $w_k$ is a lower solution of (4.5.2) and $\bar{u}_k$ is its upper solution. The hypotheses (f1)–(f4) ensure then by the dual of lemma 4.4.7 that (4.5.2) has for each fixed $k = 1,\dots,n$ the minimal solution $u_k$ in $[w_k, \bar{u}_k]$. Obviously, $u = (u_1,\dots,u_n)$ is the minimal solution of the system (4.5.2) in $[w, \bar{u}]$. Since $w \le u$, the definition of the solution and conditions (f2)–(f4) imply that

$$\begin{aligned}
&\langle \frac{\partial u_k}{\partial t}, \varphi\rangle + \ell_k(u_k,\varphi) \\
&= \int_Q f_k(x,(w_1,\dots,w_{k-1},u_k,w_{k+1},\dots,w_n),w)\varphi\, dxdt \\
&\le \int_Q f_k(x,(u_1,\dots,u_{k-1},u_k,u_{k+1},\dots,u_n),u)\varphi\, dxdt
\end{aligned}$$

for all $\varphi \in \mathcal{V}_o \cap L^2_+(Q)$ and $k = 1,\dots,n$. Denoting $u = Fw$ we then have $Fw \in P$. The last assertion of the lemma follows from lemma 4.4.4. □

The main result of this subsection is

**Theorem 4.5.1:** *If the hypotheses (f0)–(f4) are satisfied, then the system (4.5.1) has the extremal solutions in the order interval* $[\underline{u}, \bar{u}]$ *of* $\mathcal{W}^n$.

*Proof.* We shall first show that the mapping $F$ defined in lemma 4.5.1 satisfies the hypotheses of theorem 1.1.1. Let $C$ be the w.o. chain of $F$-iterations of $\underline{u}$. Since also $F[C]$ is a well-ordered chain, the $k$:th components $(Fw)_k$, $w \in C$ form a well-ordered and, by lemma 4.5.1, norm bounded chain in $\mathcal{W}_o$. By lemma 4.4.1 there is for each $k = 1, \ldots, n$ a nondecreasing sequence $(v(k)^j)_{j=o}^{\infty}$ in $F[C]$ such that the sequence $(v(k)_k^j)_{j=o}^{\infty}$ of $k$:th components of $v(k)^j$'s converges strongly in $L^2(Q)$ and weakly in $\mathcal{W}_o$ to $u_k = \sup\{(Fw)_k \mid w \in C\}$. Obviously, $u_* = (u_1, \ldots, u_n)$ is the supremum of $F[C]$ in $[\underline{u}, \bar{u}]$. Each $v(k)^j$ is by lemma 4.5.1 a lower solution of (4.5.1), and $v(k)^j \leq u_*$, whence the definition of a lower solution and conditions (f1)–(f3) imply that

$$\begin{aligned}\langle \frac{\partial v(k)_k^j}{\partial t}, \varphi\rangle + \ell_k(v(k)_k^j, \varphi) &\leq \int_Q f_k(x, t, v(k)^j, v(k)^j)\varphi\, dxdt \\ &\leq \int_Q f_k(x, v(k)^j, u_*)\varphi\, dxdt\end{aligned}$$

for each $\varphi \in \mathcal{V}_o \cap L_+^2(Q)$. As in the proof of theorem 4.4.1 it can be shown that $(v(k)_k^j)_{j=0}^{\infty}$ converges even strongly in $\mathcal{V}_o$ to $u_k$. Hence, it follows from the above inequality by (f3) and (f4), as $j \to \infty$, that

$$\langle \frac{\partial u_k}{\partial t}, \varphi\rangle + \ell_k(u_k, \varphi) \leq \int_Q f_k(x, u_*, u_*)\varphi\, dxdt$$

for all $\varphi \in \mathcal{V}_o \cap L_+^2(Q)$ and $k = 1, \ldots, n$. Because $v(k)_k^j(x, 0) = 0$ in $\Omega$ and $v(k)_k^j \rightharpoonup u_k$ in $\mathcal{W}_o$, then $u_k(x, 0) = 0$ in $\Omega$. Thus $u_*$

is by definition a lower solution of (4.5.1). Since $u_* \in [\underline{u}, \bar{u}]$, then $u_* = \sup F[C] \in P$. By definition, $Fu_* \in [u_*, \bar{u}]$ so that $\underline{u} \leq u_* \leq Fu_*$. Thus all the hypotheses of theorem 1.1.1 hold, whence $u_*$ is a fixed point of $F$. This and the definition of $F$ imply that $u_*$ is a solution of (4.5.1) in $[\underline{u}, \bar{u}]$.

If $u$ is any upper solution of (4.5.1) in $[\underline{u}, \bar{u}]$, then $\underline{u} \leq u$, which allows us to replace $\bar{u}$ by $u$ in the above reasonings, so that $u_* \leq u$. Thus $u_*$ is the minimal upper solution, and hence the minimal solution of (4.5.1) in $[\underline{u}, \bar{u}]$.

The assertions concerning the existence of the maximal solution $u^*$ can be proved by dual argumentation, replacing lower solutions by upper solutions and vice versa and applying proposition 1.2.1. □

### 4.5.2. Existence of extremal coupled quasisolutions

We say that the functions $v$, $w \in \mathcal{W}_o^n$ are *coupled quasisolutions* of the system (4.5.1) if

(i) $v_k(x,0) = w_k(x,0) = 0$ in $\Omega$;

(ii) $\langle \frac{\partial v_k}{\partial t}, \varphi\rangle + \ell_k(v_k, \varphi) = \int_Q f_k(x,t,v(x,t),w(x,t))\varphi(x,t)dx\,dt$ for all $\varphi \in \mathcal{V}_o$;

(iii) $\langle \frac{\partial w_k}{\partial t}, \varphi\rangle + \ell_k(w_k, \varphi) = \int_Q f_k(x,t,w(x,t),v(x,t))\varphi(x,t)dx\,dt$ for all $\varphi \in \mathcal{V}_o$.

The functions $v$, $w \in \mathcal{W}^n$ are said to be *coupled lower and upper quasisolutions* of the system (4.5.1), respectively, if

(i) $v_k(x,0) \leq 0 \leq w_k(x,0)$ for all $x \in \Omega$ and $v_k \leq 0 \leq w_k$ on $\Gamma$.

(ii) $\langle \frac{\partial v_k}{\partial t}, \varphi\rangle + \ell_k(v_k, \varphi) \leq \int_Q f_k(x,t,v(x,t),w(x,t))\varphi(x,t)dxdt$ for all $\varphi \in \mathcal{V}_o \cap L_+^2(Q)$;

(iii) $\langle \frac{\partial w_k}{\partial t}, \varphi\rangle + \ell_k(w_k, \varphi) \geq \int_Q f_k(x,t,w(x,t),v(x,t))\,\varphi(x,t)dxdt$ for all $\varphi \in \mathcal{V}_o \cap L_+^2(Q)$.

We shall impose the following hypotheses on the functions $f_k \colon Q \times \mathbb{R}^{2n} \to \mathbb{R}$.

(H1) The system (4.5.1) has coupled lower and upper quasisolutions $\underline{v}$, $\bar{w}$ satisfying $\underline{v} \leq \bar{w}$.

(H2) $f_k(\cdot,\cdot,u(\cdot,\cdot),v(\cdot,\cdot))$ is measurable whenever $u$, $v \in [\underline{v},\bar{w}]$, and there exist $h_k \in L^2_+(Q)$ such that $|f_k(x,t,r,s)| \leq h_k(x,t)$ for a.a. $(x,t) \in Q$ and for all $r$, $s \in [\underline{v}(x,t),\bar{w}(x,t)]$.

(H3) For each $k = 1,\ldots,n$ $f_k(x,t,r,s)$ is nondecreasing in $r$ and nonincreasing in $s$ on $[\underline{v}(x,t),\bar{w}(x,t)]$ for a.a. $(x,t) \in Q$.

**Theorem 4.5.2:** *Under the hypotheses (H1)–(H3) the system (4.5.1) possesses the extremal coupled quasisolutions in the order interval $[\underline{v},\bar{w}]$ of $\mathcal{W}^n$.*

*Proof.* The hypothesis (H2) implies that for all $v$, $w \in [\underline{v},\bar{w}]$ the functions

$$F_k(v,w)(x,t) = f_k(x,t,v(x,t),w(x,t))$$

are well-defined and belong to $L^2(Q)$. Given any $v$, $w \in [\underline{v},\bar{w}]$, consider the following system of parabolic equations.

$$\begin{aligned} &\frac{\partial u_k}{\partial t} - L_k u_k = F_k(v,w) \text{ in } Q, \\ &u_k(x,0) = 0 \text{ in } \Omega, \quad u_k = 0 \text{ in } \Gamma,\ k = 1,\ldots,n. \end{aligned} \tag{4.5.3}$$

One readily verifies that $\bar{w}_k$ and $\underline{v}_k$ are upper and lower solutions, respectively, to the IBVP (4.5.3) satisfying $\underline{v}_k \leq \bar{w}_k$ by (H1). Thus the system (4.5.3) has a unique solution $u \in W^n_o$ with $\underline{v} \leq u \leq \bar{w}$.

Define an operator $A\colon [\underline{v},\bar{w}] \times [\underline{v},\bar{w}] \to [\underline{v},\bar{w}]$ by

$$A(v,w) = u, \tag{4.5.4}$$

where $u \in \mathcal{W}^n_o$ is the unique solution of the system (4.5.3) in $[\underline{v},\bar{w}]$. By means of (H3) and differential inequality techniques (cf. Carl (1989)) it can be shown that the operator $A$ is mixed monotone. Finally, from the weak formulation of (4.5.3) we get the following estimate for $u = A(v, w)$:

$$\|u\|_{\mathcal{W}_o^n} \leq c \sum_{k=1}^{n} \|F_k(v, w)\|_{L^2(Q)}. \tag{4.5.5}$$

For any $v$, $w \in [\underline{v}, \bar{w}]$ the solution of (4.5.3) belongs to $[\underline{v}, \bar{w}]$, and the right-hand side of (4.5.5) is uniformly bounded. This implies a uniform $\mathcal{W}_o^n$-bound for the solution of (4.5.3) for all $v$, $w \in [\underline{v}, \bar{w}]$, i.e. there exists $b > 0$ such that

$$\|A(v, w)\|_{\mathcal{W}_o^n} \leq b \quad \text{for all } v,\ w \in [\underline{v}, \bar{w}].$$

Noticing also that $\mathcal{W}_o^n$ is a reflexive Banach space, it follows that $A$ satisfies the hypotheses of proposition 1.2.4. This ensures that $A$ has the extremal coupled fixed points, i.e. there exist $u_*, u^* \in [\underline{v}, \bar{w}]$ such that $u_* \leq u^*$, $u_* = A(u_*, u^*)$ and $u^* = A(u^*, u_*)$, and if $v$, $w$ are any coupled fixed points of $A$ in $[\underline{v}, \bar{w}] \times [\underline{v}, \bar{w}]$ with $v \leq w$ then $u_* \leq v \leq w \leq u^*$. This and the definition of $A$ imply that $u_*$, $u^*$ are the extremal coupled quasisolutions of the system (4.5.1). □

## 4.6. NOTES AND COMMENTS

In chapter 4 existence of extremal solutions of discontinuous elliptic and parabolic equations and systems are studied under various boundary and initial conditions. Section 4.1 is taken from Carl and Heikkilä (1994b). The preliminary results of subsection 4.1.2 extend and improve numerous recent results (see e.g., Akô (1961), Carl (1992a), Dancer and Sweers (1989), Deuel and Hess (1974/75), Diaz (1985), Frank and Wendt (1984), Keady and Kloeden (1987), Kura (1989), Mawhin and Schmitt (1984), and Schmitt (1978)). Theorem 4.1.1 is an extension of theorem 2.1 of Carl and Heikkilä (1994b) to nonautonomous case (see also Carl and Heikkilä (1994a)).

The results of section 4.2 are adapted from Carl and Heikkilä (1992a) and extended for nonautonomous equations. Example in subsection 4.2.3 is a discontinuous version of those given in Ambrosetti and Turner (1988) (see also Carl (1992b)). For special cases and related results see Ambrosetti and Badiale (1989), Carl (1988), Heikkilä (1990c), Keady and Kloeden (1987), Sattinger (1972), Stuart (1976), and Stuart (1978). Section 4.3 is extended and modified from Carl, Heikkilä and Kumpulainen (1993), and Carl and Heikkilä (1993). It also serves an extension to the results of Carl (1988a), Carl and Grossmann (1990), Khavanin and Lakshmikantham (1986), Korman and Leung (1986), and Ladde, Lakshmikantham and Vatsala (1984). Similar results are derived in Mitidieri and Sweers (1993).

Section 4.4. is taken from Carl and Heikkilä (1994c). The results of subsection 4.4.2 are based on Carl (1994), and they generalize those of Bebernes and Schmitt (1979), Deuel (1976), Lions (1965), and Sattinger (1972). The main result given in subsection 4.4.3 serves a generalization to Carl (1989), and Carl and Heikkilä (1990). Section 4.5 is based on Carl and Heikkilä (1993), and it extends corresponding results of Ladde, Lakshmikantham and Vatsala (1985) to the case when the nonlinear right hand sides involve discontinuities.

As for applications of boundary value problems with discontinuous nonlinearities see, e.g., Aris (1975), Davis and Fleishman (1986), Fleishman and Mahar (1981), and Korman and Leung (1986). Elliptic differential inclusions with discontinuous nonlinearities are studied in Carl and Heikkilä (1992).

# 5

# Differential Equations in Banach Spaces

## 5.0. INTRODUCTION

The main purpose of this chapter is to study discontinuous differential equations in ordered Banach spaces by using comparison methods, the method of upper and lower solutions and iteration methods.

We shall begin section 5.1 with existence, uniqueness and estimation results for first order differential equations with Carathéodory type nonlinearities in Banach spaces. In the rest of section 5.1 we shall derive existence and comparison results for extremal solutions of a differential equation in an ordered Banach space whose order cone has a nonempty interior. The dependent variable of the considered equation is decomposed to two parts.

Section 5.2 is devoted to the study of existence and dependence on the data of strong extremal solutions to first order semilinear initial value problems with discontinuous nonlinearities in ordered Banach spaces.

In section 5.3 we shall first apply results of section 5.1 to obtain existence, uniqueness and estimation results for strong solutions to IVP's of higher order semilinear equations with Carathéodory type nonlinear terms in Banach spaces. The rest of section 5.3 is devoted to the study of higher order semilinear differential equations, with discontinuities in the nonlinear lower order terms, in ordered Banach spaces by applying results of section 5.2.

In section 5.4 we shall derive existence and comparison results for first and second order discontinuous periodic boundary value problems in ordered Banach spaces. The cases when discontinuity in the dependent variable is of mixed monotone type are studied in section 5.5. Sufficient conditions for the existence of extremal coupled quasisolutions to the first and second order IVP's and PBVP's are derived.

Considerations of sections 5.1 and 5.2 are extended in section 5.6 to mild solutions of a first order semilinear IVP when the evolution term in the differential equation is a generator of a strongly continuous family of bounded linear operators. Applications to first and second order partial differential equations are also given.

In section 5.7 we shall study second order semilinear IVP's in Banach spaces when the linear operator in the equation is the infinitesimal generator of a strongly continuous cosine family. Existence, uniqueness and estimation results are first derived for the weak solution of the problem when the nonlinear term in the equation is of Carathéodory type. In the case when the equation contains discontinuities in the dependent variable we shall derive existence and comparison results for extremal mild solutions of the problem, assuming that the underlying Banach space is ordered. Results are applied to a second order partial differential equation of hyperbolic type.

In the study of the above mentioned discontinuous differential equations in ordered Banach spaces we shall usually require that the underlying space $E$ is ordered by a regular or fully regular order cone. If $E$ is reflexive, it suffices in some cases to assume that $E$ is ordered by any closed cone, or via continuous embedding in an ordered Banach space with fully regular order cone. In section 5.8 we shall present examples of such Banach spaces and introduce conditions which ensure above mentioned order properties for $E$. In some proofs it also suffices to assume that pointwise norm bounded or order bounded and equicontinuous chains of functions from a topological space $X$ into $E$, equipped with pointwise ordering, possess supremums and infimums. Sufficient conditions for this are introduced at the remaining part of section 5.8.

In addition to discontinuity, another characteristic feature of the obtained results is that no compactness assumptions are imposed on the nonlinearities of the considered differential equations.

## 5.1. EXISTENCE, UNIQUENESS & EXTREMALITY RESULTS

We shall begin with the study of existence, uniqueness and estimation of solutions to first order differential equations with Carathéodory type nonlinearities in Banach spaces. In subsections 5.1.3–5.1.4 we shall derive existence and comparison results for extremal solutions of a differential equation in an ordered Banach space whose order cone has a nonempty interior. We shall assume that the dependent variable in the equation is decomposed into two parts, one of them satisfying a generalized Lipschitz condition. Counter-examples are presented to illustrate the need of used hypotheses.

### 5.1.1. An existence and uniqueness result

Given a Banach space $E$ and $J = [0, T]$, $T > 0$, consider the IVP

$$x' = f(t, x, x), \qquad x(0) = x_o. \tag{5.1.1}$$

We are going to prove that (5.1.1) has a uniquely determined solution if $f: J \times E \times E \to E$ has the following properties.

(f0) $f(\cdot, x, y)$ is strongly measurable for all $x, y \in E$, and $f(\cdot, 0, 0) \in L^1(J, E)$.

(f1) There is $p \in L^1(J, I\!R_+)$ such that
$$\|f(t,x,z) - f(t,y,z)\| \le p(t)\|x - y\|$$
for all $x$, $y$, $z \in E$ and for a.a. $t \in J$,

(f2) There is $r > 0$ such that
$$\|f(t,x,y) - f(t,x,z)\| \le q(t, \|y - z\|)$$
for all $x$, $y$, $z \in E$ with $\|y - z\| < r$ and for a.a. $t \in J$, where $q: J \times I\!R_+ \to I\!R_+$ is a Carathéodory function, $q(t, \cdot)$ is nondecreasing for a.a. $t \in J$, the IVP

$$u' = p(t)u + q(t,u), \qquad u(0) = u_o, \tag{5.1.2}$$

has for some $u_o > 0$ an upper solution on $J$, and the zero-function is the only solution of (5.1.2) on $J$ when $u_o = 0$.

**Theorem 5.1.1:** *If the hypotheses (f0)–(f2) hold, then for each $x_o \in E$ the IVP (5.1.1) has a unique solution $x$ on $J$. Moreover, $x$ is the uniform limit of the sequence $(y_n)_{n=o}^{\infty}$ of the successive approximations*

$$y_{n+1}(t) = x_o + \int_o^t f(s, y_n(s), y_n(s))\, ds, \quad t \in J,\ n \in I\!N, \tag{5.1.3}$$

*for each choice of $y_o \in C(J, E)$.*

*Proof.* The hypotheses given for $q$ in (f2) imply by proposition 2.1.9 that there is $r_o > 0$ such that the IVP (5.1.2) has the minimal solution $b$ when $u_o = r_o$ and $r_o \le b(t) < r$ for each $t \in J$. Since $q$ is a Carathéodory function and $q(\cdot, w(\cdot))$ is for each $w \in [0, b]$ a.e. bounded above by $b'$, then the equation

$$Qw(t) = \int_o^t [q(s, w(s)) + p(s)w(s)]\, ds, \quad t \in J \tag{a}$$

defines a mapping $Q\colon [0,b] \to C(J, I\!R_+)$. Condition (f2) ensures that $Q$ is nondecreasing, and the choice of $r_o$ and $b$ that

$$r_o + Qb = b. \tag{b}$$

Thus $Q[0,b] \subset [0,b]$. Moreover, the sequence $(Q^n b)_{n=o}^{\infty}$ converges by corollary 2.1.4 uniformly on $J$ to the maximal solution $v$ of (5.1.2) with $u_o = 0$, whence $v(t) \equiv 0$ by (f2). Since $u(t) \equiv 0$ satisfies by (f2) the equation $u'(t) = q(t, u(t))$ for a.a. $t \in J$, then $q(t,0) = 0$ for a.a. $t \in J$. This and conditions (f0)–(f2) imply that also $f$ is a Carathéodory function. Then $f(\cdot, y(\cdot), y(\cdot))$ is by theorem 1.4.3 strongly measurable on $J$ for all $y \in C(J,E)$. Moreover, by choosing

$$y_i = \frac{i}{m} y, \qquad i = 0, \ldots, m \geq \frac{\|y\|_o}{r_o},$$

we have $\|y_i(t) - y_{i-1}(t)\| \leq r_o \leq b(t)$ on $J$ for each $i = 1, \ldots, m$, whence

$$\begin{aligned}
&\|f(t, y(t), y(t)) - f(t, 0, 0)\| \\
&\leq \sum_{i=1}^{m} \|f(t, y_i(t), y_i(t)) - f(t, y_{i-1}(t), y_{i-1}(t))\| \\
&\leq \sum_{i=1}^{m} [p(t)\|y_i(t) - y_{i-1}(t)\| + q(t, \|y_i(t) - y_{i-1}(t)\|)] \\
&\leq \sum_{i=1}^{m} [p(t)b(t) + q(t, b(t))] = m\, b'(t)
\end{aligned}$$

for a.a. $t \in J$. This and condition (f0) imply that $f(\cdot, y(\cdot), y(\cdot))$ is Bochner integrable on $J$.

Given $x_o \in E$, the equation

$$Fy(t) = x_o + \int_o^t f(s, y(s), y(s))ds, \quad t \in J, \tag{c}$$

defines a mapping $F: C(J,E) \to C(J,E)$. Denoting $|y| = \|y(\cdot)\|$, $y \in C(J,E)$, and using (f1), (f2), (a) and (c) it is easy to show that

$$|Fy - F\bar{y}| \leq Q|y - \bar{y}| \text{ for } y, \bar{y} \in C(J,E), \ |y - \bar{y}| \leq b. \quad (5.1.4)$$

The above reasoning shows that the operators $F$ and $Q$ satisfy the hypotheses of theorem 1.4.9. Thus the iteration sequence $(F^n y_o)_{n=o}^{\infty}$, which equals to the sequence $(y_n)_{n=o}^{\infty}$ defined by (5.1.3), converges uniformly in $J$ to a unique fixed point $x$ of $F$. This and the definition of $F$ imply by lemma 1.5.5 that $x$ is the uniquely determined solution of the IVP (5.1.1). □

### 5.1.2. Dependence on the initial value

We shall now show that the dependence of the solution $x$ of (5.1.1) on the initial point $x_o$ can be estimated above by the minimal solution of the comparison problem (5.1.2), where $u_o$ is the distance between values of $x_o$.

**Theorem 5.1.2:** *Let $f: J \times E \times E \to E$ satisfy conditions (f0)–(f2). If $x = x(\cdot, x_o)$ denotes the solution of the IVP (5.1.1) and $u = u(\cdot, u_o)$ the minimal solution of the IVP (5.1.2), then for all $x_o$, $\hat{x}_o \in E$, with $\|x_o - \hat{x}_o\|$ small enough,*

$$\|x(t, x_o) - x(t, \hat{x}_o))\| \leq u(t, \|x_o - \hat{x}_o\|), \qquad t \in J. \qquad (5.1.5)$$

*In particular, $x(\cdot, x_o)$ depends continuously on $x_o$ in the sense that $x(t, \hat{x}_o) \to x(t, x_o)$ uniformly over $t \in J$ as $\hat{x}_o \to x_o$.*

*Proof.* Assume that $x_o$, $\hat{x}_o \in E$, and that $\|x_o - \hat{x}_o\| \leq r_o$, where $r_o$ is chosen as in the proof of theorem 5.1.1. The solutions $x = x(\cdot, x_o)$ and $\hat{x} = x(\cdot, \hat{x}_o)$ exist by theorem 5.1.1, and they are by lemma 1.5.5 the unique fixed points of the integral operators $F$, $\hat{F}: C(J,E) \to C(J,E)$, defined by

$$Fy(t) = x_o + \int_o^t f(s, y(s), y(s))\, ds, \quad t \in J,$$

and

$$\hat{F}y(t) = \hat{x}_o + \int_o^t f(s, y(s), y(s))\, ds, \quad t \in J.$$

Moreover, $F$ satisfies the hypotheses of theorem 1.4.9 with $Q$ defined by

$$Qw(t) = \int_o^t [q(s, w(s)) + p(s)w(s)]\, ds, \quad t \in J.$$

By choosing $y_o = \hat{x}$, and noticing that

$$F\hat{x}(t) - \hat{x}(t) = F\hat{x}(t) - \hat{F}\hat{x}(t) = x_o - \hat{x}_o \quad \text{for each} \quad t \in J,$$

it follows from (1.4.19) that

$$|x - \hat{x}| \le u,$$

where $u$ the minimal solution of

$$w = \|x_o - \hat{x}_o\| + Qw.$$

Lemma 1.5.5 implies that $u = u(\cdot, \|x_o - \hat{x}_o\|)$, whence (5.1.5) holds. From proposition 2.1.9 it follows that $u(\cdot, u_o)$ is nondecreasing with respect to $u_o \in [0, r_o]$ and $u(t, \frac{1}{n}) \to 0$ uniformly over $t \in J$ as $n \to \infty$, whence (5.1.5) implies that $x(t, \hat{x}_o) \to x(t, x_o)$ uniformly over $t \in J$ as $\hat{x}_o \to x_o$. □

**Remark 5.1.1:** If $r = \infty$ in condition (f2), and if the IVP (5.1.2) has an upper solution for all $u_o \geq 0$, then (5.1.5) holds for all $x_0$, $\hat{x}_o \in E$. Moreover, it follows from (1.4.19) when we take $y_o(t) \equiv x_o$ that

$$\|x(t,x_o) - x_o\| \leq u(t, \int_o^T \|f(s,x_o,x_o)\|)\,ds), \quad t \in J. \tag{5.1.6}$$

If $f$ is defined in $J \times V \times V$, where $V \subset E$, and if $x_o \in V$ and $d = d(x_o, E \setminus V) > 0$, then the solution $x(\cdot, x_o)$ of the IVP (5.1.1) exists on the interval $[0,c]$, where $c \in (0,T]$ is so chosen that $u(c) < d$, where $u$ is the minimal solution of (5.1.2) with $u_o = \int_o^c \|f(s,x_o,x_o)\|\,ds$ (cf. the proof of theorem 1.4.9). Such a positive number $c$ exists by proposition 2.1.9.

Next we shall give some special cases to theorems 5.1.1 and 5.1.2.

**Proposition 5.1.1:** *Let $f\colon J \times E \times E \to E$ satisfy (f0), (f1) and assume there is $r > 0$ such that for all $x$, $y$, $z \in E$ with $\|y - z\| < r$ and for a.a. $t \in J$,*

$$\|f(t,x,y) - f(t,x,z)\| \leq p_1(t)\,\varphi(\|y - z\|), \tag{5.1.7}$$

*where $\varphi \in C(\mathbb{R}_+, \mathbb{R}_+)$ is nondecreasing, $\int_o^r \frac{dv}{\varphi(v)} = \infty$, and $p_1 \in L^1(J, \mathbb{R}_+)$. Then the IVP (5.1.1) has for each $x_o \in E$ a unique solution, and it depends continuously on $x_o$.*

*Proof.* It suffices to show that condition (f2) holds when the function $q\colon J \times \mathbb{R}_+ \to \mathbb{R}_+$ is defined by

$$q(t,v) = p_1(t)\varphi(v), \quad t \in J,\ v \in \mathbb{R}_+. \tag{a}$$

Since $\varphi$ is nondecreasing and $\int_o^r \frac{dv}{\varphi(v)} = \infty$, it can be shown by elementary analysis that

$$\varphi(0) = 0 \ \text{ and } \ \int_o^r \frac{dv}{\varphi(v) + v} = \infty.$$

Thus there is $r_o > 0$ such that $\int_o^T (p_1(s) + p(s))\, ds < \int_{r_o}^r \frac{dv}{\varphi(v)+v}$. By using the separation of variables it then follows that the IVP

$$u' = (p_1(t) + p(t))\,(\varphi(u) + u), \qquad u(0) = u_o \tag{b}$$

has for each $u_o \in (0, r_o]$ a unique solution on $J$. This solution is an upper solution of

$$u' = p_1(t)\,\varphi(u) + p(t)\,u, \qquad u(0) = u_o. \tag{c}$$

To show that the zero function is the only solution of (c) on $J$ when $u_o = 0$, let $u$ be such a solution. Denote $\bar{t} = \sup\{t \in J \mid u(t) = 0\}$, and make a counter-hypothesis: $\bar{t} < T$. Since $u$ is absolutely continuous and nondecreasing, then $u(\bar{t}) = 0$ and $u(t) > 0$ when $\bar{t} < t < T$. Since

$$\frac{u'(t)}{\varphi(u(t)) + u(t)} \le p_1(t) + p(t) \quad \text{for a.a. } t \in (\bar{t}, T),$$

then

$$\int_o^{u(T)} \frac{dv}{\varphi(v) + v} = \int_{\bar{t}}^T \frac{u'(s)ds}{\varphi(u(s)) + u(s)} \le \int_{\bar{t}}^T (p_1(s) + p(s))\, ds.$$

But this contradicts the fact that $\int_o^r \frac{dv}{\varphi(v)+v} = \infty$. Thus the IVP (a) has the zero function as the only solution when $u_o = 0$.

As a direct consequence of the hypotheses given for $p_1$ and $\varphi$ we see that $q$, defined by (a), is a Carathéodory function, and that $q(t, \cdot)$ is nondecreasing for all $t \in J$. Thus (f2) holds. □

Each of the functions $\varphi_m$, $m \in \mathbb{N}$, defined by (2.1.18) satisfy the hypotheses given for $\varphi$ in proposition 5.1.1. In particular, when $m = 0$ we obtain

**Corollary 5.1.1:** *Let the hypotheses (f0) and (f1) hold. Assume there is $q \in L^1(J, I\!R_+)$ such that $f$ satisfies a generalized Lipschitz condition*

$$\|f(t,x,y) - f(t,x,z)\| \leq q(t)\|y - z\|$$

*for all $x$, $y$, $z \in E$ and for a.a. $t \in J$. Then the IVP (5.1.1) has for each $x_o \in E$ exactly one solution $x$. Moreover, $x$ depends continuously on $x_o$.*

### 5.1.3. Extremal solutions of IVP's in Banach spaces

In this subsection we shall study the existence of extremal solutions of the IVP (5.1.1) in the case when $f\colon J \times E \times E \to E$, where $J = [0, T]$ and $E$ is an ordered Banach space. We shall assume that the order cone $K$ of $E$ has a nonempty interior. Denote by $\partial K$ the boundary of $K$, and by $K'$ the dual cone of $K$, defined by

$$K' = \{c \in E' \mid cx \geq 0 \text{ for each } x \in K\}.$$

Given $V \subseteq E$ and $g\colon V \to E$, we say that $g$ is *quasimonotone nondecreasing* if for all $x$, $y \in V$ with $y - x \in \partial K$ there is $c \in K'$, $c \neq 0$, such that $cx = cy$ and $cg(x) \leq cg(y)$. In the case when $E = I\!R^m$ and $K = I\!R^m_+$ the above definition is equivalent to that given in section 2.4.

We shall prove the existence of extremal solutions of the IVP (5.1.1) when $K$ is regular and $f$ satisfies condition (f1) and the following conditions.

(f3) For each $y \in AC(J, E)$ the function $f(\cdot, x, y(\cdot))$ is strongly measurable when $x \in E$, and $f(\cdot, 0, y(\cdot)) \in L^1(J, E)$.

(f4) $f(t, \cdot, y)$ is quasimonotone nondecreasing and $f(t, y, \cdot)$ is nondecreasing for all $y \in E$ and for a.a. $t \in J$.

(f5) $\|f(t, 0, y)\| \leq p_2(t)h(\|y\|)$ for all $y \in E$ and for a.a. $t \in J$, where $p_2 \in L^1(J, I\!R_+)$, $h\colon I\!R_+ \to (0, \infty)$ is nondecreasing and $\int_o^\infty \frac{dv}{h(v)} = \infty$.

In the proof we shall apply the method of generalized monotone iterations and the following result which is proved in Redheffer and Walter (1986).

**Lemma 5.1.1:** *Given $g: J \times E \to E$ and $p \in L^1(J, \mathbb{R}_+)$ assume that*

$$\|g(t,x) - g(t,y)\| \leq p(t)\|x - y\|$$

*for all $x$, $y \in E$ and for a.a. $t \in J$, and that $g(t,\cdot)$ is quasimonotone nondecreasing for a.a. $t \in J$. If the order cone of $E$ has a nonempty interior, then $x$, $y \in AC(J,E)$, $x' - g(t,x) \leq y' - g(t,y)$ a.e. on $J$ and $x(0) \leq y(0)$ imply $x(t) \leq y(t)$ for each $t \in J$.*

Assume now that the order cone $K$ of $E$ is normal and has a nonempty interior. Let $e$ be a fixed vector from the interior of $K$. It is well-known (cf. Zeidler (1985)) that the equation

$$\|x\|_e = \inf\{a > 0 \mid -ae \leq x \leq ae\} \tag{5.1.8}$$

defines a norm $\|\cdot\|_e$ in $E$ which is equivalent to the original norm $\|\cdot\|$ of $E$, i.e. there exist positive numbers $\mu$, $\nu$ such that

$$\mu\,\|x\|_e \leq \|x\| \leq \nu\,\|x\|_e \text{ for each } x \in E.$$

It is easy to see that the hypotheses (f1), (f3), (f4) and (f5) hold when the norm $\|\cdot\|$ is replaced by $\|\cdot\|_e$ and the functions $p$ and $h$ by the functions $p_e$ and $h_e$, defined by

$$p_e(t) = \frac{\nu}{\mu}p(t) \text{ and } h_e(v) = \frac{1}{\mu}h(\nu v),\ t \in J,\ v \in \mathbb{R}_+.$$

Given $x_o \in E$, denote $q = p_e + p_2$, and let $w: J \to \mathbb{R}_+$ be the solution of the IVP

$$w' = q(t)(h_e(w) + w), \qquad w(0) = \|x_o\|_e. \tag{5.1.9}$$

Condition (f5) ensures by lemma 1.5.3 that $w$ exists and is uniquely determined. Choose

$$\alpha(t) = x_o + \|x_o\|_e e - w(t)e \text{ and } \beta(t) = x_o - \|x_o\|_e e + w(t)e, \tag{5.1.10}$$

and denote $[\alpha, \beta] = \{x \in AC(J, E) \mid \alpha \le x \le \beta\}$.

**Lemma 5.1.2:** *For each choice of $y \in [\alpha, \beta]$ the IVP*

$$x' = f(t, x, y(t)), \qquad x(0) = x_o \tag{5.1.11}$$

*has a unique solution $x$, and this solution belongs to $[\alpha, \beta]$.*

*Proof.* Let $y \in [\alpha, \beta]$ be given. From the definition of the norm $\|\cdot\|_e$ and from (5.1.10) it follows that $\|y(t)\|_e \le w(t)$ for each $t \in J$. The hypotheses (f1) and (f3) ensure that the equation

$$g(t, x) = f(t, x, y(t)), \qquad t \in J, \ x \in E \tag{a}$$

defines a Carathéodory function $g : J \times E \to E$. From (f1), (f5) and (5.1.9) it follows that for each $x \in [\alpha, \beta]$

$$\|g(t, x(t))\|_e \le p(t)(h_e(w(t)) + w(t)) = w'(t) \text{ for a.a. } t \in J. \tag{b}$$

In particular, $\|g(t, x_o)\|_e \le w'(t)$ for a.a. $t \in J$, whence $g(t, x_o)$ is Bochner integrable. Condition (f1) implies that

$$\|g(t, x) - g(t, \bar{x})\|_e \le p_e(t)\|x - \bar{x}\|_e$$

for all $x$, $y \in E$ and for a.a. $t \in J$. Thus the IVP (5.1.11) has by corollary 5.1.1 a unique solution $x$. Moreover, $x$ is obtained as the uniform limit of the iteration sequence $(F^n y_o)_{n=o}^{\infty}$, where $y_o(t) \equiv x_o$ and $F$ is defined by

$$Fz(t) = x_o + \int_o^t f(s, z(s), y(s))\, ds, \ t \in J. \tag{c}$$

If $z \in [\alpha, \beta]$, it follows from (a), (b) and (c) that

$$\|Fz(t) - x_o\|_e \leq \int_o^t \|g(s, z(s))\|_e \, ds$$
$$\leq \int_o^t w'(s) \, ds = w(t) - \|x_o\|_e, \ t \in J.$$

This implies by the definition of $\|\cdot\|_e$ that

$$(\|x_o\|_e - w(t))e \leq Fz(t) - x_o \leq (w(t) - \|x_o\|_e)e, \quad t \in J.$$

This and (5.1.10) ensure that $Fz \in [\alpha, \beta]$ for all $z \in [\alpha, \beta]$. Since $y_o \in [\alpha, \beta]$, then $x = \lim_n F^n y_o \in [\alpha, \beta]$, because $x$, as the solution of (5.1.11), is absolutely continuous. □

**Lemma 5.1.3:** *Let $x = Gy$ denote the solution of the IVP (5.1.11) for given $y \in [\alpha, \beta]$. If $y_1, y_2 \in [\alpha, \beta]$, and $y_1 \leq y_2$, then $Gy_1 \leq Gy_2$.*

*Proof.* Denote $x_i = Gy_i$, $i = 1, 2$, and $g(t, x) = f(t, x, y_2(t))$, $t \in J$, $x \in E$. Conditions (f1) and (f4) ensure that $g$ satisfies the hypotheses of lemma 5.1.1. Since $f(t, x, \cdot)$ is nondecreasing by (f4), then $x_1' - g(t, x_1) \leq 0 = x_2' - g(t, x_2)$ a.e. on $J$, whence lemma 5.1.1 implies that $x_1 \leq x_2$. □

By using the results of lemma 5.1.3 and theorem 1.4.7 we shall prove.

**Theorem 5.1.3:** *Let $E$ be an ordered Banach space whose order cone $K$ is regular and has a nonempty interior. If the function $f: J \times E \times E \to E$ satisfies conditions (f1), (f3), (f4) and (f5), then for each $x_o \in E$ the IVP (5.1.1) has the extremal solutions.*

*Proof.* Given $x_o \in E$ and $e \in \text{int } K$, let $w$ be the solution of the IVP (5.1.9) on $J$. Choosing $\alpha$, $\beta$ by (5.1.10), let $G: [\alpha, \beta] \to [\alpha, \beta]$ be the operator which assigns to each $y \in [\alpha, \beta]$ the solution of the IVP (5.1.11). Lemma 5.1.3 implies that $G$ is nondecreasing. Moreover, $Gy$ is for each $y \in [\alpha, \beta]$ the only function in $[\alpha, \beta]$ which satisfies the integral equation

$$Gy(t) = x_o + \int_o^t f(s, Gy(s), y(s))\, ds. \tag{a}$$

From (f1), (f5), (5.1.9) and (5.1.10) it follows that

$$\|f(t, Gy(t), y(t))\|_e \le p_e(t)\, w(t) + p_2(t) h_e(w(t)) \le w'(t)$$

for a.a. $t \in J$ and for all $y \in [\alpha, \beta]$, where $w$ is the solution of the IVP (5.1.9). This and (a) imply that

$$\|Gy(t) - Gy(s)\|_e \le |w(t) - w(s)| \tag{b}$$

whenever $y \in [\alpha, \beta]$ and $s,\ t \in J$. Thus the hypotheses of theorem 1.4.7 are valid, whence $G$ has the least fixed point $x_*$ and the greatest fixed point $x^*$. Applying the definition of $G$ we see that $x_*$ is the minimal and $x^*$ is the maximal solution of the IVP (5.1.1) in $[\alpha, \beta]$.

If $x$ is any solution of (5.1.1), then it satisfies also the integral equation

$$x(t) = x_o + \int_o^t f(s, x(s), x(s))\, ds, \quad t \in J. \tag{c}$$

Thus

$$\begin{aligned}\|x(t)\|_e &\le \|x_o\|_e + \int_o^t \|f(s, x(s), x(s))\|_e\, ds \\ &\le \|x_o\|_e + \int_o^t p(s)(h_e(\|x(s)\|_e) + \|x(s)\|_e)\, ds\end{aligned}$$

for all $t \in J$. This implies by lemma 1.5.3 that $\|x(t)\|_e \le w(t)$ for all $t \in J$, where $w$ is the solution of the IVP (5.1.9). From (f5), (5.1.9) and (c) it then follows that

$$\begin{aligned}\|x(t) - x_o\|_e &\le \int_o^t p_e(s)\|x(s)\|_e + p_2(s)\, h_e(\|x(s)\|_e)\, ds \\ &\le \int_o^t p(s)(h_e(w(s)) + w(s))\, ds = w(t) - \|x_o\|_e\end{aligned}$$

for each $t \in J$. In particular, $x \in [\alpha, \beta]$. Because $x_*$ and $x^*$ are the minimal and the maximal solutions of (5.1.1) in $[\alpha, \beta]$, then $x \in [x_*, x^*]$. Since $x$ was an arbitrary solution of (5.1.1), then $x_*$ and $x^*$ are the extremal solutions of the IVP (5.1.1). □

As a consequence of theorem 5.1.3 we obtain

**Corollary 5.1.2:** *Given $f_i \colon J \times E \to E$, $i = 1, 2$, assume there is a null set $Z$ in $J$ such that*

(i) *$f_1(\cdot, x)$ is strongly measurable for each $x \in E$ and $f_1(\cdot, 0)$ is Bochner integrable.*
(ii) *$f_2(\cdot, y(\cdot))$ is strongly measurable for all $y \in AC(J, E)$.*
(iii) *There is $p_1 \in L^1(J, I\!R_+)$ such that*
$$\|f_1(t, x) - f_1(t, \bar{x})\| \le p_1(t)\|x - \bar{x}\|$$
*whenever $t \in J \setminus Z$ and $x, \bar{x} \in E$.*
(iv) *$f_1(t, \cdot)$ is quasimonotone nondecreasing and $f_2(t, \cdot)$ is nondecreasing for all $t \in J \setminus Z$.*
(v) *$\|f_2(t, y)\| \le p_2(t) h(\|y\|)$ for all $t \in J \setminus Z$ and $y \in E$, where $p_2 \in L^1(J, I\!R_+)$, $h \colon I\!R_+ \to (0, \infty)$ is nondecreasing and $\int_o^\infty \frac{dv}{h(v)} = \infty$.*

*If the order cone of E is regular and has a nonempty interior, then the IVP*

$$x' = f_1(t, x) + f_2(t, x), \qquad x(0) = x_o$$

*has for each $x_o \in E$ the extremal solutions.*

*Proof.* Conditions (i)–(v) imply that $f: J \times E \times E \to E$, defined by

$$f(t,x,y) = f_1(t,x) + f_2(t,y), \quad t \in J, \ x, \ y \in E,$$

has properties (f1), (f3), (f4) and (f5), whence the assertion follows from theorem 5.1.3. □

**Remark 5.1.2:** Replacing condition (f5) by the existence of a lower solution $\alpha$ and an upper solution $\beta$ of (5.1.1) we obtain existence results for extremal solutions of (5.1.1) in the order interval $[\alpha, \beta]$. Moreover, if $f$ is continuous, then the Lipschitz condition (f1) is not needed (cf. Chaljub-Simon et al. (1992)).

Given a Banach space $E$ and a closed and convex subset $B$ of $E$ which does not contain the zero-vector, and which has a nonempty interior, define

$$K = \{\lambda y \mid \lambda \geq 0, \ y \in B\}.$$

It is easy to show that $K$ is a closed positive cone with a nonempty interior. In particular, if $E$ is finite-dimensional, then each order cone of $E$ generated by any basis of $E$ is regular and has a nonempty interior. As for other examples of closed cones in Banach spaces with a nonempty interior, see section 5.8.

If $f$ is defined in $J \times V \times V$ where $V \subset E$, if $x_o \in V$, and if the hypotheses of theorem 5.1.3 hold, then the extremal solutions of the IVP (5.1.1) exist on the interval $[0, c]$, where $c \in (0, T]$ is so chosen that $\alpha(t)$, $\beta(t) \in V$ for each $t \in [0, c]$, where $\alpha$, $\beta$ are defined by (5.1.10). This holds if $w(c) - \|x_o\|_e < \inf\{\|y - x_o\|_e \mid y \notin V\}$, where $w$ is the solution of the IVP (5.1.9).

The following example shows that the above derived extremality results do not hold in general if the hypothesis: $K$ is regular is dropped.

**Example 5.1.1:** (cf. Chaljub-Simon et al. (1992)) Choose $\Omega = \{\frac{1}{i} \mid i = 1, 2, \ldots\} \cup \{0\}$, equipped with the metric $d(r,s) = |r - s|$, $r$, $s \in \Omega$. $E = C(\Omega, \mathbb{R})$ is a Banach space with respect to the supremum norm $\|x\| = \sup\{|x(s)| \mid s \in \Omega\}$, and $K = C(\Omega, \mathbb{R}_+) = \{x \in E \mid x(s) \geq 0 \text{ for each } s \in \Omega\}$ is a closed positive cone in $E$ with nonempty interior. The partial ordering defined by $K$ in $E$ is the pointwise ordering. Denote in the following $x(s) = x_s$ when $x \in E$ and $s \in \Omega$. Given $T \in (0,2]$, denote $J = [0,T]$ and define a mapping $g\colon E \to E$ by $g(y)_s = \varphi(y_s)$, $s \in \Omega$, where

$$\varphi(\xi) = \begin{cases} 2, & \xi \geq 4, \\ \sqrt{\xi}, & 0 \leq \xi \leq 4, \\ 0, & \xi < 0. \end{cases} \tag{5.1.12}$$

It is easy to see that $g$ is bounded, continuous and nondecreasing. Define $a \in E$ by

$$a_s = \begin{cases} 0, & s = 0, \\ \frac{(-1)^i}{i}, & s = \frac{1}{i},\ i = 1, 2, \ldots. \end{cases}$$

We shall show that the IVP

$$x' = g(x), \qquad x(0) = a, \tag{a}$$

does not possess any solutions. To see this, assume that $x\colon J \to E$ satisfies (a). The above definitions imply that

$$x'(t)_s = g(x(t))_s = \varphi(x(t)_s), \text{ for a.a. } t \in J, \quad x(0)_s = a_s$$

for each $s \in \Omega$. Easy calculations show that

$$x(t)_s = \begin{cases} 0, & \text{when } s = \frac{1}{2k+1},\ k = 1, 2, \ldots, \\ & \quad \text{or } s = 0, \\ \left(\frac{t}{2} + \frac{1}{\sqrt{2k}}\right)^2, & \text{when } s = \frac{1}{2k},\ k = 1, 2, \ldots. \end{cases}$$

But this implies that $s \mapsto x(t)_s$ is not continuous at 0, so that $x(t) \notin E$ when $t \in (0,T]$, a contradiction.

It is easy to see that if $g$ is as above, $f(t,x,y) = f_2(t,y) = g(y)$ and $f_1(t,x) \equiv 0$, then $f$ satisfies conditions (f1), (f3), (f4) and (f5), while conditions (i)–(v) hold for $f_1$, $f_2$, when $Z = \emptyset$, but the conclusions of theorem 5.1.3 and corollary 5.1.2 don't hold.

As shown in Chaljub-Simon et al. (1992), the above type of counter-example can be obtained in any space $E = C(\Omega, I\!R)$, where $\Omega$ is a compact metric space possessing at least one accumulation point (the case $\Omega = [0,1]$ is considered in Volkmann (1985)).

The following example shows that regularity of the order cone $K$ of $E$ is not a sufficient assumption in theorem 5.1.3 and in corollary 5.1.2 if the interior of $K$ is empty.

**Example 5.1.2:** (cf. Dieudonné (1950)) Let $E$ be a Banach space $(c_o)$ of all real-valued sequences $y = (y_i)_{i=1}^{\infty}$ with $\lim_{i\to\infty} y_i = 0$ and norm $\|y\| = \sup_i |y_i|$. The cone $K$ formed by all nonnegative-valued sequences of $E$ is regular and the norm $\|\cdot\|$ is monotone in $K$. Define a function $g\colon E \to E$ by

$$g(y) = (\varphi(y_i))_{i=1}^{\infty} \quad \text{for } y = (y_i)_{i=1}^{\infty} \in E,$$

where $\varphi$ is defined by (5.1.12). It is obvious that $g$ is bounded, continuous and nondecreasing with respect to the norm defined above and the partial ordering defined by $K$. But the IVP

$$x' = g(x), \qquad x(0) = x_o = (\frac{1}{i})_{i=1}^{\infty}$$

has no absolutely continuous solution on any interval $J = [0,T]$, $T > 0$. To see this, make a counter-hypothesis: There is an interval $J = [0,T]$, $0 < T < 2$, and an absolutely continuous function $x\colon J \to E$ such that

$$x'(t) = g(x(t)) \text{ for a.a. } t \in J, \text{ and } x(0) = x_o.$$

Denoting $x(t) = (x_i(t))_{i=1}^{\infty}$ and $x_o = (x_i^o)_{i=1}^{\infty}$, it follows by the definition of $g$ that

$$x_i'(t) = \sqrt{x_i(t)} \text{ for a.a. } t \in J \text{ and } x_i(0) = \frac{1}{i}, \quad i = 1, 2, \ldots .$$

These IVP's possess unique solutions,

$$x_i(t) = \left(\frac{t}{2} + \frac{1}{\sqrt{i}}\right)^2, \quad t \in J, \ i = 1, 2, \ldots .$$

But this implies that $x(t) = (x_i(t))_{i=1}^{\infty} \notin E$ for any $t \in (0, T]$, contradicting the assumption that $x\colon J \to E$.

Defining $f(t, x, y) = f_2(t, y) = g(y)$ and $f_1(t, x) \equiv 0$, then $f$ satisfies conditions (f1), (f3), (f4) and (f5), while conditions (i)–(v) hold for $f_1$, $f_2$, when $Z = \emptyset$, but the conclusions of theorem 5.1.3 and corollary 5.1.2 don't hold.

On the other hand, if we look for solutions of countable systems which are only componentwise differentiable, the results derived in section 2.7 show that the analogues to theorem 5.1.3 and corollary 5.1.2 hold when $E = l^p$ and $K = l_+^p$, $1 \le p \le \infty$. In the case when $p = \infty$, $K$ is not regular and $E$ is not reflexive, and when $1 \le p < \infty$, $K$ has no interior points. Moreover, in these cases the results in question hold when the Lipschitz condition of $x \mapsto f(t, x, y)$ is weakened in theorem 5.1.3 to continuity of $x_i \mapsto f_i(t, x, y)$, $i = 1, 2, \ldots$.

### 5.1.4. Dependence the data

In this subsection we shall consider the dependence of solutions of the IVP (5.1.1) on the initial value $x_o$ and on the function $f$. Our main result is

**Proposition 5.1.2:** *Let $f$, $\hat{f}\colon J \times E \times E \to E$ satisfy conditions (f1), (f3), (f4) and (f5), and assume that the order cone $K$ of $E$*

*is regular and has a nonempty interior. Given $x_o$, $\hat{x}_o \in E$, let $x_*$ denote the minimal solution of (5.1.1), and let $\hat{x}^*$ be the maximal solution of the IVP*

$$\hat{x}' = \hat{f}(t, \hat{x}, \hat{x}), \qquad \hat{x}(0) = \hat{x}_o. \tag{5.1.13}$$

*If $x_o \leq \hat{x}_o$ and $f(t, x, y) \leq \hat{f}(t, x, y)$ for a.a. $t \in J$ and for all $x$, $y \in E$, then all the solutions of (5.1.1) and (5.1.13) belong to the order interval $[x_*, \hat{x}^*]$.*

*Proof.* Let $\hat{x}$ be any solution of the IVP (5.1.13). Denoting $x = G\hat{x}$, where $G$ is defined as in lemma 5.1.3, we have

$$x' = f(t, x, \hat{x}(t)) \leq \hat{f}(t, x, \hat{x}(t)) \text{ a.e. on } J, \; x(0) = x_o \leq \hat{x}_o. \tag{a}$$

The function $g(t, y) = \hat{f}(t, y, \hat{x}(t))$, $t \in J$, $y \in E$, satisfies a generalized Lipschitz condition. From (5.1.13) and (a) it follows that

$$x' - g(t, x) \leq 0 = \hat{x}' - g(t, \hat{x}) \quad \text{and} \quad x(0) \leq \hat{x}(0).$$

This implies by lemma 5.1.1 that $x \leq \hat{x}$, i.e. $G\hat{x} \leq \hat{x}$. Because $x_*$ is the least fixed point of $G$, it follows from (1.2.1) that $x_* \leq \hat{x}$. Noticing also that $x_*$ is the minimal solution of the IVP (5.1.1), then $x_* \leq x$ for each solution $x$ of (5.1.1) or (5.1.13). Dual reasoning shows that $x \leq \hat{x}^*$ for each solution $x$ of (5.1.1) or (5.1.13). □

A an immediate consequence of proposition 5.1.2 we have

**Corollary 5.1.3:** *Under assumptions of theorem 5.1.3 the minimal and maximal solutions of the IVP (5.1.1) are nondecreasing with respect to $x_o$ and $f$.*

Denoting $u \ll v$ if $v - u$ belongs to the interior of $K$, we obtain the following generalization to proposition 2.1.5.

**Proposition 5.1.3:** *Given $f$, $\hat{f}$: $J \times E \times E \to E$, and $x_o$, $\hat{x}_o \in E$, $x_o \ll \hat{x}_o$, assume that the IVP's (5.1.8) and (5.1.13) both have solutions on $J$. Assume also that $f(t,x,x) \leq \hat{f}(t,x,x)$ for a.a. $t \in J$ and for all $x \in E$, and there is $q \in L^1(J, I\!R)$ such that $y \mapsto f(t,y,y) + q(t)y$ or $y \mapsto \hat{f}(t,y,y) + q(t)y$ is nondecreasing for a.a. $t \in J$. If $x$ is any solution of (5.1.1) and $\hat{x}$ any solution of (5.1.13) on $J$, then $x(t) \ll \hat{x}(t)$ for each $t \in J$.*

*Proof.* Let $x$ and $\hat{x}$ be solutions of (5.1.1) and (5.1.13), respectively. Denoting $Q(t) = \int_o^t q(s)\,ds$, $t \in J$, it follows from lemma 1.5.5 that

$$e^{Q(t)}x(t) = x_o + \int_o^t [e^{Q(s)}(f(s,x(s),x(s)) + q(s)x(s)]\,ds, \qquad \text{(a)}$$

and

$$e^{Q(t)}\hat{x}(t) = \hat{x}_o + \int_o^t [e^{Q(s)}(\hat{f}(s,\hat{x}(s),\hat{x}(s)) + q(s)\hat{x}(s)]\,ds. \qquad \text{(b)}$$

Denote $t_1 = \sup\{s \in J \mid x(t) \ll \hat{x}(t) \text{ for each } t \in [0,s]\}$, and make a counter-hypothesis: $t_1 < T$. The continuity of $x$ and $\hat{x}$ implies that $\hat{x}(t_1) = x(t_1) + k$ for some $k \in \partial K$. Since $x(0) \ll \hat{x}(0)$, then $t_1 > 0$. Because $x(t) \leq \hat{x}(t)$ for each $t \in [0,t_1]$, it follows from (a) and (b) by the given hypotheses that

$$x_o - \hat{x}_o + e^{Q(t_1)}k = e^{Q(t_1)}\hat{x}(t_1) - \hat{x}_o - (e^{Q(t_1)}x(t_1) - x_o) \geq 0.$$

But then $0 \ll e^{-Q(t_1)}(\hat{x}_o - x_o) \leq k$, which would imply that $0 \ll k$, contradicting the fact that $k \in \partial K$. Thus $x(t) \ll \hat{x}(t)$ for each $t \in [0,T)$. This and the above reasoning with $t_1 = T$ shows that also $x(T) \ll \hat{x}(T)$. □

## 5.2. EXTREMAL SOLUTIONS OF SEMILINEAR IVPS

In this section we shall consider the existence of extremal solutions of the semilinear IVP

$$x' = A(t)\,x + g(t, x), \qquad x(0) = x_o, \tag{5.2.1}$$

when $E$ is an ordered Banach space whose order cone does not necessarily have a nonempty interior. No continuity or compactness hypotheses are imposed on the function $g$ in (5.2.1).

### 5.2.1. On the existence of extremal solutions

We shall assume in this subsection that the order cone $K$ of $E$ is regular. Then $K$ is by proposition 1.3.4 also normal, whence there exists $\gamma > 0$ such that

$$\|y\| \leq \gamma\|z\| \text{ whenever } y,\ z \in K \text{ and } y \leq z. \tag{1.3.2}$$

We say that a bounded and linear mapping $B\colon E \to E$, denote $B \in L(E)$, is *order-preserving* if $B[K] \subseteq K$, or equivalently, if $B$ is nondecreasing.

Let us impose the following hypotheses on $A\colon J \to L(E)$ and $g\colon J \times E \to E$.

(B0) $A(\cdot)x$ is strongly measurable for each $x \in E$.
(B1) There is $\hat{p} \in L^1(J, I\!R_+)$ such that $\|A(t)\| \leq \hat{p}(t)$ and $A(t) + \hat{p}(t)I$ is order-preserving for a.a. $t \in J$.
(B2) (5.2.1) has a lower solution $\underline{x}$ and an upper solution $\bar{x}$ on $J$ such that $\underline{x} \leq \bar{x}$.
(B3) $g(\cdot, x(\cdot))$ is strongly measurable whenever $x \in AC(J, E)$.
(B4) There is $\bar{p} \in L^1(J, I\!R_+)$ such that the function $x \mapsto g(t, x) + \bar{p}(t)x$ is nondecreasing for a.a. $t \in J$.

**Theorem 5.2.1:** *If the hypotheses (B0)–(B4) hold, then the IVP (5.2.1) has the extremal solutions between* $\underline{x}$ *and* $\bar{x}$.

*Proof.* Denote $p = \bar{p} + \hat{p}$, and let $x$ belong to the order interval $[\underline{x}, \bar{x}]$ of $AC(J, E)$. Since $t \mapsto A(t)x(t) + g(t, x(t)) + p(t)x(t)$ is nondecreasing a.e. on $J$, it follows that

$$\underline{x}'(t) + p(t)\underline{x}(t) \leq A(t)x(t) + g(t, x(t)) + p(t)x(t) \leq \bar{x}'(t) + p(t)\bar{x}(t)$$

for a.a. $t \in J$. By applying this, the triangle inequality and (1.3.2) we obtain

$$\|A(t)x(t) + g(t, x(t) + p(t)x(t)\| \leq (1+\gamma)M(t) \quad \text{for a.a. } t \in J, \tag{a}$$

where

$$M(t) = p(t)(\|\underline{x}(t)\| + \|\bar{x}(t)\|) + \|\underline{x}'(t))\| + \|\bar{x}'(t)\|, \quad t \in J.$$

This and the hypotheses (B0) and (B3) imply that the equation

$$Gx(t) = e^{-P(t)}[x_o + \int_o^t e^{P(s)}(A(s)x(s) + g(s, x(s)) + p(s)x(s))\, ds], \tag{5.2.2}$$

where $P(t) = \int_o^t p(s)\, ds$, defines an absolutely continuous and a.e. differentiable mapping $Gx \colon J \to E$ for each $x \in [\underline{x}, \bar{x}]$. In view of conditions (B1) and (B4) it follows from (5.2.2) by corollary 1.4.6 that $Gx \leq Gy$ whenever $x, y \in [\underline{x}, \bar{x}]$ and $x \leq y$. Moreover, it follows from (5.2.2) by the definition of a lower solution that for each $t \in J$

$$\begin{aligned} G\underline{x}(t) &= e^{-P(t)}[x_o + \int_o^t e^{P(s)}(A(s)\underline{x}(s) + g(s, \underline{x}(s)) + p(s)\underline{x}(s))ds] \\ &\geq e^{-P(t)}[x_o + \int_o^t e^{P(s)}(\underline{x}'(s) + p(s)\underline{x}(s))ds] \\ &= e^{-P(t)}[x_o + e^{P(t)}\underline{x}(t) - \underline{x}(0)] \geq \underline{x}(t). \end{aligned}$$

Thus $\underline{x} \leq G\underline{x}$. Similarly, it can be shown that $G\bar{x} \leq \bar{x}$.

The above proof shows that $G$ is a nondecreasing mapping from $[\underline{x}, \bar{x}]$ to $[\underline{x}, \bar{x}]$.

From (5.2.2) it follows by differentiation that for each $x \in [\underline{x}, \bar{x}]$

$$(Gx)'(t) = A(t)x(t) + g(t, x(t)) + p(t)x(t) - p(t)Gx(t) \tag{b}$$

for a.a. $t \in J$. This and (a) imply that

$$\|(Gx)'(t)\| \leq (1+\gamma)N(t) \quad \text{for a.a.} \quad t \in J, \tag{c}$$

where

$$N(t) = M(t) + p(t)(\|\underline{x}(t)\| + \|\bar{x}(t)\|), \quad t \in J.$$

From (c) we then obtain

$$\|Gx(t) - Gx(\bar{t})\| \leq (1+\gamma)\left|\int_{\bar{t}}^{t} N(s)\,ds\right|, \quad t, \bar{t} \in J. \tag{d}$$

The above proof shows that $G$ satisfies the hypotheses of theorem 1.4.7, whence $G$ has the least fixed point $x_*$ and the greatest fixed point $x^*$. This and the definition (5.2.2) of $G$ imply that $x_*$ and $x^*$ are solutions of the integral equation

$$x(t) = e^{-P(t)}[x_o + \int_o^t e^{P(s)}(A(s)x(s) + g(s, x(s)) + p(s)x(s))ds],$$

for $t \in J$. Hence, $x_*$ and $x^*$ are by lemma 1.5.5 solutions of the IVP (5.2.1).

If $x$ is any solution of (5.2.1) and $\underline{x} \le x \le \bar{x}$, then $x \in [\underline{x}, \bar{x}]$, whence $x$ is by lemma 1.5.5 a fixed point of $G$. Since $x_*$ and $x^*$ are the least and the greatest ones, it follows that $x_* \le x \le x^*$. Thus $x_*$ and $x^*$ are the extremal solutions of the IVP (5.2.1) in $[\underline{x}, \bar{x}]$. □

Condition (B3) holds if $g$ is a Carathéodory function. In this case we obtain

**Proposition 5.2.1:** *Let $g\colon J \times E \to E$ be a Carathéodory function which satisfies conditions (B2) and (B4), and let $A\colon J \to L(E)$ satisfy (B0) and (B1). Then the sequence $(y_n)_{n=o}^{\infty}$ of the successive approximations*

$$e^{P(t)}y_{n+1}(t) = x_o + \int_o^t e^{P(s)}(A(s)y_n(s) + g(s, y_n(s)) + p(s)y_n(s))ds \tag{5.2.3}$$

*with $p = \bar{p} + \hat{p}$, converges uniformly on $J$ to the minimal (resp. the maximal) solution of the IVP (5.2.1) between $\underline{x}$ and $\bar{x}$, if $y_o = \underline{x}$ (resp. $y_o = \bar{x}$).*

*Proof.* Theorem 1.4.3 implies that $g(\cdot, x(\cdot))$ is strongly measurable on $J$ for each $x \in [\underline{x}, \bar{x}]$. Thus, from the proof of theorem 5.2.1 it follows that the equation (5.2.2) defines a nondecreasing mapping $G\colon [\underline{x}, \bar{x}] \to [\underline{x}, \bar{x}]$. Choosing $y_o = \underline{x}$ and denoting $y_n = G^n y_o$, we obtain a nondecreasing sequence $(y_n)_{n=o}^{\infty}$ in $[\underline{x}, \bar{x}]$, which equals to the sequence of the successive approximations defined by (5.2.3). The proof of theorem 1.4.7 implies that the sequence $(Gy_n)_{n=o}^{\infty}$ converges uniformly on $J$ to an absolutely continuous function $x_* \in [\underline{x}, \bar{x}]$. From the definition of $y_n$ it follows that

$$x_*(t) = \lim_{n\to\infty} Gy_n(t) = \lim_{n\to\infty} y_n(t), \quad t \in J. \tag{a}$$

By using (5.2.2), (a), and the dominated convergence theorem, and noticing that $g(t, \cdot)$ is continuous for a.a. $t \in J$, it is easy to show that

$$Gx_*(t) = \lim_{n\to\infty} Gy_n(t), \quad t \in J.$$

Thus $x_*$ is a fixed point of $G$. Since $y_o = \underline{x}$ is a lower bound of $G[\underline{x}, \bar{x}]$, and since $G$ is nondecreasing, it is easy to see that $x_*$ is the least fixed point of $G$, and hence, by the proof of theorem 5.2.1, the minimal solution of (5.2.1) in $[\underline{x}, \bar{x}]$. □

To ensure that (5.2.1) has extremal solutions in the whole $AC(J, E)$, we shall need the following result.

**Lemma 5.2.1:** *Let $E$ be an ordered Banach space, and let $A\colon J \to L(E)$ and $g\colon J \times E \to E$ satisfy conditions (B0), (B1), (B3), (B4) and*

(B5) *$g(\cdot, 0) \in L^1(J, E)$, and there is $r > 0$ such that $\|g(t,x) - g(t,y)\| \le q(t, \|x - y\|)$ for all $x, y \in E$ with $\|x - y\| < r$ and for a.a. $t \in J$, where $q\colon J \times I\!R_+ \to I\!R_+$ is a Carathéodory function, $q(t, \cdot)$ is nondecreasing for a.a. $t \in J$, the IVP*

$$u' = p(t)u + q(t, u), \qquad u(0) = u_o, \tag{5.1.2}$$

*has for some $u_o > 0$ and for $p = \bar{p} + 2\hat{p}$ an upper solution on $J$, and the zero-function is the only solution of (5.1.2) on $J$ when $u_o = 0$.*

*Then the IVP (5.2.1) has for each $x_o \in E$ a unique solution $x$, and it depends continuously on $x_o$. Moreover,*

a) *$x \le \bar{x}$ for each upper solution $\bar{x}$ of (5.2.1),*

b) *$\underline{x} \le x$ for each lower solution $\underline{x}$ of (5.2.1).*

*Proof.* Assume first that (B0), (B3) and (B5) hold, and let $x_o \in E$ be given. The existence and uniqueness of the solution of (5.2.1) and its continuous dependence on $x_o$ follow from theorems 5.1.1 and 5.1.2 with $f(t, x, y) = A(t)x + g(t, y)$. Assume moreover that $E$ is ordered, and that conditions (B1) and (B4) hold. Denote $p = \bar{p} + 2\hat{p}$, let $G\colon C(J, E) \to C(J, E)$ be defined by (5.2.2), and let $\underline{x}$ be a lower solution of (5.2.1) on $J$. The hypotheses given for $q$ in condition (B5) imply by the proof of theorem 5.1.1 that there exists $u_o > 0$ such that the IVP (5.1.2) has the minimal solution $b$ which satisfies $u_o \leq b(t) < r$ for each $t \in J$, that the relation

$$Qu(t) = \int_o^t ((q(s, u(s)) + p(s)u(s))\, ds, \quad t \in J \tag{a}$$

defines a nondecreasing operator $Q$: $[0, b] \to [0, b]$, that $u_o + Qb = b$, and that the sequence $(Q^n b)_{n=o}^{\infty}$ converges uniformly on $J$ to the fixed point of $Q$, i.e. to the 0-function. From (5.2.2) and (a) it follows by (B5) that

$$|Gy - G\bar{y}| \leq Q|y - \bar{y}|, \quad \text{for } \ y, \bar{y} \in C(J, E), \ |y - \bar{y}| < b.$$

The above proof implies by theorem 1.4.9 that the iteration sequence $(G^n \underline{x})_{n=o}^{\infty}$ converges uniformly on $J$ to a unique fixed point of $G$. By definition (5.2.2) of $G$ and by lemma 1.5.5 this fixed point is the solution $x$ of the IVP (5.2.1). Since conditions (B1) and (B4) imply that $G$ is nondecreasing, and since $\underline{x} \leq G\underline{x}$ by the proof of theorem 5.2.1, the sequence $(G^n \underline{x})_{n=o}^{\infty}$ is nondecreasing. Thus $\underline{x} \leq \lim_n G^n \underline{x} = x$. The proof of assertion b) is similar. □

**Proposition 5.2.2:** *Let $E$ be an ordered Banach space with regular order cone. Assume that $g$: $J \times E \to E$ satisfies conditions (B3) and (B4), that $A$: $J \to L(E)$ satisfies (B0) and (B1), and that*

$$g_1(t, x) \leq g(t, x) \leq g_2(t, x) \text{ for } x \in E \text{ and for a.a. } t \in J, \tag{5.2.4}$$

*where $g_i$: $J \times E \to E$, $i = 1, 2$, satisfy conditions (B3), (B4) and (B5). Then the IVP (5.2.1) has for each choice of $x_o \in E$ the extremal solutions.*

*Proof.* Let $x_o \in E$ be given. Lemma 5.2.1 implies that the IVP

$$x' = A(t)x + g_i(t,x), \qquad x(0) = x_o, \tag{a}$$

has a unique solution $\underline{x}$ when $i = 1$ and a unique solution $\bar{x}$ when $i = 2$. From (a) and (5.2.4) it follows that

$$\underline{x}'(t) \leq A(t)\underline{x}(t) + g_2(t, \underline{x}(t)) \quad \text{for a.a.} \quad t \in J,$$

whence $\underline{x}$ is a lower solution of (a) with $i = 2$. Thus $\underline{x} \leq \bar{x}$ by lemma 5.2.1. Moreover,

$$\underline{x}'(t) \leq A(t)\underline{x}(t) + g(t, \underline{x}(t)), \;\; A(t)\bar{x}(t) + g(t, \bar{x}(t)) \leq \bar{x}'(t),$$

for a.a. $t \in J$, whence $\underline{x}$ is a lower solution of (5.2.1) and $\bar{x}$ is its upper solution. By theorem 5.2.1 the IVP (5.2.1) has the least solution $x_*$ and the greatest solution $x^*$ between $\underline{x}$ and $\bar{x}$.

If $x$ is a solution of (5.2.1) on $J$, it follows from (5.2.1), (a) and (5.2.4) that

$$A(t)x(t)+g_1(t, x(t)) \leq x'(t) \leq A(t)x(t)+g_2(t, x(t)) \text{ for a.a. } t \in J.$$

Thus $x$ is an upper solution of (a) for $i = 1$ and a lower solution of (a) for $i = 2$, whence the results a) and b) of lemma 5.2.1 imply that $\underline{x} \leq x \leq \bar{x}$. Hence all the solutions of (5.2.1) on $J$ lie between $\underline{x}$ and $\bar{x}$, so that $x_*$ is the least and $x^*$ the greatest of all the solutions of (5.2.1) on $J$. □

### 5.2.2. Existence of minimal or maximal solution

Assume now that the order cone $K$ of $E$ is fully regular, and replace condition (B2) by the existence of an upper solution or a lower solution of the IVP (5.2.1), and by condition

(B6) $\|g(t,x)\| \leq h(t,\|x\|)$ for all $x \in E$ and for a.a. $t \in J$, where $h\colon J \times \mathbb{R}_+ \to \mathbb{R}_+$ is a standard function, the IVP

$$u' = q(t)u + h(t,u), \quad u(0) = u_o, \tag{5.2.5}$$

has for each $q \in L^1(J,\mathbb{R}_+)$ and for each $u_o \in \mathbb{R}_+$ an upper solution, and there is $p_o \in L^1(J,\mathbb{R})$ such that the function $u \mapsto h(t,u) + p_o(t)u$ is nondecreasing for a.a. $t \in J$.

This replacement ensures the existence of the minimal solution or the maximal solution of the IVP (5.2.1). This holds also when $E$ is reflexive and ordered by a closed cone, or via continuous embedding in an ordered Banach space with fully regular order cone.

**Theorem 5.2.2:** *Let $E$ be a Banach space which is ordered by a fully regular order cone, and let $A\colon J \to L(E)$ and $g\colon J \times E \to E$ satisfy conditions (B0), (B1), (B3), (B4) and (B6).*

a) *If the IVP (5.2.1) has a lower solution $\underline{x}$, then (5.2.1) has the minimal solution in $[\underline{x})$.*

b) *If the IVP (5.2.1) has an upper solution $\bar{x}$, then (5.2.1) has the maximal solution in $(\bar{x}]$.*

*Proof.* a) Assume that $\underline{x}$ is a lower solution of (5.2.1). Denote $u_o = \|\underline{x}\|_o$, $p = \bar{p} + \hat{p}$, $q = p_o + \hat{p}$, and $P(t) = \int_o^t p(s)\,ds$. Let $\bar{u}$ be an upper solution of the IVP (5.2.5). Since the zero function is a lower solution of (5.2.5), it follows from theorem 5.2.1 that the IVP (5.2.5) has the minimal solution $u_*$ on $J$ between 0 and $\bar{u}$. Moreover, lemma 1.5.5 implies that

$$e^{P(t)}u_*(t) = u_o + \int_o^t e^{P(s)}(h(s,u_*(s)) + (q(s)+p(s))u_*(s))\,ds. \tag{a}$$

Denote

$$Y = \{x \in AC(J,E) \mid \underline{x} \leq x \text{ and } \|x(t)\| \leq u_*(t) \text{ for } t \in J\}. \tag{b}$$

$Y$ is nonempty because $\underline{x} \in Y$. Let $x \in Y$ and $t, \bar{t} \in J$, $\bar{t} \leq t$, be given. As in the proof of theorem 5.2.1 one can show that the equation (5.2.2) defines a mapping $Gx \in AC(J, E)$. From (5.2.2) and (a) it follows by (B6) that $\|Gx(t)\| \leq u_*(t)$ for each $t \in J$. The hypothesis of a) implies (cf. the proof of theorem 5.2.1) that $\underline{x} \leq G\underline{x}$. This and the conditions (B1) and (B4) imply that equation (5.2.2) defines a nondecreasing mapping $G\colon Y \to Y$. By using (5.2.2), (a), (b) and (B6) we obtain

$$\|Gx(t) - Gx(\bar{t})\| \leq |u_*(t) - u_*(\bar{t})| + 2\int_{\bar{t}}^{t} p(s)u_*(s))\, ds. \qquad \text{(c)}$$

Let $C$ be a w.o. chain of $G$ iterations of $\underline{x}$, defined in theorem 1.1.1. By definition, $G[C]$ is a well-ordered chain in $Y$. Since $G[C]$ is pointwise bounded, and by (c) equicontinuous, it follows from proposition 1.3.9 that $x_* = \sup G[C]$ exists and belongs to $Y$. Thus $G$ satisfies the hypotheses of theorem 1.2.1, whence $x_*$ is the least fixed point $x_*$. From lemma 1.5.5 it follows that $x_*$ is a solution of the IVP (5.2.1).

Assume now that $x\colon J \to E$ is a solution of (5.2.1) and $\underline{x} \leq x$. By choosing in the above proof $u_o = \max\{\|\underline{x}\|_o, \|x\|_o\}$ we may assume that $x$ belongs to the set $Y$, defined by (b). From lemma 1.5.5 it follows that $x$ is a fixed point of $G$. As in the proof of theorem 1.2.1 one can show that if $C$ is the w.o. chain $C$ of $G$-iterations of $\underline{x}$, then $C \subseteq [\underline{x}, x]$, whence $x_* = \max C \leq x$. This proves that $x_*$ is the least of all the solutions of the IVP (5.2.1) in $[\underline{x})$. The proof of the case b) is dual to the above one.

□

**Corollary 5.2.1:** *If the hypotheses of theorem 5.2.2 hold with (B3) replaced by the Carathéodory conditions, then in the case a) the sequence $(y_n)_{n=o}^{\infty}$, defined in proposition 5.2.1 with $y_o = \underline{x}$, is nondecreasing and converges on $J$ uniformly to the minimal solution of the IVP (5.2.1) in $[\underline{x})$. In the case b), and when $y_o = \bar{x}$, the sequence $(y_n)_{n=o}^{\infty}$ is nonincreasing and converges uniformly on $J$ to the maximal solution of (5.2.1) in $(\bar{x}]$.*

The conclusions of theorem 5.2.2 hold also in the following cases.

**Theorem 5.2.3:** *Let $E$ be a reflexive Banach space which is ordered by a closed cone or via continuous embedding in a Banach space with fully regular order cone. If $A\colon J \to L(E)$ and $g\colon J \times E \to E$ satisfy conditions (B0), (B1), (B3), (B4) and (B6), then the conclusions of theorem 5.2.2 hold.*

*Proof.* The proof is the same as that of theorem 5.2.2 with the exception that the use of proposition 1.3.9 is replaced by the use of proposition 1.3.10. □

**Remark 5.2.1:** Example 5.1.1 shows that the results of theorem 5.2.1 and proposition 5.2.1 do not hold true in general when regularity of the order cone $K$ of $E$ is replaced by its normality, unless $E$ is reflexive. Example 5.1.2 implies in turn that regularity of the order cone $K$ of $E$ is not a sufficient assumption in theorem 5.2.2 and in corollary 5.2.1, and reflexivity of $E$ cannot be dropped out from the hypotheses of theorem 5.2.3.

### 5.2.3. Dependence on the data

Next we shall consider the dependence of the extremal solutions of the IVP (5.2.1) on the initial value $x_o$ and on the function $g$.

**Proposition 5.2.3:** *Let $E$ be an ordered Banach space with regular order cone, let $A\colon J \to L(E)$ satisfy conditions (B0) and (B1), and let $g$, $\hat{g}\colon J \times E \to E$ satisfy the hypotheses given for $g$ in proposition 5.2.2. If $x_o$, $z_o \in E$, $x_o \leq z_o$, and*

$$g(t,x) \leq \hat{g}(t,x) \textit{ for all } x \in E \textit{ and for a.a. } t \in J,$$

*then all the solutions of $x$ the IVP (5.2.1) on $J$ and the IVP*

$$x' = A(t)x + \hat{g}(t,x), \qquad x(t_o) = z_o \tag{5.2.6}$$

*on $J$ satisfy $x_* \le x \le z^*$, where $x_*$ is the minimal solution of (5.2.1) and $z^*$ is the maximal solution of (5.2.6) on $J$.*

*Proof.* By the proof of proposition 5.2.2 there exist $\underline{x}, \bar{x} \in AC(J, E)$ such that all the solutions of (5.2.1) and (5.2.6) belong to $[\underline{x}, \bar{x}]$. Let $z$ be any solution of the IVP (5.2.6). The hypotheses imply that $z$ is an upper solution of the IVP (5.2.1). Replacing $\bar{x}$ in the proof of theorem 5.2.1 by $z$, we see that the IVP (5.2.1) has the minimal solution $z_*$ between $\underline{x}$ and $z$, obtained by $z_* = \sup G[C]$, where $G$ is defined by (5.2.2) and $C$ is the w.o. chain of $G$-iterations of $\underline{x}$. Thus $z_*$ equals to the minimal solution $x_*$ of (5.2.1) between $\underline{x}$ and $\bar{x}$, which is also the least of all the solutions of (5.2.1).

The above proof and theorem 5.2.1 imply that $x_* \le x$ whenever $x$ is a solution of (5.2.1) or (5.2.6). The dual reasoning shows that $x \le z^*$ for any solution $x$ of (5.2.1) or (5.2.6). □

As special case of the above result we obtain

**Corollary 5.2.2:** *If the hypotheses of proposition 5.2.2 hold, then both the extremal solutions of the IVP (5.2.1) are nondecreasing with respect to $x_o$ and $g$.*

Similarly, it can be shown

**Corollary 5.2.3:** *If the hypotheses of theorem 5.2.2 a) hold, then the minimal solution of the IVP (5.2.1) is nondecreasing with respect to $x_o$ and $g$.*

### 5.2.4. Special cases

As a consequence of proposition 5.2.2 and corollary 5.2.2 we obtain

**Proposition 5.2.4:** *Let $E$ be an ordered Banach space with regular order cone. Assume that $g\colon J \times E \to E$ and $\bar{p} \in L^1(J, I\!R)$ satisfy (B3) and (B4), that $A\colon J \to L(E)$ satisfies (B0) and (B1),* *and that for all $x \in E$ and for a.a. $t \in J$*

$$A_1(t)x + C_1(t) \le g(t,x) \le A_2(t)x + C_2(t), \tag{5.2.7}$$

*where $A_i\colon J \to L(E)$ satisfy (B0) and (B1), and the functions $C_i\colon J \to E$, $i = 1,2$, are Bochner integrable. Then the IVP (5.2.1) has for each choice of $x_o \in E$ the extremal solutions, both of which are nondecreasing in $x_o$.*

*Proof.* It is easy to see that the hypotheses of proposition 5.2.2 hold when $g_i(t,x) = A_i(t)x + C_i(t)$, $i = 1,2$, whence the assertions follow from proposition 5.2.2 and corollary 5.2.2. □

In the following we assume that $E$ is such an ordered Banach space that its order cone $K$ is fully regular, or $E$ is reflexive.

**Proposition 5.2.5:** *Given $A\colon J \to L(E)$ and $g\colon J \times E \to K$, assume there is a null set $Z$ in $J$ such that*

(C0) *$A(\cdot)x$ is right or left continuous in $J \setminus Z$ for all $x \in E$.*
(C1) *$A(t)$ is order-preserving for all $t \in J \setminus Z$, and there is $\tilde{p} \in L^1(J, \mathbb{R}_+)$ such that $\|A(t)\| \le \tilde{p}(t)$ for all $t \in J \setminus Z$.*
(C2) *$g(t,\cdot)$ is nondecreasing for all $t \in J \setminus Z$.*
(C3) *$g(\cdot,x)$ is right or left continuous in $J \setminus Z$ for all $x \in E$.*
(C4) *There exist $p_1 \in L^1(J, \mathbb{R}_+)$, and a nondecreasing function $\varphi\colon \mathbb{R}_+ \to (0,\infty)$ such that $\int_o^\infty \frac{dv}{\varphi(v)} = \infty$, and that $\|g(t,x)\| \le p_1(t)\varphi(\|x\|)$ for all $x \in E$ and for a.a. $t \in J$.*

*Then for each choice of $x_o \in E$ the IVP (5.2.1) has the minimal solution $x_*$ on $J$. Moreover,*

$$\|x_*(t) - x_o\| \le u(t) - \|x_o\| \quad \textit{for all} \quad t \in J, \tag{5.2.8}$$

*where $u$ is the solution of the IVP*

$$u' = p(t)(\varphi(u) + u), \qquad u(0) = \|x_o\|, \tag{5.2.9}$$

*with* $p = p_1 + \tilde{p}$.

*Proof.* Let $x_o \in E$ be given. It is easy to see that $\underline{x}(t) \equiv x_o$ is a lower solution of (5.2.1). By lemma 1.5.3 the IVP (5.2.9) has a unique and nondecreasing solution $u$ on $J$, which satisfies

$$\int_{\|x_o\|}^{u(t)} \frac{dv}{\varphi(v) + v} = \int_{t_o}^{t} p(s)ds. \tag{a}$$

Denote

$$\begin{aligned} P =\{x\colon J \to E \mid x(0) = x_o, x \text{ is nondecreasing and} \\ \|x(t) - x(s)\| \le u(t) - u(s) \text{ for } s \le t\}. \end{aligned}$$

Choosing $s = 0$, the definition of $P$ implies that if $x \in P$ and $t \in J$ then

$$\|x(t) - x_o\| \le u(t) - \|x_o\|. \tag{b}$$

The hypotheses given for $g$ and $A$ can be shown to imply by theorem 1.4.1 that $g(\cdot, x(\cdot))$ and $A(\cdot)x(\cdot)$ are strongly measurable on $J$ whenever $x \in P$. This, (b), (C1) and (C4) imply that the equation

$$Gx(t) = x_o + \int_0^t [A(s)x(s) + g(s, x(s))]\, ds, \qquad t \in J, \tag{c}$$

defines a mapping $Gx\colon J \to E$. From (b) it follows also that $\|x(t)\| \le u(t)$ for each $t \in J$. Thus, for all $s, t \in J$, $s \le t$, we have

$$\begin{aligned} &\|Gx(t) - Gx(s)\| \le \int_s^t \|A(\tau)x(\tau) + g(\tau, x(\tau))\| d\tau \\ &\le \int_s^t p(\tau)(\varphi(u(\tau)) + u(\tau))\, d\tau = \int_s^t u'(\tau)d\tau = u(t) - u(s). \end{aligned}$$

Moreover, $Gx(0) = x_o$, and $Gx$ is nondecreasing by lemma 1.4.4, whence $Gx$ belongs to $P$. This holds for each $x \in P$, so that $G$ maps $P$ into $P$. Moreover, $G$ is nondecreasing by condition (C2). Since the functions of $P$ are equicontinuous, the arguments used in the proof of theorem 5.2.2 show that $G$ satisfies the hypotheses of theorem 1.2.1. Thus the integral operator $G \colon P \to P$, defined by (c), has the least fixed point $x_*$. But this means by lemma 1.5.5 that $x_*$ is the minimal solution of (5.2.1) in $P$. The proof that $x_*$ is least of all the solutions of (5.2.1) is similar to that given in the proof of theorem 5.2.2. Since $x_*$ belongs to $P$, then the estimate (5.2.8) follows from (b). □

**Remark 5.2.2:** If the order cone $K$ has a nonempty interior, it follows from corollary 5.1.3 that the result of proposition 5.2.5 holds if in condition (C2) $x \mapsto A(t)x$ is quasimonotone nondecreasing, but not necessarily nondecreasing, for all $t \in J \setminus Z$. Moreover, the IVP (5.2.1) has also the maximal solution.

From (b) it follows easily by condition (C5) that each function $x \in P$ satisfies

$$\int_o^{\|x(t)-x_o\|} \frac{dv}{\varphi(v+\|x_o\|)+v+\|x_o\|} \leq \int_{t_o}^t p(s)ds \text{ for each } t \in J.$$

This can be applied in the study of the growth and stability of the solutions of (5.2.1).

If $\int_o^\infty \frac{dv}{\varphi(v)} < \infty$, then the solution of the IVP (5.2.1) in proposition 5.2.5 can be obtained in such subinterval $[0, c]$ of $J$ that

$$\int_o^c p(s)\, ds < \int_{\|x_o\|}^\infty \frac{dv}{\varphi(v)+v}.$$

If $g$ is defined on $J \times V$, where $V \subset E$ and $d = d(x_o, E \setminus V)$, then the solution of the IVP (5.2.1) in proposition 5.2.5 can be obtained in such subinterval $[0, c]$ of $J$ that

$$\int_o^c p(s)\, ds < \int_{\|x_o\|}^{\|x_o\|+d} \frac{dv}{\varphi(v)+v}.$$

Condition (C3) can be replaced in proposition 5.2.5 by

(C5) $g(\cdot, x(\cdot))$ is strongly measurable on $J$ whenever $x \in AC(J, E)$ is nondecreasing.

By theorem 1.4.2 this holds, for instance if

(C6) for each $(t, x) \in [0, T) \times E$ there is $z \in E$ such that $\|g(t+h, x+k) - z\| \to 0$ as $h \to 0+$ and $k \to 0$ in $K$.

In particular, we have

**Corollary 5.2.4:** *Let $E$ be an ordered Banach space which is reflexive or has a fully regular order cone $K$. If $A$, $\bar{A}$: $J \to L(E)$ satisfy conditions (C0) and (C1), and if $C$: $K \to K$ is nondecreasing and maps bounded sets into bounded sets, then to each $x_o \in E$ there corresponds a $c > 0$, such that the IVP*

$$x' = A(t)x + \bar{A}(t)C(x), \qquad x(0) = x_o$$

*has the minimal solution on $J_o = [0, c]$.*

*Proof.* Let $x_o \in E$ be given Denote $g(t, x) = \bar{A}(t)C(x)$, $(t, x) \in J \times E$. Since $C$ is nondecreasing and $\bar{A}(t)$ is order-preserving for a.a. $t \in J$, then condition (C2) holds. Define $\varphi$: $\mathbb{R}_+ \to \mathbb{R}_+$ by

$$\varphi(v) = 1 + \sup\{\|C(x)\| \mid x \in E,\ \|x\| \leq v\}.$$

It is easy to see that $\varphi$ is nondecreasing and

$$\|g(t, x)\| = \|\bar{A}(t)\, C(x)\| \leq \tilde{p}(t)\, \varphi(\|x\|), \quad t \in J,\ x \in E,$$

so that condition (C4) holds, except that $\int_o^\infty \frac{dv}{\varphi(v)}$ can be finite. Moreover, if $x \in AC(J, E)$ is nondecreasing, then the function $y = C(x(\cdot))$ is nondecreasing, and hence regulated on $J$ by proposition 1.4.1. Thus $g(\cdot, x(\cdot))$ is strongly measurable on $J$, whence condition (C5) holds. The assertion follows then from remark 5.2.2 and proposition 5.2.5. □

**Example 5.2.1:** To each $i = 1, 2, \ldots$, choose an increasing sequence $(c_n^i)_{n \in \mathbb{N}}$ of rational numbers such that $c_o^i = \frac{1}{2}$ and $\sup_n c_n^i = 1$.

Given $r \in (0,1)$, let $p_i \colon J \to \mathbb{R}_+$ $i = 0, 1, \ldots$ be an a.e. left continuous function such that $0 < p_i(t) \leq \frac{p_o(t)}{i^r}$ for all $t \geq 0$ and $i = 1, 2, \ldots$, where $p_o \in L^1(J, \mathbb{R}_+)$. Denote

$$b_i(t) = \int_o^t p_i(s)\, ds, \quad t \geq 0,\ i = 1, 2, \ldots,$$

and define for each $i = 1, 2, \ldots$ a function $g_i \colon \mathbb{R}_+ \times \mathbb{R}_+ \to \mathbb{R}_+$ by

$$g_i(t,u) = \begin{cases} p_i(t), & \text{if } u \geq b_i(t), \\ c_{n+1}^i p_i(t), & \text{if } c_n^i b_i(t) \leq u < c_{n+1}^i b_i(t), \quad n \in \mathbb{N}, \\ \dfrac{p_i(t)}{2}, & \text{if } 0 \leq u < \dfrac{b_i(t)}{2}. \end{cases}$$

If $p > \frac{1}{r}$, then the equation

$$g(t,x) = (g_i(t, x_i))_{i=1}^{\infty}, \quad t \geq 0,\ x = (x_i)_{i=1}^{\infty},$$

defines a function $g \colon \mathbb{R}_+ \times l_+^p \to l_+^p$. Moreover, $g(t,x)$ is a.e. left continuous in $t$ for each $x \in l_+^p$, nondecreasing in $x$ for each $t \geq 0$, and

$$\|g(t,x)\|_p \leq p_o(t) \left( \sum_{i=1}^{\infty} \frac{1}{i^{rp}} \right)^{\frac{1}{p}} \quad \text{for each } (t,x) \in \mathbb{R}_+ \times l_+^p.$$

Thus, $g$ satisfies the hypotheses (C2)–(C4) of proposition 5.2.5 when $J = \mathbb{R}_+$, $E = l^p$ and $K = l^p_+$, whence the IVP

$$x' = g(t, x), \qquad x(t_o) = x_o \tag{a}$$

has for each $(t_o, x_o) \in \mathbb{R}_+ \times l^p_+$ the minimal solution on $J = [t_o, c]$ for each $c > t_o$. In fact, the solution $x$ of (a) is uniquely defined by $x(t) = x_o + (b_i(t))_{i=1}^{\infty}$. If $x_o = 0$, then $x$ is obtained by

$$x(t) = \sup_n \int_o^t g(s, x_n(s))\, ds, \quad t \in J,$$

where $(x_n)_{n \in \mathbb{N}}$ is the sequence of the successive approximations $x_o(t) \equiv 0$, and

$$x_{n+1}(t) = \int_o^t g(s, x_n(s))\, ds, \quad t \in J,\ n \in \mathbb{N}.$$

## 5.3. HIGHER ORDER DIFFERENTIAL EQUATIONS

In this section we shall study higher order semilinear initial value problems in Banach spaces. We shall first apply results of section 5.1 to derive existence, uniqueness and estimation results, as well as sufficient conditions for continuous dependence of the solution and its lower order derivatives on the initial values. By assuming that the underlying Banach space is ordered, we shall then prove existence and comparison results for extremal solutions of higher order IVP's, by using results of section 5.2.

### 5.3.1. Preliminaries

Given a Banach space $E$ and $J = [0, T]$, $T > 0$, consider the $m$:th order IVP

$$\begin{aligned} &y^{(m)} + A_m(t) y^{(m-1)} + \cdots + A_1(t) y = f(t, \underline{y}), \\ &y(0) = x_{o1},\ y'(0) = x_{o2},\ \ldots,\ y^{(m-1)}(0) = x_{om}, \end{aligned} \tag{5.3.1}$$

where $f\colon J \times E^m \to E$, $A_i\colon J \to L(E)$, $x_{oi} \in E$, $i = 1, \ldots, m$, and $\underline{y} = (y, y', \ldots, y^{(m-1)})$.

By a *solution* of (5.3.1) we mean a function $y\colon J \to E$ with $y^{(m-1)} \in AC(J, E)$, which satisfies the differential equation of (5.3.1) a.e. on $J$, and all the initial conditions of (5.3.1). If $y \in AC^{m-1}(J, E)$, then $\underline{y} \in AC(J, E^m)$, where $E^m$ is equipped with the norm

$$\|x\| = \|x_1\| + \cdots + \|x_m\|, \qquad x = (x_1, \ldots, x_m) \in E^m. \tag{5.3.2}$$

By using these notations and definitions the IVP (5.3.1) can be converted into the first order IVP (5.2.1) as follows:

**Lemma 5.3.1:** *$y \in AC^{m-1}(J, E)$ is a solution of the IVP (5.3.1) if and only if*

$$x = (x_1, \ldots, x_m) = \underline{y} = (y, y', \ldots, y^{(m-1)})$$

*is a solution of the IVP*

$$x' = A(t)x + g(t, x), \qquad x(0) = x_o, \tag{5.2.1}$$

*where $g\colon J \times E^m \to E^m$ and $A\colon J \to L(E^m)$ are defined by*

$$\begin{aligned} &g(t, x) = (0, \ldots, 0, f(t, x)), \\ &A(t)x = (x_2, x_3, \ldots, x_m, -\sum_{i=1}^{m} A_i(t)x_i), \end{aligned} \tag{5.3.3}$$

*when $t \in J$ and $x = (x_1, \ldots, x_m) \in E^m$.*

## 5.3.2. Existence, uniqueness and estimation results

We shall first prove that the IVP (5.3.1) has a uniquely determined solution if $A_i\colon J \to L(E)$ and $f\colon J \times E^m \to E$ have the following properties.

(A0) For each $i = 1, \ldots, m$ and $x \in E$, $A_i(\cdot)x$ is strongly measurable, and there is $p_i \in L^1(J, \mathbb{R}_+)$ such that $\|A_i(t)\| \leq p_i(t)$ for a.a. $t \in J$.

(A1) $f(\cdot, x)$ is strongly measurable for each $x \in E^m$, and $f(\cdot, 0) \in L^1(J, E)$.

(A2) There is $r > 0$ such that $\|f(t,x) - f(t,y)\| \leq q(t, \|x-y\|)$ for all $x$, $y \in E^m$ with $\|x - y\| < r$ and for a.a. $t \in J$, where $q\colon J \times \mathbb{R}_+ \to \mathbb{R}_+$ is a Carathéodory function $q(t, \cdot)$ is nondecreasing for a.a. $t \in J$, the IVP

$$u' = (1 + \sum_{i=1}^{m} p_i(t))u + q(t,u), \qquad u(0) = u_o, \quad (5.3.4)$$

has for some $u_o > 0$ an upper solution on $J$, and the zero-function is the only solution of (5.3.4) on $J$ when $u_o = 0$.

**Theorem 5.3.1:** *If the hypotheses (A0–(A2) hold, then the IVP (5.3.1) has for each $x_o = (x_{o1}, \ldots, x_{om}) \in E^m$ a unique solution $y$ on $J$. Moreover, $y$ can be obtained by the method of successive approximations.*

*Proof.* The hypotheses (A0)–(A2) imply that the function $(t, x, y) \mapsto A(t)x + g(t, y)$ satisfies the hypotheses given for $f$ in theorem 5.1.1 with $E$ replaced by $E^m$ and $p = 1 + \sum_{i=1}^{m} p_i$. With these choices the IVP (5.2.1) has by theorem 5.1.1 a unique solution $x = (x_1, \ldots, x_m)$, which can be obtained as the uniform limit of the successive approximations. This and lemma 5.3.1 imply the assertions. □

Next we shall consider the dependence of the solution of the IVP (5.3.1) on the initial values $x_{o1}, \ldots, x_{om}$.

**Theorem 5.3.2:** *Let $f\colon J \times E^m \to E$, and $A_i\colon J \to L(E)$, $i = 1, \ldots, m$, satisfy conditions (A0)–(A2). If $y = y(\cdot, (x_{o1}, \ldots, x_{om}))$ denotes the solution of the IVP (5.3.1), and $u = u(\cdot, u_o)$ denotes* *the minimal solution of the IVP (5.3.4), then for* $x_o = (x_{o1}, \ldots, x_{om})$, $\hat{x}_o = (\hat{x}_{o1}, \ldots, \hat{x}_{om}) \in E^m$, *with* $\|x_o - \hat{x}_o\|$ *sufficiently small,*

$$\|\underline{y}(t,(x_{o1}, \ldots, x_{om})) - \underline{y}(t,(\hat{x}_{o1}, \ldots, \hat{x}_{om}))\| \leq u(t, \|x_o - \hat{x}_o\|) \tag{5.3.5}$$

*for* $t \in J$, *where* $\underline{y} = (y, y', \ldots, y^{(m-1)})$. *Moreover,* $\underline{y}$ *depends continuously on* $x_{o1}, \ldots, x_{om}$.

*Proof.* The assertions are immediate consequences of lemma 5.3.1 and theorem 5.1.2. □

As a consequence of proposition 5.1.1 and lemma 5.3.1 we obtain

**Proposition 5.3.1:** *Let* $A_i \colon J \to L(E)$, $i = 1, \ldots, m$, *satisfy (A0), let* $f \colon J \times E^m \to E$ *satisfy (A1), and assume there is* $r > 0$ *and* $q \in L^1(J, \mathbb{R}_+)$ *such that*

$$\|f(t,x) - f(t,y)\| \leq q(t)\,\varphi(\|x - y\|) \tag{5.3.6}$$

*for all* $x, y \in E^m$ *with* $\|x - y\| < r$ *and for a.a.* $t \in J$, *where* $\varphi \in C(\mathbb{R}_+, \mathbb{R}_+)$ *is nondecreasing and* $\int_o^r \frac{dv}{\varphi(v)} = \infty$. *Then the IVP (5.3.1) has for each* $x_o = (x_{o1}, \ldots, x_{om}) \in E^m$ *a unique solution* $y$. *Moreover,* $y$ *and its derivatives up to the* $(m-1)$*:th order depend continuously on* $x_{o1}, \ldots, x_{om}$.

Each of the functions $\varphi_m$, $m \in \mathbb{N}$, defined by (2.1.18) satisfy the hypotheses given for $\varphi$ in proposition 5.3.1. In particular, when $m = 0$ we obtain

**Corollary 5.3.1:** *Let the hypotheses (A0) and (A1) hold. Assume there is* $r > 0$ *such that* $f$ *satisfies for all* $x, y \in E^m$ *with* $\|x - y\| < r$ *and for a.a.* $t \in J$ *a generalized Lipschitz condition*

$$\|f(t,x) - f(t,y)\| \leq q(t)\|x - y\|, \tag{5.3.7}$$

*where* $q \in L^1(J, I\!R_+)$. *Then the IVP (5.3.1) has for each* $x_o = (x_{o1}, \ldots, x_{om}) \in E^m$ *exactly one solution* $y$. *Moreover,* $y$, $y'$, ..., $y^{(m-1)}$ *depend continuously on* $x_{o1}, \ldots, x_{om}$.

In particular, we have

**Corollary 5.3.2:** *If* $A_i\colon J \to L(E)$, $i = 1, \ldots, m$, *satisfy condition (A0), and if* $C \in L^1(J, E)$, *then the IVP*

$$y^{(m)} + A_m(t)y^{(m-1)} + \cdots + A_1(t)y = C(t),$$
$$y(0) = x_{o1},\ y'(0) = x_{o2}, \ldots,\ y^{(m-1)}(0) = x_{om}$$

*has for each* $x_o = (x_{o1}, \ldots, x_{om}) \in E^m$ *a unique solution, and it, together with its first* $m-1$ *derivatives, depend continuously on* $x_{o1}, \ldots, x_{om}$.

### 5.3.3. Extremal solutions of (5.3.1) in Banach spaces

In this section we shall consider the existence of extremal solutions of the IVP (5.3.1), when $E$ is an ordered Banach space. We shall assume in this subsection that the order cone $K$ of $E$ is regular. Obviously, $K^m$ is a regular order cone in $E^m$, where the norm is defined by (5.3.2). If $y \in AC^{m-1}(J, E)$, denote, as above, $\underline{y} = (y, y', \ldots, y^{(m-1)})$.

Let us impose the following hypotheses on the mappings $A_i\colon J \to L(E)$, $i = 1, \ldots, m$, and $g\colon J \times E^m \to E$.

- (B0) $A_i(\cdot)x$ is strongly measurable for each $x \in E$.
- (B1) There is $\hat{p}_i \in L^1(J, I\!R_+)$ such that $\|A_i(t)\| \le \hat{p}_i(t)$, and that $A_i(t) + \hat{p}_i(t)I$ is order-preserving for a.a. $t \in J$.
- (B2) (5.3.1) has a lower solution $\alpha$ and an upper solution $\beta$ on $J$ such that $\underline{\alpha} \le \underline{\beta}$.
- (B3) $f(\cdot, x(\cdot))$ is strongly measurable whenever $x \in AC(J, E^m)$.
- (B4) There is $\bar{p} \in L^1(J, I\!R_+)$ such that $x \mapsto f(t, x) + \bar{p}(t)x_m$ is nondecreasing for a.a. $t \in J$.

**Theorem 5.3.3:** *If the hypotheses (B0)–(B4) hold, then the IVP (5.3.1) has the extremal solutions between* $\underline{\alpha}$ *and* $\underline{\beta}$.

*Proof.* Let $A: J \to L(E^m)$ and $g: J \times E^m \to E^m$ be defined by (5.3.3). Denoting $\hat{p} = 1 + \sum_{i=1}^{m} \hat{p}_i$, $\underline{x} = \underline{\alpha}$ and $\bar{x} = \underline{\beta}$, it is easy to see that the hypotheses (B0)–(B4) given in subsection 5.2.1 hold. Thus the IVP (5.2.1) has by theorem 5.2.1 the extremal solutions in $[\underline{x}, \bar{x}]$. This and lemma 5.3.1 imply the assertions. □

Replacing condition (B3) by Carathéodory conditions we obtain as a consequence of proposition 5.2.1 and lemma 5.3.1.

**Proposition 5.3.2:** *Let* $f: J \times E^m \to E$ *be a Carathéodory function which satisfies conditions (B2) and (B4), and let* $A_i: J \to L(E)$, $i = 1, \ldots, m$, *satisfy (B0) and (B1). Then the extremal solutions of the IVP (5.3.1) in* $[\underline{\alpha}, \underline{\beta}]$ *can be obtained by the method of successive approximations.*

By applying lemma 5.2.1 and lemma 5.3.1 we obtain

**Lemma 5.3.2:** *Let E be an ordered Banach space, let* $A_i: J \to L(E)$, $i = 1, \ldots, m$, *satisfy (B0) and (B1), and let* $f: J \times E^m \to E$ *satisfy hypotheses (A1), (A2) and (B4). Then the IVP (5.3.1) has for each* $x_o = (x_{o1}, \ldots, x_{om}) \in E^m$ *a unique solution* $y$, *and*

a) $\alpha \le y$, $\alpha' \le y', \ldots, \alpha^{(m-1)} \le y^{(m-1)}$ *for each lower solution* $\alpha$ *of (5.3.1).*

b) $y \le \beta$, $y' \le \beta', \ldots, y^{(m-1)} \le \beta^{(m-1)}$ *for each upper solution* $\beta$ *of (5.3.1).*

Lemma 5.3.1, lemma 5.3.2 and proposition 5.2.2 imply.

**Proposition 5.3.3:** *Let E be an ordered Banach space with regular order cone. Assume that* $f: J \times E^m \to E$ *satisfies conditions (B3) and (B4), that* $A_i: J \to L(E)$, $i = 1, \ldots, m$, *satisfy (B0) and (B1), and that*

$$f_1(t,x) \le f(t,x) \le f_2(t,x) \quad \textit{for all } x \in E^m \textit{ and for a.a. } t \in J,$$

*where* $f_j\colon J \times E^m \to E$, $j = 1, 2$, *satisfy conditions (A1), (A2) and (B4). Then the IVP (5.3.1) has for each choice of* $x_o = (x_{o1}, \ldots, x_{om}) \in E^m$ *the extremal solutions.*

In the case when the order cone $K$ of $E$ is fully regular, the existence of an upper or a lower solution of the IVP (5.3.1) can be replaced by the condition

(B5) $\|f(t,x)\| \le h(t, \|x\|)$ for all $x \in E^m$ and for a.a. $t \in J$, where $h\colon J \times \mathbb{R}_+ \to \mathbb{R}_+$ is a standard function, the IVP

$$u' = q(t)u + h(t,u), \quad u(0) = u_o, \tag{5.3.8}$$

has for each $q \in L^1(J, \mathbb{R}_+)$ and for each $u_o \in \mathbb{R}_+$ an upper solution, and there is $p_o\colon L^1(J, \mathbb{R})$ such that the function $u \mapsto h(t,u) + p_o(t)u$ is nondecreasing for a.a. $t \in J$.

**Theorem 5.3.4:** *Let $E$ be an ordered Banach space with fully regular order cone, let $f\colon J \times E^m \to E$ satisfy conditions (B3), (B4) and (B5), and let $A_i\colon J \to L(E)$, $i = 1, \ldots, m$, satisfy conditions (B0) and (B1).*

a) *If the IVP (5.3.1) has a lower solution $\alpha$, then (5.3.1) has the minimal solution in $[\underline{\alpha})$.*

b) *If the IVP (5.4.1) has an upper solution $\beta$, then (5.3.1) has the maximal solution in $(\underline{\beta}]$.*

*Proof.* Since $K$ is a fully regular order cone of $E$, then $K^m$ is a fully regular order cone of $E^m$. Thus the assertions follow from lemma 5.3.1 and theorem 5.2.2. □

If $E$ is reflexive, then the above result holds also when the ordering of $E$ is not induced by a fully regular order cone.

**Theorem 5.3.5:** *Let $E$ be a reflexive Banach space which is ordered by a closed cone, or via continuous embedding in an ordered Banach space with fully regular order cone. If $f\colon J \times E^m \to E$ satisfies conditions (B3), (B4) and (B5), and if $A_i\colon J \to L(E)$, $i =$*

*$1, \ldots, m$, satisfy conditions (B0) and (B1), then the conclusions of theorem 5.3.4 hold.*

*Proof.* Because $E$ is reflexive, then $E^m$ is reflexive, too. If $K$ is a closed cone in $E$, then $K^m$ is a closed cone in $E^m$, and if $E$ is continuously embedded in $E_1$, then $E^m$ is continuously embedded in $E_1^m$. Thus the assertions follow from lemma 5.3.1 and theorem 5.2.3. □

### 5.3.4. Dependence on the data

Next we shall consider the dependence of the extremal solutions of the IVP (5.3.1) on the initial condition and on the function $f$.

As a consequence of proposition 5.3.3 and lemma 5.3.1 we obtain (cf. the proof of proposition 5.2.3).

**Proposition 5.3.4:** *Let $E$ be an ordered Banach space with regular order cone, let $A_i \colon J \to L(E)$, $i = 1, \ldots, m$, satisfy conditions (B0) and (B1), and let $f$, $\hat{f} \colon J \times E^m \to E$ satisfy the hypotheses given for $f$ in proposition 5.3.3. If $x_o = (x_{o1}, \ldots, x_{om})$, $z_o = (z_{o1}, \ldots, z_{om}) \in E^m$, $x_o \leq z_o$, and*

$$f(t, x) \leq \hat{f}(t, x) \text{ for all } x \in E^m \text{ and for a.a. } t \in J, \tag{5.3.9}$$

*then all the solutions of y the IVP (5.3.1) on $J$ and the IVP*

$$\begin{aligned} &y^{(m)} + A_m(t) y^{(m-1)} + \cdots + A_1(t) y = \hat{f}(t, \underline{y}), \\ &y(0) = z_{o1}, \ y'(0) = z_{o2}, \ \ldots, \ y^{(m-1)}(0) = z_{om}, \end{aligned} \tag{5.3.10}$$

*on $J$ satisfy $\underline{y}_* \leq \underline{y} \leq \underline{z}^*$ where $y_*$ is the minimal solution of (5.4.1) and $z^*$ is the maximal solution of (5.3.10) on $J$.*

As special case of the above result we obtain

**Corollary 5.3.3:** *If the hypotheses of proposition 5.3.4 hold, then both the extremal solutions of the IVP (5.3.1) are nondecreasing with respect to* $x_{o1}, \dots, x_{om}$ *and* $f$.

Similarly, it can be shown

**Corollary 5.3.4:** *If the hypotheses of theorem 5.3.4 a) hold, then the minimal solution of the IVP (5.3.1) in* $[\underline{\alpha})$ *is nondecreasing with respect to* $x_{o1}, \dots, x_{om}$ *and* $f$.

### 5.3.5. Special cases and examples

As a consequence of proposition 5.3.1 and corollary 5.3.3 we obtain

**Proposition 5.3.5:** *Let* $E$ *be an ordered Banach space with regular order cone. Assume that* $f\colon J \times E^m \to E$ *and* $\bar{p} \in L^1(J, I\!R)$ *satisfy (B3) and (B4), that* $A_i\colon J \to L(E)$, $i = 1, \dots, m$, *satisfy (B0) and (B1), and that for all* $x \in E^m$ *and for a.a.* $t \in J$

$$\sum_{i=1}^{m} B_i^1(t)x_i + C_1(t) \le f(t,x) \le \sum_{i=1}^{m} B_i^2(t)x_i + C_2(t), \qquad (5.3.11)$$

*where* $B_i^j\colon J \to L(E)$, $i = 1, \dots, m$, *satisfy (B0) and (B1) and* $C_j\colon J \to E$, $j = 1, 2$, *are Bochner integrable. Then the IVP (5.3.1) has for each choice of* $x_{o1}, \dots, x_{om} \in E$ *the extremal solutions, both of which are nondecreasing in* $x_{o1}, \dots, x_{om}$.

In the following we assume that $E$ is an ordered Banach space with order cone $K$ such that $K$ is fully regular or $E$ is reflexive.

**Proposition 5.3.6:** *Given* $A_i\colon J \to L(E)$, $i = 1, \dots, m$ *and* $f\colon J \times E^m \to K$, *assume there is a null set* $Z$ *in* $J$ *such that*

(C0) $A_i(\cdot)x$ *is right or left continuous in* $J \setminus Z$ *for all* $x \in E$.

(C1) *$A_i(t)$ is order-preserving for all $t \in J \setminus Z$, and there is $\hat{p}_i \in L^1(J, \mathbb{R}_+)$ such that $\|A(t)\| \le \hat{p}_i(t)$ for all $t \in J \setminus Z$.*

(C2) *$f(t, \cdot)$ is nondecreasing for all $t \in J \setminus Z$.*

(C3) *$f(\cdot, x)$ is right or left continuous in $J \setminus Z$ for all $x \in E^m$.*

(C4) *There exist $q \in L^1(J, \mathbb{R}_+)$, and a nondecreasing function $\varphi\colon \mathbb{R}_+ \to (0, \infty)$ such that $\int_o^\infty \frac{dv}{\varphi(v)} = \infty$, and that $\|f(t,x)\| \le q(t)\varphi(\|x\|)$ for all $x \in E^m$ and for a.a. $t \in J$.*

*Then for each choice of $x_o = (x_{o1}, \dots, x_{om}) \in E^m$ the IVP (5.3.1) has the minimal solution $y_*$ on $J$. Moreover,*

$$\|\underline{y}_*(t) - x_o\| \le u(t) - \|x_o\| \quad \text{for all} \quad t \in J, \tag{5.3.12}$$

*where $u$ is the solution of the IVP*

$$u' = p(t)(\varphi(u) + u), \qquad u(0) = \|x_o\|,$$

*with $p = q + 1 + \sum_{i=1}^m \hat{p}_i$.*

*Proof.* The functions $A$ and $g$, defined by (5.3.3), satisfy the hypotheses of proposition 5.2.5 which, together with lemma 5.3.1, imply the assertions. □

The results derived in this section can be applied, for instance to infinite systems of second order initial value problems.

**Example 5.3.1:** Given $r \in (0,1)$ and $J = [0,T]$, $T > 0$, let $h_i\colon J \to \mathbb{R}_+$, $i = 1, 2, \dots$, be left continuous functions such that $0 < h_i(t) \le \frac{h_o(t)}{i^r}$ for all $t \in J$ and $i = 1, 2, \dots$, where $h_o \in L^1(J, \mathbb{R}_+)$. Denote

$$b_i(t) = \int_o^t \int_o^s h_i(\tau) d\tau \, ds, \quad t \in J, \ i = 1, 2, \dots,$$

and define for each $i = 1, 2, \ldots$ a function $f_i \colon J \times \mathbb{R}_+ \to \mathbb{R}_+$ by $f_i(t, 0) \equiv 0$ and

$$f_i(t,u) = \begin{cases} h_i(t), \quad \text{if } u \geq b_i(t), \\ \dfrac{i+n+1}{2i+n+1} h_i(t), \\ \text{if } \dfrac{i+n}{2i+n} b_i(t) \leq u < \dfrac{i+n+1}{2i+n+1} b_i(t), \quad n \in \mathbb{N}, \\ \dfrac{h_i(t)}{2}, \quad \text{if } 0 < u < \dfrac{b_i(t)}{2}. \end{cases}$$

Denote by $l_+^p$ the cone of nonnegative sequences of $l^p$. If $p > \frac{1}{r}$, then the equation

$$f(t,x) = (f_i(t, x_i))_{i=1}^\infty, \quad t \in J, \ x = (x_i)_{i=1}^\infty,$$

defines a function $f \colon J \times l_+^p \to l_+^p$, since

$$\|f(t,x)\|_p \leq h_o(t) \left( \sum_{i=1}^\infty \frac{1}{i^{rp}} \right)^{\frac{1}{p}} \quad \text{for each } (t,x) \in J \times l_+^p.$$

It is easy to show that $f(\cdot, x)$ is left continuous in $J$ for each $x \in l_+^p$ and $f(t, \cdot)$ is nondecreasing in $l_+^p$ for each $t \in J$. Thus $f$ satisfies the hypotheses of proposition 5.3.5 when $E = l^p$ and $K = l_+^p$. (5.3.11) holds with $B_i^j(t) = C_1(t) \equiv 0$ and $C_2(t) = h_o(t)\,(\frac{1}{i^r})_{i=1}^\infty$. If $A \in L^1(J, L(l^p))$ and $-A(t)$ is order-preserving for a.a. $t \in J$, then the IVP

$$y'' + A(t)y' = f(t,y), \qquad y(0) = y_1, \ y'(0) = y_2 \qquad (5.3.13)$$

has by proposition 5.3.5 the extremal solutions for all $y_1, y_2 \in l_+^p$.

For instance, if $y_1 = y_2 = A = 0$, then $y_* = 0$ is the least solution of (5.3.13), and $y^* = (b_i)_{i=1}^{\infty}$ is its greatest solution.

If $r \in (\frac{1}{2}, 1)$ we can choose $p = 2$. If the functions $a_j^i \colon J \to I\!R_+$, $1 \leq i, j < \infty$, are Lebesgue integrable, and

$$M = \sup\{\sum_{i=1}^{\infty} \sum_{j=1}^{\infty}(a_j^i(t))^2 \mid t \in J\} < \infty,$$

and if $A(t)$, $t \in J$, is the linear mapping whose matrix in the natural basis of $l^2$ is $(a_j^i(t))_{i,j=1}^{\infty}$, we obtain a Bochner integrable mapping $A\colon J \to L(l^2)$. If the functions $f_i$ are defined as above, the above result implies that the infinite system

$$y_i'' = \sum_{j=1}^{\infty} a_j^i(t) y_i' + f_i(t, y_i),\ y_i(0) = x_o^i,\ y_i'(0) = x_1^i, \tag{5.3.14}$$

has the least and the greatest solution whenever $\sum_{i=1}^{\infty} |x_n^i|^2 < \infty$ for $n = 0, 1$.

The full regularity of the order cone $K$ of $E$ cannot be replaced in theorem 5.3.4 by regularity, as we see from

**Example 5.3.2:** Let $E$ be a Banach space $(c_0)$ of all real-valued sequences $y = (y_i)_{i=1}^{\infty}$ with $\lim_{i\to\infty} y_i = 0$ and norm $\|y\| = \sup_i |y_i|$. The cone $K$ formed by all nonnegative-valued sequences of $E$ is regular and the norm $\|\cdot\|$ is monotone in $K$. For given positive numbers $a$ and $b$ the function $f\colon K \to K$, defined by

$$f(y) = (\frac{a}{i} + b\sqrt{y_i})_{i=1}^{\infty} \quad \text{for } y = (y_i)_{i=1}^{\infty} \in K,$$

is continuous and nondecreasing with respect to the partial ordering defined by $K$. Moreover,

$$\|f(y)\| \leq a + b\sqrt{\|y\|} \quad \text{for each } y \in K.$$

But the IVP

$$x'' = f(x'), \qquad y(0) = x_o,\ y'(0) = x_1 \tag{5.3.15}$$

has no solution on $J = [0,T]$ for any $T > 0$ and $(y_o, y_1) \in K^2$. To see this, make a counter-hypothesis: There is $(y_o, y_1) \in K^2$, $J = [0,T]$, $T > 0$ and $x \in AC^1(J,K)$ such that (5.3.15) holds. Denoting $x(t) = (x_i(t))_{i=1}^\infty$, $x_o = (x_i^o)_{i=1}^\infty$, and $x_1 = (x_i^1)_{i=1}^\infty$, then

$$x_i''(t) = \frac{a}{i} + b\sqrt{x_i'(t)} \text{ for a.e. } t \in J.$$
$$x_i(0) = x_i^o,\ x_i'(0) = x_i^1,\ i = 1, 2, \ldots .$$

By the separation of variables we obtain

$$\int_{x_i^1}^{x_i'(t)} \frac{dv}{\frac{a}{i} + b\sqrt{v}} = \int_o^t ds,\ t \in J,\ i = 1, 2, \ldots,$$

which implies that

$$\int_o^{x_i'(t)} \frac{dv}{2\sqrt{v}} \geq \frac{bt}{2},\ t \in J,\ i = 1, 2, \ldots,$$

or equivalently,

$$x_i'(t) \geq \frac{b^2t^2}{4},\ t \in J,\ i = 1, 2, \ldots .$$

Thus

$$x_i(t) \geq x_i^o + \frac{b^2t^3}{12},\ t \in J,\ i = 1, 2, \ldots,$$

which implies that $x(t) \notin K$ for any $t \in (0,T]$, contradicting the assumption that $x\colon J \to K$.

The following example shows that the hypotheses (B0)–(B3) don't guarantee the validity of theorem 5.3.1, even in the case when $E = \mathbb{R}$, when $f$ does not depend on $t$, and when $A_i(t) \equiv 0$.

**Example 5.3.3:** Define $f\colon I\!R \to I\!R$ by $f(y) = \begin{cases} 2, & \text{if } y = 0, \\ 0, & \text{if } y \neq 0. \end{cases}$

Let $J = [0,T]$, $T > 0$ be given. It is easy to see that the IVP

$$x''(t) = f(x'), \qquad x(0) = x_o,\ x'(0) = 0 \tag{a}$$

has for each $x_o \in I\!R$ an upper solution $\bar{x}(t) = x_o + t^2$ and a lower solution $\underline{x}(t) \equiv x_o$ on $J$. This implies that condition (B2) holds. If $y\colon J \to I\!R$ is continuous, then the set $U = \{t \in J \mid y(t) = 0\}$ is closed, and hence Lebesgue measurable. Thus, the set $\{t \in J \mid f(y(t)) < a\}$ is $J$ if $a > 1$, $U$ if $0 < a \leq 1$ and $\emptyset$ if $a \leq 0$, whence $f(y(\cdot))$ is measurable. This ensures that condition (B3) holds. But the IVP (a) has no solution on $J$. To see this, notice first that $x\colon J \to I\!R$ is by a solution of (a) on $J$ if and only if $x(t) = x_o + \int_o^t y(s)ds$, $t \in J$, where $y\colon J \to I\!R$ is a solution of the integral equation

$$y(t) = \int_o^t f(y(s))ds, \qquad t \in J. \tag{b}$$

Each possible solution of (b) on $J$ is absolutely continuous and nondecreasing on $J$ and vanishes at 0. If $y$ is such a function, denote $b = \sup\{a \in J \mid y(t) = 0 \text{ on } [0,a]\}$. If $b = 0$, then $y(t) > 0$ for each $t \in (0.T]$. But then $\int_o^t f(y(s))ds = 0$ for each $t \in J$, whence $y$ does not satisfy (b) on $J$. If $b > 0$, then $y(t) = 0$ for each $t \in [0,b)$, whence $\int_o^t f(y(s))ds = 2t$ for each $t \in [0,b)$, so that $y$ is not a solution of (b) on $J$. This implies that (b), and hence also (a), has no solution on any interval $J = [0,T]$, $T > 0$.

## 5.4. PERIODIC BOUNDARY VALUE PROBLEMS

In this section we shall study the existence of extremal solutions of first and second order periodic boundary value problems in ordered Banach spaces. No continuity or compactness conditions are imposed on the nonlinear terms in the considered differential equations.

### 5.4.1. First order periodic BVPs

Assume that $E$ is an ordered Banach space with regular order cone $K$. Given $J = [0,T]$, consider the PBVP

$$x' = f(t,x), \qquad x(0) = x(T), \tag{5.4.1}$$

where $f\colon J \times E \to E$. Let us impose the following assumptions for the function $f$.

(A0) There exist $\alpha, \beta \in AC(J,E)$, $\alpha \le \beta$, $c_\alpha, c_\beta \in L^1(J,E)$ and $p \in L^1(J, I\!R_+)$ with $P(t) = \int_0^t p(s)\,ds \not\equiv 0$ such that $\alpha' \le f(t,\alpha) - c_\alpha(t)$ and $\beta' \ge f(t,\beta) + c_\beta(t)$ on $J \setminus Z$, $\alpha(0) - \alpha(T) \le \int_t^T e^{P(s)-P(T)} c_\alpha(s) ds + \int_0^t e^{P(s)} c_\alpha(s)\, ds$ and $\beta(0) - \beta(T) \ge \int_t^T e^{P(s)-P(T)} c_\beta(s)\, ds + \int_0^t e^{P(s)} c_\beta(s)\, ds$, $t \in J$.

(A1) $f(\cdot, x(\cdot))$ is strongly measurable for all $x \in [\alpha, \beta]$.

(A2) $f(t,x) + p(t)\,x$ is nondecreasing in $x \in [\alpha(t), \beta(t)]$ for all $t \in J \setminus Z$.

**Theorem 5.4.1:** *If there exists a null set $Z$ in $J$ such that conditions (A0)–(A2) hold, then the extremal solutions of the PBVP (5.4.1) exist in the order interval $[\alpha, \beta]$ of $AC(J,E)$.*

*Proof.* The hypotheses (A0)–(A2) imply that if $x \in [\alpha, \beta]$, then

$$\begin{aligned}\alpha'(t) + c_\alpha(t) + p(t)\alpha(t) &\le f(t, x(t)) + p(t)x(t)\\ &\le \beta'(t) + c_\beta(t) + p(t)\beta(t)\end{aligned}$$

for each $t \in J \setminus Z$. This, the normality of the order cone $K$ of $E$, and (A1) imply that $g(t, x(\cdot)) \in L^1(J, E)$, where $g \colon J \times E \to E$ is defined by

$$g(t,x) = f(t,x) + p(t)\,x, \quad t \in J,\ x \in [\alpha(t), \beta(t)]. \tag{5.4.2}$$

Define an operator $G$ in $[\alpha, \beta]$ by

$$\begin{aligned} Gx(t) &= e^{-P(t)} \int_0^t e^{P(s)} g(s, x(s))\, ds \\ &+ \frac{e^{-P(t)}}{e^{P(T)} - 1} \int_0^T e^{P(s)} g(s, x(s))\, ds, \quad t \in J. \end{aligned} \tag{5.4.3}$$

From (5.4.3) it follows that for each $x \in [\alpha, \beta]$

$$(Gx)'(t) = g(t, x(t)) - p(t)Gx(t) \text{ a.e. on } J,\ Gx(0) = Gx(T). \tag{a}$$

This, (A0) and (5.4.2) imply that $w = G\alpha - \alpha$ satisfies the inequalities

$$w'(t) + p(t)w(t) \geq c_\alpha(t) \ \text{ for a.a. } \ t \in J, \tag{b}$$

and

$$w(T) - w(0) \leq \int_t^T e^{P(s)-P(T)} c_\alpha(s)\, ds + \int_0^t e^{P(s)} c_\alpha(s)\, ds, \ t \in J. \tag{c}$$

From (b) it follows that

$$e^{P(t)} w(t) \geq w(0) + \int_0^t e^{P(s)} c_\alpha(s)\, ds, \quad t \in J. \tag{d}$$

Inequalities (c) and (d) imply that for each $t \in J$,

$$w(0) + \int_o^t e^{P(s)} c_\alpha(s)\, ds \geq w(T) - \int_t^T e^{P(s)-P(T)} c_\alpha(s)\, ds$$
$$\geq e^{-P(T)} w(0) + \int_o^t e^{P(s)-P(T)} c_\alpha(s)\, ds.$$

This implies that

$$w(0) + \int_o^t e^{P(s)} c_\alpha(s)\, ds \geq 0, \quad t \in J. \tag{e}$$

In view of (d) and (e) we then have $w = G\alpha - \alpha \geq 0$, i.e. $\alpha \leq G\alpha$. The proof that $G\beta \leq \beta$ is similar.

From (A2) and (5.4.2) it follows that $g(t, \cdot)$ is nondecreasing in $[\alpha, \beta]$ for all $t \in J \setminus Z$. This and the definition (5.4.3) of $G$ imply that $G$ is nondecreasing. Obviously, $Gx \in AC(J, E)$ for each $x \in [\alpha, \beta]$, whence (5.4.3) defines a nondecreasing operator $G\colon [\alpha, \beta] \to [\alpha, \beta]$. This, (a) and (1.3.2) imply that

$$\|(Gx)'(t)\| \leq N(t) \text{ for all } x \in [\alpha, \beta] \text{ and for a.a. } t \in J,$$

where

$$N(t) = (1+\gamma)(\|g(t, \alpha(t)\| + \|g(t, \beta(t))\| + p(t)(\|\alpha(t)\| + \|\beta(t)\|)).$$

Since $Gx \in AC(J, E)$, we then have

$$\|Gx(t) - Gx(\bar{t})\| \leq |v(t) - v(\bar{t})|, \quad t, \bar{t} \in J,$$

where $v(t) = \int_o^t N(s)\, ds$, $t \in J$.

The above argumentation proves that $G$ satisfies the hypotheses of theorem 1.4.7, whence $G$ has the least fixed point $x_*$ and the greatest fixed point $x^*$. From the definitions of $g$ and $G$ and lemma 1.5.7 it follows that $x_*$ and $x^*$ are also solutions of the PBVP (5.4.1) in $[\alpha, \beta]$.

If $x$ is any solution of (5.4.1) in $[\alpha, \beta]$, then it is by lemma 1.5.7 a fixed point of $G$, whence $x_* \leq x \leq x^*$. Thus $x_*$ is the minimal solution and $x^*$ is the maximal solution of the PBVP (5.4.1) in $[\alpha, \beta]$. □

As for the dependence of the extremal solutions of (5.4.1) on $f$ we have

**Proposition 5.4.1:** *If the hypotheses of theorem 5.4.1 hold, then the extremal solutions of the PBVP (5.4.1) in $[\alpha, \beta]$ are nondecreasing with respect to $f$.*

*Proof.* Let $f$, $\hat{f}\colon J \times E \to E$ satisfy

$$f(t,x) \leq \hat{f}(t,x) \text{ for a.a. } t \in J \text{ and for all } x \in E. \tag{a}$$

Assume that the hypotheses of theorem 5.4.1 hold for $f$ and $\hat{f}$. Let $x_*$ be the minimal solution of (5.4.1) in $[\alpha, \beta]$, and let $\hat{x}_*$ be the minimal solution of the PBVP

$$x' = \hat{f}(t,x), \qquad x(0) = x(T). \tag{b}$$

Let $G\colon [\alpha, \beta] \to [\alpha, \beta]$ be defined by (5.4.3), where $g$ is given by (5.4.2). From lemma 1.5.7 it follows that

$$\begin{aligned} \hat{x}_*(t) &= e^{-P(t)} \int_0^t e^{P(s)} \hat{g}(s, \hat{x}_*(s))\, ds \\ &+ \frac{e^{-P(t)}}{e^{P(T)} - 1} \int_0^T e^{P(s)} \hat{g}(s, \hat{x}_*(s))\, ds, \ t \in J, \end{aligned} \tag{c}$$

where

$$\hat{g}(t,x) = \hat{f}(t,x) + p(t)\,x, \qquad t \in J,\ x \in [\alpha(t), \beta(t)]. \qquad \text{(d)}$$

From (a), (d) and (5.4.2) it follows that

$$g(t,x) \le \hat{g}(t,x), \quad \text{for a.a. } t \in J \text{ and for all } \ x \in [\alpha(t), \beta(t)],$$

whence (5.4.3) and (c) imply that $G\hat{x}_* \le \hat{x}_*$. Since $G$ satisfies the hypotheses of theorem 1.4.7 and $x_*$ is the least fixed point of $G$, it follows from (1.4.11) that $x_* \le \hat{x}_*$.

Similarly, it can be shown that $x^* \le \hat{x}^*$, where $x^*$ denotes the maximal solution of (5.4.1) in $[\alpha, \beta]$ and $\hat{x}^*$ is the maximal solution of (b) in $[\alpha, \beta]$. □

It is easy to show that condition (A0) holds if

(A3) There exist $\alpha$, $\beta \in AC(J,E)$, $\alpha \le \beta$, $h$, $k \in K$, $p \in L^1(J, I\!R_+)$, $P(t) = \int_0^t p(s)\,ds$, with $P(T) > 0$ such that
$\alpha' \le f(t,\alpha) - p(t)h$ and $\beta' \ge f(t,\beta) + p(t)k$ on $J \setminus Z$,
$\alpha(0) - \alpha(T) \le (1 - e^{-P(T)})\,h$ and
$\beta(0) - \beta(T) \ge (1 - e^{-P(T)})\,k$.

By choosing $c_\alpha = c_\beta = 0$ in (A0) we obtain

**Corollary 5.4.1:** *The conclusions of theorem 5.4.1 and proposition 5.4.1 are valid if condition (A0) is replaced by*

(A4) *There exist $\alpha$, $\beta \in AC(J,E)$, $\alpha \le \beta$, such that $\alpha' \le f(t,\alpha)$ and $\beta' \ge f(t,\beta)$ on $J \setminus Z$, $\alpha(0) \le \alpha(T)$ and $\beta(0) \ge \beta(T)$.*

As for the existence of the globally extremal solutions we have

**Proposition 5.4.2:** *Given $g\colon J \times E \to E$, assume there is a null set $Z$ of $J$ such that $g(t,\cdot)$ is nondecreasing for all $t \in J \setminus Z$,* *that $g(\cdot, x(\cdot))$ is strongly measurable for each $x \in AC(J,E)$, and there exist $a$, $b \in L^1(J,E)$ such that*

$$a(t) \le g(t,x) \le b(t) \quad \textit{for all } t \in J \setminus Z \textit{ and } x \in E. \tag{5.4.4}$$

*If $p \in L^1(J, I\!R)$ satisfies $\int_o^T p(s)\,ds > 0$, then the PBVP*

$$x' + p(t)\,x = g(t,x), \qquad x(0) = x(T) \tag{5.4.5}$$

*has the extremal solutions.*

*Proof.* By choosing $f(t,x) = g(t,x) - p(t)\,x$, $t \in J$, $x \in E$, it is easy to show that the hypotheses of corollary 5.4.1 hold with

$$\alpha(t) = e^{-P(t)} \left( \frac{\int_o^T e^{P(s)} a(s)\,ds}{e^{P(T)} - 1} + \int_o^t e^{P(s)} a(s)\,ds \right) \tag{a}$$

and

$$\beta(t) = e^{-P(t)} \left( \frac{\int_o^T e^{P(s)} b(s)\,ds}{e^{P(T)} - 1} + \int_o^t e^{P(s)} b(s)\,ds \right), \quad t \in J. \tag{b}$$

Thus the PBVP (5.4.5) has the extremal solutions $x_*$ and $x^*$ in $[\alpha, \beta]$. Moreover, if $x$ is any solution of (5.4.5), it follows from lemma 1.5.7 that for each $t \in J$,

$$x(t) = e^{-P(t)} \left( \frac{\int_o^T e^{P(s)} g(s, x(s))\,ds}{e^{P(T)} - 1} + \int_o^t e^{P(s)} g(s, x(s))\,ds \right).$$

This, (5.4.4), (a) and (b) imply that each solution of (5.4.5) belongs to $[\alpha, \beta]$, whence $x_*$ and $x^*$ are the least and the greatest of all the solutions of (5.4.5). □

### 5.4.2. Second order periodic BVPs

In this subsection we shall study the existence of extremal solutions the second order periodic boundary problem

$$-x'' = f(t,x,x'), \qquad x(0) = x(T),\ x'(0) = x'(T), \tag{5.4.6}$$

in an ordered Banach space $E$, where $f\colon J \times E^2 \to E$ and $J = [0,T]$, $T > 0$.

We say that the function $x \in AC^1(J,E)$ is a *solution* of (5.4.6) if $-x''(t) = f(t,x(t),x'(t))$ for a.a. $t \in J$, and if $x(0) = x(T)$ and $x'(0) = x'(T)$.

First we shall convert the PBVP (5.4.6) to an operator equation. The following result holds also when $E$ is not ordered.

**Lemma 5.4.1:** *If $p \in L^1(J, \mathbb{R}_+)$, $P(t) = \int_0^t p(s)ds$, $t \in J$, and $P(T) \neq 0$, then $x$ is a solution of the PBVP (5.4.6) if and only if $x$ and $y = x'$ satisfy the operator equation*

$$(x,y) = G(x,y) = (G_1(x,y), G_2(x,y)), \tag{5.4.7}$$

*where*

$$\begin{aligned} G_1(x,y)(t) = {} & \frac{e^{-P(t)}}{e^{P(T)}-1}\int_o^T e^{P(s)}[p(s)x(s)+y(s)]\,ds \\ & + e^{-P(t)}\int_o^t e^{P(s)}[p(s)x(s)+y(s)]ds, \end{aligned} \tag{5.4.8}$$

*and*

$$\begin{aligned} G_2(x,y)(t) = {} & e^{-P(t)}\int_o^t e^{P(s)}[p(s)y(s) - f(s,x(s),y(s))]ds \\ & + \frac{e^{-P(t)}}{e^{P(T)}-1}\int_o^T e^{P(s)}[p(s)y(s) - f(s,x(s),y(s))]ds. \end{aligned} \tag{5.4.9}$$

*Proof.* Assume first that $x$ is a solution of the PBVP (5.4.6). Since the function $y = x'$ is absolutely continuous on $J$, then the function $x''$ is Bochner integrable. From (5.4.6) it then follows that also $f(\cdot, x(\cdot), y(\cdot))$ is Bochner integrable, whence the integrals in the right hand side of (5.4.9) are defined. Moreover, since $x(0) = x(T)$, it follows from (5.4.8) that

$$
\begin{aligned}
e^{P(t)}G_1(x,y)(t) &= (e^{P(T)} - 1)^{-1} \int_o^T [p(s)e^{P(s)}x(s) + e^{P(s)}x'(s)]ds \\
&+ \int_o^t [p(s)e^{P(s)}x(s) + e^{P(s)}x'(s)]ds \\
&= (e^{H_i(T)} - 1)^{-1}[e^{P(T)}x(T) - x(0)] + e^{P(t)}x(t) - x(0)] \\
&= e^{P(t)}x(t), \ \ t \in J.
\end{aligned}
$$

From (5.4.6) and (5.4.9), it follows that for all $t \in J$

$$
\begin{aligned}
e^{P(t)}G_2(x,y)(t) &= \int_o^t [p(s)e^{P(s)}x'(s) - e^{P(s)}f(s,x(s),x'(s))]ds \\
&+ \frac{1}{e^{P(T)} - 1}\int_o^T [p(s)e^{P(s)}x'(s) - e^{P(s)}f(s,x(s),x'(s))]ds \\
&= \int_o^t [p(s)e^{P(s)}x'(s) + e^{P(s)}x''(s)]ds \\
&+ \frac{1}{e^{P(T)} - 1}\int_o^T [p(s)e^{P(s)}x'(s) + e^{P(s)}x''(s)]ds \\
&= e^{P(t)}x'(t) - x'(0) + \frac{1}{e^{P(T)} - 1}[e^{P(T)}x'(T) - x'(0)] \\
&= e^{P(t)}y(t).
\end{aligned}
$$

Thus $x$ and $y = x'$ satisfy the operator equation (5.4.7).

Conversely, assume that $x$, $y$ satisfy (5.4.7). In view of (5.4.8) we then have for each $t \in J$

$$
\begin{aligned}
e^{P(t)}x(t) = &\frac{1}{e^{P(T)}-1}\int_o^T e^{P(s)}[p(s)x(s)+y(s)]ds \\
&+\int_o^t e^{P(s)}[p(s)x(s)+y(s)]ds.
\end{aligned}
\tag{a}
$$

Noticing that $x$, $y \in AC(J, E)$, then (a) implies by differentiation that

$$
x'(t) = y(t), \quad t \in J. \tag{b}
$$

From (5.4.7) and (5.4.9) it follows that for each $t \in J$,

$$
\begin{aligned}
e^{P(t)}y(t) = &\int_o^t e^{P(s)}[p(s)y(s) - f(s, x(s), y(s))]ds \\
+&\frac{1}{e^{P(T)}-1}\int_o^T e^{P(s)}[p(s)y(s) - f(s, x(s), y(s))]ds.
\end{aligned}
\tag{c}
$$

Differentiating (c) with respect to $t$ we obtain for a.a. $t \in J$

$$
e^{P(t)}[p(t)y(t) + y'(t)] = e^{P(t)}[p(t)y(t) - f(t, x(t), y(t))],
$$

which together with (b) implies that

$$
x''(t) = -f(t, x(t), x'(t)) \quad \text{for a.a.} \quad t \in J.
$$

From (a) it follows that $x(T) = x(0)$, and from (c) that $x'(T) = x'(0)$, so that $x$ is a solution of the PBVP (5.4.6). □

Let us impose the following hypotheses on the function $f\colon\ J \times E^2 \to E$.

(B0) There exist $\alpha$, $\beta \in AC^1(J,E)$ with $(\alpha,\alpha') \le (\beta,\beta')$, $c_\alpha$, $c_\beta \in L^1(J,E)$ and a null set $Z$ in $J$ such that $-\beta''(t) \le f(t,\beta(t),\beta'(t)) - c_\beta(t)$, $t \in J\backslash Z$, $\beta(T) \le \beta(0)$, $-\alpha''(t) \ge f(t,\alpha(t),\alpha'(t)) + c_\alpha(t)$, $t \in J\backslash Z$, $\alpha(T) \ge \alpha(0)$, and $p \in L^1(J, I\!R_+)$ with $P(t) = \int_0^t p(s)ds \not\equiv 0$ such that $\beta'(T) - \beta'(0) \le \int_t^T e^{P(s)-P(T)} c_\beta(s)ds + \int_o^t e^{P(s)} c_\beta(s)ds$, $\alpha'(0) - \alpha'(T) \le \int_t^T e^{P(s)-P(T)} c_\alpha(s)ds + \int_o^t e^{P(s)} c_\alpha(s)ds$, $t \in J$.

(B1) $f(\cdot,x(\cdot),y(\cdot))$ is strongly measurable on $J$ when $x$, $y \in AC(J,E)$, $\alpha \le x \le \beta$, $\alpha' \le y \le \beta'$.

(B2) $f(t,x,y) - p(t)y$ is nonincreasing in $(x,y)$ on $[\alpha(t),\beta(t)] \times [\alpha'(t),\beta'(t)]$ for all $t \in J \setminus Z$.

From now on we shall assume that the order cone $K$ of $E$ is normal.

**Lemma 5.4.2:** *Let $f: J \times E^2 \to E$ satisfy conditions (B0) – (B2). Denoting*

$$a(t) = (\alpha(t),\alpha'(t)), \quad b(t) = (\beta(t),\beta'(t)), \quad t \in J, \tag{5.4.10}$$

*then $Gb \le b$ and $a \le Ga$.*

*Proof.* The hypotheses (B0)–(B2) imply that for all $x$, $y$ in $AC(J,E)$, satisfying $(\alpha,\alpha') \le (x,y) \le (\beta,\beta')$,

$$\begin{aligned}\beta''(t) + c_\beta(t) + p(t)\beta(t) &\le p(t)y(t) - f(t,x(t),y(t))\\ &\le \beta''(t) - c_\beta(t) + p(t)\beta(t), \ t \in J \setminus Z.\end{aligned}$$

This and (1.3.2) imply that $f(\cdot,x(\cdot),y(\cdot))$ is pointwise a.e. norm bounded by a function of $L^1(J,I\!R_+)$. This and (B2) ensure that $G_j a$ and $G_j b$ are defined for $j = 1,2$.

Condition (B0) implies that for each $t \in J$

$$e^{P(t))} G_1 b(t) = (e^{P(T)} - 1)^{-1} \int_0^T e^{P(s)}[p(s)\beta(s) + \beta'(s)]ds$$

$$+\int_o^t e^{P(s)}[p(s)\beta(s)+\beta'(s)]ds$$

$$=(e^{P(T)}-1)^{-1}[e^{P(T)}\beta(T)-\beta(0)]+e^{P(t)}\beta(t)-\beta(0)$$

$$\le \beta(0)+e^{P(t)}\beta(t)-\beta(0)=e^{P(t)}\beta(t).$$

From (5.4.9), (5.4.10) and (B0) it follows that for each $t\in J$

$$e^{P(t)}G_2b(t)=\int_o^t [p(s)e^{P(s)}\beta'(s)-e^{P(s)}f(s,\beta(s),\beta'(s))]ds$$

$$+\frac{1}{e^{P(T)}-1}\int_o^T [p(s)e^{P(s)}\beta'(s)-e^{P(s)}f(s,\beta(s),\beta'(s))]ds$$

$$\le \frac{1}{e^{P(T)}-1}\int_o^T [p(s)e^{P(s)}\beta'(s)+e^{P(s)}(\beta''(s)-c_\beta(s))]ds$$

$$+\int_o^t [p(s)e^{P(s)}\beta'(s)+e^{P(s)}(\beta''(s)-c_\beta(s))]ds$$

$$=\frac{1}{e^{P(T)}-1}[e^{P(T)}\beta'(T)-\beta'(0)-\int_o^T e^{P(s)}c_\beta(s)ds]$$

$$+e^{P(t)}\beta'(t)-\beta'(0)\int_o^t e^{P(s)}c_\beta(s)ds$$

$$=e^{P(t)}\beta'(t)+\frac{e^{P(T)}}{e^{P(T)}-1}[\beta'(T)-\beta'(0)]$$

$$-(e^{P(T)}-1)^{-1}\int_o^T e^{P(s)}c_\beta(s)ds-\int_o^t e^{P(s)}c_\beta(s)ds]$$

$$\le e^{P(t)}\beta'(t).$$

Thus, $G_1b\le\beta$ and $G_2b\le\beta'$, whence $Gb\le b$ The proof that $a\le Ga$ is similar. □

**Lemma 5.4.3:** *Assume that conditions (B0)–(B2) hold. Denoting*

$$[a,b] = \{(x,y) \in AC(J,E^2) \mid a \leq (x,y) \leq b\}, \tag{5.4.11}$$

*then the equations (5.4.7)–(5.4.9) define a nondecreasing mapping* $G\colon [a,b] \to [a,b]$.

*Proof.* Let $(x,y)$, $(u,v) \in [a,b]$, $(x,y) \leq (u,v)$, be given. The hypotheses (B0)–(B2) imply that for each $t \in J$

$$e^{P(t)}G_1(x,y)(t)$$

$$= \frac{1}{e^{P(T)}-1}\int_o^T [P(s)x(s)+y(s)]ds + \int_o^t [p(s)x(s)+y(s)]ds$$

$$\leq \frac{1}{e^{P(T)}-1}\int_o^T [h_i(s)u(s)+v(s)]ds + \int_o^t [p(s)u(s)+v(s)]ds$$

$$= e^{P(t)}G_1(u,v)(t),$$

and

$$e^{P(t)}G_2(x,y)(t) = \int_o^t e^{P(s)}[p(s)y(s) - f(s,x(s),y(s))]ds$$

$$+\frac{1}{e^{P(T)}-1}\int_o^T e^{P(s)}[p(s)y(s) - f(s,x(s),y(s))]ds$$

$$\leq \int_o^t e^{P(s)}[p(s)v(s) - f(s,u(s),v(s))]ds$$

$$+\frac{1}{e^{P(T)}-1}\int_o^T e^{P(s)}[p(s)v(s) - f(s,u(s),v(s))]ds$$

$$= e^{P(t)}G_2(u,v)(t).$$

Thus $G_j(x,y) \le G_j(u,v)$, $j = 1,2$. This and the result of lemma 5.4.2 imply the assertion. □

### 5.4.3. Existence of extremal solutions

Now we are ready to prove our main result concerning the existence of the extremal solutions of the periodic boundary value problem (5.4.6).

**Theorem 5.4.2:** *Let $E$ be an ordered Banach space with regular order cone $K$ and $J = [0,T]$, $T > 0$. Given $f\colon J \times E^2 \to E$, assume that conditions (B0)–(B2) hold. Then the PBVP (5.4.6) has such solutions $\underline{x}$ and $\bar{x}$ in $[\alpha,\beta]$ that $\underline{x} \le x \le \bar{x}$ and $\underline{x}' \le x' \le \bar{x}'$ for each solution $x$ of (5.4.6) in $[\alpha,\beta]$ such that $x' \in [\alpha',\beta']$.*

*Proof.* In view of lemma 5.4.3 the equations (5.4.7)–(5.4.9) define a nondecreasing mapping $G\colon [a,b] \to [a,b]$. From (5.4.8) and (5.4.9) it follows by differentiation that

$$(G_1z)'(t) = y(t) + p(t)(x(t) - G_1z(t)) \tag{a}$$

and

$$(G_2z)'(t) = p(t)(y(t) - G_2z(t)) - f(t,x(t),y(t)) \tag{b}$$

for all $z = (x,y) \in [a,b]$ and for a.a. $t \in J$. Denoting

$$v_j(t) = \int_0^t [\|\alpha^{(j)}(s)\| + \|\beta^{(j)}(s)\| + 2p(s)(\|\alpha^{(j-1)}(s)\| + \|\beta^{(j-1)}(s)\|)]ds,$$

$t \in J$, $j = 1,2$, it follows from (a), (b), (B0) and (1.3.2) that

$$\|G_jz(t) - G_jz(s)\| \le (1+\gamma)|v_j(t) - v_j(s)|,\ z \in [a,b],\ s,\ t \in J. \tag{c}$$

The above proof shows that $G$ satisfies the hypotheses of theorem 1.4.7, whence $G$ has the least fixed point $x_* = (\underline{x},\underline{y})$ and the greatest fixed point $x^* = (\bar{x}, \bar{y})$. From lemma 5.4.1 it follows that $\underline{x}$, $\bar{x}$ are solutions of the PBVP (5.4.6), and that $\underline{x}' = \underline{y}$ and $\bar{x}' = \bar{y}$.

If $x \in [\alpha, \beta]$ is a solution of (5.4.6), and if $x' \in [\alpha', \beta']$, it follows from lemma 5.4.1 that $z = (x, x')$ is a fixed point of $G$. Because of the extremality of $x_*$ and $x^*$ it follows that $x_* \le z \le x^*$, i.e. $\alpha \le x \le \beta$ and $\alpha' \le x' \le \beta'$. □

As a consequence of theorem 5.4.2 we then obtain

**Corollary 5.4.2:** *Given functions $f_1$, $f_2$: $J \times E \to E$, assume that conditions (B0) and (B1) hold for the function*

$$f(t, x, y) = f_1(t, x) + f_2(t, y), \qquad t \in J,\ x,\ y \in E.$$

*If $f_1(t, \cdot)$ is nonincreasing in $[\alpha(t), \beta(t)]$ for a.a. $t \in J$, and if $f_2(t, \cdot)$ is nonincreasing in $[\alpha'(t), \beta'(t)]$ for a.a. $t \in J$, then the PBVP (5.4.6) has the extremal solutions in $[\alpha, \beta]$.*

**Remark 5.4.1:** If $f$ is a Carathéodory function in theorem 5.4.2, then $G$ is continuous. In this case $(\underline{x}, \underline{x}')$ (resp. $(\bar{x}, \bar{x}')$) are the uniform limits of the successive approximations $(x_n, y_n)$ given by $(x_o, x_o') = (\alpha, \alpha')$ (resp. $(x_o, x_o') = (\beta, \beta')$) and

$$x_{j+1} = G_1(x_j, y_j), \quad y_{j+1} = G_2(y_j, x_j), \quad j \in \mathbb{N}.$$

Moreover, if the function $f$ is continuous, then the differential equation in (5.4.6) holds for all $t \in J$.

If $c_\alpha(t) = p(t)\, h$, $c_\beta(t) = p(t)\, k$, $h$, $k \in E$, in (B0), we obtain the condition

(B3) $-\beta''(t) \le f(t, \beta(t), \beta'(t)) - p(t)\, k$ and $-\alpha''(t) \ge f(t, \alpha(t), \alpha'(t)) + p(t)\, h$ for all $t \in J \setminus Z$, $\beta'(T) - \beta'(0) \le (1 - e^{-P(T)})k$ and $\alpha'(0) - \alpha'(T) \le (1 - e^{-P(T)})h$.

If $c_\alpha(t) = c_\beta(t) \equiv 0$ in (B0), it is reduced to

(B4) $-\beta''(t) \le f(t, \beta(t), \beta'(t))$ and $-\alpha''(t) \ge f(t, \alpha(t), \alpha'(t))$ for all $t \in J \setminus Z$, $\alpha(0) \le \alpha(T)$, $\beta(T) \le \beta(0)$, $\beta'(T) \le \beta(0)$ and $\alpha'(0) \le \alpha'(T)$.

## 5.5. MIXED MONOTONE EQUATIONS

In this section we shall consider first and second order IVP's and PBVP's of mixed monotone type in an ordered Banach space $E$. Throughout this section we shall assume that the order cone of $E$ is regular.

### 5.5.1. First order IVP of mixed monotone type

Consider first the IVP

$$x' = f(t, x, x), \qquad x(0) = x_o, \tag{5.5.1}$$

where $f\colon J \times E^2 \to E$, $J = [0, T]$.

The functions $y$, $z \in AC(J, E)$ are said to be *coupled quasisolutions* of (5.5.1) if

$$\begin{aligned} y'(t) &= f(t, y(t), z(t)) \text{ for a.a. } t \in J, \text{ and } y(0) = x_o, \\ z'(t) &= f(t, z(t), y(t)) \text{ for a.a. } t \in J, \text{ and } z(0) = x_o. \end{aligned} \tag{5.5.2}$$

We shall impose the following hypotheses on $f$.

(A0) $\alpha$, $\beta \in AC(J, E)$, $\alpha \leq \beta$ and $\alpha' \leq f(t, \alpha, \beta)$, $\beta' \geq f(t, \beta, \alpha)$ on $J \setminus Z$.

(A1) $f(\cdot, y(\cdot), z(\cdot))$ is strongly measurable for all $y$, $z \in [\alpha, \beta]$.

(A2) There is $p \in L^1(J, \mathbb{R}_+)$ such that the mapping $(y, z) \mapsto f(t, y, z) + p(t)y$ is mixed monotone in $[\alpha(t), \beta(t)] \times [\alpha(t), \beta(t)]$ for all $t \in J \setminus Z$.

We shall show that conditions (A0)–(A2) imply the existence of extremal coupled quasisolutions of (5.5.1) in the order interval $[\alpha, \beta]$.

**Theorem 5.5.1:** *If there is a null set $Z$ in $J$ so that the hypotheses (A0)–(A2) hold, then the IVP (5.5.1) has extremal* *coupled quasisolutions in the order interval* $[\alpha, \beta]$ *of* $AC(J, E)$ *for each* $x_o \in [\alpha(0), \beta(0)]$.

*Proof.* Let $x_o \in [\alpha(0), \beta(0)]$ and $y,\ z \in [\alpha, \beta]$ be given. From (A1) it follows that the function $t \mapsto f(t, y(t), z(t))$ is strongly measurable, and conditions (A0) and (A2) imply that

$$\alpha'(t) + p(t)\alpha(t) \leq f(t, y(t), z(t)) + p(t)y(t) \leq \beta'(t) + p(t)\beta(t)$$

on $J \setminus Z$. This and (1.3.2) imply that

$$\|f(t, y(t), z(t)) + p(t)y(t)\| \leq N(t) \tag{a}$$

for all $t \in J \setminus Z$, where

$$N(t) = (1+\gamma)(\|\beta'(t)\| + \|\alpha'(t)\| + p(t)(\|\alpha(t)\| + \|\beta(t)\|).$$

This ensures that we can define a map $A \colon [\alpha, \beta] \times [\alpha, \beta] \to E$ by

$$A(y, z)(t) = e^{-P(t)}[x_o + \int_o^t e^{P(s)}(f(s, y(s), z(s)) + p(s)y(s))\, ds], \tag{b}$$

where $P(t) = \int_o^t p(s)ds,\ t \in J$. From (b) it follows that for all $y,\ z \in [\alpha, \beta]$

$$\begin{aligned} &A(y, z)'(t) = f(t, y(t), z(t)) + p(t)(y(t) - A(y, z)(t)) \\ &\text{for a.a. } t \in J,\ A(y, z)(0) = x_o. \end{aligned} \tag{c}$$

This, (A0) and the choice of $x_o$ imply that $w = A(\alpha, \beta) - \alpha$ satisfies

$$w'(t) + p(t)w(t) \geq 0 \text{ for a.a. } t \in J, \quad w(0) \geq 0,$$

so that

$$w(t) \geq e^{-P(t)}w(0) \geq 0 \text{ for all } t \in J.$$

This proves that $\alpha \leq A(\alpha,\beta)$. The proof that $A(\beta,\alpha) \leq \beta$ is similar. Condition (A2) implies by corollary 1.4.6 that $A$ is mixed monotone. In particular, $A(y,z) \in [\alpha,\beta]$ for all $y$, $z \in [\alpha,\beta]$. From (a) and (c) it follows by (1.3.2) that

$$\|(A(y,z)'(t)\| \leq N(t) + (1+\gamma)(\|\alpha(t)\| + \|\beta(t)\|)$$

for all $y$, $z \in [\alpha,\beta]$ and for a.a. $t \in J$. The above proof shows that $A$ satisfies the hypotheses of proposition 1.4.6, whence $A$ has the extremal coupled fixed points $y$, $z$. From (5.5.2) and (b) it follows by lemma 1.5.5 that $u$, $v$ are coupled quasisolutions of the IVP (5.5.1) in $[\alpha,\beta]$ if and only if they are coupled fixed points of $A$. Thus $y$, $z$ are the extremal coupled quasisolutions of the IVP (5.5.1). □

Next we shall consider the existence of extremal coupled quasisolutions of the IVP

$$x' = f(t,x,x,x), \qquad x(0) = x_o. \tag{5.5.3}$$

We say that the functions $y$, $z \in AC(J,E)$ are *coupled quasisolutions* of (5.5.3) if

$$y'(t) = f(t,y(t),y(t),z(t)) \text{ for a.a. } t \in J, \text{ and } y(0) = x_o,$$
$$z'(t) = f(t,z(t),z(t),y(t)) \text{ for a.a. } t \in J, \text{ and } z(0) = x_o.$$

We shall assume that the order cone $K$ of $E$ is regular and has a nonempty interior, and that the following conditions hold.

(f0) $\alpha$, $\beta \in AC(J,E)$, $\alpha \leq \beta$ and $\alpha' \leq f(t,\alpha,\alpha,\beta)$, $\beta' \geq f(t,\beta,\beta,\alpha)$ on $J \setminus Z$.

(f1) $f(\cdot,x,y(\cdot),z(\cdot))$ strongly measurable for all $x \in E$ and $y$, $z \in [\alpha,\beta]$.

(f2) $f(\cdot, 0, \alpha(\cdot), \beta(\cdot))$ and $f(\cdot, 0, \beta(\cdot), \alpha(\cdot))$ are Bochner integrable.

(f3) There is $p \in L^1(J, I\!R_+)$ such that
$$\|f(t, x, y, z) - f(t, \bar{x}, y, z)\| \le p(t)\|x - \bar{x}\|$$
whenever $t \in J \setminus Z$ and $x$, $\bar{x}$, $y$, $z \in E$.

(f4) $f(t, \cdot, y, z)$ is quasimonotone nondecreasing, $f(t, y, \cdot, z)$ is nondecreasing and $f(t, y, z, \cdot)$ is nonincreasing for all $t \in J \setminus Z$ and $y$, $z \in E$.

**Proposition 5.5.1:** *Let $E$ be an ordered Banach space whose order cone $K$ is regular and has a nonempty interior. If there is a null set $Z$ in $J$ such that $f\colon J \times E \times E \times E \to E$ satisfies conditions (f0)–(f4), then for each $x_o \in [\alpha(0), \beta(0)]$ the IVP (5.5.3) has the extremal coupled quasisolutions in $[\alpha, \beta]$.*

*Proof.* Let $x_o \in [\alpha(0), \beta(0)]$ be given. Applying conditions (f0), (f2)–(f4) and (1.3.2) one can find a $q \in L^1(J, I\!R_+)$ such that

$$\|f(t, x, y(t), z(t))\| \le q(t) + p(t)\,\|x\| \tag{a}$$

for all $t \in J \setminus Z$, $x \in E$ and $y$, $z \in [\alpha, \beta]$. This and conditions (f1)–(f3) imply by corollary 5.1.1 that the IVP

$$x' = f(t, x, y(t), z(t)), \qquad x(0) = x_o \tag{b}$$

has for all fixed $y$, $z \in [\alpha, \beta]$ a unique solution $x = A(y, z)$. Applying lemma 5.1.1 and conditions (f0) and (f3) it is easy to show that $A(y, z) \in [\alpha, \beta]$, and that the so obtained mapping $A\colon [\alpha, \beta] \times [\alpha, \beta] \to [\alpha, \beta]$ is mixed monotone. Denoting $M = \max\{\|\alpha\|_o, \|\beta\|_o\}$, it follows from (a) that

$$\|A(y, z)'(t)\| \le q(t) + M\,p(t) \quad \text{for a.a.} \quad t \in J.$$

The above proof implies by proposition 1.4.6 that $A$ has the extremal coupled fixed points $y$, $z$, which, by the definition of

$A$, are the extremal coupled quasisolutions of the IVP (5.5.3) in $[\alpha, \beta]$. □

### 5.5.2. First order PBVP of mixed monotone type

In this subsection we shall consider the existence of the extremal coupled quasisolutions of the PBVP

$$x' = f(t, x, x), \qquad x(0) = x(T), \tag{5.5.4}$$

where $f: J \times E^2 \to E$. The functions $y$, $z \in AC(J, E)$ are called *coupled quasisolutions* of (5.5.4) if

$$\begin{aligned} y'(t) &= f(t, y(t), z(t)) \text{ for a.a. } t \in J, \text{ and } y(0) = y(T), \\ z'(t) &= f(t, z(t), y(t)) \text{ for a.a. } t \in J, \text{ and } z(0) = z(T). \end{aligned} \tag{5.5.5}$$

Assume that $E$ is an ordered Banach space with regular order cone $K$, and that the following conditions hold:

(B0) There exist $\alpha$, $\beta \in AC(J, E)$, $\alpha \le \beta$, $c_\alpha$, $c_\beta \in L^1(J, E)$, $p \in L^1(J, I\!R_+)$, $P(t) = \int_o^t p(s)\,ds$, with $P(T) > 0$
$\alpha' \le f(t, \alpha, \beta) - c_\alpha(t)$ and $\beta' \ge f(t, \beta, \alpha) + c_\beta(t)$ on $J \backslash Z$,
$\alpha(0) - \alpha(T) \le \int_t^T e^{P(s)-P(T)} c_\alpha(s)\,ds + \int_o^t e^{P(s)} c_\alpha(s)\,ds$,
$\beta(0) - \beta(T) \ge \int_t^T e^{P(s)-P(T)}\, c_\beta(s)\,ds + \int_o^t e^{P(s)} c_\beta(s) ds$,
$t \in J$.

(B1) $f(\cdot, y(\cdot), z(\cdot))$ is strongly measurable for all $y$, $z \in [\alpha, \beta]$.

(B2) $(y, z) \mapsto f(t, y, z) + p(t)\,y$ is mixed monotone in $[\alpha(t), \beta(t)] \times [\alpha(t), \beta(t)]$ for all $t \in J \setminus Z$.

**Theorem 5.5.2:** *If there exists a null set $Z$ in $J$ and such that conditions (B0)–(B2) are valid, then the PBVP (5.5.4) has the extremal coupled quasisolutions in the order interval $[\alpha, \beta]$ of $AC(J, E)$.*

*Proof.* The hypotheses (B0)–(B2) imply that if $y,\ z \in [\alpha, \beta]$, then

$$\begin{aligned} \alpha'(t) + c_\alpha(t) + p(t)\alpha(t) &\le f(t, y(t), z(t)) + p(t)y(t)) \\ &\le \beta'(t) + c_\beta(t) + p(t)\beta(t) \end{aligned}$$

for each $t \in J \setminus Z$. This, the normality of the order cone $K$ of $E$, and (B1) imply that $g(t, y(\cdot), z(\cdot)) \in L^1(J, E)$, where $g\colon J \times E^2 \to E$ is defined by

$$g(t, y, z) = f(t, y, z) + p(t)\, y, \quad t \in J,\ y,\ z \in [\alpha(t), \beta(t)]. \tag{5.5.6}$$

Define an operator $A$ in $[\alpha, \beta] \times [\alpha, \beta]$ by

$$\begin{aligned} A(y, z)(t) &= e^{-P(t)} \int_o^t e^{P(s)} g(s, y(s), z(s))\, ds \\ &+ \frac{e^{-P(t)}}{e^{P(T)} - 1} \int_o^T e^{P(s)} g(s, y(s), z(s))\, ds. \end{aligned} \tag{5.5.7}$$

From (5.5.7) it follows that for all $y,\ z \in [\alpha, \beta]$

$$\begin{aligned} &A(y, z)'(t) = g(t, y(t), z(t)) - p(t)A(y, z)(t) \\ &\text{for a.a. } \ t \in J, \quad A(y, z)(0) = A(y, z)(T). \end{aligned} \tag{a}$$

This, (B0) and (5.5.6) imply that $w = A(\alpha, \beta) - \alpha$ satisfies the inequalities

$$w'(t) + p(t)w(t) \ge c_\alpha(t) \ \text{ for a.a. } \ t \in J \tag{b}$$

and

$$w(0) - w(T) \ge -\int_t^T e^{P(s) - P(T)} c_\alpha(s)\, ds - \int_o^t e^{P(s)} c_\alpha(s)\, ds,\ t \in J. \tag{c}$$

From (b) it follows that

$$e^{P(t)}w(t) \geq w(0) + \int_o^t e^{P(s)}c_\alpha(s)\,ds, \quad t \in J. \tag{d}$$

Inequalities (c) and (d) imply that for each $t \in J$,

$$\begin{aligned} w(0) + \int_o^t e^{P(s)}c_\alpha(s)\,ds &\geq w(T) - \int_t^T e^{P(s)-P(T)}c_\alpha(s)\,ds \\ &\geq e^{-P(T)}w(0) + \int_o^t e^{P(s)-P(T)}c_\alpha(s)\,ds. \end{aligned}$$

This implies that

$$w(0) + \int_o^t e^{P(s)}c_\alpha(s)\,ds \geq 0, \quad t \in J. \tag{e}$$

In view of (d) and (e) we then have $w = A(\alpha, \beta) - \alpha \geq 0$, i.e. $\alpha \leq A(\alpha, \beta)$. The proof that $A(\beta, \alpha) \leq \beta$ is similar.

From (B2) and (5.5.6) it follows that $(y, z) \mapsto g(t, y, z)$ is mixed monotone in $[\alpha(t), \beta(t)] \times [\alpha(t), \beta(t)]$ for all $t \in J \setminus Z$. This and the definition (5.5.7) of $A$ imply that $A$ is mixed monotone. Obviously, $A(y, z) \in AC(J, E)$ for all $y$, $z \in [\alpha, \beta]$, whence (5.5.7) defines a mixed monotone operator $A\colon [\alpha, \beta] \times [\alpha, \beta] \to [\alpha, \beta]$. This, (a) and (1.3.2) ensure that

$$\|(A(y, z))'(t)\| \leq (1+\gamma)N(t) \text{ for all } y,\ z \in [\alpha, \beta] \text{ and for a.a. } t \in J,$$

where

$$N(t) = (\|g(t, \alpha(t), \beta(t))\| + \|g(t, \beta(t), \alpha(t))\| + p(t)(\|\alpha(t)\| + \|\beta(t)\|)).$$

Since $A(y,z) \in AC(J,E)$, we then have

$$\|A(y,z)(t) - A(y,z)(\bar{t})\| \le |v(t) - v(\bar{t})|, \quad t,\, \bar{t} \in J,$$

where $v(t) = \int_o^t N(s)\,ds,\ t \in J$.

The above proof shows that $A$ satisfies the hypotheses of theorem 1.4.8, whence $A$ has the extremal coupled fixed points $y$, $z$. From the definitions of $g$ and $A$ and lemma 1.5.7 it follows that $y$, $z$ are also coupled quasisolutions of the PBVP (5.5.4) in $[\alpha,\beta]$.

If $u$, $v$ are coupled quasisolutions of (5.5.4) in $[\alpha,\beta]$, then they are by lemma 1.5.7 coupled fixed points of $A$, whence $y \le u,\, v \le z$. Thus $y$, $z$ are the extremal coupled quasisolutions of the PBVP (5.5.4) in $[\alpha,\beta]$. □

Routine calculations show that condition (B0) holds if

(B3) There exist $\alpha,\, \beta \in AC(J,E)$, $\alpha \le \beta$, $h,\, k \in K$, $p \in L^1(J, I\!R_+)$, $P(t) = \int_o^t p(s)\,ds$, with $P(T) > 0$ such that
$\alpha' \le f(t,\alpha,\beta) - p(t)h$ and $\beta' \ge f(t,\beta,\alpha) + p(t)k$ on $J \setminus Z$,
$\alpha(0) - \alpha(T) \le h(1 - e^{-P(T)})$ and $\beta(0) - \beta(T) \ge k(1 - e^{-P(T)})$.

By choosing $c_\alpha = c_\beta = 0$ in (B0) we obtain

**Corollary 5.5.1:** *The conclusions of theorem 5.5.2 hold if condition (B0) is replaced by*

(B4) *There exist $\alpha,\, \beta \in AC(J,E)$, $\alpha \le \beta$, such that $\alpha' \le f(t,\alpha,\beta)$ and $\beta' \ge f(t,\beta,\alpha)$ on $J \setminus Z$, $\alpha(0) \le \alpha(T)$ and $\beta(0) \ge \beta(T)$.*

### 5.5.3. Second order IVP of mixed monotone type

Consider next the second order IVP

$$x'' = f(t, x, x, x'), \qquad x(0) = x_o,\ x'(0) = x_1, \tag{5.5.8}$$

where $f\colon J \times E^3 \to E,\ J = [0, T]$.

The functions $y,\ z \in AC^1(J, E)$ is said to be *coupled quasisolutions* of (5.5.8) if

$$\begin{aligned} &y''(t) = f(t, y(t), z(t), y'(t)) \text{ for a.a. } t \in J,\\ &z''(t) = f(t, z(t), y(t), z'(t)) \text{ for a.a. } t \in J,\\ &y(0) = x_o,\ y'(0) = x_1,\ z(0) = x_o \text{ and } z'(0) = x_1. \end{aligned} \tag{5.5.9}$$

We shall impose the following hypotheses on $f$.

(C0) $\alpha,\ \beta \in AC^1(J, E),\ \alpha \le \beta,\ \alpha' \le \beta'$ and $\alpha' \le f(t, \alpha, \beta, \alpha')$, $\beta' \ge f(t, \beta, \alpha, \beta')$ on $J \setminus Z$.

(C1) $f(\cdot, x(\cdot), y(\cdot), z(\cdot))$ is strongly measurable for all $x,\ y \in [\alpha, \beta]$ and $z \in [\alpha', \beta']$.

(C2) The mapping $(x, y) \mapsto f(t, x, y, z)$ is mixed monotone in $[\alpha(t), \beta(t)] \times [\alpha(t), \beta(t)]$ for all $z \in [\alpha'(t), \beta'(t)]$ and $t \in J \setminus Z$.

(C3) There is $p \in L^1(J, I\!R_+)$ such that the mapping $z \mapsto f(t, x, y, z) + p(t)z$ is nondecreasing in $[\alpha'(t), \beta'(t)]$ for all $x,\ y \in [\alpha(t), \beta(t)]$ and $t \in J \setminus Z$.

The main result of this subsection is

**Theorem 5.5.3:** *If there is a null set $Z$ in $J$ so that conditions (C0)–(C3) hold, then the IVP (5.5.8) has for each $(x_o, x_1) \in [\alpha(0), \beta(0)] \times [\alpha'(0), \beta'(0)]$ such coupled quasisolutions $y,\ z$ that $\alpha \le y \le z \le \beta$ and $\alpha' \le y' \le z' \le \beta'$, and if $v,\ w$ are another coupled quasisolutions of (5.5.8) such that $\alpha \le v,\ w \le \beta$ and $\alpha' \le v',\ w' \le \beta'$, then $y \le v,\ w \le z$ and $y' \le v',\ w' \le z'$.*

*Proof.* Let $x_o \in [\alpha(0), \beta(0)]$, $x_1 \in [\alpha'(0), \beta'(0)]$ be given. Denote

$$Y = \{x \in AC^1(J,E) \mid (x, x') \in [\alpha, \beta] \times [\alpha', \beta']\},$$

and define a norm and a partial ordering in $C^1(J,E)$ by

$$\|x\| = \sup\{\|x(t)\| + \|x'(t)\| \mid t \in J\},$$

and

$$x \leq y \quad \text{if and only if } x(t) \leq y(t) \text{ and } x'(t) \leq y'(t) \text{ for all } \; t \in J.$$

The given hypotheses imply that the equation

$$g(y,z)(t) = f(t, y(t), z(t), y'(t)) + p(t)y'(t), \quad t \in J, \tag{a}$$

defines a mixed monotone mapping $g\colon Y \times Y \to L^1(J,E)$. This and corollary 5.3.2 imply that the IVP

$$x'' + p(t)x' + g(y,z)(t), \qquad x(0) = x_o, \; x'(0) = x_1, \tag{b}$$

has for all $y$, $z \in Y$ a unique solution $x = A(y,z) \in AC^1(J,E)$. It is easy to see that

$$\begin{aligned} A(y,z)(t) = x_o &+ \int_o^t e^{-P(s)} x_1 \, ds \\ &+ \int_o^t \int_o^s e^{P(\tau)-P(s)} g(y,z)(\tau)\, d\tau ds, \end{aligned} \tag{5.5.10}$$

where $P(t) = \int_o^t p(s)ds$, $t \in J$. The so obtained mapping $A\colon Y \times Y \to AC(J,E)$ is mixed monotone because $g$ is. Denoting $w =$

$\alpha - A(\alpha,\beta)$, it follows from (C0), (a) and (b) and from the choice of $x_o$ and $x_1$ that $w$ is a lower solution of the IVP

$$u'' + p(t)u' = 0, \quad u(0) = 0,\ u'(0) = 0.$$

Since the zero function is the only solution of this IVP, it follows from lemma 5.3.2 that $w(t) \leq 0$ and $w'(t) \leq 0$, i.e. $\alpha(t) \leq A(\alpha,\beta)(t)$ and $\alpha'(t) \leq A(\alpha,\beta)'(t)$ on $J$. Similarly, it can be shown that $A(\beta,\alpha)(t) \leq \beta(t)$ and $A(\beta,\alpha)'(t) \leq \beta'(t)$ on $J$. The above results imply that the equation (5.5.10) defines a mixed monotone mapping $A\colon Y \times Y \to Y$.

Assume now that $(y_n)_{n=o}^{\infty}$ and $(z_n)_{n=o}^{\infty}$ are monotone sequences in $Y$, the one being nonincreasing and the other one nondecreasing. The sequences $(g(y_n, z_n)(t))_{n=o}^{\infty}$, $t \in J$ are monotone and are contained in $[g(\alpha,\beta)(t), g(\beta,\alpha)(t)]$, whence they converge, because the order cone $K$ of $E$ is regular. Denote

$$C(t) = \lim_{n\to\infty} g(y_n, z_n)(t), \quad t \in J.$$

Since $K$ is also normal, it follows by (1.3.2) that

$$\|g(y_n, z_n)(t)\| \leq (1+\gamma)(\|g(\alpha,\beta)(t)\| + \|g(\beta,\alpha)(t)\|).$$

Thus $C \in L^1(J,E)$ by the dominated convergence theorem for Bochner integrals, and

$$\lim_{n\to\infty} \int_o^t g(y_n, z_n)(s)\, ds = \int_o^t C(s)\, ds, \quad t \in J. \tag{c}$$

Defining $x\colon J \to E$ by

$$x(t) = x_o + \int_o^t e^{-P(s)} x_1\, ds + \int_o^t \int_o^s e^{P(\tau)-P(s)} C(\tau)\, d\tau ds,\ t \in J$$

then

$$x'(t) = e^{-P(t)}[x_1 + \int_o^t e^{P(s)}C(s)\,ds], \ t \in J.$$

These equations, together with (5.5.10) and (c) imply that

$$\lim_{n\to\infty} A(y_n, z_n)(t) = x(t), \ \lim_{n\to\infty} A(y_n, z_n)'(t) = x'(t), \ t \in J. \quad \text{(d)}$$

Moreover,

$$0 \le x(t) - A(y_n, z_n)(t) \le T\,e^{P(T)} \int_o^T (x(s) - g(y_n, z_n)(s))\,ds$$

and

$$0 \le x'(t) - A(y_n, z_n)'(t) \le e^{P(T)} \int_o^T (x(s) - g(y_n, z_n)(s))\,ds, \ t \in J.$$

These inequalities, (c) and the normality of $K$ imply that the convergences in (d) are uniform on $J$. Thus $A(y_n, z_n) \to x$ in $C^1(J, E)$. Obviously, $x \in Y$, whence $(A(y_n, z_n))_{n=o}^{\infty}$ converges in $Y$.

The above proof shows that $A$ satisfies the hypotheses of theorem 1.2.4, whence $A$ has the extremal coupled fixed points $y$, $z$ in $Y$. By the definitions of $Y$ and the partial ordering in $C^1(J, E)$ this means that $\alpha \le y \le z \le \beta$ and $\alpha' \le y' \le z' \le \beta'$, and if $v$, $w$ are another coupled fixed points of $A$ such that $\alpha \le v, w \le \beta$ and $\alpha' \le v', w' \le \beta'$, then $y \le v, w \le z$ and $y' \le v', w' \le z'$. From the definition of $A$ it follows that $v$, $w$ are coupled quasisolutions of (5.5.8) satisfying $\alpha \le v, w \le \beta$ and $\alpha' \le v', w' \le \beta'$; if and only if $v$, $w$ are coupled fixed points of $A$, which concludes the proof. □

### 5.5.4. Mixed monotone PBVP's in Banach spaces

In this subsection we shall study the second order mixed monotone periodic boundary value problem

$$-x'' = f(t, x, x, x'), \qquad x(0) = x(T),\ x'(0) = x'(T), \quad (5.5.11)$$

in an ordered Banach space $E$, where $f\colon J \times E^3 \to E$ and $J = [0, T],\ T > 0$.

We say that the functions $y,\ z \in AC^1(J, E)$ are *coupled quasisolutions* of (5.5.11) if

$$\begin{aligned}
&-y''(t) = f(t, y(t), z(t), y'(t)) \ \text{ for a.a. } t \in J,\\
&y(0) = y(T),\ y'(0) = y'(T),\\
&-z''(t) = f(t, z(t), y(t), z'(t)) \text{ for a.a. } t \in J,\\
&z(0) = z(T),\ z'(0) = z'(T).
\end{aligned} \quad (5.5.12)$$

In order to apply the coupled fixed point theory of mixed monotone operators to the PBVP (5.5.11) we shall convert it to a pair of operator equations.

**Lemma 5.5.1:** *If $p \in L^1(J, \mathbb{R})$, $P(t) = \int_0^t p(s)ds$, $t \in J$ and $P(T) \neq 0$, then $y$, $z$ are coupled quasisolutions of the PBVP (5.5.11) if and only if the functions*

$$\begin{aligned}
u(t) &= (u_1(t), u_2(t)) = (y(t), y'(t)),\\
v(t) &= (v_1(t), v_2(t)) = (z(t), z'(t)), \quad t \in J
\end{aligned} \quad (5.5.13)$$

*are solutions of the operator equations*

$$\begin{aligned}
u &= A(u, v) = (A_1(u, v), A_2(u, v)),\\
v &= A(v, u) = (A_1(v, u), A_2(v, u)),
\end{aligned} \quad (5.5.14)$$

*where*

$$A_1(u,v)(t) = \frac{e^{-P(t)}}{e^{P(T)}-1}\int_o^T e^{P(s)}[p(s)u_1(s)+u_2(s)]\,ds \\ + e^{-P(t)}\int_o^t e^{P(s)}[p(s)u_1(s)+u_2(s)]ds, \tag{5.5.15}$$

*and*

$$A_2(u,v)(t) = \int_o^t e^{P(s)-P(t)}[p(s)u_2(s) - f(s,u_1(s),v_1(s),u_2(s))]ds \\ + \frac{e^{-P(t)}}{e^{P(T)}-1}\int_o^T e^{P(s)}[p(s)u_2(s) - f(s,u_1(s),v_1(s),u_2(s))]ds. \tag{5.5.16}$$

*Proof.* Assume first that the functions $y$, $z$ are coupled quasisolutions of the PBVP (5.5.11). Since the functions $y'$ and $z'$ are absolutely continuous on $J$, then the functions $y''$ and $z''$ are Bochner integrable. From (5.5.11) it then follows that also $f(\cdot,y(\cdot),z(\cdot),y'(\cdot))$ and $f(\cdot,z(\cdot),y(\cdot),z'(\cdot))$ are Bochner integrable, whence the integrals in the right hand side of (5.5.15) and (5.5.16) are defined. Moreover, since $y(0)=y(T)$, it follows from (5.5.13) and (5.5.15) that

$$\begin{aligned} &e^{P(t)}A_1(u,v)(t) \\ &= (e^{P(T)}-1)^{-1}\int_o^T [p(s)e^{P(s)}y(s) + e^{P(s)}y'(s)]ds \\ &\quad + \int_o^t [p(s)e^{P(s)}y(s)+e^{P(s)}y'(s)]ds \\ &= (e^{P(T)}-1)^{-1}[e^{P(T)}y(T)-y(0)] + e^{P(t)}y(t) - y(0) \\ &= e^{P(t)}u_1(t),\ t\in J. \end{aligned}$$

From (5.5.13) and (5.5.16), it follows that for all $t \in J$

$$e^{P(t)}A_2(u,v)(t) = \int_o^t [p(s)e^{P(s)}y'(s)$$

$$-e^{P(s)}f(s,y(s),,z(s),y'(s))]ds$$

$$+\frac{1}{e^{P(T)}-1}\int_o^T [p(s)e^{P(s)}y'(s)-e^{P(s)}f(s,y(s),z(s),y'(s))]ds$$

$$=\int_o^t [p(s)e^{P(s)}y'(s)+e^{P(s)}y''(s)]ds$$

$$+\frac{1}{e^{P(T)}-1}\int_o^T [p(s)e^{P(s)}y'(s)+e^{P(s)}y''(s)]ds$$

$$=e^{P(t)}y'(t)-y'(0)+\frac{1}{e^{P(T)}-1}[e^{P(T)}y'(T)-y'(0)]=e^{P(t)}u_2(t).$$

Thus $u=A(u,v)$. Similarly, it can be shown that $v=A(v,u)$, whence the operator equations (5.5.14) are valid.

Conversely, assume that $u=(u_1,u_2)$ and $v=(v_1,v_2)$ satisfy the equations (5.5.14). In view of (5.5.15) we then have for each $t\in J$

$$e^{P(t)}u_1(t)=\frac{1}{e^{P(T)}-1}\int_o^T e^{P(s)}[p(s)u_1(s)+u_2(s)]ds + \int_o^t e^{P(s)}[p(s)u_1(s)+u_2(s)]ds. \tag{a}$$

Noticing that both $u_1$ and $u_2$ are absolutely continuous, it follows from (a) by differentiation that

$$u_1'(t)=u_2(t)\quad \text{for all } t\in J. \tag{b}$$

From (5.5.14) and (5.5.16) it follows that for each $t\in J$

$$e^{P(t)}u_2(t)=\int_o^t e^{P(s)}[p(s)u_2(s)-f(s,u_1(s),v_1(s),u_2(s))]ds + \frac{1}{e^{P(T)}-1}\int_o^T e^{P(s)}[p(s)u_2(s)-f(s,u_1(s),v_1(s),u_2(s))]ds. \tag{c}$$

Differentiating (c) with respect to $t$ we obtain

$$u_2'(t) = -f(t, u_1(t), v_1(t), u_2(t)) \text{ for a.a. } t \in J. \tag{d}$$

Denoting

$$y(t) = u_1(t), \ z(t) = v_1(t), \quad t \in J, \tag{e}$$

it follows from (b) and (d) that $y \in AC^1(J,E)$, and that

$$y''(t) = -f(t, y(t), z(t), y'(t)) \quad \text{for a.a. } t \in J.$$

From (a) it follows that $y(T) = y(0)$, and from (c) that $y'(T) = y'(0)$. Similarly, it can be shown that $z \in AC^1(J,E)$, and that

$$z''(t) = -f(t, z(t), y(t), z'(t)) \quad \text{for a.a. } t \in J,$$
$$z(0) = z(T) \text{ and } z'(0) = z'(T).$$

Thus the functions $y$, $z$ given by (e) are coupled quasisolutions of the PBVP (5.5.11). □

We are going to show that the PBVP (5.5.11) has extremal coupled quasisolutions in an order interval of $AC^1(J,E)$ if $E$ is an ordered Banach space with regular order cone $K$, and if there is a null set $Z$ in $J$ such that the function $f: J \times E^3 \to E$ has the following properties.

(D0) There exist $\alpha, \beta \in AC^1(J,E)$ with $(\alpha, \alpha') \leq (\beta, \beta')$, and $c_\alpha$, $c_\beta \in L^1(J,E)$ such that
$-\beta''(t) \leq f(t, \beta(t), \alpha(t), \beta'(t)) - c_\beta(t)$ a.e. on $J \setminus Z$,
$-\alpha''(t) \geq f(t, \alpha(t), \beta(t), \alpha'(t)) + c_\alpha(t)$ a.e. on $J \setminus Z$,
$\alpha(T) \geq \alpha(0)$ and $\beta(T) \leq \beta(0)$,
and a function $p \in L^1(J, \mathbb{R}_+)$ with $P(t) = \int_0^t p(s)ds \not\equiv 0$ such that for each $t \in J$,
$\beta'(T) - \beta'(0) \leq \int_t^T e^{P(s)-P(T)} c_\beta(s)ds + \int_o^t e^{P(s)} c_\beta(s)ds$,
$\alpha'(0) - \alpha'(T) \leq \int_t^T e^{P(s)-P(T)} c_\alpha(s)ds + \int_o^t e^{P(s)} c_\alpha(s)ds$.

(D1) $f(\cdot, x(\cdot), y(\cdot), z(\cdot))$ is strongly measurable whenever $x$, $y \in [\alpha, \beta]$ and $z \in [\alpha', \beta']$.

(D2) The function $f(t, x, y, z) - p(t)z$ is nonincreasing in $x$ on $[\alpha(t), \beta(t)]$ and in $z$ on $[\alpha'(t), \beta'(t)]$, and nondecreasing in $y$ on $[\alpha(t), \beta(t)]$ for all $t \in J \setminus Z$.

**Lemma 5.5.2:** *Assume that conditions (D0)–(D2) hold, and denote*

$$a(t) = (\alpha(t), \alpha'(t)), \quad b(t) = (\beta(t), \beta'(t)), \quad t \in J. \tag{5.5.17}$$

*Then* $a \leq A(a, b)$ *and* $A(b, a) \leq b$.

*Proof.* The hypotheses (D0)–(D2) ensure that for all $x$, $y$, $z$ in $AC(J, E)$, satisfying $\alpha \leq x$, $y \leq \beta$ and $\alpha' \leq z \leq \beta'$, and for all $t \in J \setminus Z$,

$$\begin{aligned} \alpha''(t)c_\alpha(t) + p(t)\alpha'(t) &\leq p(t)z(t) - f(t, x(t), y(t), z(t)) \\ &\leq \beta''(t) - c_\beta(t) + p(t)\beta'(t). \end{aligned} \tag{5.5.18}$$

This and (1.3.2) imply that $p(t) - f(t, x(t), y(t), z(t))$ is pointwise a.e. norm-bounded by a function of $L^1(J, I\!R_+)$. This and (D1) ensure that $A_j(a, b)$ and $A_j(b, a)$ are defined for $j = 1, 2$.

Condition (D0) implies that for each $t \in J$

$$\begin{aligned} e^{P(t))} A_1(b, a)(t) &= (e^{P(T)} - 1)^{-1} \int_0^T e^{P(s)}[p(s)\beta(s) + \beta'(s)]ds \\ &\quad + \int_o^t e^{P(s)}[p(s)\beta(s) + \beta'(s)]ds \\ &= (e^{P(T)} - 1)^{-1}[e^{P(T)}\beta(T) - \beta(0)] + e^{P(t)}\beta(t) - \beta(0) \\ &\leq \beta(0) + e^{P(t)}\beta(t) - \beta(0) = e^{P(t)}\beta(t). \end{aligned}$$

From (5.5.16), (5.5.17) and (D0) it follows that for each $t \in J$

$$e^{P(t)}A_2(b,a)(t) = \int_o^t [p(s)e^{P(s)}\beta'(s) - e^{P(s)}f(s,\beta(s),\beta'(s))]ds$$

$$+\frac{1}{e^{P(T)}-1}\int_o^T [p(s)e^{P(s)}\beta'(s) - e^{P(s)}f(s,\beta(s),\alpha(s),\beta'(s))]ds$$

$$\leq \frac{1}{e^{P(T)}-1}\int_o^T [p(s)e^{P(s)}\beta'(s) + e^{P(s)}(\beta''(s) - c_\beta(s))]ds$$

$$+\int_o^t [p(s)e^{P(s)}\beta'(s) + e^{P(s)}(\beta''(s) - c_\beta(s))]ds$$

$$= \frac{1}{e^{P(T)}-1}[e^{P(T)}\beta'(T) - \beta'(0) - \int_o^T e^{P(s)}c_\beta(s)\,ds]$$

$$+e^{P(t)}\beta'(t) - \beta'(0) - \int_o^t e^{P(s)}c_\beta(s)\,ds$$

$$= e^{P(t)}\beta'(t) + \frac{e^{P(T)}}{e^{P(T)}-1}[\beta'(T) - \beta'(0)]$$

$$-\frac{1}{e^{P(T)}-1}\int_o^T e^{P(s)}c_\beta(s)\,ds - \int_o^t e^{P(s)}c_\beta(s)\,ds \leq e^{P(t)}\beta'(t).$$

Thus, $A_1(b,a) \leq \beta$ and $A_2(b,a) \leq \beta'$, whence $A(b,a) \leq b$. The proof that $a \leq A(a,b)$ is similar. □

**Lemma 5.5.3:** *Assume that conditions (D0)–(D2) hold. Denoting*

$$[a,b] = \{(x,y) \in AC(J,E^2) \mid a \leq (x,y) \leq b\}, \qquad (5.5.19)$$

*then the equations (5.5.14)–(5.5.16) define a mixed monotone mapping* $A\colon [a,b] \times [a,b] \to [a,b]$.

*Proof.* In view of lemma 5.5.2 it suffices to show that $A$ is mixed monotone. Let $u = (u_1, u_2)$, $v = (v_1, v_2)$, $w = (w_1, w_2) \in [a, b]$, $u \leq v$, be given. The hypotheses (D0)–(D2) imply that

$$\begin{aligned}
&e^{P(t)}A_1(u,w)(t)\\
&= \frac{1}{e^{P(T)}-1}\int_o^T [p(s)u_1(s)+u_2(s)]ds + \int_o^t [p(s)u_1(s)+u_2(s)]ds\\
&\leq \frac{1}{e^{P(T)}-1}\int_o^T [p(s)v_1(s)+v_2(s)]ds + \int_o^t [p(s)v_1(s)+v_2(s)]ds\\
&= e^{P(t)}A_1(v,w)(t), \quad t \in J,
\end{aligned}$$

and

$$\begin{aligned}
&e^{P(t)}A_2(u,w)(t)\\
&= \int_o^t e^{P(s)}[p(s)u_2(s) - f(s, u_1(s), w_1(s), u_2(s))]ds\\
&+ \frac{1}{e^{P(T)}-1}\int_o^T e^{P(s)}[p(s)u_2(s) - f(s, u_1(s), w_1(s), u_2))]ds\\
&\leq \int_o^t e^{P(s)}[p(s)v_1(s) - f(s, v_1(s), w_1(s), v_2(s))]ds\\
&+ \frac{1}{e^{P(T)}-1}\int_o^T e^{P(s)}[p(s)v_1(s) - f(s, v_1(s), w_1(s), v_2(s))]ds\\
&= e^{P(t)}A_2(v,w)(t), \quad t \in J.
\end{aligned}$$

Thus $A_j(u, w) \leq A_j(v, w)$, $j = 1, 2$. Similarly, it can be shown that $A_j(w, u) \geq A_j(w, v)$, $j = 1, 2$, whence $A$ is mixed monotone. This and the result of lemma 5.5.2 imply the assertion. □

### 5.5.5. Existence of extremal coupled quasisolutions

As an application of lemmas 5.5.1–5.5.3 and theorem 1.4.8 we obtain the following existence result for the extremal coupled quasisolutions of the periodic boundary value problem (5.5.11).

**Theorem 5.5.4:** *Let $E$ be an ordered Banach space with regular order cone $K$. Given $f: J \times E^3 \to E$, assume there is a null set $Z$ in $J$ such that the conditions (D0)–(D2) hold. Then the PBVP (5.5.11) has such coupled quasisolutions $\underline{x}$, $\bar{x}$ in $[\alpha, \beta]$ that $\underline{x} \leq y$, $z \leq \bar{x}$ and $\underline{x}' \leq y'$, $z \leq \bar{x}'$ for all coupled quasisolutions $y$, $z$ of (5.5.11) in $[\alpha, \beta]$ such that $y'$, $z' \in [\alpha', \beta']$.*

*Proof.* By lemma 5.5.1 it suffices to show that the operator $A: [a, b] \times [a, b] \to [a, b]$, defined by (5.5.14)–(5.5.16), has the extremal coupled fixed points. From (5.5.15) and (5.5.16) it follows by differentiation that

$$(A_1(u,v))'(t) = u_2(t) + p(t)(u_1(t) - A_1(u,v)(t), \tag{a}$$

and

$$(A_2(u,v))'(t) = p(t)(u_2(t) - A_2(u,v)(t)) - f(t, u_1(t), v_1(t), u_2(t)), \tag{b}$$

for a.a. $t \in J$. Denoting

$$w_j(t) = \int_0^t (\|\alpha^{(j)}(s)\| + \|\beta^{(j)}(s)\| + 2p(s)(\|\alpha^{(j-1)}(s)\| + \|\beta^{(j-1)}(s)\|)),$$

it follows from (a), (b) (C0) and (1.3.2) that

$$|A_j(u,v)(t) - A_j(u,v)(s)| \leq (1+\gamma)|w_j(t) - w_j(s)|, \tag{c}$$

for $(u, v) \in [a, b]$, $s$, $t \in J$.

The above argumentation verifies that $A$ satisfies the hypotheses of theorem 1.4.8, whence $A$ has the extremal coupled fixed points $x_* = (\underline{x}, \underline{y})$ and $x^* = (\bar{x}, \bar{y})$. From lemma 5.5.1 it follows that $\underline{x}$, $\bar{x}$ are coupled quasisolutions of the PBVP (5.5.11), and that $\underline{x}' = \underline{y}$ and $\bar{x}' = \bar{y}$.

If $y$, $z \in [\alpha, \beta]$ are coupled quasisolutions of (5.5.11) such that If $y'$, $z' \in [\alpha', \beta']$, it follows from lemma 5.5.1 that $(y, y')$,

$(z, z')$ are coupled fixed points of $A$. Because of the extremality of $x_*$ and $x^*$ it follows that $x_* \le (y, y'), (z, z') \le x^*$, i.e. $\underline{x} \le y$, $z \le \bar{x}$ and $\underline{x}' \le y'$, $z' \le \bar{x}'$. □

As a consequence of theorem 5.5.4 we obtain

**Corollary 5.5.2:** *Given functions $f_j \colon J \times E \to E$, $j = 1, 2, 3$, assume that conditions (D0) and (D1) hold for the function*

$$f(t, x, y, z) = f_1(t, x) + f_2(t, y) + f_3(t, z), \quad t \in J, \; x, \; y, \; z \in I\!R.$$

*If $f_1(t, \cdot)$ and $f_2(t, \cdot)$ is nonincreasing in $[\alpha(t), \beta(t)]$ for a.a. $t \in J$, and if $f_3(t, \cdot)$ is nondecreasing in $[\alpha'(t), \beta'(t)]$ for a.a. $t \in J$, then the PBVP*

$$\begin{aligned} -x'' &= f_1(t, x) + f_2(t, x) + f_3(t, x'), \\ x(0) &= x(T), \; x'(0) = x'(T) \end{aligned}$$

*has the extremal coupled quasisolutions in $[\alpha, \beta]$.*

If the functions $f$, $f_j$ are Carathéodory functions in theorem 5.5.4 and in corollary 5.5.2, then $A$ is continuous. In this case $(\underline{x}, \underline{x}')$ and $(\bar{x}, \bar{x}'))$ are the uniform limits of the successive approximations $(x_n, y_n)$ given by $x_o = (\alpha, \alpha')$ $y_o = (\beta, \beta'))$ and

$$x_{j+1} = A(x_j, y_j), \quad y_{j+1} = A(y_j, x_j), \quad j \in I\!N.$$

Moreover, if the functions $f$, $f_j$ are continuous, then the differential equations in (5.5.11) and in (5.5.21) hold for all $t \in J$.

If $c_\alpha(t) = p(t)\, h$ and $c_\beta(t) = p(t)\, k$, $h$, $k \in E$, condition (D0) can be reduced to the form

(D3) There exist $\alpha$, $\beta \in AC^1(J, E)$ with $(\alpha, \alpha') \le (\beta, \beta')$, and $p \in L^1(J, I\!R_+)$ such that $P(t) = \int_o^t p(s)\, ds$ is positive at

$t = T$, and that
$-\beta''(t) \le f(t, \beta(t), \alpha(t), \beta'(t)) - p(t)k$ and
$-\alpha''(t) \ge f(t, \alpha(t), \beta(t), \alpha'(t)) + p(t)h$
for all $t \in J \setminus Z$, $\alpha(0) \le \alpha(T)$ and $\beta(0) \ge \beta(T)$,
$\beta'(T) - \beta'(0) \le (1 - e^{-P(T)})k$ and
$\alpha'(0) - \alpha'(T) \le (1 - e^{-P(T)})h$.

If $c_\alpha(t) = c_\beta(t) \equiv 0$ in (D0), it is reduced to

(D4) There exist $\alpha, \beta \in AC^1(J, E)$ with $(\alpha, \alpha') \le (\beta, \beta')$,
$-\beta''(t) \le f(t, \beta(t), \alpha(t), \beta'(t))$ for all $t \in J \setminus Z$,
$\beta(T) \le \beta(0)$, $\beta'(T) \le \beta'(0)$, and
$-\alpha''(t) \ge f(t, \alpha(t), \beta(t), \alpha'(t))$ for all $t \in J \setminus Z$,
$\alpha(T) \ge \alpha(0)$, $\alpha'(T) \ge \alpha'(0)$.

## 5.6. ON MILD SOLUTIONS OF FIRST ORDER IVPS

In this section we shall consider the solvability of the IVP

$$x' = A(t)x + g(t, x), \qquad x(0) = x_o. \tag{5.6.1}$$

Given a closed interval $J = [0, c]$ and a Banach space $E$, assume that $g\colon J \times E \to E$, and that each $A(t)$, $t \in J$, is a linear operator from a dense subspace $D(A(t))$ of $E$ into $E$, given by

$$A(t)x = \lim_{h \to 0+} \frac{T(t+h, t)x - x}{h}, \qquad x \in D(A(t)), \tag{5.6.2}$$

where $T(t, s) \in L(E)$, i.e. $T(t, s)$ is a bounded linear operator on $E$, for each point $(t, s)$ in $\Gamma = \{(t, s) \mid 0 \le s \le t \le c\}$, satisfying

(T0) $T(t, t) = I$ ($I$ is the identity operator in $E$);
(T1) $T(t, s)T(s, r) = T(t, r)$ for $0 \le r \le s \le t \le c$;
(T2) the mapping $(t, s) \mapsto T(t, s)x$ is strongly continuous in $\Gamma$ for each $x \in E$.

If $g$ is continuous, and if $x$ is a differentiable solution of the integral equation

$$x(t) = T(t,0)x_o + \int_o^t T(t,s)g(s,x(s))\,ds \tag{5.6.3}$$

on $J$, then $x$ is a solution of the IVP (5.6.1) with $A$ given by (5.6.2). Our purpose is to show that (5.6.3) may have continuous solutions, called *mild solutions* of (5.6.1) (cf. Ladas and Lakshmikantham (1972), Martin (1977)), also when $g$ is not continuous.

We shall first present existence and uniqueness results for mild solutions of (5.6.1) when $g$ is a Carathéodory function. In the case when $E$ is an ordered Banach space we shall study existence of extremal mild solutions of (5.6.1) by assuming that $g(t,\cdot)$ is nondecreasing, but not necessarily continuous. Dependence of mild solutions of (5.6.1) on the initial value $x_o$ and on the function $g$ is also studied. The obtained results are then applied to initial value problems of first and second order partial differential equations involving discontinuous nonlinearities.

### 5.6.1. Existence, uniqueness & continuous dependence

Let $T\colon \Gamma \to L(E)$ satisfy conditions (T0)–(T2). Condition (T2) implies by the uniform boundedness principle that

$$M = \sup\{\|T(t,s)\| \mid (t,s) \in \Gamma\} < \infty. \tag{5.6.4}$$

Assume that $g\colon J \times E \to E$ satisfies the following conditions.

(g0) $g(\cdot, z)$ is strongly measurable for each $z \in E$, and $g(\cdot,0)$ is Bochner integrable.

(g1) There is $r > 0$ such that $\|g(t,x) - g(t,y)\| \le q(t, \|x-y\|)$ for all $x$, $y \in E$ with $\|x - y\| < r$ and for a.a. $t \in J$, where $q\colon J \times \mathbb{R}_+ \to \mathbb{R}_+$ is a Carathéodory function, $q(t,\cdot)$ is nondecreasing for a.a. $t \in J$, the IVP

$$u' = M\,q(t,u), \qquad u(0) = u_o, \tag{5.6.5}$$

with $M$ given by (5.6.4), has for some $u_o > 0$ an upper solution on $J$, and the zero-function is the only solution of (5.6.5) when $u_o = 0$.

**Theorem 5.6.1:** *If the hypotheses (T0)–(T2), (g0) and (g1) hold, and if $x_o \in E$, then for each choice of $y_o \in C(J, E)$ the sequence $(y_n)_{n=o}^{\infty}$ of the successive approximations*

$$y_{n+1}(t) = T(t,0)x_o + \int_o^t T(t,s)g(s, y_n(s))\, ds, \quad t \in J, \qquad (5.6.6)$$

*converges uniformly on $J$ to a unique mild solution $x = x(\cdot, x_o)$ of (5.6.1). Moreover, if $u = u(\cdot, u_o)$ denotes the minimal solution of the IVP (5.6.5) on $J$, then for all $x_o$, $\hat{x}_o \in E$ with $\|x_o - \hat{x}_o\|$ small enough*

$$\|x(t, x_o) - x(t, \hat{x}_o)\| \leq u(t, M\|x_o - \hat{x}_o\|), \qquad t \in J. \qquad (5.6.7)$$

*In particular, $x = x(\cdot, x_o)$ depends continuously on $x_o$ in the sense that $x(t, \hat{x}_o) \to x(t, x_o)$ uniformly over $t \in J$ as $\hat{x}_o \to x_o$.*

*Proof.* The hypotheses given for $q$ in (g1) imply by proposition 2.1.9 (cf. the proof of theorem 5.1.1) that there is $r_o > 0$ such that the minimal solution $b = u(\cdot, r_o)$ exists, that $r_o \leq b(t) < r$ for each $t \in J$, that the equation

$$Qw(t) = \int_o^t M\, q(s, w(s))\, ds, \quad t \in J \qquad \text{(a)}$$

defines a nondecreasing mapping $Q\colon [0, b] \to [0, b]$, that

$$r_o + Qb(t) = b(t), \quad t \in J \qquad \text{(b)}$$

and that the sequence $(Q^n b)_{n=o}^{\infty}$ converges uniformly on $J$ to the 0-function.

Since $u(t) \equiv 0$ satisfies $u'(t) = q(t, u(t))$ for a.a. $t \in J$, then $q(t, 0) = 0$ for a.a. $t \in J$. From (g1) it then follows that also $g$ is a Carathéodory function. If $y \in C(J, E)$ is given, then $g(\cdot, y(\cdot))$ is by theorem 1.4.3 strongly measurable on $J$. This and (T2) imply by corollary 1.4.4 that $s \mapsto T(t,s)g(s, y(s))$ is strongly measurable on $[0, t]$ for each $t \in J$. Moreover, as in the proof of theorem 5.1.1 one can show that

$$\|T(t,s)g(s, y(s))\| \leq N(s) = M\,\|g(s, 0)\| + m\,b'(s), \qquad \text{(c)}$$

for all $t \in J$ and for a.a. $s \in [0, t]$, where $M$ is given by (5.6.4) and $m \geq \frac{\|y\|_o}{r_o}$. Thus $s \mapsto T(t,s)g(s, y(s))$ is Bochner integrable on $[0, t]$ for each $t \in J$. Hence, the equation

$$Fy(t) = T(t, 0)x_o + \int_o^t T(t,s)g(s, y(s))\,ds, \quad t \in J \qquad \text{(d)}$$

defines for each fixed $x_o \in E$ a mapping $Fy\colon J \to E$. If $0 \leq \bar{t} \leq t \leq c$, it follows by using (T1), (c) and (5.6.4) that

$$\begin{aligned} &\|Fy(t) - Fy(\bar{t})\| \leq \|T(t,0)x_o - T(\bar{t},0)x_o\| \\ &+ \|\int_o^{\bar{t}} (T(t,s) - T(\bar{t},s))g(s, y(s))\,ds\| + \int_{\bar{t}}^t \|T(t,s)g(s, y(s))\|\,ds \\ &\leq \|T(t,0)x_o - T(\bar{t},0)x_o\| + \|(T(t,\bar{t}) - I)\int_o^{\bar{t}} T(\bar{t},s)g(s, y(s))\,ds\| \\ &+ \int_{\bar{t}}^t N(s)\,ds. \end{aligned}$$

This implies by (T0) and (T2) that $Fy \in C(J, E)$. By using (g1), (a) and (d) it is easy to show that (5.1.4) holds for $F$ and $Q$.

The above proof shows that all the hypotheses of theorem 1.4.9 are valid, whence the iteration sequence $(F^n y_o)_{n=o}^\infty$, which equals to the sequence of the successive approximations (5.6.6), converges for each $y_o \in C(J, E)$ uniformly in $J$ to a unique fixed point $x$ of $F$. From the definition (d) of $F$ it follows that $x$ is the uniquely determined solution the integral equation (5.6.3) on $J$, and hence the only mild solution of the IVP (5.6.1).

To prove (5.6.7), let $x_o$, $\hat{x}_o \in E$ satisfy $M\|x_o - \hat{x}_o\| \le r_o$. The solutions $x = x(\cdot, x_o)$ and $\hat{x} = x(\cdot, \hat{x}_o)$ exist by the above proof, and they are the unique fixed points of the integral operators $F$, $\hat{F}\colon C(J, E) \to C(J, E)$, defined by (d) and

$$\hat{F}y(t) = T(t,0)\hat{x}_o + \int_o^t T(t,s)g(s, y(s))\, ds, \quad t \in J.$$

Corollary 2.1.4 and lemma 1.5.5 imply that $u(t, M\|x_o - \hat{x}_o\|)$ exists and is a solution of the operator equation

$$u = M\|x_o - \hat{x}_o\| + Qu,$$

with $Q$ defined by (a). Because $F$ and $Q$ satisfy also the hypotheses of theorem 1.4.9, and since

$$F\hat{x}(t) - \hat{x}(t) = F\hat{x}(t) - \hat{F}\hat{x}(t) = T(t,0)(x_o - \hat{x}_o), \quad t \in J,$$

then (5.6.7) follows from (1.4.19) when $y_o = \hat{x}$. In view of proposition 2.1.9 $u(\cdot, u_o)$ is nondecreasing with respect to $u_o \in [0, r_o]$ and $u(t, \frac{1}{n}) \to 0$ uniformly over $t \in J$ as $n \to \infty$, whence (5.6.7) implies that $x(t, \hat{x}_o) \to x(t, x_o)$ uniformly over $t \in J$ as $\hat{x}_o \to x_o$. □

**Remark 5.6.1:** If the IVP (5.6.5) has for some positive value of $M$ the zero function as the only solution when $u_o = 0$, the same does not necessarily hold for all positive $M$, as we see from the following example (cf. Carathéodory (1948), p. 676).

Define $q \in C(J \times \mathbb{R}_+, \mathbb{R}_+)$ by

$$q(t,r) = \begin{cases} 2t, & \text{for } r \ge t^2,\ t \in J, \\ \frac{2r}{t}, & \text{for } 0 \le r < t^2,\ 0 < t \le c. \end{cases}$$

It is easy to show that $u(t) \equiv 0$ is the only solution of (5.6.5) when $M = \frac{1}{2}$ and $u_o = 0$, whereas $u(t) = \gamma t^2$, $t \in J$ is for each $\gamma \in [0,1]$ a solution of (5.6.5) when $M = 1$ and $u_o = 0$.

The following result shows that the above kind of counterexample does not exist if $q$ is of Osgood type.

**Proposition 5.6.1:** *Let the hypotheses (T0)–(T2) and (g0) hold, and assume there is $r > 0$ and $p \in L^1(J, \mathbb{R}_+)$ such that for all $x, y \in E$ with $\|x - y\| < r$ and for a.a. $t \in J$,*

$$\|g(t,x) - g(t,y)\| \le p(t)\,\varphi(\|x-y\|), \tag{5.6.8}$$

*where $\varphi \in C(\mathbb{R}_+, \mathbb{R}_+)$ is nondecreasing and $\int_o^r \frac{dv}{\varphi(v)} = \infty$. Then the IVP (5.6.1) has for each $x_o \in E$ a unique mild solution $x$, and it depends continuously on $x_o$.*

*Proof.* The proof of proposition 5.1.1 with slight modifications shows that condition (g1) holds when $q\colon J \times \mathbb{R}_+ \to \mathbb{R}_+$ is defined by

$$q(t,v) = p(t)\varphi(v), \quad t \in J,\, v \in \mathbb{R}_+ \tag{a}$$

Thus the hypotheses of theorem 5.6.1 hold with $q$ given by (a).

□

Each of the functions $\varphi_m$, $m \in \mathbb{N}$, defined by (2.1.18) satisfy the hypotheses given for $\varphi$ in proposition 5.6.1. In particular, when $m = 0$ we obtain

**Corollary 5.6.1:** *Let the hypotheses (T0)–(T2) and (g0) hold. Assume there is $p \in L^1(J, \mathbb{R}_+)$ such that*

$$\|g(t,x) - g(t,y)\| \le p(t)\|x - y\| \tag{5.6.9}$$

*for all $x, y \in E$ and for a.a. $t \in J$. Then the IVP (5.6.1) has for each $x_o \in E$ exactly one mild solution $x$. Moreover, $x$ depends continuously on $x_o$.*

In particular, we have

**Corollary 5.6.2:** *If the hypotheses (T0)–(T2) hold, and if* $C \in L^1(J,E)$, *then the IVP*

$$x' = A(t)x + C(t), \qquad x(0) = x_o \tag{5.6.10}$$

*has for each* $x_o \in E$ *a unique mild solution, and it depends continuously on* $x_o$.

### 5.6.2. On extremal mild solutions

In this subsection we shall consider the existence of the extremal mild solutions of the IVP (5.6.1), when $E$ is an ordered Banach space with regular order cone. Assume now that $T\colon \Gamma \to L(E)$ satisfies conditions (T0)–(T2) and condition

(T3) $T(t,s)$ is order-preserving for all $(t,s) \in \Gamma$.

Given $x_o \in E$, we say that $x \in C(J,E)$ is a *lower mild solution* of the IVP (5.6.1) on $J$ if

$$x(t) \leq T(t,0)x_o + \int_o^t T(t,s)g(s,x(s))\,ds, \quad t \in J. \tag{5.6.11}$$

An upper mild solution of (5.6.1) is defined similarly, by reversing inequality sign in (5.6.11).

Let us impose the following hypotheses on $g\colon J \times E \to E$.

(g2) (5.6.1) has a lower mild solution $\underline{x}$ and an upper mild solution $\bar{x}$ on $J$, such that $\underline{x} \leq \bar{x}$, and that $g(\cdot,\underline{x}(\cdot))$ and $g(\cdot,\bar{x}(\cdot))$ are Bochner integrable.

(g3) $g(\cdot,x(\cdot))$ is strongly measurable whenever $x \in C(J,E)$.

(g4) $g(t,\cdot)$ is nondecreasing for a.a. $t \in J$.

**Theorem 5.6.2:** *Let $E$ be an ordered Banach space with regular order cone. If the hypotheses (T0)–(T3) and (g2)–(g4) hold, then the IVP (5.6.1) has the extremal mild solutions between* $\underline{x}$ *and* $\bar{x}$.

*Proof.* By definitions, $\underline{x}$, $\bar{x} \in C(J,E)$. If $x$ belongs to the order interval $[\underline{x}, \bar{x}]$ of $C(J,E)$, then (g3) and (T2) imply by corollary 1.4.3 that $s \mapsto T(t,s)g(s,x(s))$ is strongly measurable on $[0,s]$ for each $t \in J$. By conditions (T3) and (g4) we have

$$T(t,s)g(s,\underline{x}(s)) \leq T(t,s)g(s,x(s)) \leq T(t,s)g(s,\bar{x}(s))$$

for all $t \in J$ and for a.a. $s \in [0,t]$. Applying this, the triangle inequality, (1.3.2) and (5.6.4), it follows that for all $t \in J$ and for a.a. $s \in [0,t]$,

$$\|T(t,s)g(s,x(s))\| \leq N(s), \tag{a}$$

where

$$N(s) = (1+\gamma)M(\|g(s,\underline{x}(s))\| + \|g(s,\bar{x}(s))\|).$$

Thus $s \mapsto T(t,s)g(s,x(s))$ is Bochner integrable on $[0,t]$ for each $t \in J$, so that the equation

$$Gx(t) = T(t,0)x_o + \int_o^t T(t,s)g(s,x(s))\,ds, \quad t \in J, \tag{5.6.12}$$

defines a continuous mapping $Gx\colon J \to E$ for each $x \in [\underline{x}, \bar{x}]$. In view of conditions (T3) and (g4) it follows from (5.6.12) by corollary 1.4.6 that $Gx \leq Gy$ whenever $x, y \in [\underline{x}, \bar{x}]$ and $x \leq y$. Moreover, it follows from (5.6.12) by the definition of mild upper and lower solutions of (5.6.1) that $\underline{x} \leq G\underline{x}$ and $G\bar{x} \leq \bar{x}$. Thus $G$ is a nondecreasing mapping from $[\underline{x}, \bar{x}]$ to $[\underline{x}, \bar{x}]$.

Assume now that $(y_n)_{n=1}^{\infty}$ is a monotone sequence in $[\underline{x}, \bar{x}]$. Because the sequence $(Gy_n)_{n=1}^{\infty}$ is pointwise order bounded and monotone, and since the order cone $K$ of $E$ is regular, it follows that

$$y(t) = \lim_{n \to \infty} Gy_n(t) \tag{c}$$

exists for all $t \in J$. If $0 \le \bar{t} \le t \le c$, then (cf. the proof of theorem 5.6.1)

$$\|Gy_n(t) - Gy_n(\bar{t})\| \le \|T(t,0)x_o - T(\bar{t},0)x_o\| + \|(T(t,\bar{t}) - I)z_n\| + \int_{\bar{t}}^{t} N(s)\,ds, \tag{d}$$

where $z_n = \int_o^{\bar{t}} T(\bar{t},s)g(s,y_n(s))\,ds$, $n = 1,2,\dots$. Since $(z_n)_{n=1}^{\infty}$ is by (T3), (g4) and corollary 1.4.6 monotone and order bounded, then it converges. From (d) it then follows that the sequence $(Gy_n)_{n=o}^{\infty}$ is equicontinuous. Thus the convergence in (c) is uniform on $J$. Obviously, $y \in [\underline{x}, \bar{x}]$, whence $(Gy_n)_{n=1}^{\infty}$ converges in $[\underline{x}, \bar{x}]$.

The above proof shows that the mapping $G\colon [\underline{x}, \bar{x}] \to [\underline{x}, \bar{x}]$ satisfies the hypotheses of theorem 1.2.2, whence $G$ has the least fixed point $x_*$ and the greatest fixed point $x^*$. This and the definition (5.6.12) of $G$ imply that $x_*$ and $x^*$ are solutions of the integral equation (5.6.3). Hence, $x_*$ and $x^*$ are mild solutions of the IVP (5.6.1).

If $x$ is any mild solution of (5.6.1) and $\underline{x} \le x \le \bar{x}$, then $x \in [\underline{x}, \bar{x}]$, whence $x$ is a fixed point of $G$. Since $x_*$ and $x^*$ are the least and the greatest ones, it follows that $x_* \le x \le x^*$. Thus $x_*$ and $x^*$ are the extremal mild solutions of the IVP (5.6.1) in $[\underline{x}, \bar{x}]$. □

**Proposition 5.6.2:** *If $g\colon J \times E \to E$ in theorem 5.6.2 is a Carathéodory function, then the sequence $(y_n)_{n=o}^{\infty}$ of the successive approximations (5.6.6) converges uniformly on $J = [0,c]$ to the minimal (resp. the maximal) mild solution of the IVP (5.6.1) between $\underline{x}$ and $\bar{x}$, if $y_o = \underline{x}$ (resp. $y_o = \bar{x}$).*

*Proof.* Theorem 1.4.3 implies that $g(\cdot, x(\cdot))$ is strongly measurable on $J$ for each $x \in [\underline{x}, \bar{x}]$. From the proof of theorem 5.6.2 it follows that the equation (5.6.12) defines a nondecreasing mapping $G\colon [\underline{x}, \bar{x}] \to [\underline{x}, \bar{x}]$. Choosing $y_o = \underline{x}$ and denoting $y_n = G^n y_o$, we obtain a nondecreasing sequence $(y_n)_{n=o}^{\infty}$ in

$[\underline{x}, \bar{x}]$, which equals to the sequence of the successive approximations defined in (5.6.6). The proof of theorem 5.6.3 implies that the sequence $(Gy_n)_{n=o}^{\infty}$ converges uniformly on $J$ to a continuous function $x_* \in [\underline{x}, \bar{x}]$. From the definition of $y_n$ it follows that

$$x_*(t) = \lim_{n\to\infty} Gy_n(t) = \lim_{n\to\infty} y_n(t), \quad t \in J. \tag{5.6.13}$$

By using (5.6.12), (5.6.13), and the dominated convergence theorem, and noticing that $g(t,\cdot)$ is continuous for a.a. $t \in J$, it is easy to show that

$$Gx_*(t) = \lim_{n\to\infty} Gy_n(t), \quad t \in J.$$

Thus $x_*$ is a fixed point of $G$. Since $y_o = \underline{x}$ is a lower bound of $G[\underline{x}, \bar{x}]$, and since $G$ is nondecreasing, it is easy to see that $x_*$ is the least fixed point of $G$, and hence, by the proof of theorem 5.6.2, the minimal mild solution of (5.6.1) in $[\underline{x}, \bar{x}]$. □

**Lemma 5.6.1:** *Let $E$ be an ordered Banach space, let $T : \Gamma \to L(E)$ satisfy conditions (T0)–(T3), and let $g\colon J \times E \to E$ satisfy hypotheses (g0), (g1) and (g4). Then the IVP (5.6.1) has for each $x_o \in E$ a unique mild solution $x$ on $J$, and*

a) *$x \le \bar{x}$ for each upper mild solution $\bar{x}$ of (1.1),*
b) *$\underline{x} \le x$ for each lower mild solution $\underline{x}$ of (1.1).*

*Proof.* Let $x_o \in E$ be given and let $\underline{x}$ be a lower mild solution of (5.6.1) on $J$. The given hypotheses imply that (5.6.12) defines a nondecreasing operator $G\colon C(J,E) \to C(J,E)$. From theorem 5.6.1 it follows that the sequence $(G^n\underline{x})_{n=o}^{\infty}$ converges uniformly on $J$ to a unique mild solution $x$ the IVP (5.6.1), which is also a unique fixed point of $G$. Since $\underline{x} \le G\underline{x}$ by (5.6.11) and (5.6.12), the sequence $(G^n\underline{x})_{n=o}^{\infty}$ is nondecreasing. Thus $\underline{x} \le \lim_n G^n\underline{x} = x$. The proof of assertion b) is similar. □

**Proposition 5.6.3:** *Let $E$ be an ordered Banach space with regular order cone. Assume that $g\colon J \times E \to E$ satisfies conditions* *(g3) and (g4), that $T\colon \Gamma \to L(E)$ satisfies conditions (T0)–(T3), and that*

$$g_1(t,x) \le g(t,x) \le g_2(t,x) \text{ for all } x \in E \text{ and for a.a. } t \in J, \tag{5.6.14}$$

*where $g_i\colon J \times E \to E$, $i = 1, 2$, satisfy conditions (g0), (g1) and (g4). Then the IVP (5.6.1) has for each choice of $x_o \in E$ the extremal mild solutions.*

*Proof.* Let $x_o \in E$ be given. Theorem 5.6.1 implies that that the integral equation

$$x(t) = T(t,0)x_o + \int_o^t T(t,s)g_i(t,x(s))\,ds \tag{a}$$

has a unique solution $\underline{x}$ when $i = 1$ and $\bar{x}$ when $i = 2$. From (a) and (5.6.14) it follows that $\underline{x}$ is a lower solution of (a) with $i = 2$. Thus $\underline{x} \le \bar{x}$ by lemma 5.6.1. Applying, (g4), the triangle inequality, (1.3.2) and (5.6.14), it follows that if $x \in [\underline{x}, \bar{x}]$, then

$$\|g(t,x(t))\| \le N(t) \quad \text{for a.a. } \; t \in J, \tag{a}$$

where

$$N(t) = (1+\gamma)(\|g_1(t,\underline{x}(t))\| + \|g_2(t,\bar{x}(t))\|).$$

This and (g3) imply that the functions $g(\cdot,\underline{x}(\cdot))$ and $g(\cdot,\bar{x}(\cdot))$ are Bochner integrable. Moreover,

$$\underline{x}(t) \le T(t,0)x_o + \int_o^t T(t,s)g(s,\underline{x}(s))\,ds \;\text{ for a.a. } t \in J,$$

whence $\underline{x}$ is a mild lower solution of (5.6.1). Similarly, it can be shown that $\bar{x}$ is a mild upper solution of (5.6.1). Thus $g$

satisfies also condition (g2). By theorem 5.6.2 the IVP (5.6.1) has the minimal mild solution $x_*$ and the maximal mild solution $x^*$ between $\underline{x}$ and $\bar{x}$.

If $x$ is a mild solution of (5.6.1) on $J$, it follows from (5.6.3), (5.6.14) and (a) that $x$ is a mild upper solution of (a) for $i = 1$ and a mild lower solution of (a) for $i = 2$, whence the results a) and b) of lemma 5.6.1 imply that $\underline{x} \le x \le \bar{x}$. Hence all the mild solutions of (5.6.1) on $J$ lie between $\underline{x}$ and $\bar{x}$, whence $x_*$ is the least and $x^*$ the greatest of all the mild solutions of (5.6.1) on $J$.

□

### 5.6.3. Existence of minimal or maximal mild solution

In the case when the order cone $K$ of $E$ is fully regular, the existence of mild upper or lower solutions of the IVP (5.6.1) can be replaced by the condition

(g5) $\|g(t,x)\| \le h(t, \|x\|)$ for all $x \in E$ and for a.a. $t \in J$, where $h\colon J \times I\!R_+ \to I\!R_+$ is a standard function, $h(t,\cdot)$ is nondecreasing for a.a. $t \in J$, and the IVP

$$u' = M\,h(t,u), \quad u(t_o) = u_o, \tag{5.6.15}$$

with $M$ given by (5.6.4), has for each $u_o \in I\!R_+$ an upper solution.

**Theorem 5.6.3:** *Given an ordered Banach space $E$ with fully regular order cone, assume that $T\colon \Gamma \to L(E)$ and $g\colon J \times E \to E$ satisfy conditions (T0)–(T3) and (g3)–(g5).*

a) *If the IVP (5.6.1) has a mild lower solution $\underline{x}$, then (5.6.1) has the minimal mild solution in $[\underline{x})$.*

b) *If the IVP (5.6.1) has a mild upper solution $\bar{x}$, then (5.6.1) has the maximal mild solution in $(\bar{x}]$.*

*Proof.* a) Let $\underline{x}$ be a mild lower solution of (5.6.1). Denote $u_o = \max\{M\|x_o\|, \|\underline{x}\|_o\}$, and let $\bar{u}$ be an upper solution of the IVP

(5.6.15). Since the zero function is a lower solution of (5.6.15), it follows from theorem 2.1.3 that (5.6.15) has the minimal solution $u_*$ on $J$ between 0 and $\bar{u}$. Moreover, lemma 1.5.5 implies that

$$u_*(t) = u_o + \int_o^t M\, h(s, u_*(s))\, ds, \quad t \in J. \tag{a}$$

Denote by $Y = \{y \in C(J,E) \mid \underline{x} \leq x \text{ and } |y| \leq u_*\}$. $Y$ is nonempty since $\underline{x} \in Y$. If $y \in Y$, it is easy to see that the equation (5.6.12) defines a mapping $Gy \in C(J,E)$. Moreover,

$$\begin{aligned} \|Gy(t)\| &\leq \|T(t,0)x_o\| + \int_o^t \|T(t,s)g(s,y(s))\|\, ds \\ &\leq M\,\|x_o\| + \int_o^t M\,\|g(s,y(s))\|\, ds \\ &\leq M\,\|x_o\| + \int_o^t M\, h(s, \|y(s)\|)\, ds \\ &\leq M\,\|x_o\| + \int_o^t M\, h(s, u_*(s))\, ds \leq u_*(t) \end{aligned}$$

for each $t \in J$. The definition (5.6.11) of a mild lower solution of (5.6.1), and (5.6.12) imply that $\underline{x} \leq G\underline{x}$. Conditions (T3) and (g4) imply that $G$ is nondecreasing in $Y$. Thus the equation (5.6.12) defines a nondecreasing mapping $G\colon Y \to Y$. As in the proof of theorem 5.6.1 it can be shown that if $(x_n)_{n=o}^{\infty}$ is a nondecreasing sequence in $Y$, then $(Gx_n)_{n=o}^{\infty}$ converges uniformly on $J$ to a function $y \in Y$. Since $\underline{x}$ is the least element of $Y$, then $G$ has by proposition 1.2.2 b) the least fixed point $x_*$ in $Y$. By definition, $x_*$ is a mild solution of the IVP (5.6.1) in $Y$.

Assume now that $x\colon J \to E$ is a mild solution of (5.6.1) in $[\underline{x})$. By choosing in the above proof $u_o = \max\{M\|x_o\|, \|\underline{x}\|_o, \|x\|_o\}$ one can assume that $x \in Y$. By the definition (5.6.12) of $G$, $x = Gx$. To prove that $x_* \leq x$, let $C$ be the w.o. chain of $G$ iterations of $\underline{x}$ defined in theorem 1.1.1. To show that $y \leq x$ for each $y \in C$ make a counter-hypothesis: there is the least element $y \in C$ such that $y \not\leq x$. Since $\underline{x} \leq x$, then $\underline{x} < y$. If $z \in C$ and $z < y$ then $z \leq x$. Since $G$ is nondecreasing, then $Gz \leq Gx = x$. Thus $y = \sup G\{z \in C \mid z < y\} \leq x$, contradicting with the choice of $y$. Consequently $x_* = \max C \leq x$, which proves that $x_*$ is the least of all the mild solutions of the IVP (1.1) in $[\underline{x})$.

The proof of the case b) is similar. □

**Corollary 5.6.3:** *Let the hypotheses of theorem 5.6.3 hold, and let $g$ be a Carathéodory function.*

a) *If the IVP (5.6.1) has a mild lower solution $\underline{x}$, then the sequence $(y_n)_{n=o}^{\infty}$, defined by (5.6.6) with $y_o = \underline{x}$, is nondecreasing and converges on $J$ uniformly to the minimal mild solution of the IVP (5.6.1) in $[\underline{x})$.*

b) *If the IVP (5.6.1) has a mild upper solution $\bar{x}$, then the sequence $(y_n)_{n=o}^{\infty}$, defined by (5.6.6) with $y_o = \bar{x}$, is nondecreasing and converges on $J$ uniformly to the maximal mild solution of the IVP (5.6.1) in $(\bar{x}]$.*

### 5.6.4. Dependence on the data

Next we shall consider the dependence of the extremal mild solutions of the IVP (5.6.1) on $x_o$ and $g$.

**Proposition 5.6.4:** *Let $E$ be an ordered Banach space with regular order cone, let $T\colon \Gamma \to L(E)$ satisfy conditions (T0)–(T2), and let $g$, $\hat{g}\colon J \times E \to E$ satisfy the hypotheses given for $g$ in proposition 5.6.3. If $x_o$, $z_o \in E$, $x_o \leq z_o$, and if*

$$g(t,x) \leq \hat{g}(t,x) \text{ for all } x \in E \text{ and for a.a. } t \in J,$$

*then all the mild solutions of $x$ the IVP (5.6.1) and the IVP (5.2.6) satisfy $x_* \leq x \leq z^*$, where $x_*$ is the minimal mild solution of (5.6.1) and $z^*$ is the maximal mild solution of (5.2.6).*

*Proof.* Similar to that of proposition 5.2.3. □

As special case of the above result we obtain

**Corollary 5.6.4:** *If the hypotheses of proposition 5.6.4 hold, then the extremal mild solutions of the IVP (5.6.1) are nondecreasing with respect to* $x_o$ *and* $g$.

Similarly, it can be shown

**Corollary 5.6.5:** *If the hypotheses of theorem 5.6.3 a) hold, then the minimal mild solution of the IVP (5.6.1) in* $[\underline{x})$ *is nondecreasing with respect to* $x_o$ *and* $g$.

**Corollary 5.6.6:** *Let* $E$ *be an ordered Banach space with regular order cone. Assume that* $g\colon J \times E \to E$ *satisfies (g3) and (g4), that* $T\colon \Gamma \to L(E)$ *satisfy (T0)–(T3), and that for all* $x \in E$ *and for a.a.* $t \in J$

$$C_1(t) \leq g(t,x) \leq C_2(t),$$

*where* $C_i \in L^1(J,E)$, $i = 1,2$. *Then the IVP (5.6.1) has for each choice of* $x_o \in E$ *the extremal mild solutions, both of which are nondecreasing in* $x_o$.

*Proof.* It is easy to see that the hypotheses of proposition 5.6.4 hold when $g_i(t,x) = C_i(t)$, $i = 1,2$, whence the assertions follow from proposition 5.6.4 and corollary 5.6.4. □

### 5.6.5. Applications to PDEs

In this subsection we demonstrate how to apply the preceding results to partial differential equations. Consider first the following IVP of the first order partial differential equation

$$\frac{\partial u}{\partial t} = k\frac{\partial u}{\partial \xi} + f(\xi,t,u), \qquad u(\xi,0) = x_o(\xi), \tag{5.6.16}$$

where $k$ is a given positive constant. If $J = [0, c]$, $c > 0$, and $f: \mathbb{R} \times J \times \mathbb{R} \to \mathbb{R}$ is continuously differentiable, it is easy to see that a continuously differentiable function $u: \mathbb{R} \times J \to \mathbb{R}$ is a solution of the IVP (5.6.16) if and only if it satisfies the integral equation

$$u(\xi, t) = x_o(\xi + kt) + \int_o^t f(\xi + k(t-s), s, u(\xi + k(t-s), s))\, ds. \tag{5.6.17}$$

On the other hand, (5.6.17) may have solutions also when $f$ is not even continuous. Given $p \in [0, \infty)$ we shall now study existence of solutions of (5.6.17) in the set $\mathcal{U}_p$ of those measurable functions $u: \mathbb{R} \times J \to \mathbb{R}$ for which $u(\cdot, t) \in L^p(\mathbb{R})$ for each $t \in J$ and $\lim_{t \to t_o} \int_{-\infty}^{\infty} |(u(\xi, t) - u(\xi, t_o)|^p d\xi = 0$ for each $t_o \in J$. Such solutions of (5.6.17) are called *mild solutions* of the IVP (5.6.16). Because a.e. equal measurable functions are identified, then a mild solution of (5.6.16) is required to satisfy the integral equation (5.6.17) for all $t \in J$ and for a.a. $\xi \in \mathbb{R}$.

Choose $E = L^p(\mathbb{R})$, and define for each $x \in E$

$$T(t, s)x(\xi) = x(\xi + k(t-s)),\ 0 \le s \le t \le c,\ \xi \in \mathbb{R}. \tag{5.6.18}$$

It can be shown (cf. Dunford and Schwartz (1958)) that (5.6.18) defines a mapping $T: \Gamma \to L(E)$ which has properties (T0)–(T2).

Assume first that $f: \mathbb{R} \times J \times \mathbb{R} \to \mathbb{R}$ has the following properties:

(f0) $f(\cdot, \cdot, z)$ is measurable on $\mathbb{R} \times J$ for each $z \in \mathbb{R}$ and $t \mapsto f(\cdot, t, 0)$ is a Bochner integrable mapping from $J$ to $L^p(\mathbb{R})$.

(f1) There is $q \in L^p_+(\mathbb{R})$ such that
$$|f(\xi, t, y) - f(\xi, t, z)| \le q(t)|y - z|$$
for all $t \in J$, $y$, $z \in \mathbb{R}$ and for a.a. $\xi \in \mathbb{R}$.

If $x_o \in L^p(\mathbb{R})$ and $u \in \mathcal{U}_p$, it follows that the right hand side of (5.6.17) is defined. In view of (5.6.18) the equation (5.6.17) can be rewritten as

$$u(\xi,t) = T(t,0)x_o(\xi) + \int_o^t T(t,s)f(\xi,s,u(\xi,s))\,ds. \qquad (5.6.19)$$

If $u \in \mathcal{U}_p$ is a mild solution of (5.6.16), i.e. a solution of (5.6.19), then denoting

$$x(t) = u(\cdot,t),\ g(t,y) = f(\cdot,t,y(\cdot)),\ t \in J,\ y \in L^p(\mathbb{R}), \qquad (5.6.20)$$

$x$ belongs to $C(J, L^p(\mathbb{R}))$, and $x$ is a solution of the integral equation

$$x(t) = T(t,0)x_o + \int_o^t T(t,s)g(s,x(s))\,ds, \quad t \in J. \qquad (5.6.21)$$

Conversely, if $x \in C(J, L^p(\mathbb{R}))$ is a solution of (5.6.21), there is $u \in \mathcal{U}_p$ such that $x(t) = u(\cdot,t)$ for all $t \in J$, and $u$ is a solution of (5.6.19), and thus a mild solution of (5.6.16) (cf. Hille and Phillips (1957)).

Conditions (f0) and (f1) imply that (5.6.20) defines a mapping $g\colon J \times L^p(\mathbb{R}) \to L^p(\mathbb{R})$ which satisfies condition (g0) and the Lipschitz condition (5.6.9). Thus all the hypotheses of corollary 5.6.1 hold, whence the integral equation (5.6.21) has for each $x_o \in L^p(\mathbb{R})$ a unique solution $x$ in $C(J, L^p(\mathbb{R}))$, and it depends continuously on $x_o$. According to the above stated correspondence between solutions of (5.6.21) in $C(J, L^p(\mathbb{R}))$ and mild solutions of (5.6.16) and identification of a.e. equal functions we then obtain

**Proposition 5.6.5:** *If conditions (f0) and (f1) hold, then the IVP (5.6.16) has for each $x_o \in L^p(\mathbb{R})$ a unique mild solution $u = u(\cdot,\cdot;x_o) \in \mathcal{U}_p$, which depends continuously on $x_o$ in the sense that $\|u(\cdot,t;\bar{x}_o) - u(\cdot,t;x_o)\|_p \to 0$ uniformly over $t \in J$ as $\bar{x}_o \to x_o$ in $L^p(\mathbb{R})$.*

Assume next that $f$ has the following properties.

(f2) $f$ is a standard function from $(\mathbb{R} \times J) \times \mathbb{R}$ to $\mathbb{R}$.

(f3) $f(\xi, t, \cdot)$ is nondecreasing for all $t \in J$ and for a.a. $\xi \in \mathbb{R}$.

(f4) There exist $f_1, f_2 \colon \mathbb{R} \times J \times \mathbb{R} \to \mathbb{R}$ which satisfy conditions (f0), (f1) and (f3) such that for all $(\xi, t, v) \in \mathbb{R} \times J \times \mathbb{R}$, $f_1(\xi, t, v) \leq f(\xi, t, v) \leq f_2(\xi, t, v)$.

Assume also that $L^p(\mathbb{R})$ is ordered by the cone $L^p_+(\mathbb{R})$ of its a.e. nonnegative-valued elements. From (5.6.18) it follows that $T(t, s)$ is order-preserving for each $(t, s) \in \Gamma$, so that conditions (T0)–(T3) hold. In view of conditions (f2)–(f4) and definition (5.6.20) of $g$ it is easy to see that conditions (g3) and (g4) hold, and that $g_1(t, y) \leq g(t, y) \leq g_2(t, y)$ for all $t \in J$ and $y \in L^p(\mathbb{R})$, where $g_i$, $i = 1, 2$, is defined by (5.6.20) with $f = f_i$. Noticing also that the order cone $L^p_+(\mathbb{R})$ of $L^p(\mathbb{R})$ is regular, then all the hypotheses of proposition 5.6.3 hold, whence (5.6.21) has for each $x_o \in L^p(\mathbb{R})$ extremal solutions $x_*$ and $x^*$, and they are nondecreasing in $x_o$. According to the above discussion there exist $u_*, u^* \in \mathcal{U}_p$ such that $x_*(t)(\xi) = u_*(\xi, t)$ and $x^*(t)(\xi) = u^*(\xi, t)$ for all $t \in J$ and for a.a. $\xi \in \mathbb{R}$, that $u_*$ and $u^*$ are mild solutions of the IVP (5.6.16). Moreover, defining a partial ordering in $\mathcal{U}_p$ by

$$u \leq v \quad \text{if } u(\xi, t) \leq v(\xi, t) \text{ for all } t \in J \text{ and for a.a. } \xi \in \mathbb{R}, \tag{5.6.22}$$

it is easy to see that if $x, y \in C(J, L^p(\mathbb{R}))$ and if $u$, $v$ are their representatives in $\mathcal{U}_p$, respectively, then $x \leq y$ if and only if $u \leq v$. Thus $u_*$ and $u^*$ are extremal mild solutions of (5.6.16), and they are nondecreasing in $x_o$. Thus we have proved the following result.

**Proposition 5.6.6:** *If conditions (f2)–(f4) hold, then the IVP (5.6.16) has for each $x_o \in L^p(\mathbb{R})$ the extremal mild solutions in $\mathcal{U}_p$, and they are nondecreasing in $x_o$.*

Consider next the IVP of the second order partial differential equation of the form

$$\frac{\partial u}{\partial t} = k\,\frac{\partial^2 u}{\partial \xi^2} + f(\xi, t, u), \qquad u(\xi, 0) = x_o(\xi). \tag{5.6.23}$$

If $J = [0, c]$, $c > 0$, if $f\colon \mathbb{R} \times J \times \mathbb{R} \to \mathbb{R}$ is continuous, and if $k > 0$, then each solution $u\colon \mathbb{R} \times J \to \mathbb{R}$ of the IVP (5.6.23) satisfies the integral equation

$$\begin{aligned} u(\xi, t) = &\int_{-\infty}^{\infty} K(\xi - z, t) x_o(z)\, dz \\ &+ \int_o^t \int_{-\infty}^{\infty} K(\xi - z, t - s) f(z, s, u(z, s))\, dz\, ds, \end{aligned} \tag{5.6.24}$$

where $K(z, t) = \frac{k}{\sqrt{4\pi t}} e^{\frac{-z^2}{4kt}}$, $z \in \mathbb{R}$, $t > 0$ (cf. DuChateau and Zachmann (1987)). Given $p \in [1, \infty)$ we say that $u \in \mathcal{U}_p$ is a *mild solution* of the IVP (5.6.23) if $u$ satisfies the integral equation (5.6.24) for all $t \in J$ and for a.a. $\xi \in \mathbb{R}$.

Define for $y \in E = L^p(\mathbb{R})$, $0 \le s \le t \le c$, and $\xi \in \mathbb{R}$,

$$T(t, s) y(\xi) = \int_{-\infty}^{\infty} K(\xi - z, t - s) y(z)\, dz. \tag{5.6.25}$$

It can be shown (cf. Yoshida (1974)) that (5.6.25) defines a mapping $T\colon \Gamma \to L(E)$ which has properties (T0)–(T2). Because $K$ is positive-valued, then $T(t, s)$ is order-preserving for each $(t, s) \in \Gamma$, so that condition (T3) holds.

Assume that $f\colon \mathbb{R} \times J \times \mathbb{R} \to \mathbb{R}$ has properties (f0)–(f1) or (f2)–(f4). If $x_o \in L^p(\mathbb{R})$ and $u \in \mathcal{U}_p$, it follows that the right hand side of (5.6.24) is defined. In view of (5.6.25) the equation (5.6.24) can be rewritten in the form (5.6.19). Using the definitions (5.6.20) it follows from the above discussion that it suffices to consider the integral equation (5.6.21). Because $f$ satisfies the same hypotheses as in our first application, we obtain

**Proposition 5.6.7:** *If conditions (f0) and (f1) hold, then the IVP (5.6.23) has for each $x_o \in L^p(\mathbb{R})$ a unique mild solution $u$ in $\mathcal{U}_p$ which depends continuously on $x_o$ in the sense defined in proposition 5.6.5. If conditions (f2)–(f4) hold, then the IVP (5.6.23) has for each $x_o \in L^p(\mathbb{R})$ the extremal mild solutions in $\mathcal{U}_p$, and they are nondecreasing in $x_o$.*

## 5.7. SECOND ORDER SEMILINEAR IVPS

In this section we shall consider the second order semilinear initial value problem

$$x'' = Ax + g(t, x, x'), \qquad x(0) = x_o,\ x'(0) = x_1, \tag{5.7.1}$$

where $A$ is the infinitesimal generator of a strongly continuous cosine family $\{C(t) \mid t \in \mathbb{R}\}$ in a Banach space $E$, $g\colon J \times E^2 \to E$ and $J = [0, T]$, $T > 0$.

We shall first prove existence, uniqueness and estimation results for weak solutions of the IVP (5.7.1), by using a comparison method and assuming that $g$ satisfies Carathéodory conditions. The existence of the extremal mild solutions of (5.7.1) is then studied when $E$ is an ordered Banach space, and when $g$ does not depend on $x'$. We shall also discuss the dependence of these solutions on the initial values and on $g$. A characteristic feature of the results concerning extremal solutions is that $g$ is not assumed to be continuous in any of its arguments. Moreover, no compactness assumptions are imposed on $g$. The obtained results are then applied to a second order partial differential equation of hyperbolic type.

### 5.7.1. Preliminaries

Given a Banach space $E$, we say that a family $\{C(t) \mid t \in \mathbb{R}\}$ of operators in $L(E)$ is a *strongly continuous cosine family* if

(i) $C(0) = I$;
(ii) $t \mapsto C(t)x$ is strongly continuous for each fixed $x \in E$;
(iii) $C(t+s) + C(t-s) = 2C(t)C(s)$ for all $s, t \in \mathbb{R}$.

The *strongly continuous sine family* $\{S(t) \mid t \in \mathbb{R}\}$, associated to the given strongly continuous cosine family $\{C(t) \mid t \in \mathbb{R}\}$, is defined by

$$S(t)x = \int_o^t C(s)\,x\,ds, \quad x \in E, \ t \in \mathbb{R}. \tag{5.7.2}$$

Denote

$$\begin{aligned} E_1 &= \{x \in E \mid C(\cdot)x \in C^1(\mathbb{R}, E)\}, \\ E_2 &= \{x \in E \mid C(\cdot)x \in C^2(\mathbb{R}, E)\}. \end{aligned} \tag{5.7.3}$$

It can be shown that $\overline{E}_2 = E$. Obviously, $E_2$ is a subspace of $E_1$. As for the properties of strongly continuous cosine and sine families, see Fattorini (1968, 1969), Goldstein (1985), Travis and Webb (1978a, b).

The *infinitesimal generator* $A\colon E_2 \to E$ of a cosine family $\{C(t) \mid t \in \mathbb{R}\}$ is defined by

$$Ax = \frac{d^2}{dt^2} C(0)x. \tag{5.7.4}$$

Assume now that $A$ is an infinitesimal generator of a given strongly continuous cosine family $\{C(t) \mid t \in \mathbb{R}\}$.

By a *strong solution* of the IVP (5.7.1) on $J$ we mean a function $x\colon J \to E$ with absolutely continuous first derivative, whose second derivative $x''(t)$ exists and equals to $Ax(t) + g(t, x(t), x'(t))$ for a.a. $t \in J$, and which satisfies the initial conditions $x(0) =$

$x_o$, $x'(0) = x_1$. Given $(x_o, x_1) \in E_2 \times E$, we say that $x \in C^1(J, E)$ is a *weak solution* of (5.7.1) if there is $y \in C(J, E)$ such that

$$x(t) = x_o + \int_o^t y(s)\, ds,$$

$$y(t) = S(t)Ax_o + C(t)x_1 + \int_o^t C(t-s)g(s, x(s), y(s))ds. \tag{5.7.5}$$

By the reasoning used in the proofs of proposition 1.2 and theorem 1.3 in Kusano and Oharu (1992) (see also Travis and Webb (1978b)) one can show that

(a) a strong solution $x$ of (5.7.1) is also its weak solution if $g(\cdot, x(\cdot), x'(\cdot))$ is continuous;
(b) a weak solution $x$ of (5.7.1) is also its strong solution if $g(\cdot, x(\cdot), x'(\cdot))$ is absolutely continuous and a.e. differentiable;
(c) if $x \in C^1(J, E)$ is a weak solution of the IVP (5.7.1), then it satisfies the integral equation

$$x(t) = C(t)x_o + S(t)x_1 + \int_o^t S(t-s)g(s, x(s), x'(s))\, ds. \tag{5.7.6}$$

### 5.7.2. Existence and uniqueness results

In this subsection we shall assume that

(C0) $\{C(t) \mid t \in \mathbb{R}\}$ is a strongly continuous cosine family, that $A$ is its infinitesimal generator, and that $\{S(t) \mid t \in \mathbb{R}\}$ is the associated sine family.

Condition (ii), definition (5.7.2) and the uniform boundedness principle imply that

$$M = \sup\{\|C(t)\| \mid t \in J\} < \infty, \tag{5.7.7}$$

and that

$$\|S(t)x - S(\bar{t})x\| \le M|t - \bar{t}|\|x\|, \; x \in E, \quad t, \bar{t} \in J. \tag{5.7.8}$$

Assume also that $g\colon J \times E^2 \to E$ satisfies the following conditions.

(C1) $g(\cdot, x, y)$ is strongly measurable for all $x$, $y \in E$, and $g(t, 0, 0) \in L^1(J, E)$.

(C2) For all $x$, $y$, $h$, $k \in E$ and for a.a. $t \in J$,

$$\|g(t, x + h, y + k) - g(t, x, y)\| \le q(t, \|h\|, \|k\|), \tag{5.7.9}$$

where $q\colon J \times I\!R_+^2 \to I\!R_+$ is a Carathéodory function, $q(t, \cdot, \cdot)$ is nondecreasing for a.a. $t \in J$, the IVP

$$u'' = M\, q(t, u, u'), \qquad u(0) = u_o, \; u'(0) = u_1, \tag{5.7.10}$$

with $M$ given by (5.7.7), has for each $(u_o, u_1) \in I\!R_+^2$ an upper solution on $J$, and the zero-function is the only solution of (5.7.10) when $u_o = u_1 = 0$.

**Theorem 5.7.1:** *If the hypotheses (C0)–(C2) hold, then for each $(x_o, x_1) \in E_2 \times E$ the IVP (5.7.1) has a unique weak solution $x$ on $J$. Moreover, $x$ is of the form $x(t) = x_o + \int_o^t y(s)\, ds$, $t \in J$, where $y$ is the uniform limit of the sequence $(y_n)_{n=o}^{\infty}$ of the successive approximations*

$$\begin{aligned} y_{n+1}(t) &= S(t)Ax_o + C(t)x_1 \\ &\quad + \int_o^t C(t-s)g(s, x_o + \int_o^s y_n(\tau))d\tau, y_n(s))\, ds, \end{aligned} \tag{5.7.11}$$

*$t \in J$, $n \in I\!N$, and with arbitrarily chosen $y_o \in C(J, E)$.*

*Proof.* Let $(x_o, x_1) \in E_2 \times E$ be given. The function $s \mapsto q(s, v(s), v'(s))$ is for each $v \in C^1(J, I\!R_+)$ bounded above by $\frac{1}{M}u''$, where $u$ is an upper solution of the IVP (5.7.10) with $u(0) = \|v\|_o$ and $u'(0) = \|v'\|_o$. The hypotheses given for $q$ in (C2) imply that the equation

$$Qw(t) = \int_o^t M\, q(s, \int_o^s w(\tau)d\tau, w(s))\, ds, \quad t \in J \tag{5.7.12}$$

defines a nondecreasing mapping $Q\colon C(J, \mathbb{R}_+) \to C(J, \mathbb{R}_+)$. Since $u(t) \equiv 0$ satisfies $u''(t) = q(t, u(t), u'(t))$ for a.a. $t \in J$, then $q(t, 0, 0) = 0$ for a.a. $t \in J$. From (C2) it then follows that also $g$ is a Carathéodory function. Thus $g(\cdot, x(\cdot), y(\cdot))$ is by theorem 1.4.3 strongly measurable in $J$ for all $x$, $y \in C(J, E)$. From (C2) it also follows that

$$\|g(t, x(t), y(t))\| \le \|g(t, 0, 0)\| + q(t, \|x\|_o, \|y)\|_o), \; t \in J$$

whence $g(\cdot, x(\cdot), y(\cdot))$ is Bochner integrable. This implies that the equation

$$\begin{aligned} Fz(t) &= S(t)Ax_o + C(t)x_1 \\ &+ \int_o^t C(t-s)g(s, x_o + \int_o^s z(\tau)d\tau, z(s))\, ds \end{aligned} \tag{5.7.13}$$

defines a mapping $F\colon C(J, E) \to C(J, E)$. By using (5.7.9), (5.7.12) and (5.7.13) it is easy to show that

$$|Fy - F\bar{y}| \le Q|y - \bar{y}|, \quad y, \bar{y} \in C(J, E).$$

Condition (C2) ensures that the operator equation

$$u_1 + Qv(t) = v(t), \quad t \in J \tag{a}$$

has an upper solution for each $u_1 \in \mathbb{R}_+$. Moreover, for any such an upper solution $v$ the sequence $(Q^n v)_{n=o}^{\infty}$ converges by proposition 5.2.1 and lemma 5.3.1 uniformly on $J$ to the maximal solution of (5.7.10) with $u_o = u_1 = 0$, i.e. to the 0-function. Thus all the hypotheses of proposition 1.4.8 are valid, whence the iteration sequence $(F^n y_o)_{n=o}^{\infty}$, which equals to the sequence of the successive approximations (5.7.11), converges for each choice of $y_o \in C(J, E)$ uniformly in $J$ to a unique fixed point $y$ of $F$. From the definition (5.7.13) of $F$ it follows that $y$ is the uniquely determined solution of the second integral equation of (5.7.5) when $x(t) = x_o + \int_o^t y(s)\,ds$, $t \in J$, so that $x$ is a unique weak solution of the IVP (5.7.1) on $J$. □

### 5.7.3. Dependence on the initial value

The dependence of the weak solution of the IVP (5.7.1) on the initial values $x_o$ and $x_1$ can be estimated by the minimal solutions of the comparison problem (5.7.10) in the following manner.

**Theorem 5.7.2:** *Let the hypotheses (C0)–(C2) hold, and let $x = x(\cdot, x_o, x_1)$ denote the weak solution of the IVP (5.7.1) on $J$, and $u = u(\cdot, u_o, u_1)$ the minimal solution of the IVP (5.7.10) on $J$. Then for all $x_o$, $\bar{x}_o \in E_2$, $x_1$, $\bar{x}_1 \in E$ and $t \in J$,*

$$\|x(t) - \bar{x}(t)\| \le u(t), \quad \textit{and} \quad \|x'(t) - \bar{x}'(t)\| \le u'(t), \tag{5.7.14}$$

*where $x = x(\cdot, x_o, x_1)$, $\bar{x} = x(\cdot, \bar{x}_o, \bar{x}_1)$ and*
$$u = u(\cdot, \|x_o - \bar{x}_o\|, MT\|A(x_o - \bar{x}_o)\| + M\|x_1 - \bar{x}_1\|).$$
*Moreover, both $x$ and $x'$ depend continuously on $x_1$.*

*Proof.* Let $x_o$, $\bar{x}_o \in E$ be given. The solutions $x = x(\cdot, x_o)$ and $\bar{x} = x(\cdot, \bar{x}_o)$ exist by theorem 5.7.1, and

$$x(t) = x_o + \int_o^t y(s)\,ds, \quad \text{and} \quad \bar{x}(t) = \bar{x}_o + \int_o^t \bar{y}(s)\,ds,$$

where $y$, $\bar{y}$ are the unique fixed points of the integral operators $F$, $\bar{F}: C(J,E) \to C(J,E)$, defined by

$$Fz(t) = S(t)Ax_o + C(t)x_1 + \int_o^t C(t-s)g(s, x_o + \int_o^s z(\tau)d\tau, z(s))\, ds,$$

and

$$\bar{F}z(t) = S(t)A\bar{x}_o + C(t)\bar{x}_1 + \int_o^t C(t-s)g(s, \bar{x}_o + \int_o^s z(\tau)d\tau, z(s))\, ds.$$

Moreover,

$$u(t) = \|x_o - \bar{x}_o\| + \int_o^t v(s)\, ds, \qquad t \in J,$$

where $v$ is the least fixed point of the operator $G: C(J, \mathbb{R}_+) \to C(J, \mathbb{R}_+)$, given by

$$\begin{aligned} Gw(t) = {} & MT\|A(x_o - \bar{x}_o)\| + M\|x_1 - \bar{x}_1\| \\ & + \int_o^t M\, q(s, \|x_o - \bar{x}_o\| + \int_o^s w(\tau)d\tau, w(s))\, ds. \end{aligned}$$

These definitions and condition (C2) imply that

$$|Fy - \bar{F}\bar{y}| \le G|y - \bar{y}| \quad \text{for} \quad y, \bar{y} \in C(J,E). \tag{a}$$

Denoting $y_o(t) \equiv x_1$ and $\bar{y}_o(t) \equiv \bar{x}_1$, then equality holds in

$$|F^n y_o - \bar{F}^n \bar{y}_o| \le G^n |y_o - \bar{y}_o| \tag{b}$$

when $n = 0$. Since $G$ is nondecreasing, it follows from (a) and (b) that

$$|F^{n+1}y_o - \bar{F}^{n+1}\bar{y}_o| \le G|F^n y_o - \hat{F}^n \bar{y}_o| \le G^{n+1}|y_o - \bar{y}_o|,$$

whence (b) holds for all $n \in \mathbb{N}$. Theorem 5.7.1 and proposition 5.3.2 imply that

$$y = \lim_{n\to\infty} F^n y_o,\ \bar{y} = \lim_{n\to\infty} \bar{F}^n \bar{y}_o \text{ and } v = \lim_{n\to\infty} G^n|y_o - \bar{y}_o|.$$

From this and (b) it follows, when $n \to \infty$, that

$$\|y(t) - \bar{y}(t)\| \le v(t) \text{ for each } t \in J.$$

This and the definitions of $y$, $\bar{y}$ and $v$ imply that the estimates (5.7.14) hold. The hypotheses given for $q$ in (C2) ensure (cf. the proof of proposition 2.1.9) that the minimal solution $u$ of the IVP (5.7.10) and its derivative $u'$ are nondecreasing with respect to $u_o$ and $u_1$, and that both of them tend to zero uniformly over $t \in J$ as $u_o \to 0$ and $u_1 \to 0$. This implies by (5.7.14) that the weak solution $x$ of (5.7.1) and its derivative depend continuously on $x_1$. □

As a consequence of theorem 5.7.2 and corollary 5.3.1 we obtain

**Corollary 5.7.1:** *Let the hypotheses (C0) and (C1) hold. Assume moreover that for all $x$, $y$, $h$, $k \in E$ and for a.a. $t \in J$,*

$$\|g(t, x + k, y + k) - g(t, x, y)\| \le p(t)\|h\| + q(t)\|k\|,$$

*where $p$, $q \in L^1(J, \mathbb{R}_+)$. Then the IVP (5.7.1) has for each $(x_o, x_1) \in E_2 \times E$ exactly one weak solution $x$. Moreover, $x$ and $x'$ depend continuously on $x_1$.*

*Proof.* From corollary 5.3.1 it follows that conditions (C2) holds when $q(t,u,v) = p(t)u + q(t)v$. Thus the relations (5.7.14) hold. Corollary 5.3.2 implies moreover that the right hand sides of the inequalities (5.7.14) tend to 0 uniformly on $J$ as $\bar{x}_o = x_o$ and $\bar{x}_1 \to x_1$ in $E$, which implies the last conclusion of corollary. □

In particular, we have

**Corollary 5.7.2:** *If the hypotheses (C0) and (C1) hold, and if $A_1$, $A_2 \in L^1(J, L(E))$ and $A_3 \in L^1(J, E)$, then the IVP*

$$x'' = Ax + A_1(t)x + A_2(t)x' + A_3(t), \qquad x(0) = x_o,\ x'(0) = x_1$$

*has for each $(x_o, x_1) \in E_2 \times E$ a unique weak solution, and it, as well as its derivative, depend continuously on $x_1$.*

### 5.7.4. On extremal mild solutions

In this subsection we shall consider the existence of extremal mild solutions of the IVP

$$x'' = Ax + g(t,x), \qquad x(0) = x_o,\ x'(0) = x_1, \tag{5.7.15}$$

between assumed upper and lower mild solutions, when $E$ is an ordered Banach space with regular order cone and $g\colon J \times E \to E$.

Given $x_o$, $x_1 \in E \times E$, we say that $x \in C(J,E)$ is a *lower mild solution* of the IVP (5.7.15) on $J$ if

$$x(t) \le C(t)x_o + S(t)x_1 + \int_o^t S(t-s)g(s,x(s))\,ds \tag{5.7.16}$$

for each $t \in J$. An upper mild solution of (5.7.15) is defined similarly, by reversing the inequality sign in (5.7.16). If equality holds in (5.7.16), we say that $x$ is a *mild solution* of (5.7.15).

Compared with the notion of weak solution we now don't require the differentiability of the solution. In the case when $x_o \in E_2$ and $g$ is continuous it can be shown (cf. Kusano and Oharu (1992)) that (5.7.15) has the same weak and mild solutions.

Let us impose the following hypotheses on the mappings $g: J \times E \to E$ and $C: J \to L(E)$.

(C3) (5.7.15) has a lower mild solution $\underline{x}$ and an upper mild solution $\bar{x}$ such that $\underline{x} \leq \bar{x}$, and that the functions $g(\cdot, \underline{x}(\cdot))$ and $g(\cdot, \bar{x}(\cdot))$ are Bochner integrable.

(C4) $g(\cdot, x(\cdot))$ is strongly measurable whenever $x \in C(J, E)$.

(C5) $g(t, \cdot)$ is nondecreasing for a.a. $t \in J$.

(C6) $C(t)$ is order-preserving for all $t \in J$.

If (C0) and (C6) hold, it follows from (5.7.2) that $S(t)$ is also order-preserving for each $t \in J$.

**Theorem 5.7.3:** *If the hypotheses (C0) and (C3)-(C6) hold, then the IVP (5.7.15) has the extremal mild solutions between $\underline{x}$ and $\bar{x}$.*

*Proof.* By definitions $\underline{x}$, $\bar{x} \in C(J, E)$. If $x$ belongs to the order interval $[\underline{x}, \bar{x}]$ of $C(J, E)$, it follows from (C4) and (5.7.2) by corollary 1.4.3 that the mapping $s \mapsto S(t-s)g(s, x(s))$ is strongly measurable on $[0, t]$ for all $t \in J$. Conditions (C5) and (C6) imply that for all $t \in J$ and for a.a. $s \in [0, t]$,

$$S(t-s)g(s, \underline{x}(s)) \leq S(t-s)g(s, x(s)) \leq S(t-s)g(s, \bar{x}(s)).$$

Applying this, the triangle inequality, (1.3.2) and (5.7.8), it follows that

$$\|S(t-s)g(s, x(s)\| \leq MT\, N(s) \tag{a}$$

for all $t \in J$ and for a.a. $s \in [0, t]$, where

$$N(s) = (1+\gamma)(\|g(s, \underline{x}(s))\| + \|g(s, \bar{x}(s))\|).$$

This and the hypotheses (C3) and (C4) imply that the function $s \mapsto S(t-s)g(s,x(s))$ is Bochner integrable on $[0,t]$ for each $t \in J$. Thus the equation

$$Gx(t) = C(t)x_o + S(t)x_1 + \int_o^t S(t-s)g(s,x(s))\,ds \qquad (5.7.17)$$

defines a mapping $Gx\colon J \to E$ for each $x \in [\underline{x}, \bar{x}]$. Moreover, if $0 \le \bar{t} \le t \le T$ and $x \in [\underline{x}, \bar{x}]$, it follows from (5.7.2), (5.7.7), (5.7.17) and (a) that

$$\begin{aligned}
&\|Gx(t) - Gx(\bar{t})\| \le \|C(t)x_o - C(\bar{t})x_o\| + \|S(t)x_1 - S(\bar{t})x_1\| \\
&+ \int_o^{\bar{t}} \|S(t-s)g(s,x(s)) - S(\bar{t}-s)g(s,x(s))\|ds \\
&+ \int_{\bar{t}}^t \|S(t-s)g(s,x(s))\|ds \\
&\le \|C(t)x_o - C(\bar{t})x_o\| + \|S(t)x_1 - S(\bar{t})x_1\| \\
&+ M|t-\bar{t}| \int_o^T N(s)\,ds + MT \int_{\bar{t}}^t N(s)\,ds.
\end{aligned}$$

Thus the family of functions $Gx$, $x \in [\underline{x}, \bar{x}]$, is equicontinuous. In view of conditions (C5) and (C6) it follows from (5.7.17) by corollary 1.4.6 that $Gx \le Gy$ whenever $x, y \in [\underline{x}, \bar{x}]$ and $x \le y$. Moreover, it follows from (5.7.17) by the definition of mild upper and lower solutions of (5.7.15) that $\underline{x} \le G\underline{x}$ and $G\bar{x} \le \bar{x}$. Thus $G$ is a nondecreasing mapping from $[\underline{x}, \bar{x}]$ to $[\underline{x}, \bar{x}]$.

Assume now that $(x_n)_{n=1}^{\infty}$ is a monotone sequence in $[\underline{x}, \bar{x}]$. From the monotonicity of $G$ it follows that the sequence $(Gx_n)_{n=1}^{\infty}$ is also monotone. Since the order cone $K$ of $E$ is regular, it follows that

$$y(t) = \lim_{n \to \infty} Gx_n(t) \qquad \text{(c)}$$

exists for all $t \in J$. Because the functions $Gx_n$ are equicontinuous, the convergence in (c) is uniform on $J$, so that $Gx_n \to y$

in $C(J,E)$ with respect to the supremum norm. Obviously, $y$ belongs to $[\underline{x},\bar{x}]$, whence $(Gx_n)_{n=1}^{\infty}$ converges in $[\underline{x},\bar{x}]$.

The above proof shows that the mapping $G\colon [\underline{x},\bar{x}] \to [\underline{x},\bar{x}]$ satisfies the hypotheses of theorem 1.2.2, whence $G$ has the least fixed point $x_*$ and the greatest fixed point $x^*$. This and the definition (5.7.17) of $G$ imply that $x_*$ and $x^*$ are mild solutions of the IVP (5.7.15).

If $x$ is any mild solution of (5.7.15) and $\underline{x} \le x \le \bar{x}$, then $x \in [\underline{x},\bar{x}]$, whence $x$ is a fixed point of $G$. Since $x_*$ and $x^*$ are the least and the greatest ones, it follows that $x_* \le x \le x^*$. Thus $x_*$ and $x^*$ are the extremal solutions of the IVP (5.7.15) in $[\underline{x},\bar{x}]$. □

**Proposition 5.7.1:** *If $g\colon J \times E \to E$ in theorem 5.7.3 is a Carathéodory function, then the sequence $(y_n)_{n=o}^{\infty}$ of the successive approximations*

$$y_{n+1}(t) = C(t)x_o + S(t)x_1 + \int_o^t S(t-s)g(s,y_n(s))\,ds,$$

*$n \in \mathbb{N}$, $t \in J$, converges uniformly on $J$ to the minimal (resp. the maximal) mild solution of the IVP (5.7.15) between $\underline{x}$ and $\bar{x}$, if $y_o = \underline{x}$ (resp. $y_o = \bar{x}$).*

*Proof.* Cf. the proof of proposition 5.6.2. □

**Lemma 5.7.1:** *Let the hypotheses (C0) and (C6) hold, and assume that $g\colon J \times E \to E$ satisfies condition (C5) and conditions*

(C7) *$g(\cdot,z)$ is strongly measurable for each $z \in E$ and $g(\cdot,0) \in L^1(J,E)$.*

(C8) *There is $r > 0$ such that $\|g(t,x)-g(t,y)\| \le q(t,\|x-y\|)$ for all $x$, $y \in E$ with $\|x-y\| < r$ and for a.a. $t \in J$, where $q\colon J \times \mathbb{R}_+ \to \mathbb{R}_+$ is a Carathéodory function, $q(t,\cdot)$ is nondecreasing for a.a. $t \in J$, the IVP*

$$u' = MT\,q(t,u), \qquad u(0) = u_o, \tag{5.7.18}$$

*with $M$ given by (5.7.7), has for some $u_o \in I\!R_+$ an upper solution on $J$, and the zero-function is the only solution of (5.7.18) when $u_o = 0$.*

*Then the IVP (5.7.15) has for each $(x_o, x_1) \in E \times E$ a unique mild solution $x$ on $J$, and it depends continuously on $(x_o, x_1)$. Moreover,*

a) *$x \leq \bar{x}$ for each upper mild solution $\bar{x}$ of (5.7.15),*

b) *$\underline{x} \leq x$ for each lower mild solution $\underline{x}$ of (5.7.15).*

*Proof.* Let $x_o$, $x_1 \in E$ be given. The existence and uniqueness of the mild solution $x = x(\cdot, x_o, x_1)$ of (5.7.15) follows by the reasoning used in the proof of theorem 5.6.1. Following the proof of theorem 5.6.1 it is easy to show that if $\|x_i - \bar{x}_i\|$, $i = 0, 1$, are small enough, then

$$\|x(t, x_o, x_1) - x(t, \bar{x}_o, \bar{x}_1)\| \leq u(t, M\|x_o - \bar{x}_o\| + MT\|x_1 - \bar{x}_1\|), \tag{a}$$

where $u = u(\cdot, u_o)$ denotes the minimal solution of the IVP (5.7.18). Moreover, $u(\cdot, u_o)$ tends by proposition 2.1.9 to 0 uniformly over $t \in J$ as $u_o \to 0$. This and (a) imply that $x(\cdot, x_o, x_1)$ depends continuously on $(x_o, x_1)$. The proof of lemma 5.6.1 can then be applied to reach the conclusion a) and b). □

**Proposition 5.7.2:** *Let $E$ be an ordered Banach space with regular order cone. Assume that $g\colon J \times E \to E$ satisfies conditions (C4) and (C5), and that*

$$g_1(t, x) \leq g(t, x) \leq g_2(t, x) \tag{5.7.19}$$

*for all $x \in E$ and for a.a. $t \in J$, where $g_i\colon J \times E \to E$, $i = 1, 2$ satisfy conditions (C5), (C7) and (C8). Then the IVP (5.7.15) has for each choice of $x_o \in E$ the extremal mild solutions.*

*Proof.* Let $x_o$, $x_1 \in E$ be given. Lemma 5.7.1 implies that the integral equation

$$x(t) = C(t)x_o + S(t)x_1 + \int_o^t S(t-s)g_i(t, x(s))\, ds \tag{a}$$

has unique mild solutions $\underline{x}$ when $i = 1$ and $\bar{x}$ when $i = 2$. From (a) and (5.7.19) it follows that $\underline{x}$ is a lower solution of (a) with $i = 2$. Thus $\underline{x} \leq \bar{x}$ by lemma 5.7.1. Applying, (C5), the triangle inequality, (1.3.2) and (5.7.19), it follows that if $x \in [\underline{x}, \bar{x}]$, then

$$\|g(t, x(t))\| \leq N(t) \quad \text{for a.a. } \ t \in J,$$

where

$$N(t) = (1 + \gamma)(\|g_1(t, \underline{x}(t))\| + \|g_2(t, \bar{x}(t))\|).$$

This and (C4) imply that the functions $g(\cdot, \underline{x}(\cdot))$ and $g(\cdot, \bar{x}(\cdot))$ are Bochner integrable. Moreover,

$$\underline{x}(t) \leq C(t)x_o + S(t)x_1 + \int_o^t S(t - s)g(s, \underline{x}(s))\, ds \quad \text{for a.a. } t \in J,$$

whence $\underline{x}$ is a mild lower solution of (5.7.15). Similarly, it can be shown that $\bar{x}$ is a mild upper solution. Thus $g$ satisfies also condition (C3). By theorem 5.7.3 the IVP (5.7.15) has the minimal mild solution $x_*$ and the maximal mild solution $x^*$ between $\underline{x}$ and $\bar{x}$.

If $x$ is a mild solution of (5.7.15) on $J$, it follows from (5.7.19) that $x$ is a mild upper solution of (a) for $i = 1$ and a mild lower solution of (a) for $i = 2$, whence the results a) and b) of lemma 5.7.1 imply that $\underline{x} \leq x \leq \bar{x}$. Hence all the mild solutions of (5.7.15) on $J$ lie between $\underline{x}$ and $\bar{x}$, whence $x_*$ is the least and $x^*$ the greatest of all the mild solutions of (5.7.15) on $J$. □

### 5.7.4. Existence of minimal or maximal mild solution

In the case when the order cone of $E$ is fully regular, the existence of upper or lower mild solutions of the IVP (5.7.15) can be replaced by condition

(C9) $\|g(t,x)\| \le h(t,\|x\|)$ for all $x \in E$ and for a.a. $t \in J$, where $h\colon J \times \mathbb{R}_+ \to \mathbb{R}_+$ is a standard function, $h(t,\cdot)$ is nondecreasing for a.a. $t \in J$, and the IVP

$$u' = MT\,h(t,u), \quad u(t_o) = u_o, \tag{5.7.20}$$

with $M$ given by (5.7.7), has for each $u_o \in \mathbb{R}_+$ an upper solution.

**Theorem 5.7.4:** *Given an ordered Banach space $E$ with fully regular order cone, assume that $g\colon J \times E \to E$ satisfies conditions (C4), (C5) and (C9) and that (C0) and (C6) hold.*
*a) If the IVP (5.7.15) has a lower mild solution $\underline{x}$, then (5.7.15) has the minimal mild solution in $[\underline{x})$.*
*b) If the IVP (5.7.15) has an upper mild solution $\bar{x}$, then (5.7.15) has the maximal mild solution in $(\bar{x}]$.*

*Proof.* a) Assume that $\underline{x}$ is a lower mild solution of the IVP (5.7.15). Denote $u_o = M\,\|x_o\| + MT\,\|x_1\| + \|\underline{x}\|_o$, and let $\bar{u}$ be an upper solution of the IVP (5.7.20). Since the zero function is a lower solution of (5.7.20), it follows from theorem 2.1.3 that (5.7.20) has the minimal solution $u_*$ on $J$ between 0 and $\bar{u}$. Moreover, lemma 1.5.5 implies that

$$u_*(t) = u_o + \int_o^t MT\,h(s,u_*(s))\,ds, \quad t \in J. \tag{a}$$

Denote $Y = \{y \in C(J,E) \mid \underline{x} \le y \text{ and } |y| \le u_*\}$. $Y$ is nonempty because $\underline{x} \in Y$. If $y \in Y$, it is easy to see that the equation (5.7.17) defines a mapping $Gy\colon J \to E$. From (5.7.7), (5.7.8) and (C9) it follows that

$$\|Gy(t)\| \le \|C(t)x_o\| + \|S(t)x_1\| + \int_o^t \|S(t-s)g(s,y(s))\|\,ds$$

$$\leq M\,\|x_o\| + MT\,\|x_1\| + MT\int_o^t \|g(s,y(s))\|\,ds$$

$$\leq u_o + \int_o^t MT\,h(s,\|y(s)\|)\,ds \leq u_o + \int_o^t MT\,h(s,u_*(s))\,ds$$

$$= u_*(t)$$

for each $t \in J$. The definition of a mild lower solution of (5.7.15) implies that $\underline{x} \leq G\underline{x}$. These properties, together with conditions (C5) and (C6) imply that the equation (5.7.17) defines a nondecreasing mapping $G\colon Y \to Y$. Moreover, if $y \in Y$ and $0 \leq \bar{t} \leq t \leq T$, then

$$\|Gy(t) - Gy(\bar{t})\| \leq \|C(t)x_o - C(\bar{t})x_o\| + \|S(t)x_1 - S(\bar{t})x_1\|$$
$$+ M|t-\bar{t}|\int_o^T h(s,u_*(s))\,ds + MT\int_{\bar{t}}^t h(s,u_*(s))\,ds. \qquad \text{(b)}$$

Let $C$ be a w.o. chain of $G$-iterations of $\underline{x}$. By definition, $G[C]$ is a well-ordered chain in $Y$, and hence pointwise bounded, and by (b) equicontinuous, whence proposition 1.3.9 implies that $x_* = \sup G[C]$ exists. Thus $G$ satisfies the hypotheses of theorem 1.2.1, so that $G$ has the least fixed point $x_*$. By definition, $x_*$ is a mild solution of the IVP (5.7.15) in $Y$.

Assume now that $x\colon J \to E$ is a mild solution of (5.7.15). By choosing $u_o = M\,\|x_o\| + MT\,\|x_1\| + \|\underline{x}\|_o + \|x\|_o$ in the above proof we may assume that $x \in Y$. Thus the reasoning used in the proof of theorem 5.6.3 shows that $x_* \leq x$, whence $x_*$ is the least of all the mild solutions of the IVP (5.7.15) in $[\underline{x})$.

The proof of the case b) is similar. □

**Corollary 5.7.3:** *Let the hypotheses of theorem 5.7.4 hold, let $g$ be a Carathéodory function, and let $G$ be defined by (5.7.17).*

a) *If the IVP (5.7.15) has a lower mild solution $\underline{x}$, then the sequence $(G^n\underline{x})_{n=o}^{\infty}$ is nondecreasing and converges on $J$ uniformly to the minimal mild solution of (5.7.15) in $[\underline{x})$.*

b) *If the IVP (5.7.15) has an upper mild solution* $\bar{x}$, *then the sequence* $(G^n\bar{x})_{n=o}^{\infty}$, *is nondecreasing and converges on* $J$ *uniformly to the maximal mild solution of (5.7.15) in* $(\bar{x}]$.

In the case when the order cone of $E$ is not fully regular we can use proposition 1.3.10, instead of proposition 1.3.9, in the proof of theorem 5.7.4 to obtain the following result.

**Theorem 5.7.5:** *If* $E$ *is a reflexive Banach space, ordered by a closed cone or via continuous embedding in an ordered Banach space with fully regular order cone, then conditions (C0), (C4), (C5), (C6) and (C9) ensure that the conclusions of theorem 5.7.4 hold.*

### 5.7.5. Dependence on the data

Next we shall consider the dependence of the extremal mild solutions of the IVP (5.7.15) on the initial condition and on the function $g$.

**Proposition 5.7.3:** *If the hypotheses of proposition 5.7.2 hold, then the extremal mild solutions of the IVP (5.7.15) are nondecreasing with respect to* $x_o$, $x_1$ *and* $g$.

*Proof.* Cf. the proof of proposition 5.2.3. □

Similarly, it can be shown

**Corollary 5.7.4:** *If the hypotheses of theorem 5.7.4 a) hold, then the minimal mild solution of the IVP (5.7.15) in* $[\underline{x})$ *is nondecreasing with respect to* $x_o$ *and* $g$.

**Corollary 5.7.5:** *Let* $E$ *be an ordered Banach space with regular order cone. Assume that conditions (C0) and (C4)–(C6) hold, that for all* $x \in E$ *and for a.a.* $t \in J$, $C_1(t) \leq g(t,x) \leq C_2(t)$, *where* $C_i \in L^1(J,E)$, $i = 1,2$. *Then the IVP (5.7.15) has for each choice of* $x_o$, $x_1 \in E$ *the extremal mild solutions, both of which are nondecreasing in* $x_o$ *and in* $x_1$.

*Proof.* It is easy to see that the hypotheses of proposition 5.7.3 hold when $g_i(t,x) = C_i(t)$, $i = 1,2$, whence the assertions follow from proposition 5.7.3. □

### 5.7.6. Application to a PDE

Consider the IVP of the second order hyperbolic partial differential equation

$$\frac{\partial^2 u}{\partial t^2} = k^2 \frac{\partial^2 u}{\partial \xi^2} + f(\xi,t,u), \quad u(\xi,0) = x_o(\xi), \ \frac{\partial}{\partial t}u(\xi,0) = x_1(\xi), \tag{5.7.21}$$

where $k$ is a given positive constant. If $J = [0,T]$, $T > 0$, and $f\colon \mathbb{R} \times J \times \mathbb{R} \to \mathbb{R}$ is continuously differentiable, it is easy to see that a twice continuously differentiable function $u\colon \mathbb{R} \times J \to \mathbb{R}$ is a solution of the IVP (5.7.21) if and only if it satisfies the integral equation

$$\begin{aligned} u(\xi,t) =& \frac{1}{2}(x_o(\xi+kt) + x_o(\xi-kt)) + \frac{1}{2k}\int_{\xi-kt}^{\xi+kt} x_1(z)dz \\ &+ \frac{1}{2k}\int_o^t \int_{\xi-kt}^{\xi+kt} f(z,s,u(z,s))dz\,ds. \end{aligned} \tag{5.7.22}$$

On the other hand, (5.7.22) may have solutions also when $f$ is not even continuous. Given $p \in [1,\infty)$, we shall now study existence of solutions of (5.7.22) in the set $\mathcal{U}_p$ of those measurable functions $u\colon \mathbb{R} \times J \to \mathbb{R}$ for which $u(\cdot,t) \in L^p(\mathbb{R})$ for each $t \in J$ and $\lim_{t\to t_o} \int_{-\infty}^{\infty} |(u(\xi,t) - u(\xi,t_o)|^p d\xi = 0$ for each $t_o \in J$. We say that $u \in \mathcal{U}_p$ is a *mild solution* of the IVP (5.7.21) if (5.7.22) holds for all $t \in J$ and for a.a. $\xi \in \mathbb{R}$.

Choose $E = L^p(\mathbb{R})$, and define for each $x \in E$

$$C(t)x(\xi) = \frac{1}{2}(x(\xi + kt) + x(\xi - kt)), \ t \in \mathbb{R}, \ \xi \in \mathbb{R}. \quad (5.7.23)$$

To show that (5.7.23) defines a strongly continuous cosine family $\{C(t) \mid t \in \mathbb{R}\}$ of operators in $L(L^p(\mathbb{R}))$, note first that properties $C(0) = I$ and $C(t+s) + C(t-s) = 2C(t)C(s)$ are trivially verified. The equation $T(t)x(\xi) = x(\xi + kt)$ defines a strongly continuous semigroup $T\colon \mathbb{R} \to L(L^p(\mathbb{R}))$ (cf. Dunford and Schwartz (1958)) and $C(t) = \frac{1}{2}(T(t) + T(-t))$, which implies strong continuity of the mapping $t \mapsto C(t)x$. The corresponding sine family, defined in (5.7.2), can be given by

$$S(t)x(\xi) = \frac{1}{2k}\int_{\xi - kt}^{\xi + kt} x(z)\,dz. \quad (5.7.24)$$

Assume first that $f\colon \mathbb{R} \times J \times \mathbb{R} \to \mathbb{R}$ has the following properties:

(f0) $f(\cdot,\cdot,z)$ is measurable on $\mathbb{R} \times J$ for each $z \in \mathbb{R}$ and $t \mapsto f(\cdot,t,0)$ is a Bochner integrable mapping from $J$ to $L^p(\mathbb{R})$.

(f1) There is $q \in L^p_+(\mathbb{R})$ such that
$$|f(\xi,t,y) - f(\xi,t,z)| \le q(t)|y - z|$$
for all $t \in J$, $y, z \in \mathbb{R}$ and for a.a. $\xi \in \mathbb{R}$.

If $x_o \in L^p(\mathbb{R})$ and $u \in \mathcal{U}_p$, it follows that the right hand side of (5.7.22) is defined. In view of (5.7.23) and (5.7.24) the equation (5.7.22) can be rewritten as (cf. DuChateau and Zachmann (1987))

$$u(\xi,t) = C(t)x_o(\xi) + S(t)x_1(\xi) + \int_o^t S(t-s)f(\xi,s,u(\xi,s))\,ds. \quad (5.7.25)$$

If $u \in \mathcal{U}_p$ is a mild solution of (5.7.21), i.e. a solution of (5.7.25), then denoting

$$x(t) = u(\cdot, t), \; g(t, y) = f(\cdot, t, y(\cdot)), \; t \in J, \; y \in L^p(\mathbb{R}), \quad (5.7.26)$$

then $x \in C(J, L^p(\mathbb{R}))$, and $x$ is a solution of the integral equation

$$x(t) = C(t)x_o + S(t)x_1 + \int_o^t S(t-s)g(s, x(s))\, ds. \quad (5.7.27)$$

Conversely, if $x \in C(J, L^p(\mathbb{R}))$ is a solution of (5.7.27), there is $u \in \mathcal{U}_p$ such that $x(t) = u(\cdot, t)$ for all $t \in J$, and $u$ is a solution of (5.7.25), and thus a mild solution of (5.7.21) (cf. Hille and Phillips (1957)).

Conditions (f0) and (f1) imply that (5.7.26) defines a mapping $g\colon J \times L^p(\mathbb{R}) \to L^p(\mathbb{R})$ which satisfies conditions (C7) and (C8). Thus the hypotheses of lemma 5.7.1 hold, whence the integral equation (5.7.27) has for each choice of $x_o$, $x_1 \in L^p(\mathbb{R})$ a unique solution $x$ in $C(J, L^p(\mathbb{R}))$. Moreover, $x$ depends continuously on $x_o$ and on $x_1$. According to the above stated correspondence between solutions of (5.7.27) in $C(J, L^p(\mathbb{R}))$ and mild solutions of (5.7.21) and identification of a.e. equal functions we then obtain

**Proposition 5.7.4:** *If conditions (f0) and (f1) hold, then the IVP (5.7.21) has for each $x_o \in L^p(\mathbb{R})$ a unique mild solution $u = u(\cdot, \cdot; x_o, x_1)$ in $\mathcal{U}_p$, which depends continuously on $x_o$ and on $x_1$ in the sense that $\|u(\cdot, t; \bar{x}_o, \bar{x}_1) - u(\cdot, t; x_o, x_1)\|_p \to 0$, uniformly over $t \in J$ as $\bar{x}_o \to x_o$ and $\bar{x}_1 \to x_1$ in $L^p(\mathbb{R})$.*

Assume next that $f$ has the following properties.

(f2) $f$ is a standard function from $(\mathbb{R} \times J) \times \mathbb{R}$ to $\mathbb{R}$.
(f3) $f(\xi, t, \cdot)$ is nondecreasing for all $(\xi, t) \in J \times \mathbb{R}$.
(f4) There exist $f_1$, $f_2\colon \mathbb{R} \times J \times \mathbb{R} \to \mathbb{R}$ which satisfy conditions (f0), (f1) and (f3) such that for all $(\xi, t, v) \in$

$I\!R \times J \times I\!R$,
$f_1(\xi, t, v) \le f(\xi, t, v) \le f_2(\xi, t, v)$.

Assume that $L^p(I\!R)$ is partially ordered by the cone $L^p_+(I\!R)$ of a.e. nonnegative-valued elements of $L^p(I\!R)$. This and (5.7.23) imply that $C(t)$ is order-preserving for each $t \in I\!R$. In view of conditions (f2)–(f4) and definition (5.7.26) of $g$ it is easy to see that (C4) and (C5) hold, and that

$$g_1(t, y) \le g(t, y) \le g_2(t, y) \quad \text{for all } t \in J \text{ and } \; y \in L^p(I\!R),$$

where $g_i$, $i = 1, 2$, is defined by (5.7.26) with $f = f_i$. Noticing also that the order cone $L^p_+(I\!R)$ of $L^p(I\!R)$ is a regular, then all the hypotheses of proposition 5.7.2 hold, whence (5.7.27) has for each $x_o \in L^p(I\!R)$ extremal solutions $x_*$ and $x^*$, and they are nondecreasing in $x_o$ and in $x_1$. According to the above discussion there exist $u_*$, $u^* \in \mathcal{U}_p$ such that $x_*(t) = u_*(\cdot, t)$ and $x^*(t) = u^*(\cdot, t)$ for all $t \in J$ so that $u_*$ and $u^*$ are mild solutions of the IVP (5.7.21). Moreover, defining a partial ordering in $\mathcal{U}_p$ by

$$u \le v \quad \text{if } u(\xi, t) \le v(\xi, t) \text{ for all } t \in J \text{ and for a.a. } \; \xi \in I\!R,$$

it is easy to see that if $x$, $y \in C(J, L^p(I\!R))$ and if $u$, $v$ are their representatives in $\mathcal{U}_p$, respectively, then $x \le y$ if and only if $u \le v$. Thus $u_*$ and $u^*$ are the extremal mild solutions of (5.7.21) and they are nondecreasing in $x_o$ and in $x_1$. Thus we have proved the following result.

**Proposition 5.7.5:** *If conditions (f2)–(f4) hold, then the IVP (5.7.21) has for each choice of $x_o$, $x_1 \in L^p(I\!R)$ the extremal mild solutions in $\mathcal{U}_p$, and they are nondecreasing in $x_o$ and in $x_1$.*

## 5.8. ORDERED BANACH & FUNCTION SPACES

In the theory of discontinuous differential equations in an ordered Banach space $E$, presented above, we assumed that the order cone of $E$ is regular or fully regular, or that $E$ is reflexive and ordered by a closed cone or via continuous embedding in an ordered Banach space with fully regular order cone. In this section we shall introduce examples of such ordered Banach spaces, and derive sufficient conditions to ensure such kind of properties.

In some of the proofs we have used the property that a pointwise order bounded or norm bounded and equicontinuous chain in a space of continuous functions from a topological space $X$ to $E$, equipped with pointwise ordering, should posses the supremum and the infimum. Sufficient conditions to ensure this are derived in the last subsection.

### 5.8.1. Ordered Banach spaces & Hilbert spaces

In this subsection we shall derive conditions under which the order cone of an ordered Banach space $E$ is regular or fully regular. The following examples show that a regular cone need not to be fully regular and a normal cone need not be regular.

**Example 5.8.1:** Let $E$ be the Banach space $(c_o)$ of real-valued sequences $y = (y_i)_{i=1}^{\infty}$ with $\lim_i y_i = 0$ and norm $\|y\| = \sup_i |y_i|$. The order cone $K$ of $(c_o)$, formed by all the nonnegative-valued sequences of $E$ is regular, but not fully regular. For instance, if $x_n$ is a sequence whose $n$ first coordinates equal to 1, and the remaining are zero, then $(x_n)_{n=1}^{\infty}$ is a norm bounded and increasing sequence in $K$ which does not converge strongly in $(c_o)$. □

**Example 5.8.2:** Let $l^\infty$ be the Banach space of bounded and real-valued sequences $y = (y_i)_{i=1}^{\infty}$ with the norm $\|y\|_\infty = \sup_i |y_i|$. The cone $l_+^\infty$ formed by all the nonnegative-valued sequences of $l^\infty$ is normal, since the norm $\|\cdot\|_\infty$ is monotone in $l_+^\infty$. On the other hand, $l_+^\infty$ is not regular. For instance, if $x_n$ is as in example 5.8.1, then $(x_n)_{n=1}^\infty$ is an order (and also norm) bounded and increasing sequence in $l_+^\infty$, which does not converge strongly in $l^\infty$, since $\|x_{n+1} - x_n\|_\infty = 1$ for each $n = 1, 2, \ldots$ . However, $l_+^\infty$ is *strongly minihedral*, i.e. its each nonempty and bounded subset has the supremum. □

We say that an ordered Banach space $E$ is a *UMB-space* if its norm is *uniformly monotone* in the order cone $K$ of $E$, i.e. (cf. Birkhoff (1973)) if to each $\epsilon > 0$ there corresponds such a $\delta > 0$ that

$$y, z \in K, \ \|y\| = 1 \text{ and } \|y + z\| \le 1 + \delta \text{ imply } \|z\| \le \epsilon. \tag{5.8.1}$$

The following result shows that the order cone of a UMB-space is normal.

**Lemma 5.8.1:** *If $E = (E, K, \|\cdot\|)$ is a UMB-space, then the norm $\|\cdot\|$ of $E$ is strictly monotone in $K$, i.e.*

$$y, x \in K \text{ and } y < x \quad \text{imply } \|y\| < \|x\|. \tag{5.8.2}$$

*Proof.* Assume there exist such $x, y \in K$, $y < x$, that $\|y\| \ge \|x\|$. Dividing the last inequality by the norm of $y$, we may assume that $\|y\| = 1$. Denoting $z = x - y$, we then have $\|y + z\| \le 1$, so that condition (5.8.1) does not hold if one chooses $\epsilon = \|z\|/2$, contradicting the fact that $E$ is a UMB-space. □

**Proposition 5.8.1:** *The order cone $K$ of a UMB-space $E$ is regular and fully regular.*

*Proof.* Let $(x_n)_{n=o}^\infty$ be a bounded and nondecreasing sequence in $K$. The sequence $(\|x_n\|)_{n=o}^\infty$ is nondecreasing by lemma 5.8.1 whence $M = \lim_n \|x_n\|$ exists and is finite. If $M = 0$, then $(x_n)_{n=o}^\infty$ converges to 0. Assume next that $M > 0$. Given $\epsilon > 0$, choose $\delta > 0$ so that (5.8.1) holds. Choose $k \in I\!N$ such that $M - \|x_k\| \le \frac{M\delta}{2}$ and $\|x_k\| \ge \frac{M}{2}$. If $m \ge n \ge k$, then

$$0 \le \|x_m\| - \|x_n\| \le M - \|x_k\| \le \frac{M\delta}{2} \le \|x_n\|\delta.$$

This implies by (5.8.1) that $\|x_n - x_m\| \le \|x_n\|\epsilon \le M\epsilon$. Thus $(x_n)_{n=o}^{\infty}$ is a Cauchy sequence in $E$, whence $(x_n)_{n=o}^{\infty}$ converges in $E$.

Lemma 5.8.1 implies that the norm of $E$ is monotone in $K$. Thus the order bounded subsets of $E$ are bounded, whence the above proof holds also for order bounded sequences of $K$. □

**Corollary 5.8.1:** *An ordered Banach space $E$ is a UMB-space if there exist $c$, $p > 0$ such that*

$$\|x - y\|^p \le c\,(\|x\|^p - \|y\|^p) \quad \textit{whenever}\; 0 \le y \le x. \tag{5.8.3}$$

*Proof.* Given $\epsilon > 0$ choose $\delta = (1 + \frac{\epsilon^p}{c})^{\frac{1}{p}} - 1$. If $y, z \in K$, $\|y\| = 1$ and $\|y + z\| \le 1 + \delta$, it follows from (5.8.3) with $x = y + z$ that

$$\|z\|^p = \|x - y\|^p \le c(\|x\|^p - \|y\|^p) \le c((1 + \frac{\epsilon^p}{c}) - \|y\|^p) = \epsilon^p.$$

Thus $\|z\| \le \epsilon$, which proves that (5.8.1) is valid. □

**Example 5.8.3:** Given $1 \le p < \infty$, let $l^p$ be the Banach space of those real-valued sequences $y = (y_i)_{i=1}^{\infty}$, for which $\sum_{i=1}^{\infty} |y_i|^p < \infty$, and with the norm

$$\|y\|_p = (\sum_{i=1}^{\infty} |y_i|^p)^{\frac{1}{p}}.$$

If $x = (x_i)_{i=1}^{\infty}$, $y = (y_i)_{i=1}^{\infty} \in l_+^p$ and $y \leq x$, then $x_i - y_i \geq 0$ for all $i = 1, 2, \dots$, whence

$$\|x\|_p^p = \sum_{i=1}^{\infty}(y_i+(x_i-y_i))^p \geq \sum_{i=1}^{\infty}(y_i^p+(x_i-y_i)^p) = \|y\|_p^p+\|x-y\|_p^p.$$

Thus condition (5.8.3) holds when $c = 1$, whence $l_+^p$ is regular and fully regular, and the norm $\|\cdot\|_p$ is strictly monotone and uniformly monotone in $l_+^p$.

We say that a vector space $E$, equipped with an inner product $(\cdot|\cdot)$, is a *Hilbert space* if $E$ is a Banach space with respect to the norm $\|x\|_2 = (x|x)^{\frac{1}{2}}$. A Hilbert space, ordered by any cone which is closed in the topology determined by $\|\cdot\|_2$, is called an *ordered Hilbert space.*

**Proposition 5.8.2:** *If $E = (E, (\cdot|\cdot))$ is an ordered Hilbert space and $K$ its order cone, then the following conditions are equivalent:*

(i) *$\|\cdot\|_2$ is uniformly monotone in $K$.*
(ii) *$\|\cdot\|_2$ is strictly monotone in $K$.*
(iii) *$(y|z) \geq 0$ for all $y, z \in K$.*

*Proof.* (i) implies (ii) by lemma 5.8.1. To prove that (iii) follows from (ii), assume that (iii) does not hold. Then there exist such $y, z_o \in K$, that $\|y\|_2 = \|z_o\|_2 = 1$ and $(y|z_o) < 0$. By choosing $z = -2(y|z_o)z_o$, then $\|y + z\|_2 = 1$, so that $0 \leq y < z$. But $\|y\|_2 = \|z\|_2 = 1$, whence (ii) is not valid.

If (iii) holds, and if $y, z \in K$ and $\|y\|_2 = 1$, then condition $\|y+z\|_2 \leq 1+\delta$ implies by (iii) that $\|z\|_2^2 \leq 2\delta+\delta^2$. Thus (5.8.1) holds when $\delta = \sqrt{1+\epsilon} - 1$, whence (i) is valid. □

If $K$ is a cone in a Hilbert space $E$ such that $(x|y) < 0$ for some $x, y \in K$, it follows from proposition 5.8.2 that $(E, K, \|\cdot\|_2)$ is not a UMB-space. However, in certain cases it is possible to construct a new inner product in $E$ such that it generates a norm $\|\cdot\|$, equivalent to $\|\cdot\|_2$, such that $(E, K, \|\cdot\|)$ is a UMB-space. For instance, we have

**Proposition 5.8.3:** *Given* $e \in E$, $\|e\|_2 = 1$, *and* $c \in (0,1)$, *define*

$$K = \{x \in E \mid (x|e) \geq c\|x\|_2\}. \tag{5.8.4}$$

*Then there exists a norm* $\|\cdot\|$ *of* $E$, *equivalent to* $\|\cdot\|_2$, *such that* $(E, K, \|\cdot\|)$ *is a UMB-space.*

Proof. a) Assume first that $0 < c < 1/\sqrt{2}$, and denote

$$H = \{x \in E \mid (x|e) = 0\}.$$

Each vector $x$ of $E$ has a unique representation in the form $x = (x|e)e + P_H x$, where $P_H x$ is the orthogonal projection of $x$ on $H$. If $y = (y|e)e + P_H y$ is another vector of $E$, then

$$(x|y) = (x|e)(y|e) + (P_H x|P_H y),$$

so that

$$\|x\|_2^2 = (x|e)^2 + \|P_H x\|_2^2. \tag{a}$$

Define in $E$ a new inner product by

$$x \cdot y = \frac{1 - c^2}{c^2}(x|e)(y|e) + (P_H x|P_H y),$$

and let $\|\cdot\|$ be the norm of $E$, induced by this inner product. Thus

$$\|x\|^2 = \frac{1 - c^2}{c^2}(x|e)^2 + \|P_H x\|^2. \tag{b}$$

Since $0 < c < \frac{1}{\sqrt{2}}$, it follows from (a) and (b) that

$$\|x\|_2^2 \leq \|x\|^2 \leq \frac{1 - c^2}{c^2}\|x\|_2^2,$$

so that the norms $\|\cdot\|_2$ and $\|\cdot\|$ are equivalent. Since

$$\|e\|^2 = \frac{1-c^2}{c^2} \quad \text{and } x \cdot e = \frac{1-c^2}{c^2}(x|e),$$

then $e_o = \frac{c}{\sqrt{1-c^2}} e$ is a unit vector with respect to $\|\cdot\|$. Moreover, we have

$$\|x\|_2^2 = \|x\|^2 + \frac{2c^2-1}{c^2}(x|e)^2 = \|x\|^2 + \frac{2c^2-1}{1-c^2}(x \cdot e_o)^2.$$

Thus for all $x \in K$,

$$\begin{aligned}(x \cdot e_o)^2 &= \frac{c^2}{1-c^2}(x \cdot e)^2 = \frac{1-c^2}{c^2}(x|e)^2 \geq (1-c^2)\|x\|_2^2 \\ &= (1-c^2)\|x\|^2 + (2c^2-1)(x \cdot e_o)^2,\end{aligned}$$

which implies that

$$(x \cdot e_o)^2 \geq \|x\|^2/2.$$

Since $x \cdot e_o \geq 0$ for each $x \in K$, we then have

$$K = \{x \in E \mid x \cdot e_o \geq \frac{\|x\|}{\sqrt{2}}\}.$$

b) If $c \in [\frac{1}{\sqrt{2}}, 1)$, choose $x \cdot y = (x|y)$ for $x, y \in E$. The above reasoning implies an existence of such a norm $\|\cdot\|$, equivalent to $\|\cdot\|_2$ and induced by an inner product $(x, y) \to x \cdot y$ of $E$, such a vector $e_o$ of $E$ for which $\|e_o\| = 1$, and such a constant $c \in [\frac{1}{\sqrt{2}}, 1)$ that the set $K$ given by (5.8.4) can be represented as

$$K = \{x \in E \mid x \cdot e_o \geq c\|x\|\}. \tag{c}$$

It is easy to see that $K$ is a closed cone in $E$.

It remains to show that the norm $\|\cdot\|$ is uniformly monotone in $K$. By proposition 5.8.2 it is tantamount to show that $x \cdot y \geq 0$ for all $x, y \in K$. If $x = 0$ or $y = 0$, then $x \cdot y = 0$. Assume in the following that $x \neq 0$ and $y \neq 0$. Denote

$$u = \frac{x}{\|x\|}, \quad v = \frac{y}{\|y\|}, \quad \text{and } z = \frac{e_o}{\sqrt{2}}.$$

Since $\|u\| = \|v\| = 1$, $\|z\| = 1/\sqrt{2}$ and $x \cdot e_o/\|x\| \geq c \geq \frac{1}{\sqrt{2}}$, then

$$\|u - z\|^2 = \|u\|^2 - 2u \cdot z + \|z\|^2 = 1 - \frac{2x \cdot e_o}{\|x\|\sqrt{2}} + \frac{1}{2} \leq \frac{1}{2},$$

so that $\|u - z\| \leq \frac{1}{\sqrt{2}}$. Similarly, one can show that $\|v - z\| \leq \frac{1}{\sqrt{2}}$. Since

$$\|u - v\|^2 = \|u\|^2 - 2u \cdot v + \|v\|^2 = 2(1 - u \cdot v),$$

we then have

$$\sqrt{2}\sqrt{1 - u \cdot v} = \|u - v\| \leq \|u - z\| + \|z - v\| \leq 1/\sqrt{2} + 1/\sqrt{2} = \sqrt{2},$$

i.e. $1 - u \cdot v \leq 1$, so that $u \cdot v \geq 0$. Thus

$$x \cdot y = \|x\|\|y\| \; u \cdot v \geq 0.$$

From proposition 5.8.2 it then follows that $(E, K, \|\cdot\|)$ is a UMB-space. □

Assume now that $E$ is separable Hilbert space. As a consequence of proposition 5.8.2 we obtain

**Corollary 5.8.2:** *If $\{e_i\}_{i\in I\!N}$ is an orthonormal basis of a separable Hilbert space $E$, and*

$$K = \{x = \sum_{o}^{\infty} x_i e_i \in E \mid x_i \geq 0 \text{ for each } i \in N\}, \tag{5.8.5}$$

*then $(E, K, \|\cdot\|_2)$ is a UMB-space.*

*Proof.* Since

$$(x|y) = \sum_{o}^{\infty} x_i y_i, \quad \text{for } x = \sum_{o}^{\infty} x_i e_i, \ y = \sum_{0}^{\infty} y_i e_i \in E,$$

it follows from (5.8.5) that $(x|y) \geq 0$ for all $x, y \in K$. Moreover, it is easy to see that $K$ is a closed cone in $E$, whence the assertion follows from proposition 5.8.2. □

As a special case of proposition 5.8.3 we obtain

**Corollary 5.8.3:** *Let $\{e_i\}_{i\in I\!N}$ be an orthonormal basis of a separable Hilbert space $E$, $c > 0$, and*

$$K = \{x = \sum_{o}^{\infty} x_i e_i \mid x_o \geq c(\sum_{1}^{\infty} x_i^2)^{\frac{1}{2}}\}. \tag{5.8.6}$$

*If $c \geq 1$, then $(E, K, \|\cdot\|_2)$ is a UMB-space. If $0 < c < 1$, there exists such a norm $\|\cdot\|$ of $E$, equivalent to $\|\cdot\|_2$, that $(E, K, \|\cdot\|)$ is a UMB-space.*

*Proof.* If $x = \sum_o^\infty x_i e_i$, then $\|x\|_2^2 = (x|e_o)^2 + \sum_1^\infty x_i^2$ and $(x|e_o) = x_o$. Thus

$$x \in K \Leftrightarrow (x|e_o) \geq c(\sum_{1}^{\infty} x_i^2)^{\frac{1}{2}} \Leftrightarrow (x|e_o)^2 \geq c^2\|x\|_2^2 - c^2(x|e_o)^2,$$

which implies that

$$K = \{x \in E \mid (x|e_o) \geq \frac{c}{\sqrt{1+c^2}}\|x\|_2\}.$$

The assertion follows then from proposition 5.8.3 and its proof.□

**Remark 5.8.1:** The cone defined by (5.8.6) has nonempty interior, whereas the cone given by (5.8.5) does not have any interior points.

## 5.8.2. Weak and strong convergence in normed spaces

Recall, that a sequence $(y_n)_{n=o}^{\infty}$ *converges weakly* to $y$ in a normed space $E$, denote $y_n \rightharpoonup y$ or w-$\lim_n y_n = y$, if $\lim_n f(y_n) = f(y)$ for each $f \in E'$. $y$ is said to be a *weak limit* of $(y_n)_{n=o}^{\infty}$.

If a weak limit exists, it is uniquely determined, and each subsequence of a weakly convergent sequence posses the same weak limit. Moreover, the following conditions hold (cf. Yoshida (1974)).

If $x_n \rightharpoonup x$ and $y_n \rightharpoonup y$, then $x_n + y_n \rightharpoonup x + y$.

If $c_n \to c$ in $\mathbb{R}$ and $y_n \rightharpoonup y$, then $c_n y_n \rightharpoonup cy$.

If $y_n \to y$ strongly, then $y_n \rightharpoonup y$. Reverse holds if dim $E < \infty$.

If $y_n \rightharpoonup y$, then $(y_n)$ is bounded, i.e. $\sup_n \|y_n\| < \infty$, and

$$\|y\| \leq \liminf_{n\to\infty} \|y_n\|. \tag{5.8.7}$$

**Mazur's theorem:** (cf. Yoshida (1974)) *If $(y_n)_{n=o}^{\infty}$ converges weakly in a normed space $E$ to $y$, then to each $\epsilon > 0$ there is a convex combination $\sum_{j=o}^{m} c_j y_j$, i.e. $c_j \geq 0$, $j = 0, \ldots m$ and $\sum_{j=o}^{m} c_j = 1$ such that $\|\sum_{j=o}^{m} c_j y_j - y\| \leq \epsilon$.*

Mazur's theorem implies that the order cone of an ordered normed space is weakly closed in the following sense.

**Proposition 5.8.4:** *Let $K$ be the order cone of an ordered normed space $E$. If a sequence $(y_n)_{n=o}^{\infty}$ of $K$ converges weakly to $y$ in $E$, then $y \in K$.*

*Proof.* By Mazur's theorem there corresponds to each $k = 1, 2, \ldots$ a convex combination $x_k$ of the elements of $(y_n)_{n=o}^{\infty}$ such that $\|x_k - y\| \leq \frac{1}{k}$. Thus $(x_k)_{k=o}^{\infty}$ converges strongly to $y$. This implies that $y \in K$, since $K$ is closed and each $x_k$ belongs to $K$ by the convexity of $K$. □

**Corollary 5.8.4:** *Let $E$ be an ordered normed space. If $x_n \rightharpoonup x$ and $y_n \rightharpoonup y$ in $E$ and $x_n \leq y_n$ for sufficiently large $n$, then $x \leq y$.*

*Proof.* a) $(y_n - x_n)_{n \in \mathbf{N}}$ is a sequence which belongs (eventually) to $K$ which converges weakly to $y - x$, whence $y - x \in K$, i.e. $x \leq y$ by proposition 5.8.4. □

**Proposition 5.8.5:** *Let $(y_n)_{n=o}^{\infty}$ be a nondecreasing sequence in an ordered normed space $E$ with normal order cone. If each subsequence of $(y_n)_{n=o}^{\infty}$ has a subsequence possessing a weak limit, then $(y_n)_{n=o}^{\infty}$ converges strongly to its supremum.*

*Proof.* Corollary 1.1.3 implies that $(y_n)_{n=o}^{\infty}$ converges weakly to $y$ in $E$. To show that the convergence is strong, let $\epsilon > 0$ be given. By Mazur's theorem there is a linear combination $\sum_{j=o}^{m} c_j y_j$ with $c_j \geq 0$, $j = 0, \ldots m$, and $\sum_{j=o}^{m} c_j = 1$ such that
$\|\sum_{j=o}^{m} c_j y_j - y\| \leq \epsilon$. Since $y_j \leq y_n \leq y$ for $j \leq n$, then
$\sum_{j=o}^{m} c_j y_j \leq y_n \leq y$ for $n \geq m$. Hence,

$$0 \leq y - y_n \leq y - \sum_{j=o}^{m} c_j y_j \text{ for each } n \geq m.$$

Since the order cone of $E$ is normal, the above inequality and (1.3.2) imply that

$$\|y - y_n\| \leq \gamma \|y - \sum_{j=0}^{m} c_j y_j\| \leq \gamma \epsilon \text{ for each } n \geq m.$$

The above proof shows that $(y_n)_{n=o}^{\infty}$ converges strongly to $y$.

□

As a consequence of proposition 5.8.5, and the fact that the strong limit of a sequence is also the weak limit, we obtain

**Corollary 5.8.5:** *If the order cone of an ordered normed space $E$ is normal, then a nondecreasing sequence of $E$ has a weak limit if and only if it has a strong limit, and both equal to the supremum of the sequence.*

The results derived in this section for nondecreasing sequences have their dual counterparts for nonincreasing sequences.

Examples 5.8.1 and 5.8.2 imply that the regular order cone of an ordered Banach space $E$ is not necessarily fully regular, and the normal order cone is not necessarily regular. However, such counter-examples don't exist if $E$ is reflexive, as follows from proposition 1.3.4. Moreover, we have

**Proposition 5.8.6:** *If $E = (E, \|\cdot\|)$ is a weakly complete ordered normed space, then each of its bounded chains have both the supremum and the infimum. The conclusion holds also for order bounded chains if the order cone of $E$ is normal.*

*Proof.* If $E$ is weakly complete, it follows that (cf. Narici and Beckenstein (1985)) that bounded subsets of $E$ are relatively compact. Thus the first assertion follows from proposition 1.1.4. If the order cone of $E$ is normal, then each order bounded chain of $E$ is bounded by (1.3.2), so that the above reasoning proves also the second assertion. □

### 5.8.3. $L^p$–spaces of vector valued functions

Let $(\Omega, \mathcal{A}, \mu)$ be a measure space, $E = (E, \|\cdot\|)$ a Banach space, and $1 \leq p < \infty$. Denote by $L^p(\Omega, E)$ the space of (the equivalence classes of a.e. equal) $\mu$-measurable functions $x : \Omega \to E$, for which $|x|^p = t \mapsto \|x(t)\|^p$ is $\mu$-integrable, i.e. $|x| \in L^p(\Omega)$. $L^p(\Omega, E)$ is a vector space with respect to pointwise addition and scalar multiplication, and a Banach space with respect to the norm

$$\|x\|_p = \left(\int_\Omega |x|^p \, d\mu\right)^{\frac{1}{p}}.$$

If $1 < p, q < \infty$ and $\frac{1}{p} + \frac{1}{q} = 1$, and if $x \in L^p(\Omega, E)$ and $y \in L^q(\Omega, E)$, then $|x||y|$ belongs to $L^1_+(\Omega)$, and Hölder's inequality holds, i.e.

$$\int_\Omega |x||y| d\mu \leq \|x\|_p \|y\|_q.$$

The above results, as well as the following two convergence results are proved in Lang (1969).

**Lemma 5.8.2:** *If $(x_n)_{n=o}^\infty$ converges strongly to $x$ in $L^p(\Omega, E)$, there is a subsequence of $(x_n)_{n=o}^\infty$ which converges pointwise a.e. to $x$.*

**$L^p$–dominated convergence theorem:** *Let $(x_n)_{n=o}^\infty$ be a sequence in $L^p(\Omega, E)$, which converges pointwise a.e. to a function $x: \Omega \to E$. If there is a function $v \in L^p_+(\Omega)$ such that $|x_n| \leq v$ for each $n \in \mathbb{N}$, then $(x_n)_{n=o}^\infty$ converges strongly to $x$ in $L^p(\Omega, E)$.*

Next we shall consider $L^p(\Omega, E)$ as an ordered Banach space.

**Lemma 5.8.3:** *If $K$ is a closed cone in $E$, then*

$$L^p(\Omega, K) = \{x \in L^p(\Omega, E) \mid x(t) \in K \quad \textit{for a.a. } t \in \Omega\}$$

*is a closed cone in $L^p(\Omega, E)$.*

*Proof.* It is easy to see that $L^p(\Omega, K)$ is a cone in $L^p(\Omega, E)$. To show that it is closed, let $(x_n)_{n=o}^\infty$ be a sequence in $L^p(\Omega, K)$ which converges strongly to $x$ in $L^p(\Omega, E)$. By lemma 5.8.2, $(x_n)_{n=o}^\infty$ has a subsequence $(x_{n_k})_{k=o}^\infty$, which converges pointwise a.e. to $x$. Since $x_{n_k}(t) \in K$ for a.a. $t \in \Omega$, and since $K$ is closed, then $x(t) = \lim_k x_{n_k}(t) \in K$ for a.a. $t \in \Omega$, whence $x \in L^p(\Omega, K)$. □

If $E$ is an ordered Banach space and $K$ its order cone, then $L^p(\Omega, E)$ is an ordered Banach space with respect to the partial ordering defined by $L^p(\Omega, K)$. This ordering $\leq$ can also be defined as

$$x \leq y \text{ if and only if } x(t) \leq y(t) \text{ for a.a. } t \in \Omega.$$

**Proposition 5.8.7:** *Let $E$ be an ordered Banach space, $K$ its order cone and $p \in [1, \infty)$. If $K$ is normal, regular or fully regular, then $L^p(\Omega, K)$ is normal, regular or fully regular, respectively.*

*Proof.* Assume first that $K$ is normal, and let $x, y \in L^p(\Omega, K)$, $x \leq y$ be given. Since $K$ is normal, there is $M > 0$ such that

$$\|x(t)\| \leq M\,\|y(t)\| \qquad \text{for a.a. } t \in \Omega.$$

Thus

$$\int_\Omega \|x(t)\|^p d\mu \leq M^p \int_\Omega \|y(t)\|^p d\mu,$$

which implies that $\|x\|_p \leq M\,\|y\|_p$. This proves the normality of $L^p(\Omega, K)$.

Assume next that $K$ is regular, and let $(x_n)_{n=o}^\infty$ be a nondecreasing and order bounded sequence in $L^p(\Omega, K)$. By definition there is $b \in L^p(\Omega, K)$ such that for all $n \in \mathbb{N}$,

$$0 \leq x_n(t) \leq b(t) \text{ for a.a. } t \in \Omega.$$

Thus there is a null set $Z$ in $\Omega$ such that $(x_n(t))_{n=o}^\infty$ is a nondecreasing and order bounded sequence in $K$ for each $t \in \Omega \setminus Z$. Since $K$ is regular, then $(x_n(t))_{n=o}^\infty$ has the strong limit $x(t)$ for each $t \in \Omega \setminus Z$. Defining $x(t) = 0$ for $t \in Z$, then $(x_n)_{n=o}^\infty$ converges pointwise a.e. in $\Omega$ to $x$. Since $K$ is also normal, there is $M > 0$ such that $|x_n| \leq M|b|$ for each $n \in \mathbb{N}$. Since $M|b| \in L_+^p(\Omega)$, it follows from the $L^p$–dominated convergence theorem that $x \in L^p(\Omega, E)$, and that $(x_n)_{n=o}^\infty$ converges strongly in $L^p(\Omega, E)$ to $x$. Lemma 5.8.3 implies that $x \in L^p(\Omega, K)$. This proves that the cone $L^p(\Omega, K)$ is regular.

Assume finally that $K$ is fully regular, and let $(x_n)_{n=o}^\infty$ be a nondecreasing and bounded sequence in $L^p(\Omega, K)$. We may also assume (replacing the norm of $E$ by an equivalent one by lemma 1.3.2) that the norm $\|\cdot\|$ of $E$ is monotone in $K$. Thus the functions $|x_n|^p$ form a nondecreasing sequence in $L^1_+(\Omega)$, and the integrals $\int_\Omega |x_n|^p d\mu$ form a bounded sequence. This implies by the monotone convergence theorem that $(|x_n|^p)_{n=o}^\infty$ converges pointwise a.e. in $\Omega$ to a function $w \in L^1_+(\Omega)$. Consequently $v = w^{\frac{1}{p}}$ belongs to $L^p_+(\Omega)$ and

$$\|x_n(t)\| \leq v(t) \quad \text{for a.a. } t \in \Omega. \tag{a}$$

Thus $(x_n(t))_{n=o}^\infty$ is a bounded and nondecreasing sequence in $K$ for a.a. $t \in \Omega$. Since $K$ is fully regular, then $x(t) = \lim_n x_n(t)$ exists for a.a. $t \in \Omega$. Defining $x(t) = 0$ for the remaining $t \in \Omega$, then $(x_n)_{n=o}^\infty$ converges pointwise a.e. in $\Omega$ to $x$. This and (a) imply by the $L^p$–dominated convergence theorem that $x \in L^p(\Omega, E)$, and that $(x_n)_{n=o}^\infty$ converges strongly to $x$ in $L^p(\Omega, E)$. This shows that $L^p(\Omega, K)$ is fully regular. □

**Example 5.8.4:** Let $(\Omega_j, \mathcal{A}_j, \mu_j)$ be measure spaces, $E_o = (E_o, \|\cdot\|_o)$ a Banach space, and $K_o$ a closed cone in $E_o$. Given a sequence $(p_j)_{j=o}^\infty$ of numbers from $[1, \infty)$, denote

$$E_{j+1} = (L^{p_j}(\Omega_j, E_j), \|\cdot\|_{p_j}) \quad \text{and} \quad K_{j+1} = L^{p_j}(\Omega_j, K_j), \; j \in I\!N.$$

$E_j$ is a Banach space and $K_j$ is by lemma 5.8.3 a closed cone in $E_j$ for each $j = 1, 2, \ldots$. From proposition 5.8.7 it follows that each $K_j$ is normal, regular or fully regular whenever $K_o$ has the corresponding property.

Consider next the case when $p = \infty$. Let $(\Omega, \mathcal{A}, \mu)$ be a measure space, and $E = (E, \|\cdot\|)$ a Banach space. Denote by $L^\infty(\Omega, E)$ the space of (the equivalence classes of) those $\mu$-measurable functions $x : \Omega \to E$ for which $|x| \in L^\infty_+(\Omega)$. $L^\infty(\Omega, E)$ is a Banach space with respect to the norm

$$\|x\|_\infty = \inf\{c \geq 0 \mid \|x(t)\| \leq c \;\text{ for a.a. } t \in \Omega\}.$$

If $K$ is a closed cone in $E$, then the set $L^\infty(\Omega, K)$ of a.e. $K$-valued elements of $L^\infty(\Omega, E)$ is a closed cone in $L^\infty(\Omega, E)$. It is also easy to see that if $K$ is normal, then also $L^\infty(\Omega, K)$ is normal.

**Lemma 5.8.4:** *If $0 < \mu(\Omega) < \infty$, then $L^\infty(\Omega, E) \subseteq L^p(\Omega, E) \subseteq L^1(\Omega, E)$ for each $p \in (1, \infty)$, and*

$$\|x\|_\infty = \lim_{p \to \infty} \|x\|_p \quad \textit{for each } x \in L^\infty(\Omega, E). \tag{5.8.8}$$

*Moreover, if $E$ is an ordered Banach space and if the order cone of $E$ is regular (resp. fully regular), then each order bounded (resp. bounded) chain of $L^\infty(\Omega, E)$ has the supremum and the infimum in $L^\infty(\Omega, E)$.*

*Proof.* Let $x \in L^\infty(\Omega, E)$ be given. Since $\|x(t)\| \le \|x\|_\infty$ for a.e. $t \in \Omega$, then

$$\|x\|_p \le \mu(\Omega)^{\frac{1}{p}} \|x\|_\infty \tag{5.8.9}$$

for each $p \in [1, \infty)$, which shows that $L^\infty(\Omega, E) \subseteq L^p(\Omega, E)$.

By definition of $\|x\|_\infty$ there corresponds to each $\epsilon > 0$ a subset $B$ of $\Omega$ with $\mu(B) > 0$ such that $\|x(t)\| \ge \|x\|_\infty - \epsilon$ for each $t \in B$. Thus

$$\mu(B)^{\frac{1}{p}}(\|x\|_\infty - \epsilon) \le \|x\|_p. \tag{a}$$

From (5.8.9) and (a) it follows that for each $\epsilon > 0$

$$\|x\|_\infty - \epsilon \le \liminf_{p \to \infty} \|x\|_p \le \limsup_{p \to \infty} \|x\|_p \le \|x\|_\infty,$$

which implies (5.8.8).

Choosing a unit vector $e \in E$ and $y(t) \equiv e$ in Hölder's inequality, we have for each $x \in L^p(\Omega, E)$, $1 < p < \infty$,

$$\|x\|_1 \le \mu(\Omega)^{\frac{1}{q}} \|x\|_p, \tag{5.8.10}$$

whence $L^p(\Omega, E) \subseteq L^1(\Omega, E)$.

Assume now that $E$ is an ordered Banach space with regular (resp. fully regular) order cone $K$. Let $C$ be an order bounded (resp. bounded) and chain in $L^\infty(\Omega, E)$, and let $p \in [1, \infty)$ be given. From the first part of this lemma it follows that $C$ is an order bounded (resp. a bounded) chain in $L^p(\Omega, E)$, whose order cone $L^p(\Omega, K)$ is by proposition 5.8.7 regular (resp. fully regular). This and propositions 1.3.3 and 1.3.4 imply that $C$ has the supremum $x^*$ in $L^p(\Omega, E)$, and there is a nondecreasing sequence $(x_n)_{n=o}^\infty$ in $C$ which converges strongly to $x^*$ in $L^p(\Omega, E)$. By lemma 5.8.2 $(x_n)_{n=o}^\infty$ has a subsequence which converges pointwise a.e. in $\Omega$ to $x^*$. Since $(x_n(t))_{n=o}^\infty$ is nondecreasing for a.a. $t \in \Omega$, then $(x_n)_{n=o}^\infty$ itself converges pointwise a.e. in $\Omega$ to $x^*$. Moreover, there is $M > 0$ such that $\|x_n(t)\| \leq M$ for a.a. $t \in \Omega$. Thus

$$x^*(t) = \lim_{n \to \infty} x_n(t) = \sup_n x_n(t),$$

and $\|x^*(t)\| \leq M$ for a.a. $t \in \Omega$. This implies that $x^*$ belongs to $L^\infty(\Omega, E)$ and it is easy to see that $x^* = \sup C$ also in $L^\infty(\Omega, E)$. The existence of $\inf C$ follows from the above proof, since $\inf C = -\sup(-C)$. □

**Proposition 5.8.8:** *If $(\Omega, \mathcal{A}, \mu)$ is a $\sigma$-finite measure space and $E$ is an ordered Banach space with regular (resp. fully regular) order cone, then each order bounded (resp. bounded) chain of $L^\infty(\Omega, E)$ has the supremum and the infimum in $L^\infty(\Omega, E)$.*

*Proof.* Since $\Omega$ is $\sigma$-finite, then $\Omega = \bigcup_{n=o}^\infty A_n$, where $A_n \subseteq A_{n+1}$ and $\mu(A_n) < \infty$ for each $n \in \mathbb{N}$. Let $C$ be an order bounded (resp. bounded) chain in $L^\infty(\Omega, K)$. Choose $M > 0$ such that

$$\|x(t)\| \leq M \text{ for a.a. } t \in \Omega \text{ and for each } x \in C.$$

The set $C|A_n = \{x|A_n \mid x \in C\}$ is for each $n \in \mathbb{N}$ an order bounded (resp. bounded) chain in $L^\infty(A_n, K)$. Lemma 5.8.4 implies then that $x_n = \sup(C|A_n)$ exists in $L^\infty(A_n, K)$. Defining $x_n(t) = 0$ for $t \in \Omega \setminus A_n$, we obtain a sequence of $\mu$-measurable functions $x_n \colon \Omega \to K$. This sequence is nondecreasing, since

$A_n \subseteq A_{n+1}$, order bounded (resp. bounded) and
$\|x_n(t)\| \leq M$ for a.a. $t \in \Omega$ and for each $n \in \mathbb{N}$.
Thus

$$x^*(t) = \lim_{n\to\infty} x_n(t) = \sup_{n\in N} x_n(t) \tag{a}$$

exists for a.a. $t \in \Omega$. Defining $x^*(t) = 0$ for the remaining $t \in \Omega$, we obtain a $\mu$-measurable function $x^* : \Omega \to K$, which satisfies

$$\|x^*(t)\| \leq M \quad \text{for a.a. } t \in \Omega.$$

Thus $x^* \in L^\infty(\Omega, K)$.

If $x \in C$, then $x|A_n \leq x_n$, so that
$x(t) \leq x_n(t) \leq x^*(t)$ for a.a. $t \in A_n$ and for each $n \in \mathbb{N}$.
Thus $x \leq x^*$ for each $x \in C$, so that $x^*$ is an upper bound of $C$. If $y \in L^\infty(\Omega, K)$ is another upper bound of $C$, then
$x(t) \leq y(t)$ for a.a. $t \in \Omega$ and for each $x \in C$.
Thus $x|A_n \leq y|A_n$ for all $n \in \mathbb{N}$ and $x \in C$, whence
$x_n(t) \leq y(t)$ for a.a. $t \in \Omega$ and for each $n \in \mathbb{N}$.
This and (a) imply that $x^* \leq y$, whence $x^* = \sup C$ in $L^\infty(\Omega, K)$.

If $C$ is a chain in $L^\infty(\Omega, E)$ and $x_o \in C$, then $C_o = \{x - x_o \mid x \in C, x_o \leq x\}$ is a chain in $L^\infty(\Omega, K)$, whence $\sup C_o$ exists in $L^\infty(\Omega, K)$. But then $x_o + \sup C_o$ is the supremum of $C$ in $L^\infty(\Omega, E)$. The above proof implies also that $\inf C = -\sup(-C)$ exists in $L^\infty(\Omega, E)$. □

### 5.8.4. Special cases

If $E = (E, (\cdot|\cdot))$ a Hilbert space, then $L^2(\Omega, E)$ is a Hilbert space with respect to the inner product

$$\langle x, y \rangle = \int_\Omega (x|y) d\mu.$$

**Lemma 5.8.5:** *If $K$ is a closed cone in a Hilbert space $E$, and if $(u|v) \geq 0$ for all $u, v \in K$, then $L^2(\Omega, K)$ is a regular and fully regular order cone in $L^2(\Omega, E)$.*

*Proof.* If $x, y \in L^2(\Omega, K)$, then $(x(t)|y(t)) \geq 0$ for a.a. $t \in \Omega$, whence

$$\langle x, y\rangle = \int_{\Omega} (x|y) d\mu \geq 0.$$

This and propositions 5.8.1 and 5.8.2 imply the assertion. □

**Example 5.8.5:** Let $(\Omega_j, \mathcal{A}_j, \mu_j)$ be measure spaces, $E_o = (E_o, (\cdot|\cdot)_o)$ a Hilbert space, and $K_o$ a closed cone in $E_o$. Denoting

$$E_{j+1} = L^2(\Omega_j, E_j),\ K_{j+1} = L^2(\Omega_j, K_j),\ j \in \mathbb{N},$$

$$(x|y)_{j+1} = \int_{\Omega_j} (x|y)_j d\mu_j \ \text{ and } \|x\|_j = (x|x)_j^{\frac{1}{2}},$$

we obtain a sequence of Hilbert spaces $E_j$, $j \in \mathbb{N}$. From proposition 5.8.7 and lemma 5.8.5 it follows that if $(x|y)_o \geq 0$ for all $x, y \in K_o$, then $K_j$ is a regular and fully regular order cone in $(E_j, \|\cdot\|_j)$ for each $j \in \mathbb{N}$.

As a Hilbert space $L^2(\Omega, E)$ is reflexive whenever $E$ is a Hilbert space. Moreover, we have (cf. Lang (1969)).

**Lemma 5.8.6:** *If $E$ is a Hilbert space and $(\Omega, \mathcal{A}, \mu)$ is a measure space with $\sigma$-finite measure $\mu$, then $L^p(\Omega, E)$ is reflexive for each $p \in (1, \infty)$ and $L^p(\Omega, E)' = L^{\frac{p}{p-1}}(\Omega, E)$.*

In the case when $\Omega = [a, b] \subset \mathbb{R}$ we have (cf. Kufner, John and Fučic (1977), Lang (1969)).

**Lemma 5.8.7:** *Let $E$ be a Banach space and $a, b \in \mathbb{R}$, $a < b$.*

a) *$L^p([a, b], E)$ is separable if $E$ is separable and $1 \leq p < \infty$.*
b) *$L^p([a, b], E)$ is uniformly convex if $E$ is uniformly convex and $1 < p < \infty$.*
c) *$L^p([a, b], E)$ is reflexive if $E$ is reflexive and $1 < p < \infty$.*
d) *$L^p([a, b], E)$ is a Hilbert space if $E$ is a Hilbert space and $p = 2$.*

e) *If $E$ is continuously embedded in the Banach space $F$ and $1 \leq p \leq q \leq \infty$, then the embedding $L^q([a,b],E) \subseteq L^p([a,b],F)$ is continuous.*

### 5.8.5. Other function spaces

In this subsection we shall define and introduce some properties of Orlicz spaces, Sobolev spaces and Orlicz-Sobolev spaces.

**Orlicz spaces:** (cf. Krasnosel'skii and Rutickii (1961)) We say that $M\colon \mathbb{R} \to \mathbb{R}_+$ is an *N-function* if it can be represented in the form

$$M(t) = \int_0^t \varphi(\tau)d\tau, \qquad t \in \mathbb{R},$$

where $\varphi\colon \mathbb{R} \to \mathbb{R}$ is an odd, increasing and right continuous in $\mathbb{R}_+$, and satisfies

$$\varphi(t) = 0 \text{ if and only if } t = 0, \text{ and } \varphi(t) \to +\infty \text{ when } t \to +\infty.$$

The *conjugate function* $\bar{M}$ of a $N$-function $M$ is defined by

$$\bar{M}(t) = \int_0^t \sup\{s \mid \varphi(s) \leq \tau\}d\tau, \qquad t \in \mathbb{R}.$$

We say that $M$ satisfies the $\Delta_2$-*condition* if there exist constants $k > 0$ and $u_o \geq 0$ such that

$$M(2u) \leq k\,M(u) \text{ for } u \geq u_o.$$

Let $\Omega$ be a bounded domain in $\mathbb{R}^m$, and $M\colon \mathbb{R} \to \mathbb{R}_+$ an N-function. The function space

$$\begin{aligned} L_M(\Omega) = \{u\colon \Omega \to \mathbb{R} \mid\ & u \text{ is measurable and} \\ & \int_\Omega M\left(\frac{u(x)}{k}\right)dx < \infty \text{ for some } k > 0\} \end{aligned}$$

is called an *Orlicz space*. It can be shown that $L_M(\Omega)$ is a linear subspace of $L^1(\Omega)$, and that condition

$$\|u\|_{(M)} = \inf\{k > 0 \mid \int_\Omega M(\frac{u(x)}{k})dx \le 1\} \tag{5.8.11}$$

defines a Banach norm, called a *Luxemburg norm*, in $L_M(\Omega)$. If $M(t) = \frac{1}{p}|t|^p$, $t \in I\!R$, when $1 < p < \infty$, then $L_M(\Omega) = L^p(\Omega)$ and $\|\cdot\|_{(M)} = p^{-1/p}\|\cdot\|_p$.

**Lemma 5.8.8:** *If $M$ satisfies the $\Delta_2$-condition, then*

$$L_M^+(\Omega) = \{u \in L_M(\Omega) \mid u(x) \ge 0 \ \ \textit{for a.a. } x \in \Omega\}$$

*is a fully regular cone in $L_M(\Omega)$. If the $\Delta_2$-condition does not hold, then $L_M^+(\Omega)$ is not even regular.*

On separability and reflexivity of Orlicz spaces we have

**Lemma 5.8.9:** *$L_M(\Omega)$ is separable if and only if $M$ satisfies the $\Delta_2$-condition, and reflexive if and only if both $M$ and $\bar{M}$ satisfy the $\Delta_2$-condition.*

Denote by $E_M(\Omega)$ the closure of $L^\infty(\Omega)$ in $L_M(\Omega)$. The so obtained Banach space $E_M(\Omega)$ is separable and its nonnegative elements form a regular order cone in $E_M(\Omega)$. Moreover, $E_M(\Omega) = L_M(\Omega)$ if and only if $M$ satisfies the $\Delta_2$-condition.

<u>Sobolev spaces:</u> (cf. Kufner, John and Fučic (1977)) Let $\Omega$ be an open subset of $I\!R^m$. When $k \in I\!N$ and $1 \le p < \infty$, define

$$W^{k,p}(\Omega) = \{w \in L^p(\Omega) \mid D^j w \in L^p(\Omega) \text{ for } |j| \le k\},$$

where $j = (k_1, \dots, k_m)$, $|j| = k_1 + \cdots + k_m$, and $D^j w$ denotes a *generalized derivative*, i.e. $u = D^j w$ if

$$\int_\Omega u(x)\varphi(x)\,dx = (-1)^{|j|}\int_\Omega w(x)D^j\varphi(x)\,dx$$

for each $\varphi \in C_o^\infty(\Omega) = \bigcap_{k=o}^\infty C_o^k(\Omega)$, where $C_o^k(\Omega)$ is the set of those functions $\varphi\colon \Omega \to I\!R$ whose supports are compact and lie in $\Omega$ and the partial derivative

$$D^j\varphi = \frac{\partial^{k_1+\cdots+k_m}\varphi}{\partial x_1^{k_1}\ldots\partial x_m^{k_m}}$$

is continuous in $\Omega$ whenever $|j| \le k$.

The Sobolev space $W^{k,p}(\Omega)$ is a Banach space with respect to the norm

$$\|w\|_{k,p} = \sum_{|j|\le k} \|D^j w\|_p.$$

Denote by $W_o^{k,p}(\Omega)$ the closure of $C_o^\infty(\Omega)$ relative to this norm. The nonnegative-valued elements of $W^{k,p}(\Omega)$ form its closed cone $W_+^{k,p}(\Omega)$. The corresponding result holds also for $W_o^{k,p}(\Omega)$.

**Lemma 5.8.10:** *The Sobolev spaces $W^{k,p}(\Omega)$ and $W_o^{k,p}(\Omega)$ are separable for each $p \in [1,\infty)$ and reflexive for each $p \in (1,\infty)$. If $n > 1$, $p > 1$ and $kp \le m$, and if $\Omega$ is a bounded domain with the Lipschitz-continuous boundary $\partial\Omega$, then $W^{k,p}(\Omega)$ is continuously embedded in $L^p(\Omega)$, i.e. there exists $c > 0$ such that $\|u\|_p \le c\,\|u\|_{W^{k,p}(\Omega)}$ for each $u \in W^{k,p}(\Omega)$.*

<u>**Orlicz-Sobolev spaces:**</u> Let $\Omega$ be a bounded domain in $I\!R^m$. When $M$ is an $N$-function which satisfies the $\Delta_2$-condition, and $k \in I\!N$, define

$$W^kL_M(\Omega) = \{w \in L_M(\Omega) \mid D^jw \in L_M(\Omega) \text{ for } |j| \le k\},$$

where $j = (k_1, \ldots, k_m)$, $|j| = k_1 + \cdots + k_m$, and $D^j w$ denotes the generalized derivative. The nonnegative-valued elements of $W^k L_M(\Omega)$ form its closed cone $W^k L_M(\Omega)_+$.

**Lemma 5.8.11:** *$W^k L_M(\Omega)$ is a Banach space with respect to the norm*

$$\|w\| = \Big(\sum_{|j|\le k} \|D^j w\|^2_{(M)}\Big)^{\frac{1}{2}},$$

*and is reflexive if and only if $M$ and its conjugate $\bar{M}$ satisfy the $\Delta_2$-condition.*

**Spaces of bounded functions:** Given a nonempty set $\Omega$ and a Banach space $E$, denote

$$B(\Omega, E) = \{x \colon \Omega \to E \mid \sup_{\omega \in \Omega} \|x(\omega)\| < \infty\}. \tag{5.8.12}$$

Define a norm in $B(\Omega, E)$ by $\|x\|_o = \sup_{\omega\in\Omega} \|x(\omega)\|$. If $E$ is ordered by a closed positive cone $K$, then $B(\Omega, K) = \{x \in B(\Omega, E) \mid x(\omega) \in K \text{ for all } \omega \in \Omega\}$ is a closed positive cone in $B(\Omega, E)$, which induces the pointwise partial ordering in $B(\Omega, E)$. Moreover, it is easy to see that $B(\Omega, K)$ is normal if $K$ is, and has a nonempty interior whenever $K$ has.

**Lemma 5.8.12:** *Let $E$ be an ordered Banach space with order cone $K$. A chain $C$ of $B(\Omega, E)$ has the infimum and the supremum in the following cases.*

a) *$C$ is order bounded and $K$ is regular.*
b) *$C$ is bounded and $K$ is fully regular.*
c) *$C$ is bounded and $E$ is reflexive.*

*Proof.* a) Let $C$ be an order bounded chain in $B(\Omega, E)$, and let $a$, $b \in B(\Omega, E)$ be so chosen that $a(\omega) \le x(\omega) \le b(\omega)$ for each $x \in C$ and $\omega \in \Omega$. If $K$ is regular, it follows from proposition

1.3.2 that

$$y(\omega) = \inf_{x \in C} x(\omega) \quad \text{and} \quad z(\omega) = \sup_{x \in C} x(\omega) \tag{a}$$

exist for each $\omega \in \Omega$, and that $a(\omega) \leq y(\omega) \leq z(\omega) \leq b(\omega)$ for each $\omega \in \Omega$. This and the normality of $K$ imply that $y$, $z \in B(\Omega, E)$. Obviously, $y = \inf C$ and $z = \sup C$.

b) Let $C$ be a bounded chain in $B(\Omega, E)$. Then there is a positive number $M$ so that

$$\|x(\omega)\| \leq M \quad \text{for each } x \in C \text{ and } \omega \in \Omega. \tag{b}$$

Hence, if $K$ is fully regular, the relations (a) define by proposition 1.3.3 mappings $y$, $z : \Omega \to E$, and for each fixed $\omega \in \Omega$ there exist monotone sequences $(y_n)$ and $(z_n)$ in $C$ such that $y_n(\omega) \to y(\omega)$ and $z_n(\omega) \to z(\omega)$. In view of (b) we then see that $y$, $z \in B(\Omega, E)$. From (a) it follows that $y = \inf C$ and $z = \sup C$.

c) Let $C$ be a bounded chain in $B(\Omega, E)$, and let $M > 0$ be so chosen that (b) holds. If $E$ is reflexive, it follows from proposition 1.3.6 that the mappings $y$, $z : \Omega \to E$ are well-defined by (a), and that for each fixed $\omega \in \Omega$ there exist monotone sequences $(y_n)$ and $(z_n)$ in $C$ such that $y_n(\omega) \rightharpoonup y(\omega)$ and $z_n(\omega) \rightharpoonup z(\omega)$. In view of (b) and (5.8.7) we then have

$$\|y(\omega)\| \leq \lim\inf_{n \to \infty} \|y_n(\omega)\| \leq M$$

and

$$\|z(\omega)\| \leq \lim\inf_{n \to \infty} \|z_n(\omega)\| \leq M.$$

Thus $y$, $z \in B(\Omega, E)$, and (a) implies that $y = \inf C$ and $z = \sup C$. □

The above results imply.

**Theorem 5.8.1:** *The bounded chains have supremums and infimums in the following ordered function spaces:*

a) $L^p(\Omega)$, *ordered by* $L^p_+(\Omega)$, $1 \le p \le \infty$, *where* $(\Omega, \mathcal{A}, \mu)$ *is a measure space with* $\sigma$*-finite* $\mu$ *if* $p = \infty$.

b) $L^p(\Omega, E)$, *ordered by* $L^p(\Omega, K))$, $1 \le p \le \infty$ *where* $E$ *is an ordered Banach space with fully regular order cone* $K$, *and* $(\Omega, \mathcal{A}, \mu)$ *is a measure space with* $\sigma$*-finite* $\mu$ *if* $p = \infty$.

c) $B(\Omega, E)$, *ordered by* $B(\Omega, K)$, *where* $\Omega$ *is a nonempty set and* $E$ *is an ordered Banach space which is reflexive or its order cone* $K$ *is fully regular.*

d) $L_M(\Omega)$, *ordered by* $L_M(\Omega)_+$, *if* $\Omega$ *is a bounded domain in* $\mathbb{R}^m$ *and* $M$ *satisfies the* $\Delta_2$*-condition.*

e) $W^{k,p}(\Omega)$, *ordered by* $W^{k,p}_+(\Omega)$, $\Omega$ *being an open subset of* $\mathbb{R}^m$, $1 < p < \infty$ *and* $k \in \mathbb{N}$.

f) $W^k L_M(\Omega)$, *ordered by* $W^k L_M(\Omega)_+$, *where* $\Omega$ *is a bounded domain in* $\mathbb{R}^m$, $k \in \mathbb{N}$ *and both* $M$ *and* $\bar{M}$ *satisfy the* $\Delta_2$*-condition.*

g) $C_o^\infty(\Omega)$, *ordered by* $C_o^\infty(\Omega)_+$, *if* $\Omega$ *is an open subset of* $\mathbb{R}^m$.

**Proposition 5.8.9:** *Order bounded chains have supremums and infimums in the ordered function spaces given in theorem 5.8.1 a), b) and c), also when* $K$ *in b) and c) is regular, and in* $E_M(\Omega)$, *ordered by its non-negative elements,* $\Omega$ *being a bounded domain in* $\mathbb{R}^m$.

### 5.8.6. Ordered spaces of continuous functions

In this section we shall study the existence of supremums and infimums of chains in the spaces $C(X, L^p(\Omega, E))$, where $X$ is a topological space, $(\Omega, \mathcal{A}, \mu)$ a measure space and $E$ an ordered Banach space, the partial ordering of $C(X, L^p(\Omega, E))$ being defined by

$$x \le y \text{ if and only if } x(t) \le y(t) \text{ for each } t \in X. \tag{5.8.13}$$

**Lemma 5.8.13:** *Let the order cone of $E$ be regular (resp. fully regular), and $\mu(\Omega) < \infty$. Then each pointwise order bounded (resp. bounded), and equicontinuous chain of $C(X, L^\infty(\Omega, E))$ has the supremum and the infimum in $C(X, L^\infty(\Omega, E))$.*

*Proof.* Let $C$ be a pointwise bounded and equicontinuous chain in $C(X, L^\infty(\Omega, E))$. From lemma 5.8.7 it follows that for each $t \in X$ $x^*(t) = \sup_{x\in C} x(t)$ exists in $L^\infty(\Omega, E)$ and equals to that in $L^p(\Omega, E)$ for each $p \in [1, \infty)$. To prove that the so obtained function $x^*: X \to L^\infty(\Omega, E)$ is continuous, let $t \in X$ and $\epsilon > 0$ be given. Since $C$ is equicontinuous, there is a neighborhood $U$ of $t$ such that

$$\|x(s) - x(t)\|_\infty \le \epsilon \ \text{ whenever } s \in U \text{ and } x \in C.$$

This implies by (5.8.9) that, for each $p \in [1, \infty)$

$$\|x(s) - x(t)\|_p \le \mu(\Omega)^{\frac{1}{p}}\epsilon \ \text{ whenever } s \in U \text{ and } x \in C.$$

This, lemma 5.8.4 and the proof of proposition 5.8.8 imply that

$$\|x^*(s) - x^*(t)\|_p \le \mu(\Omega)^{\frac{1}{p}}\epsilon \ \text{ whenever } s \in U\ . \tag{a}$$

Since $x^*(s) - x^*(t) \in L^\infty(\Omega, E)$ and $\mu(\Omega) < \infty$, it follows from lemma 5.8.4 that

$$\|x^*(s) - x^*(t)\|_\infty = \lim_{p\to\infty} \|x^*(s) - x^*(t)\|_p,$$

which, together with (a), implies that

$$\|x^*(s) - x^*(t)\|_\infty \le \epsilon \ \text{ whenever } s \in U\ .$$

Thus $x^*$ is continuous at each $t \in X$, so that $x^* \in C(X, L^\infty(\Omega, E))$. From the definition of $x^*$ it follows that $x^*$ is the supremum of $C$ in $C(X, L^\infty(\Omega, E))$. The above proof and $\inf C = -\inf(-C)$ imply that $\inf C$ exists in $C(X, L^\infty(\Omega, E))$. □

**Proposition 5.8.10:** *Let $(\Omega, \mathcal{A}, \mu)$ be a $\sigma$-finite measure space and $E = (E, \|\cdot\|)$ an ordered Banach space with regular (resp. fully regular) order cone. If a chain $C$ in $C(X, L^\infty(\Omega, E))$ is pointwise order bounded (resp. bounded) and equicontinuous, then* $\sup C$ *and* $\inf C$ *exist in* $C(X, L^\infty(\Omega, E))$.

*Proof.* Assume that $\Omega = \bigcup_{n=o}^{\infty} A_n$, where $A_n \subset A_{n+1}$ and $\mu(A_n) < \infty$ for each $n \in I\!N$. From lemma 5.8.4 it follows that $x^*(t) = \sup_{x \in C} x(t)$ exists in $L^\infty(\Omega, E)$ for each $t \in X$. To prove that $x^* = \sup C$, it suffices to show that $x^*$ is continuous.

Assume that $t \in X$, and let $\epsilon > 0$ be given. Since $C$ is equicontinuous, there is a neighborhood $U$ of $t$ such that

$$\|x(s) - x(t)\|_\infty \leq \epsilon \quad \text{whenever } s \in U \text{ and } x \in C.$$

This implies that

$$\|x(s)(\omega) - x(t)(\omega)\| \leq \epsilon \ \text{ for a.a. } \omega \in \Omega \text{ and for } \ s \in U,\ x \in C.$$

Denoting by $\|\cdot\|_\infty^n$ the norm of $L^\infty(A_n, E)$, we then have

$$\|x(s)|A_n - x(t)|A_n\|_\infty^n \leq \epsilon \ s \in U,\ x \in C.$$

Since $x^*(t)|A_n = \sup_{x \in C} x(t)|A_n$ for each $n \in I\!N$ and $t \in X$, it follows from the proof of lemma 5.8.12 that

$$\|x^*(s)|A_n - x^*(t)|A_n\|_\infty^n \leq \epsilon,\ j \geq k,\ n \in I\!N$$

This means that

$$\|x^*(s)(\omega) - x^*(t)(\omega)\| \leq \epsilon \text{ for a.a. } \omega \in A_n$$

whenever $s \in U$ and $n \in \mathbb{N}$. Since $\Omega = \bigcup_{n=o}^{\infty} A_n$, we then have

$$\|x^*(s) - x^*(t)\|_\infty \leq \epsilon \text{ for } s \in U.$$

This proves that $x^*$ is continuous at $t$. Thus $x^* \in C(X, L^\infty(\Omega, E))$, so that $x^* = \sup C$ in $C(X, L^\infty(\Omega, E))$. The proof of the existence of $\inf C$ is similar. □

**Example 5.8.6:** The set $\mathbb{N}$ is $\sigma$-finite with respect to the counting measure $\mu$, and $L^p(\mathbb{N}, E)$ can be identified for $1 \leq p < \infty$ with the space $l^p(E)$ of sequences $x = (x_n)_{n=o}^{\infty}$ of the elements of the Banach space $E = (E, \|\cdot\|)$, for which $\|x\|_p = (\sum_{n=o}^{\infty} \|x_n\|^p)^{\frac{1}{p}}$ is finite. It can be shown that for $p > 1$ the space $l^p(E)$ is reflexive if and only if $E$ is reflexive.

$L^\infty(\mathbb{N}, E)$ is equal to the space $l^\infty(E)$ of all bounded sequences $x = (x_n)_{n=o}^{\infty}$ of $E$, and $\|x\|_\infty = \sup_{n \in \mathbb{N}} \|x_n\|$. If $E$ is ordered, and $K$ its order cone, then the partial ordering of $L^p(\mathbb{N}, E)$, defined by $L^p(\mathbb{N}, K)$ equals to that, defined in $l^p(E)$ by

$(x_n)_{n=o}^{\infty} \leq (y_n)_{n=o}^{\infty}$ if and only if $y_n - x_n \in K$ for each $n \in \mathbb{N}$.
Thus, the results of propositions 1.3.9 and 5.8.10 imply.

**Corollary 5.8.6:** *Let $E$ be an ordered normed space with regular (resp. fully regular) order cone and $1 \leq p \leq \infty$. If a chain $C$ in $C(X, l^p(E))$ is pointwise order bounded (resp. bounded) and equicontinuous, then* $\sup C$ *and* $\inf C$ *exist in $C(X, l^p(E))$.*

**Proposition 5.8.11:** *Let $E$ be an ordered Banach space with order cone $K$, and $X$ a topological space. An equicontinuous chain $C$ of $C(X, B(\Omega, E))$ has the infimum and the supremum in the following cases.*

a) *$C$ is pointwise order bounded and $K$ is regular.*
b) *$C$ is pointwise bounded and $K$ is fully regular.*
c) *$C$ is pointwise bounded and $E$ is reflexive.*

*Proof.* a) Let $C$ be an equicontinuous and pointwise order bounded chain in $C(X, B(\Omega, E))$. Given $t \in X$, there exist $a$, $b \in$

$B(\Omega, E)$ so that $a(\omega) \leq x(t)(\omega) \leq b(\omega)$ for each $x \in C$ and $\omega \in \Omega$. If $K$ is regular, it follows from proposition 1.3.2 that

$$z(t)(\omega) = \sup_{x \in C} x(t)(\omega) \tag{a}$$

exist for each $\omega \in \Omega$, and that $a(\omega) \leq z(t)(\omega) \leq b(\omega)$ for each $\omega \in \Omega$. This and the normality of $K$ imply that $z(t) \in B(\Omega, E)$.

Let $\epsilon > 0$ be given. Because $C$ is equicontinuous, there is such a neighborhood $U$ of $t$ that

$$\|x(s) - x(t)\|_o \leq \epsilon \text{ for all } x \in C \text{ and } s \in U. \tag{b}$$

By the definition of $\|\cdot\|_o$ this is tantamount to

$$\|x(s)(\omega) - x(t)(\omega)\| \leq \epsilon \text{ for } x \in C, s \in U, \omega \in \Omega. \tag{c}$$

If $s \in U$ and $\omega \in \Omega$ are also fixed, it follows from (a) by proposition 1.3.2 the existence of nondecreasing sequences $(x_n)$ and $(y_n)$ in $C$ such that $x_n(t)(\omega) \to z(t)(\omega)$ and $y_n(s)(\omega) \to z(s)(\omega)$. Denoting $z_n = \max\{x_n, y_n\}$, $n \in \mathbb{N}$, we obtain a nondecreasing sequence $(z_n)$ of $C$. Because

$$\begin{aligned} &x_o(t)(\omega) \leq x_n(t)(\omega) \leq z_n(t)(\omega) \leq z(t)(\omega), \text{ and} \\ &y_o(s)(\omega) \leq y_n(t)(\omega) \leq z_n(s)(\omega) \leq z(s)(\omega), \end{aligned} \tag{d}$$

for each $n \in \mathbb{N}$, it follows that $z_n(t)(\omega) \to z(t)(\omega)$ and $z_n(s)(\omega) \to z(s)(\omega)$. In view of this and (c) we have

$$\|z(s)(\omega) - z(t)(\omega)\| \leq \epsilon.$$

This holds for each $s \in U$ and $\omega \in \Omega$, whence $z$ is continuous at $t$. Because $t$ was an arbitrarily chosen point of $X$, then $z \in C(X, B(\Omega, E))$. This and (a) imply that $z = \sup C$.

b) Let $C$ be an equicontinuous and pointwise bounded chain in $C(X, B(\Omega, E))$. Given $t \in X$, there is a positive number $M(t)$ so that

$$\|x(t)(\omega)\| \le M(t) \text{ for each } x \in C \text{ and } \omega \in \Omega. \tag{e}$$

Hence, if $K$ is fully regular, the relation (a) defines by proposition 1.3.3 for each fixed $t \in X$ a mapping $z(t) \colon \Omega \to E$, and for each fixed $\omega \in \Omega$ there exist a nondecreasing sequence $(x_n)$ in $C$ such that $x_n(t)(\omega) \to z(t)(\omega)$. This holds for each $\omega \in \Omega$ whence (d) implies that $z(t) \in B(\Omega, E)$. The reasoning similar to that in case a) shows that (a) defines a mapping $z \in C(X, B(\Omega, E))$, and that $z = \sup C$.

c) Assume next that $E$ is reflexive. Let $C$ be an equicontinuous and pointwise bounded chain in $C(X, B(\Omega, E))$, let $t \in X$ be fixed, and choose $M(t) > 0$ such that (e) holds. From proposition 1.3.6 it follows that (a) defines a mapping $(t) \ : \Omega \to E$.

Given $\epsilon > 0$ choose a neighborhood $U$ of $t$ so that (c) holds. If $s \in U$ and $\omega \in \Omega$ are also fixed, there exist by (a) and proposition 1.3.6 nondecreasing sequences $(x_n)$ and $(y_n)$ in $C$ such that $x_n(t)(\omega) \rightharpoonup z(t)(\omega)$ and $y_n(s)(\omega) \rightharpoonup z(s)(\omega)$. In view of (e) and (5.8.7) we have

$$\|z(t)(\omega)\| \le \liminf_{n\to\infty} \|x_n(\omega)\| \le M(t) \text{ and}$$
$$\|z(s)(\omega)\| \le \liminf_{n\to\infty} \|y_n(s)(\omega)\| \le M(s).$$

Thus $z(t)$, $z(s) \in B(\Omega, E)$. Denoting $z_n = \max\{x_n, y_n\}$, $n \in \mathbb{N}$, we obtain a nondecreasing sequence $(z_n)$ of $C$. Because (d) holds for each $n \in \mathbb{N}$, it follows from (a), (d), (e), proposition 1.3.6 and corollary 5.8.4 that $z_n(t)(\omega) \rightharpoonup z(t)(\omega)$ and $z_n(s)(\omega) \rightharpoonup z(s)(\omega)$. In view of this, (5.8.7) and (c) we see that

$$\|z(s)(\omega) - z(t)(\omega)\| \le \liminf_{n\to\infty} \|z_n(s)(\omega) - z_n(t)(\omega)\| \le \epsilon.$$

This holds for all $s \in U$ and $\omega \in \Omega$, whence $z$ is continuous at $t$, which was arbitrarily chosen from $X$. Thus $z \in C(X, B(\Omega))$, and the definition (a) of $z$ implies that $z = \sup C$.

The existence of $\inf C$ can be proved similarly. □

### 5.8.7. Summary

As a summary of the above results and the results of section 1.3 we obtain

**Theorem 5.8.2:** *Each pointwise bounded and equicontinuous chain of $C(X,Y)$, ordered by the pointwise ordering (5.8.13), has the supremum and the infimum in $C(X,Y)$ if $X$ is a topological space and $Y$ is*

a) *an ordered normed space with fully regular order cone $K$,*
b) *a reflexive Banach space, ordered by any closed cone or via continuous embedding in an ordered Banach space with fully regular order cone.*
c) *an ordered Hilbert space,*
d) *$\mathbb{R}^m$ with any norm and ordered by any closed cone,*
e) *$B(\Omega, E)$, ordered by $B(\Omega, K)$, where $\Omega$ is a nonempty set and $E$ is an ordered Banach space which is reflexive or its order cone $K$ is fully regular,*
f) *$L^p(\Omega, E)$, ordered by $L^p(\Omega, K)$, where $E$ is an ordered Banach space with fully regular order cone $K$, $1 \le p \le \infty$, and $(\Omega, \mathcal{A}, \mu)$ is a measure space with $\sigma$-finite $\mu$ if $p = \infty$,*
g) *$L^p([a,b], E)$, ordered by $L^p([a,b], K)$, where $E$ is an ordered reflexive Banach space with order cone $K$ and $1 \le p < \infty$,*
h) *$l^p(E)$, ordered by $l^p(K)$, where $E$ is an ordered Banach space with fully regular order cone $K$ and $1 \le p \le \infty$,*
i) *$l^p(E)$, ordered by $l^p(K)$, where $E$ is an ordered reflexive Banach space, $K$ the order cone of $E$ and $1 < p < \infty$,*

j) $L_M(\Omega)$, *ordered by* $L_M(\Omega)_+$,*where* $\Omega$ *is a bounded domain in* $\mathbb{R}^m$ *and* $M$ *satisfies the* $\Delta_2$*-condition,*
k) $W^{k,p}(\Omega)$, *ordered by* $W_+^{k,p}(\Omega)$,*where* $\Omega$ *is an open subset of* $\mathbb{R}^m$, $k \in \mathbb{N}$ *and* $1 < p < \infty$,
l) $W^k L_M(\Omega)$, *ordered by* $W^k L_M(\Omega)_+$, *where* $\Omega$ *is a bounded domain in* $\mathbb{R}^m$, $k \in \mathbb{N}$, *and both* $M$ *and* $\bar{M}$ *satisfy the* $\Delta_2$*-condition.*

**Proposition 5.8.12:** *Each pointwise order bounded and equicontinuous chain of* $C(X,Y)$, *ordered by the pointwise ordering (5.8.13), has the supremum and the infimum in* $C(X,Y)$ *when* $X$ *is a topological space and* $Y$ *is as in theorem 5.8.2 a), e), f), g) (also when* $K$ *is regular), c) with* $(x|y) \geq 0$ *for all* $x$, $y \in K$, *d) and when* $Y = E_M(\Omega)$, *where* $\Omega$ *is a bounded domain in* $\mathbb{R}^N$.

Replacing $\epsilon$ by $\psi(s,t)$ in the above proofs we obtain the following extension to propositions 1.3.9 and 1.3.10.

**Proposition 5.8.13:** *If* $X$ *is a topological space, and if* $Y$ *is any of the spaces given in theorem 5.8.2 (resp. in proposition 5.8.12), then each pointwise bounded (resp. ordered bounded) chain* $C$ *of* $C(X,Y)$ *has the supremum* $x^*$ *the infimum* $x_*$ *in* $C(X,Y)$. *Moreover, if there is a function* $\psi\colon X \times X \to \mathbb{R}_+$, *such that*

$$\|x(t) - x(s)\| \leq \psi(t,s) \quad \textit{for all } t,\ s \in X \ \textit{ and } \ x \in C, \tag{5.8.14}$$

*then for all* $t$, $s \in X$,

$$\|x^*(t)-x^*(s)\| \leq \psi(t,s) \quad \textit{and} \quad \|x_*(t)-x_*(s)\| \leq \psi(t,s). \tag{5.8.15}$$

**Remark 5.8.2:** In view of proposition 5.8.12 the conclusions of theorems 5.2.2, 5.3.4, 5.6.4 and 5.7.4, and their consequences hold when the space $E$ is replaced by any of the spaces $Y$ given in theorem 5.8.2. Thus the results of these theorems can be applied, for instance, to stochastic differential equations and to countable systems of differential equations in ordered Banach spaces (cf. f) and h) in theorem 5.8.2). In fact, almost all the results of earlier subsections are applicable to these classes of differential equations, since the fixed point results derived in subsection 1.4.8 hold also when the space $E$ is replaced by any of the spaces $Y$ given in proposition 5.8.12. The proofs, (use theorem 1.2.1 and propositions 1.2.1, 1.2.3 and 5.8.12), and applications to differential equations, are left to the reader.

The case when $Y = B(\Omega, E)$ seems to be worth of further comments. Firstly, it yields to even uncountable systems of differential equations in ordered Banach spaces, when $\Omega$ is considered as an index set. These systems can be generalized further by associating to each $\omega \in \Omega$ a different Banach space $E_\omega$ (cf. Chaljub-Simon et al. (1992)). Secondly, the space $B(\Omega, E)$ equals to $C(\Omega, E)$ when $\Omega$ is endowed with the discrete topology. As the reader can verify, most of the results derived in sections 5.1–5.7 hold when $E$ is replaced by the so defined $C(\Omega, E)$. On the other hand, this is no longer true if $\Omega$ has at least one accumulation point, even when $E = \mathbb{R}$. In fact, example 5.1.1 and its generalizations (see Chaljub-Simon et al. (1992), Satz 3) imply that in such a case differential equations in question are nonsolvable even when they are of nondecreasing type and contain no discontinuities. Consequently, in dealing with infinite systems of differential equations one cannot expect continuity of solutions with respect to indexes in any nondiscrete topology.

## 5.9. NOTES AND COMMENTS

In chapter 5 we have considered discontinuous differential equations in Banach spaces. The material of subsections 5.1.1 and 5.1.2 is based on Heikkilä (1989b), and Seikkala (1978). The results of subsections 5.1.3 and 5.1.4 are from Heikkilä and Lakshmikantham (1994d). As for the related results for continuous or Carathéodory type quasimonotone differential equations and inequalities, see e.g., Chaljub-Simon et al. (1992), Deimling (1977),

Deimling and Lakshmikantham (1979a, b), Guo and Lakshmikantham (1988), Lemmert (1989), Lemmert, Schmidt and Volkmann (1991), Redheffer (1973), Redheffer and Walter (1986), Volkmann (1972, 1986), Walter (1970, 1971), and Wazewski (1950).

Section 5.2 generalizes results of Heikkilä (1989d, 1990a). Section 5.3 is new, some special cases for second order IVP's are considered in Heikkilä and Leela (1992). Section 5.4 is adapted from Heikkilä (1991), and Heikkilä, Kumpulainen and Lakshmikantham (1993), special cases being treated in Heikkilä, Lakshmikantham and Leela (1988), and Heikkilä, Lakshmikantham and Sun (1992). Section 5.5 is based on Heikkilä, Kumpulainen and Lakshmikantham (1992), and Heikkilä and Lakshmikantham (1993). As for special cases see e.g., Lakshmikantham (1989), Lakshmikantham and Leela (1983), and Nieto (1988, 1989). Section 5.6 is taken from Heikkilä and Lakshmikantham (1994c), the definitions in subsection 5.6.1 being from Lakshmikantham and Leela (1983), and Martin (1977). The contents of section 5.7 is taken from Heikkilä and Leela (1994). The definitions and basic properties of ordered Banach spaces and function spaces introduced in section 5.8 are taken from Birkhoff (1973), Deimling (1985), Heikkilä (1989c), Krasnosel'skii and Rutickii (1961), Krasnosel'skii (1964), Krein and Rutman (1950), Kufner, John and Fučic (1977), Lang (1969), Yoshida (1974), and Zeidler (1985). Some of the results concerning the existence of supremums and infimums of chains in ordered Banach spaces and function spaces are given in Heikkilä (1990d, 1992a) (see also Dunford and Schwartz (1958)).

Discontinuous differential equations through conversion to set-valued inclusions are studied, for instance, in Deimling (1992), Filippov (1964, 1988), and Hu (1988, 1990). As for discontinuous functional differential equations and integral equations in ordered Banach spaces see Heikkilä (1989b, 1990d).

# References


Abian, S. and Brown, A. B. (1961). A theorem on partially ordered sets, with applications to fixed point theorems, Can J. Math., 13: 78–82.

Akô, K. (1961). On the Dirichlet problem for quasilinear elliptic differential equations of the second order, J. Math. Soc. Japan, 13: 45–62.

Amann, H. (1976).Fixed point equations and nonlinear eigenvalue problems in ordered Banach spaces, SIAM Rev., 18: 620–709.

Amann, H. (1977). Order structures and fixed points, ATTI 2° Sem. Anal. Funz. Appl.

Ambrosetti, A. and Badiale, M. (1989). The dual variational principle and elliptic problems with discontinuous nonlinearities. J. Math. Anal. Appl., 140: 363–373.

Ambrosetti, A. and Turner, R. E. L. (1988). Some discontinuous variational problems. Diff. and Int. Eqns., 1: 341–349.

Appel, J. (1988). The superposition operator in function spaces, Expo. Math, 6: 209–270.

Aris, F. (1975). The mathematical theory of diffusion and reaction in permeable catalysts. Clarendon Press, Oxford, 112.

Bakhtin, I. A. (1972). Existence of common fixed points of Abelian families of discontinuous operators, Siberian Math. J., 3: 167–172.

Bebernes, J. W. and Schmitt, K. (1979). "On the existence of maximal and minimal solutions of parabolic partial differential equations," Proc. Amer. Math. Soc., pp. 211–218.

Bernfeld, S. and Lakshmikantham, V. (1974). An Introduction to Nonlinear Boundary Value Problems, Acad. Press, New York.

Berkovits, J. and Mustonen, V. (1992). Topological degree for perturbations of linear maximal monotone mappings and applications to a class of parabolic problems, Rendiconti di Matematica, Serie VII 12: 597–621.

Birkhoff, G. (1973). Lattice Theory, Amer. Math. Soc. Coll. Publ., Vol. XXV, Providence, Rhode Island.

Bourbaki, N. (1949–1950). Sur le Theoreme de Zorn, Archiv der Mathematik, 2: 434–437.

Bressan, A. (1988). "Unique solutions for a class of discontinuous differential equations," Proc. Amer. Math. Soc., Vol. 104, pp. 772-778.

Browder, F. (1976). Nonlinear Operators and Nonlinear Equations of Evolution in Banach Spaces, Proc. Symp. Pure Math., Vol 18/2, Amer. Math. Soc.

Brown, A. L. and Page, A. (1970). Elements of Functional Analysis, Van Nostrand Reinhold Company, London.

Carathéodory, C. (1948). Vorlesungen über Reelle Funktionen, Chelsea Publishing Company, New York.

Carl, S. (1988a). Ein konstruktiver Existenzsatz für Randwertprobleme elliptischer Differentialgleichungen zweiter ordnung mit unstetiger Nichtlinearität, Math. Nachr., 138: 55–65.

Carl, S. (1988b). A monotone iterative scheme for nonlinear reaction-diffusion systems having nonmonotone reaction terms, J. Math. Anal. Appl., 134: 81–93.

Carl, S. (1989). The monotone iterative technique for a parabolic boundary value problem with discontinuous nonlinearity, Nonlinear Anal., 13: 1399–1407.

Carl, S. (1992a). "An existence result for discontinuous elliptic equations under discontinuous nonlinear flux condition," Proceedings of the First European Conference on Elliptic and Parabolic Problems, Pont-a-Mousson (France), June 1991, Pitman Research Notes in Mathematics Series 266, pp. 120–130.

Carl, S. (1992b). A combined variational–monotone iterative method for elliptic boundary value problems with discontinuous nonlinearity, Appl. Anal., 43:21–45.

Carl, S. (1994). On the existence of weak extremal solutions for a class of quasilinear parabolic problems, Diff. Int. Eqns., to appear.

Carl, S. and Grossman, C. (1990). Monotone enclosure for elliptic and parabolic systems with nonmonotone nonlinearities. J. Math. Anal. Appl., 151: 190–202.

Carl, S. and Heikkilä, S. (1990). On a parabolic boundary value problem with discontinuous nonlinearity, Nonlinear Anal., 15: 1091-1095.

Carl, S. and Heikkilä, S. (1992a). On extremal solutions of an elliptic boundary value problem involving discontinuous nonlinearities, Diff. Integral Eqns., 5: 581–589.

Carl, S. and Heikkilä, S. (1992b). An existence result for elliptic differential inclusions with discontinuous nonlinearity, Nonlinear Anal., 18 (5): 471–479.

Carl, S. and Heikkilä, S. (1992c). Theorems of Peano type for first order ODE:s involving discontinuous nonlinearities, Preprint, Univ. of Oulu.

Carl, S. and Heikkilä, S. (1993). On extremal coupled fixed points of mixed monotone operators with applications to systems of partial differential equations, Preprint, Univ. of Oulu.

Carl, S. and Heikkilä, S. (1994a). On the existence of extremal solutions for discontinuous elliptic equations under discontinuous flux conditions, Nonlinear Anal., to appear.

Carl, S. and Heikkilä, S. (1994b). Extremal solutions of quasilinear elliptic equations with discontinuous lower order term under discontinuous boundary conditions, (submitted).

Carl, S. and Heikkilä, S. (1994c). Extremal solutions of quasilinear parabolic boundary value problems with discontinuous nonlinearities, (submitted).

Carl, S. Heikkilä, S. and Kumpulainen, M. (1993). On a generalized iteration method with applications to fixed point theorems and elliptic systems involving discontinuities. Nonlinear Anal., 20 (2), 157–167.

Chaljub-Simon, A., Lemmert, R., Schmidt, S. and Volkmann, P., (1992). Gewöhnliche Differentialgleichungen mit quasimonoton wachsenden rechten Seiten in geordneten Banachräumen, Int. Ser. Num. Math., 103, 307–320.

Chen, Y. C. (1991). Existence theorems of coupled fixed points, J. Math. Anal. Appl., 154, 142–150.

Chipot, M. and Rodrigues, J. F. (1988). "Comparison and stability of solutions to a class of quasilinear parabolic problems." Proc. Royal Soc., Edinburgh, 110 A, pp. 275–285.

Coddington, E. A. and Levinson, N. (1955). Theory of Ordinary Differential Equations, McGraw-Hill, New York-Toronto-London.

Dancer, E. N. and Sweers, G. (1989). On the existence of a maximal weak solution for a semilinear elliptic equation, Differential and Integral Equations, 2: 533–540.

Davis, E. N. and Fleishman, B. A. (1986). A discontinuous nonlinear problem: stability and convergence of iterates in a finite number of steps Appl. Anal., 23: 139–157.

Deimling, K. (1977). Ordinary Differential Equations in Banach Spaces, Lecture notes 596, Springer-Verlag.

Deimling, K. (1985). Nonlinear Functional Analysis, Springer, New York.

Deimling, K. (1992). Multivalued Differential Equations, Walter de Gruyter, Berlin-New York.

Deimling, K., Ladde, G. S. and Lakshmikantham, V. (1985). Sample solutions of stochastic boundary value problems, Stochastic Analysis and Applications, 3(2): 153-162.

Deimling, K. and Lakshmikantham, V. (1979a). On the existence of extremal solutions of differential equations in Banach spaces, Nonlinear Anal., 3: 563–568.

Deimling, K. and Lakshmikantham, V. (1979b). Existence and comparison results for differential equations in Banach spaces, Nonlinear Anal., 3: 569–575.

Deuel, J. (1976). Nichtlineare parabolische Randwertprobleme mit Unter- und Oberlösungen, ETH Diss. Nr. 5750, Zürich.

Deuel, J. and Hess, P. (1974/75). "A criterion for the existence of solutions of non–linear elliptic boundary value problems," Proc. Royal Soc. Edinburgh, Vol. 74A, pp. 49–54.

Deuel, J. and Hess, P. (1978). Nonlinear parabolic boundary value problems with upper and lower solutions, Israel J. Math., 29: 92–104.

Dhage, B. C. and Heikkilä, S. (1993). On nonlinear boundary value problems with deviating arguments and discontinuous right hand side, J. Appl. Math. and Stochastic Anal., 6 (1): 00–00.

Dia, J. I. (1985). Nonlinear Partial Differential Equations and Free Boundaries Vol. **1**, Elliptic Equations, Research Notes in Math. 106, Pitman, London.

Dieudonné, J. (1950). Deux exemples singuliers d'équations différentielles, Acta Sci. Math., 12 B: 38–40.

Dieudonné, J. (1960). Foundations of Modern Analysis, Academic Press, New York-London.

DuChateau, P. and Zachmann, D. (1987). Applied Partial Differential Equations, Harper & Row Publishers, New York.

Dunford, N. and Schwartz, J. (1958). Linear operators I, Interscience, New York-London.

Fattorini, H.O. (1968). Ordinary differential equations in linear topological spaces, I, J. Diff. Eq., 5: 72–105.

Fattorini, H.O. (1969). Ordinary differential equations in linear topological spaces, II, J. Diff. Eq., 6: 50–70.

Filippov, A. F. (1964). Differential equations with discontinuous right-hand side, Trans. Amer. Math. Soc., 42: 199-231.

Filippov, A. F. (1988). Differential Equations with Discontinuous Righthand Sides, Kluwer Acad. Publ., The Netherlands.

Fleishman, B. A. and Mahar, T. J. (1981). A step-function model in chemical reactor theory: multiplicity and stability of solutions, Nonlinear Anal., 5: 645–654.

Frank, L. S. and Wendt, W. D. (1984). On an elliptic operator with discontinuous nonlinearity, J. Differential Equations, 54: 1–18.

Gilbarg, D. and Trudinger, N. S. (1983). Elliptic Partial Differential Equations of Second Order, Springer-Verlag, Berlin-Heidelberg-New York-Tokyo.

Goldstein, J. A. (1985). Semigroups of Linear Operators and Applications, Oxford, New York.

Gossez, J. P. and Mustonen, V. (1993). Pseudo-monotonicity and the Leray-Lions condition, Differential and Integral Equations, 6: 37–45.

Guo, D. and Lakshmikantham, V. (1988). Nonlinear Problems in Abstract Cones, Academic Press, New York-London.

Hájek, O. (1979). Discontinuous differential equations I/II, J. Diff. Eqs., 32: 149-170 and 171-185.

Heikkilä, S. (1975). On the method of successive approximations for Volterra integral equations, Ann. Acad Sci. Fenn A, 1: 39-47.

Heikkilä, S. (1988). A fixed point lemma with applications, Appl. Anal., 30: 165-174.

Heikkilä, S. (1989a). On fixed points through iteratively generated chains with applications to differential equations, J. Math. Anal. Appl., 138 (2): 397–417.

Heikkilä, S. (1989b). On operator and integral equations with discontinuous right hand side, J. Math. Anal. Appl., 140 (1): 200–217.

Heikkilä, S. (1989c). On UMB-spaces and equations, Report ISBN 951-42-2722-0, Math., Univ. of Oulu, 50 pp.

Heikkilä, S. (1989d). On the Cauchy problem for ordinary differential equations in ordered Banach spaces, Report ISBN 951-42-2820-0 Math., Univ. of Oulu, 29 pp.

Heikkilä, S. (1990a). On differential equations in ordered Banach spaces with applications to differential systems and random equations Diff. and Int. Eqns., 3 (3): 589–600.

Heikkilä, S. (1990b). On fixed points through a generalized iteration method with applications to differential and integral equations involving discontinuities, Nonlinear Anal., 14 (5): 413–426.

Heikkilä, S. (1990c). On an elliptic boundary value problem with discontinuous nonlinearity, Appl. Anal., 34:183-189.

Heikkilä, S. (1990d). On functional differential equations with discontinuous right hand side in ordered Banach spaces, Funkcialaj Ekvacioij, 33: 519–526.

Heikkilä, S. (1991). On first order mixed quasimonotone periodic boundary value systems involving discontinuities, Preprint, Univ. of Oulu, 8 pp.

Heikkilä, S. (1992). On extremal solutions of operator equations in ordered normed spaces, Appl. Anal.. 44: 77–97.

Heikkilä, S. (1994a). "Monotone methods with applications to nonlinear analysis," Proceedings of the First Congress of Nonlinear Analysts, Tampa (U.S.A) 1992, Walter de Gruyter Publishers, Berlin, (to appear), 12 pp.

Heikkilä, S. (1994b). "On extremal solutions of infinite systems of discontinuous differential equations," Proceedings of Dynamical Systems and Applications, Atlanta 1993, (submitted).

Heikkilä, S. and Hu, S. (1994). On fixed points of multifunctions in ordered spaces, Appl. Anal. (to appear).

Heikkilä, S., Kumpulainen, M. and Lakshmikantham, V. (1992). On solvability of mixed monotone operator equations with applications to mixed quasimonotone differential systems involving discontinuities, J. Appl. Math. and Stochastic Anal., 5 (1): 1–18.

Heikkilä, S. and Lakshmikantham, V. (1993). On the second order mixed quasimonotone periodic boundary value systems in ordered Banach spaces, Nonlinear Anal., 20 (9): 1135–1144.

Heikkilä, S. and Lakshmikantham, V. (1994a). Extension of the method of upper and lower solutions for discontinuous differential equations, Diff. Eqns. and Dyn. Syst., to appear.

Heikkilä, S. and Lakshmikantham, V. (1994b). On the method of upper and lower solutions for discontinuous boundary value problems, Nonlinear Anal., to appear.

Heikkilä, S. and Lakshmikantham, V. (1994c). On mild solutions of first order discontinuous semilinear differential equations in Banach spaces, (submitted).

Heikkilä, S. and Lakshmikantham, V. (1994d). On extremal solutions of ordinary differential equations in ordered Banach spaces, (submitted).

Heikkilä, S., Lakshmikantham, V. and Leela, S. (1988). Applications of monotone techniques to differential equations with discontinuous right hand side, Diff. and Int. Eqns., 1 (3).

Heikkilä, S., Lakshmikantham, V. and Sun, Y. (1992). Fixed point results in ordered normed spaces with applications to abstract and differential equations, J. Math. Anal. Appl., 163 (2): 422–437.

Heikkilä, S. and Leela, S. (1992). On the solvability of the second order initial value problems in Banach spaces, Dynamic Systems and Applications, 1 (2): 141–170.

Heikkilä, S. and Leela, S. (1994). On weak and mild solutions of second order discontinuous semilinear differential equations in Banach spaces, (submitted).

Hille, E. (1972). Methods in Classical and Functional Analysis, Addison-Wesley, Reading, Massachusetts.

Hille, E. and Phillips, R. S. (1957). Functional Analysis and Semigroups, Amer. Math. Soc. Coll. Publ. 31.

Howell, G.W. and Lakshmikantham, V. (1990). A new approach to monotone iterative technique for solution of differential equations, Appl. Anal., 39 (2-3): 113-117.

Hrbacek, K. and Jech, T. (1978). Introduction to Set Theory, Marcel Dekker, Inc., New York-Basel.

Hu, S. C. (1988). "Differential equations with discontinuous nonlinearities", Proc. Int. Conf. Columbus, Vol. I, pp. 450–454.

Hu, S. C. (1991). Differential equations with discontinuous right-hand sides, J. Math. Anal. Appl., 154: 377-390.

Höft, H. and Höft, M. (1976). Some fixed point theorems for partially ordered sets, Can. J. Math., XXVII (5): 992-997.

Inkmann, F. (1982). Existence and multiplicity theorems for semilinear elliptic equations with nonlinear boundary conditions, Indiana Univ. Math. J., 31: 213–221.

Keady, G. and Kloeden, P. E. (1987). "An elliptic boundary value problem with discontinuous nonlinearity II," Proc. Roy. Soc. Edinburg 105 A, pp. 23–36.

Khavanin, M. and Lakshmikantham, V. (1986). The method of mixed monotony and second order boundary value problems, J. Math. Anal. Appl., 120: 737–744.

Kolibiar, M. (1982). Fixed point theorems in ordered sets, Studia Sci. Math., Hungary, 17.

Korman, Ph. and Leung, A. W. (1986). "A general monotone scheme for elliptic systems with applications to ecological models," Proc. Royal Soc. Edinburgh 102 A, pp. 315–325.

Krasnosel'skii, M. A. (1964). Positive Solutions of Operator Equations, Noordhoff-Groningen, The Netherlands.

Krasnosel'skii, M. A. and Rutickii, Y. (1961). Convex Functions and Orlicz Spaces, Noordhoff-Groningen, The Netherlands.

Krein, M. and Rutman, M. (1950). Linear operators leaving invariant a cone in a Banach space, Amer. Math. Soc. Transl., 26.

Krivine, J. -L. (1971). Introduction to Axiomatic Set Theory, Reidel, Dordrecht-Holland.

Kufner, A., John, O. and Fučik, S. (1977). Function Spaces, Noordhoff International Publishing, Leyden.

Kuiper, H. J. (1973). Eigenvalue problems for noncontinuous operators associated with quasilinear elliptic equations, Arch. Rat. Mech. Anal., 53: 178–186.

Kura, T. (1989). The weak supersolution–subsolution method for second order quasilinear elliptic equations, Hiroshima Math. J., 1–36.

Kusano, T. and Oharu, S. (1992). Semilinear evolution equations with singularities in ordered Banach spaces, Diff. and Int. Eqns., 5 (6): 1383-1405.

Ladas, G. and Lakshmikantham, V. (1972). Differential Equations in Abstract Spaces, Academic Press, New York-London.

Ladde, G. S., Lakshmikantham, V. and Vatsala, A. S. (1984). Existence of coupled quasisolutions of systems of nonlinear elliptic boundary value problems, Nonlinear Analysis, TMA, 8: 501–515.

Ladde, G. S., Lakshmikantham, V. and Vatsala, A. S. (1985). Monotone Iterative Techniques for Nonlinear Differential Equations, Pitman.

Lakshmikantham, V. (1989). Periodic boundary value problems of first and second order differential equations, J. of Appl. Math. and Simulation, 2 (3): 131–138.

Lakshmikantham, V. and Leela, S. (1969). Differential and Integral Inequalities I, Academic Press, New York-London.

Lakshmikantham, V. and Leela, S. (1981). Nonlinear Differential Equations in Abstract Spaces, Pergamon Press, Oxford, United Kingdom.

Lakshmikantham, V. and Leela, S. (1983). Existence and monotone method for periodic solutions of first-order differential equations, J. Math. Anal. Appl., 91: 237-243.

Lakshmikantham, V., Nieto, J. J. and Sun, Y. (1991). An existence result about periodic boundary value problems of second order differential equations, Appl. Anal., 40 (1): 1-10.

Lang, S. (1969). Real Analysis, Addison-Wesley, Reading, Massachusetts.

Lemmert, R. (1989). Existenzsätze für gewöhnliche Differentialgleichungen in geordneten Banachräumen, Funkcial. Ekvac., Ser. Internac., 32: 243–249.

Lemmert, R., Schmidt, S. and Volkmann, P. (1991). Ein Existenzsatz für gewöhnliche Differentialgleichungen in geordneten Banachräumen, Math. Nachr., 153: 349-351.

Levy, A. (1979). Basic Set Theory, Springer-Verlag, Berlin-Heidelberg-New York.

Lewin, J. (1991). A simple proof of Zorn's lemma, Amer. Math. Monthly, 98 (4): 353–354.

Lions, J. L. (1965). Sur certaines equations paraboliques non linéaires, Bull. Soc. Math. France, 93: 155-175.

Lions, J. L. (1969). Quelques Méthodes de Résolution des Problémes aux Limites Nonlinéaires, Dunod Gauthier-Villars, Paris, France. (Russian transl.: Mir, Moscow 1972).

Martin, Jr., R. H. (1977). Nonlinear perturbations of linear evolution systems, J. Math. Soc. Japan, 29 (2): 233–252.

Mawhin, J. and Schmitt, K. (1984). "Upper and lower solutions and semilinear second order elliptic equations with non-linear boundary conditions," Proc. Royal Soc. Edinburgh 97 A, pp. 199–207.

Matrosov, V. M. (1967). On differential equations and inequalities with discontinuous right-hand sides I/II, Differential Eqs., 3: 432-437.

McShane, E. J. (1974). Integration, Princeton Univ. Press, Princeton, New Jersey.

Mendelson, E. (1987). Introduction to Mathematical Logic, Third Edition, Wadsworth & Brooks/Cole Advanced Books & Software, Monterey, California.

Mikusinski, J. (1978). The Bochner Integral, Academic Press, New York.

Mitidieri, E. and Sweers, G. (1993). Existence of a maximal solution for quasimonotone elliptic system Report 93-30, Delft Univ. of Technology, pp. 22–29.

Mlak, W. (1958). Differential inequalities in linear spaces, Ann. Polon. Math. Soc., 5: 95–101.

Mlak, W. and Olech, C. (1963). Integration of infinite systems of differential inequalities, Ann. Polon. Math. Soc., 13: 105–122.

Munroe, M. E. (1959). Introduction to Measure and Integration, Addison-Wesley, Reading, Massachusetts.

Mustonen, V. (1990). "Mappings of monotone type: Theory and Applications," Proc. Int. Spring School "Nonlinear Analysis, Function Spaces and Applications IV", B. G. Teubner, Leipzig, pp. 104–126.

Nagata, J. (1974). Modern General Topology, North-Holland, Amsterdam.

Narici, L. and Beckenstein, E. (1985). Topological Vector Spaces, Marcel Dekker, Inc., New York–Basel.

Naumann, J. (1984). Einführung in die Theorie parabolischer Variationsungleichungen, BSB B.G. Teubner Verlagsgesellschaft, Leipzig, Germany.

Nieto, J. J. (1988). Nonlinear second order periodic boundary value problem, J. Math. Anal. Appl., 130: 22–29.

Nieto, J. J. (1989.) Nonlinear second order periodic boundary value problem with Carathéodory functions, Appl. Anal., 34: 111–128.

Protter, M. H. and Weinberger, H. F. (1967). Maximum Principles in Differential Equations, Prentice-Hall, Inc., Englewood Cliffs, New Jersey.

Redheffer, R. M. (1973). Gewöhnliche Differentialungleichungen mit quasimonotonen Funktionen in normierten linearen Räumen, Arc. Rational Mech. Analysis, 52: 121–133.

Redheffer, R. M. and Walter, W. (1986). Remarks on ordinary differential equations in ordered Banach spaces, Monatshefte Math., 102: 237–249.

Royden, H. L. (1968). Real Analysis, The MacMillan Company, London, England.

Sattinger, D. H. (1972). Monotone methods in nonlinear elliptic and parabolic boundary value problems, Indiana Univ. Math. J., 21: 979-1000.

Schaefer, H. (1966). Topological Vector Spaces, The MacMillan Company, New York.

Schmitt, K. (1978). Boundary value problems for quasilinear second order elliptic equations, Nonlinear Anal., 2: 263–309.

Seikkala, S. (1978). On the method of successive approximations for nonlinear equations in spaces of continuous functions, Acta Univ. Oulu A, 76.

Sentis, R. (1978). Equations differentielles à second membre measurable, Boll. Un. Mat. Ital., 15-B: 724-742.

Shragin, I. V. (1979). On the Carathéodory conditions, Russian Math Surveys, 34 (3): 183–189.

Smithson, R. E. (1971). "Fixed points of order preserving multifunctions," Proc. A.M.S. 28, pp. 304–310.

Stuart, C. A. (1976). Differential equations with discontinuous nonlinearities, Arch. Rat. Mech. Anal., 63: 59-75

Stuart, C. A. (1978). Maximal and minimal solutions of elliptic differential equations with discontinuous nonlinearities, Math. Z., 163: 239-249.

Sun, J. (1984). On the equivalence of normal cones and fully regular cones in reflexive Banach spaces (in Chinese), Kexue Tongbao, 28: 382.

Sun, J. and Sun, Y. (1986). Some fixed point theorems of increasing operators, Appl. Anal., 23: 23–27.

Sun, J. and Sun, Y. (1989). A general principle on ordered sets and its applications to fixed point theory, Appl. Anal., 34 (1+2): 129–137.

Sun, Y. (1991). A fixed point theorem for mixed monotone operators with applications, J. Math. Anal. Appl., 156: 240–252.

Tarski, A. (1955). A lattice-theoretical fixpoint theorem and its applications, Pacific J. Math., 5: 285–309.

Travis, C. C. and Webb, G. F. (1978a). "Second order differential equations in Banach space," Proc. Int. Symp. on Nonlinear Equations in Abstract Spaces, Academic Press, New York, pp. 331-361.

Travis, C. C. and Webb, G. F. (1978b). Cosine families and abstract nonlinear second order differential equations, Acta Math. Acad. Sci. Hung., ul32: 75–96.

Troianiello, G. M. (1987). Elliptic Differential Equations and Obstacle Problems, Plenum Press, New York.

Volkman, P. (1972). Gewöhnliche Differentialungleichungen mit quasimonoton wachsenden Funktionen in topologischen Vektor räumen, Math. Z, 127: 157-164.

Volkmann, P. (1985). Équations différentielles ordinairies dans les espaces des fonctions bornées, Czechoslovak Math J., 35 (110): 201-211.

Walter, W. (1965). Über sukzessive Approximation bei Volterra-Integralungleichungen in mehreren Veränderlichen, Ann. Acad. Sci. Fenn. ser. AI, 345: 1-32.

Walter, W. (1970). Differential and Integral Inequalities, Springer-Verlag, Berlin-Heidelberg-New York.

Walter, W. (1971). Ordinary differential inequalities in ordered Banach spaces, J. Diff. Eqns., 9: 253-261.

Wazewski, T. (1950). Systémes des équations et des inégalitiés différentielles ordinaires aux deuxiémes membres monotones et leurs applications, Ann. Polon. Math. Soc., 23: 112-156.

Yoshida, K. (1974). Functional Analysis, Springer-Verlag, Berlin-Heidelberg-New York.

Zeidler, E. (1985). Functional Analysis and its Applications, Vol. I, Springer-Verlag, Berlin-Heidelberg-New York.

Zeidler, E. (1990a). Nonlinear Functional Analysis and its Applications, Vol. II/A, Springer, Berlin.

Zeidler, E. (1990b). Nonlinear Functional Analysis and its Applications, Vol. II/B, Springer, Berlin.

# Index